Ewing Buchan

Buchan's Exchange Tables

Embracing simple, reliable and accurate Forms for ordinary use

Ewing Buchan

Buchan's Exchange Tables
Embracing simple, reliable and accurate Forms for ordinary use

ISBN/EAN: 9783337189327

Printed in Europe, USA, Canada, Australia, Japan

Cover: Foto ©Andreas Hilbeck / pixelio.de

More available books at **www.hansebooks.com**

BUCHAN'S
EXCHANGE TABLES,

IN THREE PARTS,

EMBRACING

SIMPLE, RELIABLE AND ACCURATE FORMS FOR ORDINARY USE,

IN THE CONVERSION OF

STERLING INTO CANADIAN CURRENCY,

(AND VICE-VERSA)

ADVANCING BY EIGHTHS;

ALSO A COMPLETE SET,

SPECIALLY ADAPTED FOR LARGE TRANSACTIONS AT CLOSE RATES,

ADVANCING BY SIXTEENTHS,

AND

CALCULATIONS FOR TRANSACTIONS IN

AMERICAN OR DOMESTIC EXCHANGE,

AT BOTH DISCOUNT AND PREMIUM;

ALSO CONTAINING

TABLE OF BROKERAGES AND STERLING EQUIVALENTS OF LONG AND SHORT EXCHANGE
IN CANADIAN AND NEW YORK SYSTEMS OF QUOTATIONS.

———

COMPILED BY

EWING BUCHAN

———

STERLING EQUIVALENTS.

CANADIAN AND UNITED STATES RATES

Advancing by 64ths from $7\frac{1}{2}$ to $11\frac{3}{64}$ per cent.

CANADA.	U. S.	CANADA.	U. S.	CANADA.	U. S.	CANADA.	U. S.
	$ cts.		$ cts.		$ cts.		$ cts.
7½	4.777777	8¼	4.817361	9½	4.858044	10¼	4.898527

EXPLANATIONS TO PART I.

STERLING EQUIVALENTS

AND

STERLING EXCHANGE TABLES,

ADVANCING BY EIGHTHS.

The intimate relations existing between the CANADIAN and NEW YORK markets for Sterling Exchange necessitate a convenient table showing at a glance the relative values according to the different systems of quotations. Such a table will be found on next page.

The Table of EQUIVALENTS of long and short Exchange, on the next succeeding page, will at once commend itself to the practical banker, and the Explanations at the head of the page will be found to thoroughly explain its use.

The STERLING CONVERSION TABLES, which follow, are pronounced to be the most complete yet issued. The system provides for conversion of any odd number of pounds up to £1000 at one glance, the units and tens being found in right and left-hand columns, and the hundreds at the top of the page. The same system is carried out in regard to shillings and pence, which will be found at the foot of the page, the pence in the right and left-hand columns, and the shillings at the top. The decimals are extended beyond the cents to allow for large amounts, so that £1000 or £10000 can be found from £100, £1500 from £150, £9700 from £970, &c., &c. A table of differences at $\frac{1}{16}$ precedes these tables, by the use of which values can be found at fractions of $\frac{1}{16}$ higher or lower than the given rate, if necessary. The range of quotations, viz., $6\frac{1}{2}$ to $12\frac{1}{2}$ % premium—advancing by 8ths from $7\frac{1}{4}$ to 11 %, and by $\frac{1}{2}$'s above and below these rates—will commend itself as being sufficient for all purposes, from the fact that the lowest average posted rate in New York in thirteen years was $7\frac{3}{4}$ %, and the highest $10\frac{1}{2}$ %.

Example : To ascertain the value of £978 11s. 2d. at $7\frac{7}{8}$ %, see pp. 22 and 23, and find column of 900 and line of 78 ; thus

| | £978 | = | $4688.96 |

Then under head of Shillings and Pence, at foot of page, find column of 11 and line of 2 ; thus,

| | 11 | 2 | = | 2.67 |

Result : £978 11 2 = $4691.63

For conversion of Canadian Currency into Sterling, it is only necessary to glance over the page until the eye lights on the amount nearest below the sum desired to convert, which will give the full amount in pounds, and the difference subtracted will be the shillings and pence.

Example: To ascertain the value of 4753\frac{19}{100}$ at $7\frac{1}{8}$, see pp. 22 and 23. The next lowest amount will be found to be

	$4751.29	=	£991	0	0
and the difference	1.90	=		7	11
Result :	$4753.19	=	£991	7	11

Nothing can be simpler than this ; and a short acquaintance with the form will secure great speed in this mode of conversion.

STERLING EQUIVALENTS.

Canadian into New York Rates, and Vice-Versa.

(ADVANCING BY 16ths)

Canada into New York.

Old Par		Present Par :—	
Old Par	= $4.44.4		
6 7/16 % pr.	= 4.73.0	9 1/2 % pr.	= $4.86.6
6 1/2	= 4.73.3	9 9/16	= 4.86.9
6 9/16	= 4.73.6	9 5/8	= 4.87.2
6 5/8	= 4.73.8	9 11/16	= 4.87.5
6 11/16	= 4.74.1	9 3/4	= 4.87.7
6 3/4	= 4.74.4	9 13/16	= 4.88.0
6 13/16	= 4.74.7	9 7/8	= 4.88.3
6 7/8	= 4.75.0	9 15/16	= 4.88.6
6 15/16	= 4.75.2	10	= 4.88.8
7	= 4.75.5	10 1/16	= 4.89.1
7 1/16	= 4.75.8	10 1/8	= 4.89.4
7 1/8	= 4.76.1	10 3/16	= 4.89.7
7 3/16	= 4.76.3	10 1/4	= 4.90.0
7 1/4	= 4.76.6	10 5/16	= 4.90.2
7 5/16	= 4.76.9	10 3/8	= 4.90.5
7 3/8	= 4.77.2	10 7/16	= 4.90.8
7 7/16	= 4.77.5	10 1/2	= 4.91.1
7 1/2	= 4.77.7	10 9/16	= 4.91.3
7 9/16	= 4.78.0	10 5/8	= 4.91.6
7 5/8	= 4.78.3	10 11/16	= 4.91.9
7 11/16	= 4.78.6	10 3/4	= 4.92.2
7 3/4	= 4.78.8	10 13/16	= 4.92.5
7 13/16	= 4.79.1	10 7/8	= 4.92.7
7 7/8	= 4.79.4	10 15/16	= 4.93.0
7 15/16	= 4.79.7	11	= 4.93.3
8	= 4.80.0	11 1/16	= 4.93.6
8 1/16	= 4.80.2	11 1/8	= 4.93.8
8 1/8	= 4.80.5	11 3/16	= 4.94.1
8 3/16	= 4.80.8	11 1/4	= 4.94.4
8 1/4	= 4.81.1	11 5/16	= 4.94.7
8 5/16	= 4.81.3	11 3/8	= 4.95.0
8 3/8	= 4.81.6	11 7/16	= 4.95.2
8 7/16	= 4.81.9	11 1/2	= 4.95.5
8 1/2	= 4.82.2	11 9/16	= 4.95.8
8 9/16	= 4.82.5	11 5/8	= 4.96.1
8 5/8	= 4.82.7	11 11/16	= 4.96.3
8 11/16	= 4.83.0	11 3/4	= 4.96.6
8 3/4	= 4.83.3	11 13/16	= 4.96.9
8 13/16	= 4.83.6	11 7/8	= 4.97.2
8 7/8	= 4.83.8	11 15/16	= 4.97.5
8 15/16	= 4.84.1	12	= 4.97.7
9	= 4.84.4	12 1/16	= 4.98.0
9 1/16	= 4.84.7	12 1/8	= 4.98.3
9 1/8	= 4.85.0	12 3/16	= 4.98.6
9 3/16	= 4.85.2	12 1/4	= 4.98.8
9 1/4	= 4.85.5	12 5/16	= 4.99.1
9 5/16	= 4.85.8	12 3/8	= 4.99.4
9 3/8	= 4.86.1	12 7/16	= 4.99.7
9 7/16	= 4.86.3	12 1/2	= 5.00.0

New York into Canada.

$4.75 1/2	= 6.98 7/8 % pr.	$4.87 3/4	= 9.74 3/8 % pr.
4.75 3/4	= 7.04 1/2	4.88	= 9.80
4.76	= 7.10	4.88 1/4	= 9.85 5/8
4.76 1/4	= 7.15 5/8	4.88 1/2	= 9.91 1/4
4.76 1/2	= 7.21 1/4	4.88 3/4	= 9.96 7/8
4.76 3/4	= 7.26 7/8	4.89	= 10.02 1/2
4.77	= 7.32 1/2	4.89 1/4	= 10.08 1/8
4.77 1/4	= 7.38 1/8	4.89 1/2	= 10.13 3/4
4.77 1/2	= 7.43 3/4	4.89 3/4	= 10.19 3/8
4.77 3/4	= 7.49 3/8	4.90	= 10 1/4
4.78	= 7.55	4.90 1/4	= 10.30 5/8
4.78 1/4	= 7.60 5/8	4.90 1/2	= 10.36 1/4
4.78 1/2	= 7.66 1/4	4.90 3/4	= 10.41 7/8
4.78 3/4	= 7.71 7/8	4.91	= 10.47 1/2
4.79	= 7.77 1/2	4.91 1/4	= 10.53 1/8
4.79 1/4	= 7.83 1/8	4.91 1/2	= 10.58 3/4
4.79 1/2	= 7.88 3/4	4.91 3/4	= 10.64 3/8
4.79 3/4	= 7.94 3/8	4.92	= 10.70
4.80	= 8	4.92 1/4	= 10.75 5/8
4.80 1/4	= 8.05 5/8	4.92 1/2	= 10.81 1/4
4.80 1/2	= 8.11 1/4	4.92 3/4	= 10.86 7/8
4.80 3/4	= 8.16 7/8	4.93	= 10.92 1/2
4.81	= 8.22 1/2	4.93 1/4	= 10.98 1/8
4.81 1/4	= 8.28 1/8	4.93 1/2	= 11.03 3/4
4.81 1/2	= 8.33 3/4	4.93 3/4	= 11.09 3/8
4.81 3/4	= 8.39 3/8	4.94	= 11.15
4.82	= 8.45	4.94 1/4	= 11.20 5/8
4.82 1/4	= 8.50 5/8	4.94 1/2	= 11.26 1/4
4.82 1/2	= 8.56 1/4	4.94 3/4	= 11.31 7/8
4.82 3/4	= 8.61 7/8	4.95	= 11 3/8
4.83	= 8.67 1/2	4.95 1/4	= 11.43 1/8
4.83 1/4	= 8.73 1/8	4.95 1/2	= 11.48 3/4
4.83 1/2	= 8.78 3/4	4.95 3/4	= 11.54 3/8
4.83 3/4	= 8.84 3/8	4.96	= 11.60
4.84	= 8.90	4.96 1/4	= 11.65 5/8
4.84 1/4	= 8.95 5/8	4.96 1/2	= 11.71 1/4
4.84 1/2	= 9.01 1/4	4.96 3/4	= 11.76 7/8
4.84 3/4	= 9.06 7/8	4.97	= 11.82
4.85	= 9 1/8	4.97 1/4	= 11.88 1/8
4.85 1/4	= 9.18 1/8	4.97 1/2	= 11.93 3/4
4.85 1/2	= 9.23 3/4	4.97 3/4	= 11.99 3/8
4.85 3/4	= 9.29 3/8	4.98	= 12.05
4.86	= 9.35	4.98 1/4	= 12.10 5/8
4.86 1/4	= 9.40 5/8	4.98 1/2	= 12.16 1/4
4.86 1/2	= 9.46 1/4	4.98 3/4	= 12.21 7/8
4.86.65	= U. S. Par.	4.99	= 12.27 1/2
4.86 3/4	= 9.51 7/8	4.99 1/4	= 12.33 1/8
4.87	= 9.57 1/2	4.99 1/2	= 12.38 3/4
4.87 1/4	= 9.63 1/8	4.99 3/4	= 12.44 3/8
4.87 1/2	= 9.68 3/4	5.00	= 12 1/2

STERLING EQUIVALENTS.

DISCOUNT to be added to 60 DAY RATE

TO ASCERTAIN VALUE OF

SHORT EXCHANGE or CABLE TRANSFERS.

EXAMPLE.—To find the value of a 30 days sight bill when 60 days exchange is quoted 4.86, and rate of discount in London is 3%, find the line of 3% at left hand side of table below, and add the decimal found in 30 days sight column, thus—60 days sight............................... 4.86

30 " " .... .0120

Value of 30 days sight bill... .... 4.8720

In the same way 45 days will be 4.8660, 3 days, 4.8828, cable 4.8900, etc., etc.

Should the rate of discount be at a smaller fraction than ¼ of 1%, add or deduct the decimal for discount rates of ⅛ and ¹⁄₁₆ as the case may be, as found in first two lines of the table.

12 days are allowed for remittance and acceptance ; allowance is also made for 3 days' grace.

Rate of Discount in London	45 Days Sight.	30 Days Sight.	15 Days Sight.	3 Days Sight.	On Demand.	Cable Transfers.
¹⁄₁₆	.0001	.0003	.0004	.0005	.0005	.0006
⅛	.0002	.0005	.0007	.0009	.0010	.0012
1 %	.0020	.0040	.0060	.0076	.0084	.0100
1¼ "	.0025	.0050	.0075	.0095	.0105	.0125
1½ "	.0030	.0060	.0090	.0114	.0126	.0150
1¾ "	.0035	.0070	.0105	.0133	.0147	.0175
2 "	.0040	.0080	.0120	.0152	.0168	.0200
2¼ "	.0045	.0090	.0135	.0171	.0189	.0225
2½ "	.0050	.0100	.0150	.0190	.0210	.0250
2¾ "	.0055	.0110	.0165	.0209	.0231	.0275
3 "	.0060	.0120	.0180	.0228	.0252	.0300
3¼ "	.0065	.0130	.0195	.0247	.0273	.0325
3½ "	.0070	.0140	.0210	.0266	.0294	.0350
3¾ "	.0075	.0150	.0225	.0285	.0315	.0375
4 "	.0080	.0160	.0240	.0304	.0336	.0400
4¼ "	.0085	.0170	.0255	.0323	.0357	.0425
4½ "	.0090	.0180	.0270	.0342	.0378	.0450
4¾ "	.0095	.0190	.0285	.0361	.0399	.0475
5 "	.0100	.0200	.0300	.0380	.0420	.0500
5¼ "	.0105	.0210	.0315	.0399	.0441	.0525
5½ "	.0110	.0220	.0330	.0418	.0462	.0550
5¾ "	.0115	.0230	.0345	.0437	.0483	.0575
6 "	.0120	.0240	.0360	.0456	.0504	.0600
6¼ "	.0125	.0250	.0375	.0475	.0525	.0625
6½ "	.0130	.0260	.0390	.0494	.0546	.0650
6¾ "	.0135	.0270	.0405	.0513	.0567	.0675
7 "	.0140	.0280	.0420	.0532	.0588	.0700

1/16

OF ONE PER CENT PREMIUM.

STERLING INTO DOLLARS AND CENTS, AND VICE-VERSA.

£	0	100	200	300	400	500	600	700	800	900	£
£	$ cts.	$ cts.	$ cts.	$ cts.	$ cts.	$ cts.	$ cts.	$ cts.	$ cts.	$ cts.	£
0	......	.27.7	.55.5	.83.3	1.11.1	1.38.8	1.66.6	1.94.4	2.22.2	2.50.0	0
1	.00.2	.28.0	.55.8	.83.6	1.11.3	1.39.1	1.66.9	1.94.7	2.22.5	2.50.2	1
2	.00.5	.28.3	.56.1	.83.8	1.11.6	1.39.4	1.67.2	1.95.0	2.22.7	2.50.5	2
3	.00.8	.28.6	.56.3	.84.1	1.11.9	1.39.7	1.67.5	1.95.2	2.23.0	2.50.8	3
4	.01.1	.28.8	.56.6	.84.4	1.12.2	1.40.0	1.67.7	1.95.5	2.23.3	2.51.1	4
5	.01.3	.29.1	.56.9	.84.7	1.12.5	1.40.2	1.68.0	1.95.8	2.23.6	2.51.3	5
6	.01.6	.29.4	.57.2	.85.0	1.12.7	1.40.5	1.68.3	1.96.1	.2.23.8	2.51.6	6
7	.01.9	.29.7	.57.5	.85.2	1.13.0	1.40.8	1.68.6	1.96.3	2.24.1	2.51.9	7
8	.02.2	.30.0	.57.7	.85.5	1.13.3	1.41.1	1.68.8	1.96.6	2.24.4	2.52.2	8
9	.02.5	.30.2	.58.0	.85.8	1.13.6	1.41.3	1.69.1	1.96.9	2.24.7	2.52.5	9
10	.02.7	.30.5	.58.3	.86.1	1.13.8	1.41.6	1.69.4	1.97.2	2.25.0	2.52.7	10
11	.03.0	.30.8	.58.6	.86.3	1.14.1	1.41.9	1.69.7	1.97.5	2.25.2	2.53.0	11
12	.03 3	.31.1	.58.8	.86.6	1.14.4	1.42.2	1.70.0	1.97.7	2.25.5	2.53.3	12
13	.03.6	.31.3	.59.1	.86 9	1.14.7	1.42.5	1.70.2	1.98.0	2.25.8	2.53.6	13
14	.03.8	.31.6	.59.4	.87.2	1.15.0	1.42.7	1.70.5	1.98.3	2.26.1	2.53.8	14
15	.04.1	.31.9	.59.7	.87.5	1.15.2	1.43.0	1.70.8	1.98.6	2.26.3	2.54.1	15
16	.04.4	.32.2	.60.0	.87.7	1.15.5	1.43.3	1.71.1	1.98.8	2.26.6	2.54.4	16
17	.04.7	.32.5	.60.2	.88.0	1.15.8	1.43.6	1.71.3	1.99.1	2.26.9	2.54.7	17
18	.05.0	.32.7	.60.5	.88.3	1.16.1	1.43.8	1.71.6	1.99.4	2.27.2	2.55.0	18
19	.05.2	.33.0	.60.8	.88.6	1.16.3	1.44.1	1.71.9	1.99.7	2.27.5	2.55.2	19
20	.05.5	.33.3	.61.1	.88.8	1.16.6	1.44.4	1.72.2	2.00.0	2.27.7	2.55.5	20
21	.05.8	.33.6	.61.3	.89.1	1.16.9	1.44.7	1.72.5	2.00.2	2.28.0	2.55.8	21
22	.06.1	.33.8	.61.6	.89.4	1.17.2	1.45.0	1.72.7	2.00.5	2.28.3	2.56.1	22
23	.06.3	.34.1	.61.9	.89.7	1.17.5	1.45.2	1.73.0	2.00.8	2.28.6	2.56.3	23
24	.06.6	.34.4	.62.2	.90.0	1.17.7	1.45.5	1.73.3	2.01.1	2.28.8	2.56.6	24
25	.06.9	.34.7	.62.5	.90.2	1.18.0	1.45.8	1.73.6	2.01.3	2.29.1	2.56.9	25
26	.07.2	.35.0	.62.7	.90.5	1.18.3	1.46.1	1.73.8	2.01.6	2.29.4	2.57.2	26
27	.07.5	.35.2	.63.0	.90.8	1.18.6	1.46.3	1.74.1	2.01.9	2.29.7	2.57.5	27
28	.07.7	.35.5	.63.3	.91.1	1.18.8	1.46.6	1.74.4	2.02.2	2.30.0	2.57.7	28
29	.08.0	.35.8	.63.6	.91.3	1.19.1	1.46.9	1.74.7	2.02.5	2.30.2	2.58.0	29
30	.08.3	.36.1	.63.8	.91.6	1.19.4	1.47.2	1.75.0	2.02.7	2.30.5	2.58.3	30
31	.08.6	.36.3	.64.1	.91.9	1.19.7	1.47.5	1.75 2	2 03.0	2.30.8	2.58.6	31
32	.08.8	.36.6	.64.4	.92.2	1.20.0	1.47.7	1.75.5	2.03.3	2.31.1	2.58.8	32
33	.09.1	.36.9	.64.7	.92.5	1.20.2	1.48.0	1.75.8	2.03.6	2.31.3	2.59.1	33
34	.09.4	.37.2	.65.0	.92.7	1.20.5	1.48.3	1.76.1	2.03.8	2.31.6	2.59.4	34
35	.09.7	.37.5	.65.2	.93.0	1.20.8	1.48.6	1.76.3	2.04.1	2.31.9	2.59.7	35
36	.10.0	.37.7	.65.5	.93.3	1.21.1	1.48.8	1.76.6	2.04.4	2.32.2	2.60.0	36
37	.10.2	.38.0	.65.8	.93.6	1.21.3	1.49.1	1.76.9	2.04.7	2.32.5	2.60.2	37
38	.10.5	.38.3	.66.1	.93.8	1.21.6	1.49.4	1.77.2	2.05.0	2.32.7	2.60.5	38
39	.10.8	.38.6	.66.3	.94.1	1.21.9	1 49.7	1.77.5	2.05.2	2.33.0	2.60.8	39
40	.11.1	.38 8	.66.6	.94.4	1.22.2	1.50.0	1.77.7	2.05.5	2.33.3	2.61.1	40
41	.11.3	.39.1	.66.9	.94.7	1.22.5	1.50.2	1.78.0	2.05.8	2.33.6	2.61.3	41
42	.11.6	.39.4	.67.2	.95.0	1.22.7	1.50.5	1.78.3	2.06.1	2.33.8	2.61.6	42
43	.11.9	.39.7	.67.5	.95.2	1.23.0	1.50.8	1.78.6	2.06.3	2.34.1	2.61.9	43
44	.12 2	.40.0	.67.7	.95.5	1.23.3	1.51.1	1.78.8	2.06.6	2.34.4	2.62.2	44
45	.12.5	.40.2	.68.0	.95.8	1.23.6	1.51.3	1.79.1	2.06.9	2.34.7	2.62.5	45
46	.12.7	.40.5	.68.3	.96.1	1.23.8	1.51 6	1.79.4	2.07.2	2.35.C	2.62.7	46
47	.13.0	.40.8	.68.6	.96.3	1.24.1	1.51.9	1.79.7	2.07.5	2.35.2	2.63.0	47
48	.13.3	.41.1	.68.8	.96.6	1.24.4	1.52.2	1.80.0	2.07.7	2.35.5	2.63.3	48
49	.13.6	.41.3	.69.1	.96.9	1.24.7	1.52.5	1.80.2	2.08.0	2.35.8	2.63.6	49

Shillings and pence not required in this table on account of the smallness of the values, that of 19 shillings, being less than 2 mills—or .00.1 of a cent.

$\frac{1}{16}$

STERLING INTO DOLLARS AND CENTS, AND VICE-VERSA.

£	0	100	200	300	400	500	600	700	800	900	£
£	$ cts.	$ cts.	$ cts.	$ cts.	$ cts.	$ cts.	$ cts.	$ cts.	$ cts.	$ cts.	£
50	.13.8	.41.6	.69.4	.97.2	1.25.0	1.52.7	1.80.5	2.08 3	2.36.1	2.63.8	50
51	.14.1	.41.9	.69.7	.97.5	1.25 2	1.53.0	1.80.8	2.08.6	2.36.3	2.64.1	51
52	.14.4	.42.2	.70.0	.97.7	1.25.5	1.53.3	1.81.1	2.08.8	2 36.6	2.64.4	52
53	.14.7	.42.5	.70.2	.98.0	1.25.8	1.53.6	1.81.3	2.09.1	2.36.9	2.64.7	53
54	.15.0	.42.7	.70.5	.98.3	1.26.1	1.53.8	1.81.6	2.09.4	2.37.2	2.65.0	54
55	.15.2	.43.0	.70.8	.98.6	1.26.3	1.54.1	1.81.9	2 09.7	2.37.5	2.65.2	55
56	.15.5	.43.3	.71.1	.98.8	1.26.6	1.54.4	1.82.2	2.10.0	2.37.7	2.65.5	56
57	.15.8	.43.6	.71.3	.99.1	1.26.9	1.54.7	1.82.5	2.10 2	2.38.0	2.65.8	57
58	.16.1	.43.8	.71.6	.99.4	1.27.2	1.55.0	1.82 7	2.10.5	3.38.3	2.66.1	58
59	.16.3	.44.1	.71.9	.99.7	1.27.5	1.55.2	1.83.0	2.10.8	2.38.6	2.66.3	59
60	.16.6	.44.4	.72.2	1.00.0	1.27.7	1.55.5	1.83.3	2.11.1	2.38.8	2.66.6	60
61	.16.9	.44.7	.72.5	1.00.2	1.28.0	1.55.8	1.83.6	2.11.3	2.39.1	2.66 9	61
62	.17.2	.45.0	.72.7	1.00.5	1.28.3	1.56.1	1.83.8	2.11.6	2.39 4	2 67.2	62
63	.17.5	.45.2	.73.0	1.00.8	1.28.6	1.56.3	1.84.1	2.11.9	2.39 7	2.67.5	63
64	.17.7	.45.5	.73.3	1.01.1	1.28.8	1.56.6	1.84.4	2.12.2	2 40 0	2.67.7	64
65	.18.0	.45.8	.73.6	1.01.3	1.29.1	1.56.9	1.84.7	2.12.5	2.40.2	2.68.0	65
66	.18.3	.46.1	.73.8	1.01.6	1.29.4	1.57.2	1.85.0	2.12.7	2 40.5	2.68.3	66
67	.18.6	.46.3	.74.1	1.01.9	1.29 7	1.57.5	1.85.2	2.13.0	2 40.8	2.68.6	67
68	.18 8	.46.6	.74.4	1.02.2	1.30.0	1.57.7	1.85.5	2.13.3	2.41.1	2.68 8	68
69	.19.1	.46.9	.74.7	1.02.5	1.30.2	1.58.0	1.85.8	2.13.6	2.41.3	2 69.1	69
70	.19.4	.47.2	.75.0	1.02.7	1.30.5	1.58.3	1.86.1	2.13.8	2.41.6	2 69.4	70
71	.19.7	.47.5	.75.2	1.03.0	1.30.8	1.58.6	1.86.3	2.14.1	2.41.9	2.69.7	71
72	.20.0	.47.7	.75.5	1.03.3	1.31.1	1.58 8	1.86.6	2.14.4	2.42.3	2.70.0	72
73	.20.2	.48.0	.75.8	1.03.6	1.31.3	1.59.1	1.86.9	2.14.7	2 42.5	2.70.2	73
74	.20.5	.48.3	.76.1	1.03.8	1.31.6	1.59.4	1.87.2	2.15.0	2.42.7	2.70 5	74
75	.20.8	.48.6	.76.3	1.04.1	1.31.9	1.59.7	1.87.5	2.15.2	2.43.0	2.70.8	75
76	.21.1	.48.8	.76.6	1.04.4	1.32.2	1.60.0	1.87.7	2.15.5	2.43.3	2.71.1	76
77	.21.3	.49.1	.76.9	1.04.7	1.32.5	1 60 2	1.88.0	2.15 8	2.43.6	2.71.3	77
78	.21.6	.49.4	.77.2	1.05.0	1.32.7	1.60.5	1.88.3	2.16.1	2.43.8	2.71.6	78
79	.21.9	.49.7	.77.5	1.05.2	1.33.0	1.60.8	1.88.6	2.16.3	2.44.1	2.71.9	79
80	.22.2	.50.0	.77.7	1.05.5	1.33.3	1 61.1	1.88.8	2.16.6	2.44.4	2.72.2	80
81	.22.5	.50.2	.78.0	1.05.8	1.33.6	1.61.3	1.89.1	2.16.9	2.44.7	2.72.5	81
82	.22.7	.50 5	.78.3	1.06.1	1.33.8	1.61.6	1.89.4	2.17.2	2.45.0	2.72.7	82
83	.23.0	.50.8	.78.6	1.06.3	1.34.1	1.61.9	1.89.7	2.17.5	2.45.2	2.73.0	83
84	.23.3	.51.1	.78.8	1.06.6	1.34.4	1.62.2	1.90.0	2.17.7	2.45 5	2.73.3	84
85	.23.6	.51.3	.79.1	1.06.9	1.34.7	1.62.5	1.90.2	2.18.0	2.45.8	2.73.6	85
86	.23.8	.51.6	.79.4	1.07.2	1.35.0	1.62.7	1.90 5	2.18.3	2.46.1	2.73.8	86
87	.24.1	.51.9	.79.7	1.07.5	1.35.2	1.63.0	1.90.8	2.18.6	2 46.3	2.74.1	87
88	.24.4	.52.2	.80.0	1.07.7	1.35 5	1.63.3	1.91.1	2.18.8	2.46.6	2.74.4	88
89	.24.7	·52.5	.80.2	1.08.0	1.35.8	1.63.6	1.91.3	2.19.1	2.46.9	2.74.7	89
90	.25.0	.52.7	.80.5	1.08.3	1.36 1	1.63.8	1.91.6	2.19.4	2.47.2	2 75 0	90
91	.25.2	.53.0	.80.8	1.08.6	1.36.3	1.64.1	1.91.9	2.19.7	2.47.5	2.75 2	91
92	.25.5	.53.3	.81.1	1.08.8	1.36.6	1.64.4	1.92.2	2.20.0	2.47.7	2.75.5	92
93	.25.8	.53.6	.81.3	1.09.1	1.36.9	1.64.7	1.92.5	2.20.2	2.48.0	2.75.8	93
94	.26.1	.53.8	.81.6	1.09.4	1.37.2	1.65.0	1 92.7	2.20.5	2.48.3	2.76.1	94
95	.26.3	.54.1	.81.9	1.09.7	1.37.5	1.65.2	1.93.0	2.20.8	2.48.6	2.76.3	95
96	.26.6	.54.4	.82.2	1.10.0	1.37.7	1.65.5	1.93.3	2.21.1	2.48.8	2.76.6	96
97	.26.9	.54.7	.82.5	1.10.2	1.38.0	1.65.8	1.93.6	2.21.3	2.49.1	2.76.9	97
98	.27.2	.55.0	.82.7	1.10.5	1.38.3	1.66.1	1.93.8	2.21.6	2.49.4	2.77.2	98
99	.27.5	.55.2	.83 0	1.10.8	1.38.6	1.66.3	1.94.1	2.21.9	2.49.7	2.77.5	99

Shillings and pence not required in this table on account of the smallness of the values, that of 19 shilling being less than 2 mills—or .00.1 of a cent.

$4.73¹¹⁄₁₀₀ TO THE POUND.

6½ %

PREMIUM.

STERLING INTO DOLLARS AND CENTS, AND VICE-VERSA.

£	0	100	200	300	400	500	600	700	800	900	£
£	$ cts.	$ cts.	$ cts.	$ cts.	$ cts.	$ cts.	$ cts.	$ cts.	$ cts.	$ cts.	£
0	......	473.33.3	946.66.6	1420.00.0	1893.33.3	2366.66.6	2840.00.0	3313.33.3	3786.66.6	4260.00.0	0
1	4.73.3	478.06.6	951.40.0	1424.73.3	1898.06.6	2371.40.0	2844.73.3	3318.06.6	3791.40.0	4264.73.3	1
2	9.46.6	482.80.0	956.13.3	1429.46.6	1902.80.0	2376.13.3	2849.46.6	3322.80.0	3796.13.3	4269.46.6	2
3	14.20.0	487.53.3	960.86.6	1434.20.0	1907.53.3	2380.86.6	2854.20.0	3327.53.3	3800.86.6	4274.20.0	3
4	18.93.3	492.26.6	965.60.0	1438.93.3	1912.26.6	2385.60.0	2858.93.3	3832.26.6	3805.60.0	4278.93.3	4
5	23.66.6	497.00.0	970.33.3	1443.66.6	1917.00.0	2390.33.3	2863.66.6	3337.00.0	3810.33.3	4283.66.6	5
6	28.40.0	501.73.3	975.06.6	1448.40.0	1921.73.3	2395.06.6	2868.40.0	3341.73.3	3815.06.6	4288.40.0	6
7	33.13.3	506.46.6	979.80.0	1453.13.3	1926.46.6	2399.80.0	2873.13.3	3346.46.6	3819.80.0	4293.13.3	7
8	37.86.6	511.20.0	984.53.3	1457.86.6	1931.20.0	2404.53.3	2877.86.6	3351.20.0	3824.53.3	4297.86.6	8
9	42.60.0	515.93.3	989.26.6	1462.60.0	1935.93.3	2409.26.6	2882.60.0	3355.93.3	3829.26.6	4302.60.0	9
10	47.33.3	520.66.6	994.00.0	1467.33.3	1940.66.6	2414.00.0	2887.33.3	3360.66.6	3834.00.0	4307.33.3	10
11	52.06.6	525.40.0	998.73.3	1472.06.6	1945.40.0	2418.73.3	2892.06.6	3365.40.0	3838.73.3	4312.06.6	11
12	56.80.0	530.13.3	1003.46.6	1476.80.0	1950.13.3	2423.46.6	2896.80.0	3370.13.3	3843.46.6	4316.80.0	12
13	61.53.3	534.86.8	1008.20.0	1481.53.3	1954.86.6	2428.20.0	2901.53.3	3374.86.6	3848.20.0	4321.53.3	13
14	66.26.6	539.60.0	1012.93.3	1186.26.6	1959.60.0	2432.93.3	2906.26.6	3379.60.0	3852.93.3	4326.26.6	14
15	71.00.0	544.33.3	1017.66.6	1491.00.0	1964.33.3	2437.66.6	2911.00.0	3384.33.3	3857.66.6	4331.00.0	15
16	75.73.3	549.06.6	1022.40.0	1495.73.3	1969.06.6	2442.40.0	2915.73.3	3389.06.6	3862.40.0	4335.73.3	16
17	80.46.6	553.80.0	1027.13.3	1500.46.6	1973.80.0	2447.13.3	2920.46.6	3393.80.0	3867.13.3	4340.46.6	17
18	85.20.0	558.53.3	1031.86.6	1505.20.0	1978.53.3	2451.86.6	2925.20.0	3398.53.3	3871.86.6	4345.20.0	18
19	89.93.3	563.26.6	1036.60.0	1509.93.3	1983.26.6	2456.60.0	2929.93.3	3403.26.6	3876.60.0	4349.93.3	19
20	94.66.6	568.00.0	1041.33.3	1514.66.6	1988.00.0	2461.33.3	2934.66.6	3408.00.0	3881.33.3	4354.66.6	20
21	99.40.0	572.73.3	1046.06.6	1519.40.0	1992.73.3	2466.06.6	2939.40.0	3412.73.3	3886.06.6	4359.40.0	21
22	104.13.3	577.46.6	1050.80.0	1524.13.3	1997.46.6	2470.80.0	2944.13.3	3417.46.6	3890.80.0	4364.13.3	22
23	108.86.6	582.20.0	1055.53.3	1528.86.6	2002.20.0	2475.53.3	2948.86.6	3422.20.0	3895.53.3	4368.86.6	23
24	113.60.0	586.93.3	1060.26.6	1533.60.0	2006.93.3	2480.26.6	2953.60.0	3426.93.3	3900.26.6	4373.60.0	24
25	118.33.3	591.66.6	1065.00.0	1538.33.3	2011.66.6	2485.00.0	2958.33.3	3431.66.6	3905.00.0	4378.33.3	25
26	123.06.6	596.40.0	1069.73.3	1543.06.6	2016.40.0	2489.73.3	2963.06.6	3436.40.0	3909.73.3	4383.06.6	26
27	127.80.0	601.13.3	1074.46.6	1547.80.0	2021.13.3	2494.46.6	2967.80.0	3441.13.3	3914.46.6	4387.80.0	27
28	132.53.3	605.86.6	1079.20.0	1552.53.3	2025.86.6	2499.20.0	2972.53.3	3445.86.6	3919.20.0	4392.53.3	28
29	137.26.6	610.60.0	1083.93.3	1557.26.6	2030.60.0	2503.93.3	2977.26.6	3450.60.0	3923.93.3	4397.26.6	29
30	142.00.0	615.33.3	1088.66.6	1562.00.0	2035.33.3	2508.66.6	2982.00.0	3455.33.3	3928.66.6	4402.00.0	30
31	146.73.3	620.06.6	1093.40.0	1566.73.3	2040.06.6	2513.40.0	2986.73.3	3460.06.6	3933.40.0	4406.73.3	31
32	151.46.6	624.80.0	1098.13.3	1571.46.6	2044.80.0	2518.13.3	2991.46.6	3464.80.0	3938.13.3	4411.46.6	32
33	156.20.0	629.53.3	1102.86.6	1576.20.0	2049.53.3	2522.86.6	2996.20.0	3469.53.3	3942.86.6	4416.20.0	33
34	160.93.3	634.26.6	1107.60.0	1580.93.3	2054.26.6	2527.60.0	3000.93.3	3474.26.6	3947.60.0	4420.93.3	34
35	165.66.6	639.00.0	1112.33.3	1585.66.6	2059.00.0	2532.33.3	3005.66.6	3479.00.0	3952.33.3	4425.66.6	35
36	170.40.0	643.73.3	1117.06.6	1590.40.0	2063.73.3	2537.06.6	3010.40.0	3483.73.3	3957.06.6	4430.40.0	36
37	175.13.3	648.46.6	1121.80.0	1595.13.3	2068.46.6	2541.80.0	3015.13.3	3488.46.6	3961.80.0	4435.13.3	37
38	179.86.6	653.20.0	1126.53.3	1599.86.6	2073.20.0	2546.53.3	3019.86.6	3493.20.0	3966.53.3	4439.86.6	38
39	184.60.0	657.93.3	1131.26.6	1604.60.0	2077.93.3	2551.26.6	3024.60.0	3497.93.3	3971.26.6	4444.60.0	39
40	189.33.3	662.66.6	1136.00.0	1609.33.3	2082.66.6	2556.00.0	3029.33.3	3502.66.6	3976.00.0	4449.33.3	40
41	194.06.6	667.40.0	1140.73.3	1614.06.6	2087.40.0	2560.73.3	3034.06.6	3507.40.0	3980.73.3	4454.06.6	41
42	198.80.0	672.13.3	1145.46.6	1618.80.0	2092.13.3	2565.46.6	3038.80.0	3512.13.3	3985.46.6	4458.80.0	42
43	203.53.3	676.86.6	1150.20.0	1623.53.3	2096.86.6	2570.20.0	3043.53.3	3516.86.6	3990.20.0	4463.53.3	43
44	208.26.6	681.60.0	1154.93.3	1628.26.6	2101.60.0	2574.93.3	3048.26.6	3521.60.0	3994.93.3	4468.26.6	44
45	213.00.0	686.33.3	1159.66.6	1632.00.0	2106.33.3	2579.66.6	3053.00.0	3526.33.3	3999.66.6	4473.00.0	45
46	217.73.3	691.06.6	1164.40.0	1637.73.3	2111.06.6	2584.40.0	3057.73.3	3531.06.6	4004.40.0	4477.73.3	46
47	222.46.6	695.80.0	1169.13.3	1642.46.6	2115.80.0	2589.13.3	3062.46.6	3535.80.0	4009.13.3	4482.46.6	47
48	227.20.0	700.53.3	1173.86.6	1647.20.0	2120.53.3	2593.86.6	3067.20.0	3540.53.3	4013.86.6	4487.20.0	48
49	231.93.3	705.26.6	1178.60.0	1651.93.3	2125.26.6	2598.60.0	3071.93.3	3545.26.6	4018.60.0	4491.93.3	49

s.	0	1	2	3	4	5	6	7	8	9	s.
d.	$ cts.	$ cts.	$ cts.	$ cts.	$ cts.	$ cts.	$ cts.	$ cts.	$ cts.	$ cts.	d.
0	......	.23.6	.47.3	.71.0	.94.6	1.18.3	1.42.0	1.65.6	1.89.3	2.13.0	0
1	.01.9	.25.6	.49.3	.72.9	.96.6	1.20.3	1.43.9	1.67.6	1.91.3	2.14.9	1
2	.03.9	.27.6	.51.2	.74.9	.98.6	1.22.2	1.45.9	1.60.6	1.93.2	2.16.9	2
3	.05.9	.29.6	.53.2	.76.9	1.00.6	1.24.2	1.47.9	1.71.6	1.95.2	2.18.9	3
4	.07.9	.31.5	.55.2	.78.9	1.02.5	1.26.2	1.49.9	1.73.5	1.97.2	2.20.9	4
5	.09.8	.33.5	.57.2	.80.8	1.04.5	1.28.2	1.51.8	1.75.5	1.99.2	2.22.8	5
6	.11.8	.35.5	.50.2	.82.8	1.06.5	1.30.2	1.53.8	1.77.5	2.01.2	2.24.8	6
7	.13.8	.37.5	.61.1	.84.8	1.08.5	1.32.1	1.55.8	1.79.5	2.03.1	2.26.8	7
8	.15.8	.39.4	.63.1	.86.8	1.10.4	1.34.1	1.57.8	1.81.4	2.05.1	2.28.8	8
9	.17.8	.41.4	.65.1	.88.8	1.12.4	1.36.1	1.59.8	1.83.4	2.07.1	2.30.8	9
10	.19.7	.43.4	.67.1	.90.7	1.14.4	1.38.1	1.61.7	1.85.4	2.09.1	2.32.7	10
11	.21.7	.45.4	.69.0	.92.7	1.16.4	1.40.0	1.63.7	1.87.4	2.11.0	2.34.7	11

$4.73 7⁄100 TO THE POUND.

6½%
PREMIUM.

STERLING INTO DOLLARS AND CENTS, AND VICE-VERSA.

£	0	100	200	300	400	500	600	700	800	900	£
	$ cts.	$ cts.	$ cts.	$ cts.	$ cts.	$ cts.	$ cts.	$ cts.	$ cts.	$ cts.	
50	236.66.6	710.00.0	1183.33.3	1656.66.6	2130.00.0	2603.33.3	3076.66.6	3550.00.0	4023.33.3	4496.66.6	50
51	241.40.0	714.73.3	1188.06.6	1661.40.0	2134.73.3	2608.06.6	3081.40.0	3554.73.3	4028.06.6	4501.40.0	51
52	246.13.3	719.46.6	1192.80.0	1666.13.3	2139.46.6	2612.80.0	3086.13.3	3559.46.6	4032.80.0	4506.13.3	52
53	250.86.6	724.20.0	1197.53.3	1670.86.6	2144.20.0	2617.53.3	3090.86.6	3564.20.0	4037.53.3	4510.86.6	53
54	255.60.0	728.93.3	1202.26.6	1675.60.0	2148.93.3	2622.26.6	3095.60.0	3568.93.3	4042.26.6	4515.60.0	54
55	260.33.3	733.66.6	1207.00.0	1680.33.3	2153.66.6	2627.00.0	3100.33.3	3573.66.6	4047.00.0	4520.33.3	55
56	265.06.6	738.40.0	1211.73.3	1685.06.6	2158.40.0	2631.73.3	3105.06.6	3578.40.0	4051.73.3	4525.06.6	56
57	269.80.0	743.13.3	1216.46.6	1689.80.0	2163.13.3	2636.46.6	3109.80.0	3583.13.3	4056.46.6	4529.80.0	57
58	274.53.3	747.86.6	1221.20.0	1694.53.3	2167.86.6	2641.20.0	3114.53.3	3587.86.6	4461.20.0	4534.53.3	58
59	279.26.6	752.60.0	1225.93.3	1699.26.6	2172.60.0	2645.93.3	3119.26.6	3592.60.0	4065.93.3	4539.26.6	59
60	284.00.0	757.33.3	1230.66.6	1704.00.0	2177.33.3	2650.66.6	3124.00.0	3597.33.3	4070.66.6	4544.00.0	60
61	288.73.3	762.06.6	1235.40.0	1708.73.3	2182.06.6	2655.40.0	3128.73.3	3602.06.6	4075.40.0	4548.73.3	61
62	293.46.6	766.80.0	1240.13.3	1713.46.6	2186.80.0	2660.13.3	3133.46.6	3606.80.0	4080.13.3	4553.46.6	62
63	298.20.0	771.53.3	1244.86.6	1718.20.0	2191.53.3	2664.86.6	3138.20.0	3611.53.3	4084.86.6	4558.20.0	63
64	302.93.3	776.26.6	1249.60.0	1722.93.3	2196.26.6	2669.60.0	3142.93.3	3616.26.6	4089.60.0	4562.93.3	64
65	307.66.6	781.00.0	1254.33.3	1727.66.6	2201.00.0	2674.33.3	3147.66.6	3621.00.0	4094.33.3	4567.66.6	65
66	312.40.0	785.73.3	1259.06.6	1732.40.0	2205.73.3	2679.06.6	3152.40.0	3625.73.3	4099.06.6	4572.40.0	66
67	317.13.3	790.46.6	1263.80.0	1737.13.3	2210.46.6	2683.80.0	3157.13.3	3630.46.6	4103.80.0	4577.13.3	67
68	321.86.6	795.20.0	1268.53.3	1741.86.6	2215.20.0	2688.53.3	3161.86.6	3635.20.0	4108.53.3	4581.86.6	68
69	326.60.0	799.93.3	1273.26.6	1746.60.0	2219.93.3	2693.26.6	3166.60.0	3639.93.3	4113.26.6	4586.60.0	69
70	331.33.3	804.66.6	1278.00.0	1751.33.3	2224.66.6	2698.00.0	3171.33.3	3644.66.6	4118.00.0	4591.33.3	70
71	336.06.6	809.40.0	1282.73.3	1756.06.6	2229.40.0	2702.73.3	3176.06.6	3649.40.0	4122.73.3	4596.06.6	71
72	340.80.0	814.13.3	1287.46.6	1760.80.0	2234.13.3	2707.46.6	3180.80.0	3654.13.3	4127.46.6	4600.80.0	72
73	345.53.3	818.86.6	1292.20.0	1765.53.3	2238.86.6	2712.20.0	3185.53.3	3658.86.6	4132.20.0	4605.53.3	73
74	350.26.6	823.60.0	1296.93.3	1770.26.6	2243.60.0	2716.93.3	3190.26.6	3663.60.0	4136.93.3	4610.26.6	74
75	355.00.0	828.33.3	1301.66.6	1775.00.0	2248.33.3	2721.66.6	3195.00.0	3668.33.3	4141.66.6	4615.00.0	75
76	359.73.3	833.06.6	1306.40.0	1779.73.3	2253.06.6	2726.40.0	3199.73.3	3673.06.6	4146.40.0	4619.73.3	76
77	364.46.6	837.80.0	1311.13.3	1784.46.6	2257.80.0	2731.13.3	3204.46.6	3677.80.0	4151.13.3	4624.46.6	77
78	369.20.0	842.53.3	1315.86.6	1789.20.0	2262.53.3	2735.86.6	3209.20.0	3682.53.3	4155.86.6	4629.20.0	78
79	373.93.3	847.26.6	1320.60.0	1793.93.3	2267.26.6	2740.60.0	3213.93.3	3687.26.6	4160.60.0	4633.93.3	79
80	378.66.6	852.00.0	1325.33.3	1798.66.6	2272.00.0	2745.33.3	3218.66.6	3692.00.0	4165.33.3	4638.66.6	80
81	383.40.0	856.73.3	1330.06.6	1803.40.0	2276.73.3	2750.06.6	3223.40.0	3696.73.3	4170.06.6	4643.40.0	81
82	388.13.3	861.46.6	1334.80.0	1808.13.3	2281.46.6	2754.80.0	3228.13.3	3701.46.6	4174.80.0	4618.13.3	82
83	392.86.6	866.20.0	1339.53.3	1812.86.6	2286.20.0	2759.53.3	3232.86.6	3706.20.0	4179.53.3	4652.86.6	83
84	397.60.0	870.93.3	1344.26.6	1817.60.0	2290.93.3	2764.26.6	3237.60.0	3710.93.3	4184.26.6	4657.60.0	84
85	402.33.3	875.66.6	1349.00.0	1822.33.3	2295.66.6	2769.00.0	3242.33.3	3715.66.6	4189.00.0	4662.33.3	85
86	407.06.6	880.40.0	1353.73.3	1827.06.6	2300.40.0	2773.73.3	3247.06.6	3720.40.0	4193.73.3	4667.06.6	86
87	411.80.0	885.13.3	1358.46.6	1831.80.0	2305.13.3	2778.46.6	3251.80.0	3725.13.3	4198.46.6	4671.80.0	87
88	416.53.3	889.86.6	1363.20.0	1836.53.3	2309.86.6	2783.20.0	3256.53.3	3729.86.6	4203.20.0	4676.53.3	88
89	421.26.6	894.60.0	1367.93.3	1841.26.6	2314.60.0	2787.93.3	3261.26.6	3734.60.0	4207.93.3	4681.26.6	89
90	426.00.0	899.33.3	1372.66.6	1846.00.0	2319.33.3	2792.66.6	3266.00.0	3739.33.3	4212.66.6	4686.00.0	90
91	430.73.3	904.06.6	1377.40.0	1850.73.3	2324.06.6	2797.40.0	3270.73.3	3744.06.6	4217.40.0	4690.73.3	91
92	435.46.6	908.80.0	1382.13.3	1855.46.6	2328.80.0	2802.13.3	3275.46.6	3748.80.0	4222.13.3	4695.46.6	92
93	440.20.0	913.53.3	1386.86.6	1860.20.0	2333.53.3	2806.86.6	3280.20.0	3753.53.3	4226.86.6	4700.20.0	93
94	444.93.3	918.26.6	1391.60.0	1864.93.3	2338.26.6	2811.60.0	3284.93.3	3758.26.6	4231.60.0	4704.93.3	94
95	449.66.6	923.00.0	1396.33.3	1869.66.6	2343.00.0	2816.33.3	3289.66.6	3763.00.0	4236.33.3	4709.66.6	95
96	454.40.0	927.73.3	1401.06.6	1874.40.0	2347.73.3	2821.06.6	3294.40.0	3767.73.3	4241.06.6	4714.40.0	96
97	459.13.3	932.46.6	1405.80.0	1879.13.3	2352.46.6	2825.80.0	3299.13.3	3772.46.6	4245.80.0	4719.13.3	97
98	463.86.6	937.20.0	1410.53.3	1883.86.6	2357.20.0	2830.53.3	3303.86.6	3777.20.0	4250.53.3	4723.86.6	98
99	468.60.0	941.93.3	1415.26.6	1888.60.0	2361.93.3	2835.26.6	3308.60.0	3781.93.3	4255.26.6	4728.60.0	99

S.	10	11	12	13	14	15	16	17	18	19	S.
d.	$ cts.	$ cts.	$ cts.	$ cts.	$ cts.	$ cts.	$ cts.	$ cts.	$ cts	$ cts.	d.
0	2.36.6	2.60.3	2.84.0	3.07.6	3.31.3	3.55.0	3.78.6	4.02.3	4.26.0	4.49.6	0
1	2.38.6	2.62.3	2.85.9	3.09.6	3.33.3	3.56.9	3.80.6	4.04.3	4.27.9	4.51.6	1
2	2.40.6	2.64.2	2.87.9	3.11.6	3.35.2	3.58.9	3.82.6	4.06.2	4.29.9	4.53.6	2
3	2.42.6	2.66.2	2.89.9	3.13.6	3.37.2	3.60.9	3.84.6	4.08.2	4.31.9	4.55.6	3
4	2.44.5	2.68.2	2.91.9	3.15.5	3.39.2	3.62.9	3.86.5	4.10.2	4.33.9	4.57.5	4
5	2.46.5	2.70.2	2.93.8	3.17.5	3.41.2	3.64.8	3.88.5	4.12.2	4.35.8	4.59.5	5
6	2.48.5	2.72.2	2.95.8	3.19.5	3.43.2	3.66.8	3.90.5	4.14.2	4.37.8	4.61.5	6
7	2.50.5	2.74.1	2.97.8	3.21.5	3.45.1	3.68.8	3.92.5	4.16.1	4.39.8	4.63.5	7
8	2.52.4	2.76.1	2.99.8	3.23.4	3.47.1	3.70.8	3.94.4	4.18.1	4.41.8	4.65.4	8
9	2.54.4	2.78.1	3.01.8	3.25.4	3.49.1	3.72.8	3.96.4	4.20.1	4.43.8	4.67.4	9
10	2.56.4	2.80.1	3.03.7	3.27.4	3.51.1	3.74.7	3.98.4	4.22.1	4.45.7	4.69.4	10
11	2.58.4	2.82.0	3.05.7	3.29.4	3.53.0	3.76.7	4.00.4	4.24.0	4.47.7	4.71.4	11

$4.75 1⅖ TO THE POUND.

7%

PREMIUM.

STERLING INTO DOLLARS AND CENTS, AND VICE-VERSA.

£	0	100	200	300	400	500	600	700	800	900	£
£	$ cts.	$ cts.	$ cts.	$ cts.	$ cts.	$ cts.	$ cts.	$ cts.	$ cts.	$ cts.	£
0	......	475.55.5	951.11.1	1426.66.6	1902.22.2	2377.77.7	2853.33.3	3328.88.8	3804.44.4	4280.00.0	0
1	4.75.5	480.31.1	955.86.6	1431.42.2	1906.97.7	2382.53.3	2858.08.8	3333.64.4	3809.20.0	4284.75.5	1
2	9.51.1	485.06.6	960 62.2	1436 17.7	1911.73.3	2387.28.8	2862.84.4	3338 40.0	3813.95.5	4289.51.1	2
3	14.26.6	489.82.2	965.37.7	1440.93.3	1916.48.8	2392.04.4	2867.60.0	3343.15.5	3818.71.1	4294.26.6	3
4	19.02.2	494.57.7	970.13.3	1445.68.8	1921.24.4	2396.80.0	2872.35.5	3347.91.1	3823 46.6	4299.02.2	4
5	23.77.7	499.33.3	974.88.8	1450.44.4	1926.00.0	2401 55.5	2877.11.1	3352.66.6	3828.22.2	4303.77.7	5
6	28.53.3	504.08.8	979.64.4	1455.20.0	1930.75.5	2406 31.1	2881.86.6	3357.42.2	3832.97.7	4308.53.3	6
7	33.28.8	508.84.4	984.40.0	1459.95.5	1935.51.1	2411.06.6	2886.62.2	3362.17.7	3837.73.3	4313.28.8	7
8	38.04.4	513.60.0	989.15.5	1464.71.1	1940.26.6	2415.82.2	2891.37.7	3366.93.3	3842.48.8	4318.04.4	8
9	42.80.0	518.35.5	993.91.1	1469.46.6	1945.02.2	2420.57.7	2896.13.3	3371.68.8	3847.24.4	4322.80.0	9
10	47.55.5	523.11.1	998.66.6	1474.22.2	1949.77.7	2425.33.3	2900.88.8	3376.44.4	3852.00.0	4327.55.5	10
11	52.31.1	527.86.6	1003.42.2	1478.97.7	1954.53.3	2430.08.8	2905.64.4	3381.20.0	3856.75.5	4332.31.1	11
12	57 06 6	532.62.2	1008.17.7	1483.73.3	1959.28.8	2434.84.4	2910.40.0	3385.95.5	3861.51.1	4337.06.6	12
13	61.82.2	537.37.7	1012.93.3	1488.48.8	1964.04.4	2439.60.0	2915.15.5	3390.71.1	3866.26.6	4341.82.2	13
14	66.57.7	542.13.3	1017.68.8	1493.24.4	1968.80.0	2444.35.5	2919.91.1	3395.46.6	3871.02.2	4346.57.7	14
15	71.33.3	546.88.8	1022.44.4	1498.00.0	1973.55.5	2449.11.1	2924.66.6	3400.22.2	3875.77.7	4351.33.3	15
16	76.08.8	551.64.4	1027.20 0	1502.75.5	1978.31.1	2453.86.6	2929.42.2	3404.97.7	3880.53.3	4356.08.8	16
17	80.84.4	556.40.0	1031.95.5	1507.51.1	1983.06.6	2458.62.2	2934.17.7	3409 73.3	3885.28.8	4360.84.4	17
18	85.60.0	561.15.5	1036.71.1	1512.26 6	1987.82.2	2463.37.7	2938.93 3	3414.48.8	3890.04.4	4365.60.0	18
19	90.35.5	565.91.1	1041.46.6	1517.02.2	1992.57.7	2468.13.3	2943.68.8	3419.24.4	3894.80.0	4370.35.5	19
20	95.11.1	570.66.6	1046.22.2	1521.77.7	1997.33.3	2472.88.8	2948.44.4	3424.00.0	3899.55.5	4375.11.1	20
21	99.86 6	575.42.2	1050.97.7	1526.53.3	2002.08.8	2477.64.4	2953.20.0	3428.75.5	3904.31.1	4379.86.6	21
22	104.62.2	580.17.7	1055.73.3	1531.28.8	2006.84.4	2482.40.0	2957.95.5	3433 51.1	3909.06.6	4384.62.2	22
23	109.37.7	584.93.3	1060.48.8	1536.04.4	2011.60.0	2487.15.5	2962.71 1	3438.26.6	3913 82.2	4389.37.7	23
24	114.13.3	589.68.8	1065.24.4	1540.80.0	2016.35.5	2491.91.1	2967.46.6	3443.02.2	3918.57.7	4394.13.3	24
25	118.88.8	594.44.4	1070.00.0	1545.55.5	2021.11.1	2496.66.6	2972.22.2	3447.77.7	3923.33.3	4398.88.8	25
26	123.64.4	599 20.0	1074.75.5	1550.31.1	2025.86.6	2501.42.2	2976.97.7	3452.53.3	3928.08.8	4403.64.4	26
27	128.40.0	603.95.5	1079.51.1	1555.06.6	2030.62.2	2506.17.7	2981.73.3	3457.28.8	3932.84.4	4408.40.0	27
28	133.15.5	608.71.1	1084.26.6	1559.82.2	2035.37.7	2510.93.3	2986.48.8	3462.04.4	3937.60.0	4413.15.5	28
29	137.91.1	613.46.6	1089.02.2	1564.57.7	2040.13.3	2515.68.8	2991.24.4	3466.80.0	3942.35.5	4417.91.1	29
30	142.66.6	618.22.2	1093.77.7	1569.33.3	2044.88.8	2520.44.4	2996.00.0	3471.55.5	3947.11.1	4422.66.6	30
31	147.42.2	622.97.7	1098.53.3	1574.08.8	2049.64.4	2525.20.0	3000.75.5	3476.31.1	3951.86.6	4427.42.2	31
32	152.17.7	627.73 3	1103.28.8	1578.84.4	2054.40.0	2529.95.5	3005.51.1	3481.06.6	3956.62.2	4432.17.7	32
33	156.93.3	632.48.8	1108.04.4	1583.60.0	2059.15.5	2534.71.1	3010.26.6	3485.82.2	3961.37.7	4436.93.3	33
34	161.68.8	637.24.4	1112.80.0	1588.35.5	2063.91.1	2539.46.6	3015.02.2	3490.57.7	3966.13.3	4441.68.8	34
35	166.44.4	642.00.0	1117.55.5	1593.11.1	2068.66.6	2544.22.2	3019.77.7	3495.33.3	3970.88.8	4446.44.4	35
36	171.20.0	646.75.5	1122.31.1	1597.86.6	2073.42.2	2548.97.7	3024.53.3	3500.08.8	3975.64.4	4451.20.0	36
37	175.95.5	651.51.1	1127.06.6	1602.62.2	2078.17.7	2553.73.3	3029.28.8	3504.84.4	3980.40.0	4455.95.5	37
38	180.71.1	656.26.6	1131.82.2	1607.37.7	2082.93.3	2558.48.8	3034.04.4	3509.60.0	3985.15.5	4460.71.1	38
39	185.46.6	661.02.2	1136.57.7	1612.13.3	2087.68.8	2563.24.4	3038.80.0	3514.35.5	3989.91.1	4465.46.6	39
40	190.22.2	665.77.7	1141 33.3	1616.88.8	2092.44.4	2568.00.0	3043.55.5	3519.11.1	3994.66.6	4470 22.2	40
41	194.97.7	670.53.3	1146.08.8	1621.64.4	2097.20.0	2572.75.5	3048.31.1	3523.86.6	3999.42.2	4474.97.7	41
42	199.73.3	675.28.8	1150.84.4	1626.40.0	2101.95.5	2577.51.1	3053 06.6	3528.62.2	4004.17.7	4479.73.3	42
43	204.48.8	680.04.4	1155.60.0	1631.15.5	2106.71.1	2582.26.6	3057.82.2	3533.37.7	4008.93.3	4484.48.8	43
44	209.24.4	684.80.0	1160.35.5	1635.91.1	2111.46.6	2587.02.2	3062.57.7	3538.13.3	4013.68.8	4489.24.4	44
45	214.00.0	689.55.5	1165.11.1	1640.66.6	2116.22.2	2591.77.7	3067.33.3	3542.88.8	4018.44.4	4494.00.0	45
46	218.75.5	694.31.1	1169.86.6	1645.42.2	2120.97.7	2596.53.3	3072.08.8	3547.64.4	4023.20.0	4498.75.5	46
47	223.51.1	699 06.6	1174.62 2	1650.17.7	2125.73.3	2601.28.8	3076.84.4	3552.40.0	4027.95 5	4503.51.1	47
48	228.26.6	703.82.2	1179.37.7	1654.93.3	2130.48.8	2606.04.4	3081.60.0	3557.15.5	4032.71.1	4508.26.6	48
49	233 02.2	708.57.7	1184.13.3	1659.68.8	2135.24.4	2610.80.0	3086.35.5	3561.91.1	4037.46.6	4513.02.2	49

S.	0	1	2	3	4	5	6	7	8	9	S.
d.	$ cts.	$ cts.	$ cts.	$ cts.	$ cts.	$ cts.	$ cts.	$ cts.	$ cts.	$ cts.	d.
0	......	.23.7	.47.5	.71.3	.95.1	1.18.8	1.42.6	1.66.4	1.90.2	2.14.0	0
1	.01.9	.25.7	.49.5	.73.3	.97.1	1.20.8	1.44.6	1.68.4	1.92.2	2.15.9	1
2	.03.9	.27.7	.51.5	.75.3	.99.0	1.22.8	1.46.6	1.70.4	1.94.2	2.17.9	2
3	.05.9	.29.7	.53.5	.77.3	1.01.0	1.24.8	1.48.6	1.72.4	1.96.1	2.19.9	3
4	.07.9	.31.7	.55.5	.79.2	1.03.0	1.26.8	1.50.6	1.74.4	1.98.1	2.21.9	4
5	.09.9	.33.7	.57.5	.81.2	1.05.0	1.28.8	1.52.6	1.76.3	2.00.1	2.23.9	5
6	.11.9	.35.7	.59.4	.83.2	1.07.0	1.30.8	1.54.6	1.78.3	2.02.1	2.25.9	6
7	.13.9	.37.7	.61.4	.85.2	1.09.0	1.32.8	1.56.5	1.80.3	2.04.1	2.27.9	7
8	.15.9	.39.6	.63.4	.87.2	1.11.0	1.34.7	1.58.5	1.82.3	2.06.1	2.29.9	8
9	.17.8	.41.6	.65.4	.89.2	1.12.9	1.36.7	1.60.5	1.84.3	2.08.1	2.31.8	9
10	.19.8	.43.6	.67.4	.91.1	1.14.9	1.38.7	1.62.5	1.86.3	2.10.0	2.33.8	10
11	.21.8	.45.6	.69.3	.93.1	1.16.9	1.40.7	1.64.5	1.88.2	2.12.0	2.35.8	11

$4.76⁴⁴ TO THE POUND.

7%

PREMIUM.

STERLING INTO DOLLARS AND CENTS, AND VICE-VERSA.

£	0	100	200	300	400	500	600	700	800	900	£
	$ cts.	$ cts.	$ cts.	$ cts.	$ cts.	$ cts.	$ cts.	$ cts.	$ cts.	$ cts.	
50	237.77.7	713.33.3	1188.88.8	1664.44.4	2140.00.0	2615.55.5	3091.11.1	3566.66.6	4042.22.2	4517.77.7	50
51	242.53.3	718.08.8	1193.64.4	1669.20.0	2144.75.5	2620.31.1	3095.86.6	3571.42.2	4046.97.7	4522.53.3	51
52	247.28.8	722.84.4	1198.40.0	1673.95.5	2149.51.1	2625.06.6	3100.62.2	3576.17.7	4051.73.3	4527.28.8	52
53	252.04.4	727.60.0	1203.15.5	1678.71.1	2154.26.6	2629.82.2	3105.37.7	3580.93.3	4056.48.8	4532.04.4	53
54	256.80.0	732.35.5	1207.91.1	1683.46.6	2159.02.2	2634.57.7	3110.13.3	3585.68.8	4061.24.4	4536.80.0	54
55	261.55.5	737.11.1	1212.66.6	1688.22.2	2163.77.7	2639.33.3	3114.88.8	3590.44.4	4066.00.0	4541.55.5	55
56	266.31.1	741.86.6	1217.42.2	1692.97.7	2168.53.3	2644.08.8	3119.64.4	3595.20.0	4070.75.5	4546.31.1	56
57	271.06.6	746.62.2	1222.17.7	1697.73.3	2173.28.8	2648.84.4	3124.40.0	3599.95.5	4075.51.1	4551.06.6	57
58	275.82.2	751.37.7	1226.93.3	1702.48.8	2178.04.4	2653.60.0	3129.15.5	3604.71.1	4080.26.6	4555.82.2	58
59	280.57.7	756.13.3	1231.68.8	1707.24.4	2182.80.0	2658.35.5	3133.91.1	3609.46.6	4085.02.2	4560.57.7	59
60	285.33.3	760.88.8	1236.44.4	1712.00.0	2187.55.5	2663.11.1	3138.66.6	3614.22.2	4089.77.7	4565.33.3	60
61	290.08.8	765.64.4	1241.20.0	1716.75.5	2192.31.1	2667.86.6	3143.42.2	3618.97.7	4094.53.3	4570.08.8	61
62	294.84.4	770.40.0	1245.95.5	1721.51.1	2197.06.6	2672.62.2	3148.17.7	3623.73.3	4099.28.8	4574.84.4	62
63	299.60.0	775.15.5	1250.71.1	1726.26.6	2201.82.2	2677.37.7	3152.93.3	3628.48.8	4104.04.4	4579.60.0	63
64	304.35.5	779.91.1	1255.46.6	1731.02.2	2206.57.7	2682.13.3	3157.68.8	3633.24.4	4108.80.0	4584.35.5	64
65	309.11.1	784.66.6	1260.22.2	1735.77.7	2211.33.3	2686.88.8	3162.44.4	3638.00.0	4113.55.5	4589.11.1	65
66	313.86.6	789.42.2	1264.97.7	1740.53.3	2216.08.8	2691.64.4	3167.20.0	3642.75.5	4118.31.1	4593.86.6	66
67	318.62.2	794.17.7	1269.73.3	1745.28.8	2220.84.4	2696.40.0	3171.95.5	3647.51.1	4123.06.6	4598.62.2	67
68	323.37.7	798.93.3	1274.48.8	1750.04.4	2225.60.0	2701.15.5	3176.71.1	3652.26.6	4127.82.2	4603.37.7	68
69	328.13.3	803.68.8	1279.24.4	1754.80.0	2230.35.5	2705.91.1	3181.46.6	3657.02.2	4132.57.7	4608.13.3	69
70	332.88.8	808.44.4	1284.00.0	1759.55.5	2235.11.1	2710.66.6	3186.22.2	3661.77.7	4137.33.3	4612.88.8	70
71	337.64.4	813.20.0	1288.75.5	1764.31.1	2239.86.6	2715.42.2	3190.97.7	3666.53.3	4142.08.8	4617.64.4	71
72	342.40.0	817.95.5	1293.51.1	1769.06.6	2244.62.2	2720.17.7	3195.73.3	3671.28.8	4146.84.4	4622.40.0	72
73	347.15.5	822.71.1	1298.26.6	1773.82.2	2249.37.7	2724.93.3	3200.48.8	3676.04.4	4151.60.0	4627.15.5	73
74	351.91.1	827.46.6	1303.02.2	1778.57.7	2254.13.3	2729.68.8	3205.24.4	3680.80.0	4156.35.5	4631.91.1	74
75	356.66.6	832.22.2	1307.77.7	1783.33.3	2258.88.8	2734.44.4	3210.00.0	3685.55.5	4161.11.1	4636.66.6	75
76	361.42.2	836.97.7	1312.53.3	1788.08.8	2263.64.4	2739.20.0	3214.75.5	3690.31.1	4165.86.6	4641.42.2	76
77	366.17.7	841.73.3	1317.28.8	1792.84.4	2268.40.0	2743.95.5	3219.51.1	3695.06.6	4170.62.2	4646.17.7	77
78	370.93.3	846.48.8	1322.04.4	1797.60.0	2273.15.5	2748.71.1	3224.26.6	3699.82.2	4175.37.7	4650.93.3	78
79	375.68.8	851.24.4	1326.80.0	1802.35.5	2277.91.1	2753.46.6	3229.02.2	3704.57.7	4180.13.3	4655.68.8	79
80	380.44.4	856.00.0	1331.55.5	1807.11.1	2282.66.6	2758.22.2	3233.77.7	3709.33.3	4184.88.8	4660.44.4	80
81	385.20.0	860.75.5	1336.31.1	1811.86.6	2287.42.2	2762.97.7	3238.53.3	3714.08.8	4189.64.4	4665.20.0	81
82	389.95.5	865.51.1	1341.06.6	1816.62.2	2292.17.7	2767.73.3	3243.28.8	3718.84.4	4194.40.0	4669.95.5	82
83	394.71.1	870.26.6	1345.82.2	1821.37.7	2296.93.3	2772.48.8	3248.04.4	3723.60.0	4199.15.5	4674.71.1	83
84	399.46.6	875.02.2	1350.57.7	1826.13.3	2301.68.8	2777.24.4	3252.80.0	3728.35.5	4203.91.1	4679.46.6	84
85	404.22.2	879.77.7	1355.33.3	1830.88.8	2306.44.4	2782.00.0	3257.55.5	3733.11.1	4208.66.6	4684.22.2	85
86	408.97.7	884.53.3	1360.08.8	1835.64.4	2311.20.0	2786.75.5	3262.31.1	3737.86.6	4213.42.2	4688.97.7	86
87	413.73.3	889.28.8	1364.84.4	1840.40.0	2315.95.5	2791.51.1	3267.06.6	3742.62.2	4218.17.7	4693.73.3	87
88	418.48.8	894.04.4	1369.60.0	1845.15.5	2320.71.1	2796.26.6	3271.82.2	3747.37.7	4222.93.3	4698.48.8	88
89	423.24.4	898.80.0	1374.35.5	1849.91.1	2325.46.6	2801.02.2	3276.57.7	3752.13.3	4227.68.8	4703.24.4	89
90	428.00.0	903.55.5	1379.11.1	1854.66.6	2330.22.2	2805.77.7	3281.33.3	3756.88.8	4232.44.4	4708.00.0	90
91	432.75.5	908.31.1	1383.86.6	1859.42.2	2334.97.7	2810.53.3	3286.08.8	3761.64.4	4237.20.0	4712.75.5	91
92	437.51.1	913.06.6	1388.62.2	1864.17.7	2339.73.3	2815.28.8	3290.84.4	3766.40.0	4241.95.5	4717.51.1	92
93	442.26.6	917.82.2	1393.37.7	1868.93.3	2344.48.8	2820.04.4	3295.60.0	3771.15.5	4246.71.1	4722.26.6	93
94	447.02.2	922.57.7	1398.13.3	1873.68.8	2349.24.4	2824.80.0	3300.35.5	3775.91.1	4251.46.6	4727.02.2	94
95	451.77.7	927.33.3	1402.88.8	1878.44.4	2354.00.0	2829.55.5	3305.11.1	3780.66.6	4256.22.2	4731.77.7	95
96	456.53.3	932.08.8	1407.64.4	1883.20.0	2358.75.5	2834.31.1	3309.86.6	3785.42.2	4260.97.7	4736.53.3	96
97	461.28.8	936.84.4	1412.40.0	1887.95.5	2363.51.1	2839.06.6	3314.62.2	3790.17.7	4265.73.3	4741.28.8	97
98	466.04.4	941.60.0	1417.15.5	1892.71.1	2368.26.6	2843.82.2	3319.37.7	3794.93.3	4270.48.8	4746.04.4	98
99	470.80.0	946.35.5	1421.91.1	1897.46.6	2373.02.2	2848.57.7	3324.13.3	3799.68.8	4275.24.4	4750.80.0	99

S.	10	11	12	13	14	15	16	17	18	19	S.
d.	$ cts.	$ cts.	$ cts.	$ cts.	$ cts.	$ cts.	$ cts.	$ cts.	$ cts.	$ cts.	d.
0	2.37.7	2.61.5	2.85.3	3.09.1	3.32.8	3.56.6	3.80.4	4.04.2	4.28.0	4.51.7	0
1	2.39.7	2.63.5	2.87.3	3.11.1	3.34.8	3.58.6	3.82.4	4.06.2	4.29.9	4.53.7	1
2	2.41.7	2.65.5	2.89.3	3.13.0	3.36.8	3.60.6	3.84.4	4.08.2	4.31.9	4.55.7	2
3	2.43.7	2.67.5	2.91.3	3.15.0	3.38.8	3.62.6	3.86.4	4.10.1	4.33.9	4.57.7	3
4	2.45.7	2.69.5	2.93.2	3.17.0	3.40.8	3.64.6	3.88.4	4.12.1	4.35.9	4.59.7	4
5	2.47.7	2.71.5	2.95.2	3.19.0	3.42.8	3.66.6	3.90.3	4.14.1	4.37.9	4.61.7	5
6	2.49.7	2.73.4	2.97.2	3.21.0	3.44.8	3.68.6	3.92.3	4.16.1	4.39.9	4.63.7	6
7	2.51.7	2.75.4	2.99.2	3.23.0	3.46.8	3.70.5	3.94.3	4.18.1	4.41.9	4.65.7	7
8	2.53.6	2.77.4	3.01.2	3.25.0	3.48.7	3.72.5	3.96.3	4.20.1	4.43.9	4.67.8	8
9	2.55.6	2.79.4	3.03.2	3.26.9	3.50.7	3.74.5	3.98.3	4.22.1	4.45.8	4.69.6	9
10	2.57.6	2.81.4	3.05.1	3.28.9	3.52.7	3.76.5	4.00.3	4.24.0	4.47.8	4.71.6	10
11	2.59.6	2.83.3	3.07.1	3.30.9	3.54.7	3.78.5	4.02.2	4.26.0	4.49.8	4.73.6	11

B4.77 11/110 TO THE POUND.

7½ %

PREMIUM.

STERLING INTO DOLLARS AND CENTS, AND VICE-VERSA.

£	0	100	200	300	400	500	600	700	800	900	£
	$ cts.	$ cts.	$ cts.	$ cts.	$ cts.	$ cts.	$ cts.	$ cts.	$ cts.	$ cts.	
0		477.77.7	955.55.5	1433.33.3	1911.11.1	2388.88.8	2866.66.6	3344.44 4	3822.22.2	4300.00.0	0
1	4.77.7	482.55.5	960.33.3	1438.11.1	1915.88 8	2393.66.6	2871.44.4	3349.22.2	3827.00.0	4304.77.7	1
2	9.55.5	487.33.3	965.11.1	1442.88.8	1920.66.6	2398.44.4	2876.22.2	3354.00.0	3831.77.7	4309.55.5	2
3	14.33.3	492.11.1	969.88.8	1447.66.6	1925.44.4	2403.22.2	2881.00.0	3358.77.7	3836.55.5	4314.33.3	3
4	19.11.1	496.88.8	974.66.6	1452.44.4	1930.22.2	2408 00.0	2885.77.7	3363.55.5	3841.33.3	4319.11.1	4
5	23.88.8	501.66.6	979.44.4	1457.22.2	1935.00.0	2412.77.7	2890.55.5	3368.33.3	3846.11.1	4323.88.8	5
6	28.66.6	506.44.4	984.22.2	1462.00.0	1939.77.7	2417.55.5	2895.33.3	3373.11.1	3850.88.8	4328.66.6	6
7	33.44.4	511.22 2	989.00.0	1466.77.7	1944.55.5	2422.33.3	2900.11.1	3377.88.8	3855.66.6	4333.44.4	7
8	38.22.2	516 00 0	993.77.7	1471.55.5	1949.33.3	2427.11.1	2904.88.8	3382 66 6	3860 44.4	4338.22.2	8
9	43.00.0	520.77.7	998.55.5	1476.33.3	1954.11.1	2431.88.8	2909.66.6	3387.44.4	3865.22 2	4343.00.0	9
10	47.77.7	525.55.5	1003.33.3	1481.11.1	1958.88.8	2436.66.6	2914.44.4	3392.22.2	3870.00.0	4317.77.7	10
11	52.55.5	530.33.3	1008.11.1	1485.88.8	1963.66.6	2441.44 4	2919.22.2	3397.00.0	3874.77.7	4352.55.5	11
12	57.33.3	535 11.1	1012.88.8	1490.66.6	1968.44.4	2446.22.2	2924.00.0	3401.77.7	3879.55.5	4357.33.3	12
13	62.11.1	539.88.8	1017.66.6	1495.44.4	1973.22.2	2451.00.0	2928.77.7	3406.55.5	3884.33.3	4302.11.1	13
14	66.88.8	544.66.6	1022.44.4	1500.22.2	1978.00.0	2455.77.7	2933 55.5	3411.33.3	3889.11.1	4366.88.8	14
15	71.66.6	549.44.4	1027.22.2	1505.00.0	1982.77.7	2460.55.5	2938.33.3	3416.11.1	3893.88.8	4371.66.6	15
16	76.44.4	554.22.2	1032.00.0	1509.77.7	1987.55.5	2465.33.3	2943.11.1	3420.88.8	3898.66.6	4376 44.4	16
17	81.22.2	559.00.0	1036.77.7	1514.55.5	1992.33.3	2470.11.1	2947.88 8	3425.66.6	3903.44.4	4381.22.2	17
18	86.00.0	563.77.7	1041.55.5	1519.33.3	1997.11.1	2474.88.8	2952.66.6	3430.44.4	3908.22 2	4386.00.0	18
19	90.77.7	568.55.5	1046.33.3	1524.11.1	2001.88.8	2479 66.6	2957.44.4	3435.22.2	3913.00.0	4390.77.7	19
20	95.55.5	573.33.3	1051.11.1	1528.88.8	2006.66.6	2484.44 4	2962.22.2	3440.00.0	3917.77.7	4395 55.5	20
21	100.33.3	578.11.1	1055.88.8	1533.66.6	2011.44.4	2489.22.2	2967.00.0	3444.77.7	3922.55.5	4400.33.3	21
22	105.11.1	582.88.8	1060.66.6	1538.44.4	2016.22.2	2494.00.0	2971.77.7	3449.55.5	3927.33.3	4405.11.1	22
23	109.88.8	587.66.6	1065.44.4	1543.22.2	2021.00.0	2498.77.7	2976.55.5	3454.33.3	3932.11.1	4409.88.8	23
24	114.66.6	592.44.4	1070.22.2	1548.00.0	2025.77.7	2503.55.5	2981.33.3	3459.11.1	3936.88.8	4414.66.6	24
25	119.44.4	597.22.2	1075.00.0	1552.77.7	2030.55.5	2508.33.3	2986.11.1	3463.88.8	3941.66.6	4419.41.4	25
26	124.22.2	602.00.0	1079.77.7	1557.55.5	2035.33.3	2513.11 1	2990.88 8	3468.66.6	3946.44.4	4424.22.2	26
27	129.00.0	606.77.7	1084 55.5	1562.33.3	2040.11.1	2517.88.8	2995.66.6	3473.44.4	3951.22.2	4429.00.0	27
28	133.77.7	611.55.5	1089.33 3	1567.11.1	2044.88.8	2522 66.6	3000.44.4	3478.22.2	3956.00.0	4433.77.7	28
29	138.55.5	616.33 3	1094.11.1	1571.88.8	2049.66.6	2527.44.4	3005.22.2	3483.00.0	3960.77.7	4438.55.5	29
30	143.33.3	621.11.1	1098 88.8	1576.66.6	2054.44.4	2532.22.2	3010.00.0	3487.77.7	3965.55.5	4443.33.3	30
31	148.11.1	625.88.8	1103.66.6	1581.44.4	2059.22.2	2537.00.0	3014.77.7	3492.55.5	3970.33.3	4448.11.1	31
32	152.88.8	630.66.6	1108.44.4	1586.22.2	2064.00 0	2541.77.7	3019.55.5	3497.33.3	3975.11.1	4452.88 8	32
33	157.66.6	635.44.4	1113.22.2	1591.00.0	2068.77.7	2546.55.5	3024 33.3	3502.11.1	3979.88.8	4457.66.6	33
34	162.44.4	640 22.2	1118.00.0	1595.77.7	2073.55.5	2551.33.3	3029.11.1	3506.88.8	3984.66 6	4462.44.4	34
35	167.22.2	645.00.0	1122.77.7	1600.55.5	2078.33 3	2556.11.1	3033.88.8	3511.66.6	3989.44.4	4467.22.2	35
36	172.00.0	649.77.7	1127.55.5	1605.33.3	2083.11.1	2560.88.8	3038 66.6	3516.44.4	3991.22.2	4472.00.0	36
37	176.77.7	654.55.5	1132.33.3	1610.11.1	2087.88.8	2565.66.6	3043.44.4	3521.22.2	3999.00.0	4476.77.7	37
38	181.55.5	659.33.3	1137.11.1	1614.88.8	2092.66.6	2570.44.4	3048.22.2	3526.00.0	4003.77.7	4481.55.5	38
39	186.33.3	664.11.1	1141.88 8	1619.66.6	2097.44.4	2575.22.2	3053.00.0	3530.77.7	4008.55.5	4486.33.3	39
40	191.11.1	668.88 8	1146.66.6	1624.44.4	2102.22.2	2580.00.0	3057.77.7	3535.55.5	4013.33.3	4491.11.1	40
41	195.88.8	673.66.6	1151.44.4	1629.22.2	2107.00.0	2584.77.7	3062.55.5	3540.33.3	4018.11.1	4495.88.8	41
42	200.66.6	678.44.4	1156.22.2	1634.00.0	2111.77.7	2589.55.5	3067.33.3	3545.11.1	4022.88.8	4500.66.6	42
43	205.44.4	683.22.2	1161.00.0	1638.77.7	2116.55.5	2594.33.3	3072.11.1	3549.88.8	4027.66.6	4505.44.4	43
44	210.22.2	688.00.0	1165.77.7	1643.55.5	2121.33.3	2599.11.1	3076.88.8	3554.66.6	4032.44.4	4510.22.2	44
45	215.00.0	692.77.7	1170.55.5	1648.33.3	2126.11.1	2603.88.8	3081.66.6	3559.44.4	4037.22.2	4515.00.0	45
46	219.77.7	697.55.5	1175.33.3	1653.11.1	2130.88.8	2608.66.6	3086.44.4	3564.22.2	4042.00.0	4519.77.7	46
47	224.55.5	702.33.3	1180.11.1	1657.88.8	2135.66.6	2613.44.4	3091.22.2	3569.00.0	4046.77.7	4524.55.5	47
48	229.33.3	707.11.1	1184.88.8	1662.66.6	2140.44.4	2618.22.2	3096.00.0	3573.77.7	4051.55.5	4529.33.3	48
49	234.11.1	711.88.8	1189.66.6	1667.44.4	2145.22.2	2623.00.0	3100.77.7	3578.55.5	4056.33.3	4534.11.1	49

S.	0.	1.	2.	3.	4.	5.	6.	7.	8.	9.	S.
d.	$ cts.	$ cts.	$ cts.	$ cts.	$ cts.	$ cts.	$ cts.	$ cts.	$ cts.	$ cts.	d.
0		.23.8	.47.7	.71.6	.95.5	1.19.4	1.43.3	1.67.2	1.91.1	2.14.9	0
1	.01.9	.25.8	.49.7	.73.6	.97.5	1.21.4	1.45.3	1.69.2	1.93.1	2.16.9	1
2	.03.9	.27.8	.51.7	.75.6	.99.5	1.23.4	1.47.3	1.71.2	1.95.0	2.18.9	2
3	.05.9	.29 8	.53 7	.77.6	1.01.5	1.25.4	1.49.3	1.73.1	1.97.0	2.20 9	3
4	.07.9	.31.8	.55.7	.79.6	1.03.5	1.27.4	1.51.2	1.75.1	1.99.0	2.22.9	4
5	.09.9	.33.8	.57.7	.81.6	1.05.5	1.29.3	1.53.2	1.77.1	2.01.0	2.24 9	5
6	.11.9	.35.8	.59.7	.83.6	1.07.4	1.31.3	1.55.2	1.79.1	2.03.0	2.26.9	6
7	.13.9	.37.8	.61.7	.85.5	1.09.4	1.33 3	1.57.2	1.81.1	2.05.0	2.28.9	7
8	.15.9	.39.8	.63.6	.87.5	1.11.4	1.35.3	1.59.2	1.83.1	2.07.0	2.30.9	8
9	.17.9	.41.7	.65.6	.89.5	1.13.4	1.37.3	1.61.2	1.85.1	2.09.0	2.32.9	9
10	.19.9	.43.7	.67.6	.91.5	1.15.4	1.39.3	1.63.1	1.87.1	2.11.0	2.34.8	10
11	.21.8	.45.7	.69.6	.93.5	1.17.4	1.41.3	1.65.1	1.89.1	2.13.0	2.36.8	11

$4.77 17/100 TO THE POUND.
7½%
PREMIUM.
STERLING INTO DOLLARS AND CENTS, AND VICE-VERSA.

£	0	100	200	300	400	500	600	700	800	900	£
	$ cts.	$ cts.	$ cts.	$ cts.	$ cts.	$ cts.	$ cts.	$ cts.	$ cts.	$ cts.	
50	238,88.8	716,66.6	1191.44.4	1672.22.2	2150.00.0	2627.77.7	3105.55.5	3583.33.3	4061.11.1	4538.88.8	50
51	243,66.6	721.44.4	1199.22.2	1677.00.0	2154.77.7	2632,55.5	3110.33.3	3588.11.1	4065.88.8	4543.66.6	51
52	248.44.4	726.22.2	1204.00.0	1681.77.7	2159.55.5	2637.33.3	3115.11.1	3592.88.8	4070.66.6	4548.44.4	52
53	253.22.2	731.00.0	1208.77.7	1686.55.5	2164.33.3	2642.11.1	3119.88.8	3597.66.6	4075.44.4	4553.22.2	53
54	258.00.0	735.77.7	1213.55.5	1691.33.3	2169.11.1	2646.88.8	3124.66.6	3602.44.4	4080.22.2	4558.00.0	54
55	262.77.7	740.55.5	1218.33.3	1696.11.1	2173.88.8	2651.66.6	3129.44.4	3607.22.2	4085.00.0	4562.77.7	55
56	267.55.5	745.33.3	1223.11.1	1700.88.8	2178.66.6	2656.44.4	3134.22.2	3612.00.0	4089.77.7	4567.55.5	56
57	272.33.3	750.11.1	1227.88.8	1705.66.6	2183.44.4	2661.22.2	3139.00.0	3616.77.7	4094.55.5	4572.33.3	57
58	277.11.1	754.88.8	1232.66.6	1710.44.4	2188.22.2	2666.00.0	3143.77.7	3621.55.5	4099.33.3	4577.11.1	58
59	281.88.8	759.66.6	1237.44.4	1715.22.2	2193.00.0	2670.77.7	3148.55.5	3626.33.3	4104.11.1	4581.88.8	59
60	286.66.6	764.44.4	1242.22.2	1720.00.0	2197.77.7	2675.55.5	3153.33.3	3631.11.1	4108.88.8	4586.66.6	60
61	291.44.4	769.22.2	1247.00.0	1724.77.7	2202.55.5	2680.33.3	3158.11.1	3635.88.8	4113.66.6	4591.44.4	61
62	296.22.2	774.00.0	1251.77.7	1729.55.5	2207.33.3	2685.11.1	3162.88.8	3640.66.6	4118.44.4	4596.22.2	62
63	301.00.0	778.77.7	1256.55.5	1734.33.3	2212.11.1	2689.88.8	3167.66.6	3645.44.4	4123.22.2	4601.00.0	63
64	305.77.7	783.55.5	1261.33.3	1739.11.1	2216.88.8	2694.66.6	3172.44.4	3650.22.2	4128.00.0	4605.77.7	64
65	310.55.5	788.33.3	1266.11.1	1743.88.8	2221.66.6	2699.44.4	3177.22.2	3655.00.0	4132.77.7	4610.55.5	65
66	315.33.3	793.11.1	1270.88.8	1748.66.6	2226.44.4	2704.22.2	3182.00.0	3659.77.7	4137.55.5	4615.33.3	66
67	320.11.1	797.88.8	1275.66.6	1753.44.4	2231.22.2	2709.00.0	3186.77.7	3664.55.5	4142.33.3	4620.11.1	67
68	324.88.8	802.66.6	1280.44.4	1758.22.2	2236.00.0	2713.77.7	3191.55.5	3669.33.3	4147.11.1	4624.88.8	68
69	329.66.6	807.44.4	1285.22.2	1763.00.0	2240.77.7	2718.55.5	3196.33.3	3674.11.1	4151.88.8	4629.66.6	69
70	334.44.4	812.22.2	1290.00.0	1767.77.7	2245.55.5	2723.33.3	3201.11.1	3678.88.8	4156.66.6	4634.44.4	70
71	339.22.2	817.00.0	1294.77.7	1772.55.5	2250.33.3	2728.11.1	3205.88.8	3683.66.6	4161.44.4	4639.22.2	71
72	344.00.0	821.77.7	1299.55.5	1777.33.3	2255.11.1	2732.88.8	3210.66.6	3688.44.4	4166.22.2	4644.00.0	72
73	348.77.7	826.55.5	1304.33.3	1782.11.1	2259.88.8	2737.66.6	3215.44.4	3693.22.2	4171.00.0	4648.77.7	73
74	353.55.5	831.33.3	1309.11.1	1786.88.8	2264.66.6	2742.44.4	3220.22.2	3698.00.0	4175.77.7	4653.55.5	74
75	358.33.3	836.11.1	1313.88.8	1791.66.6	2269.44.4	2747.22.2	3225.00.0	3702.77.7	4180.55.5	4658.33.3	75
76	363.11.1	840.88.8	1318.66.6	1796.44.4	2274.22.2	2752.00.0	3229.77.7	3707.55.5	4185.33.3	4663.11.1	76
77	367.88.8	845.66.6	1323.44.4	1801.22.2	2279.00.0	2756.77.7	3234.55.5	3712.33.3	4190.11.1	4667.88.8	77
78	372.66.6	850.44.4	1328.22.2	1806.00.0	2283.77.7	2761.55.5	3239.33.3	3717.11.1	4194.88.8	4672.66.6	78
79	377.44.4	855.22.2	1333.00.0	1810.77.7	2288.55.5	2766.33.3	3244.11.1	3721.88.8	4199.66.6	4677.44.4	79
80	382.22.2	860.00.0	1337.77.7	1815.55.5	2293.33.3	2771.11.1	3248.88.8	3726.66.6	4204.44.4	4682.22.2	80
81	387.00.0	864.77.7	1342.55.5	1820.33.3	2298.11.1	2775.88.8	3253.66.6	3731.44.4	4209.22.2	4687.00.0	81
82	391.77.7	869.55.5	1347.33.3	1825.11.1	2302.88.8	2780.66.6	3258.44.4	3736.22.2	4214.00.0	4691.77.7	82
83	396.55.5	874.33.3	1352.11.1	1829.88.8	2307.66.6	2785.44.4	3263.22.2	3741.00.0	4218.77.7	4696.55.5	83
84	401.33.3	879.11.1	1356.88.8	1834.66.6	2312.44.4	2790.22.2	3268.00.0	3745.77.7	4223.55.5	4701.33.3	84
85	406.11.1	883.88.8	1361.66.6	1839.44.4	2317.22.2	2795.00.0	3272.77.7	3750.55.5	4228.33.3	4706.11.1	85
86	410.88.8	888.66.6	1366.44.4	1844.22.2	2322.00.0	2799.77.7	3277.55.5	3755.33.3	4233.11.1	4710.88.8	86
87	415.66.6	893.44.4	1371.22.2	1849.00.0	2326.77.7	2804.55.5	3282.33.3	3760.11.1	4237.88.8	4715.66.6	87
88	420.44.4	898.22.2	1376.00.0	1853.77.7	2331.55.5	2809.33.3	3287.11.1	3764.88.8	4242.66.6	4720.44.4	88
89	425.22.2	903.00.0	1380.77.7	1858.55.5	2336.33.3	2814.11.1	3291.88.8	3769.66.6	4247.44.4	4725.22.2	89
90	430.00.0	907.77.7	1385.55.5	1863.33.3	2341.11.1	2818.88.8	3296.66.6	3774.44.4	4252.22.2	4730.00.0	90
91	434.77.7	912.55.5	1390.33.3	1868.11.1	2345.88.8	2823.66.6	3301.44.4	3779.22.2	4257.00.0	4734.77.7	91
92	439.55.5	917.33.3	1395.11.1	1872.88.8	2350.66.6	2828.44.4	3306.22.2	3784.00.0	4261.77.7	4739.55.5	92
93	444.33.3	922.11.1	1399.88.8	1877.66.6	2355.44.4	2833.22.2	3311.00.0	3788.77.7	4266.55.5	4744.33.3	93
94	449.11.1	926.88.8	1404.66.6	1882.44.4	2360.22.2	2838.00.0	3315.77.7	3793.55.5	4271.33.3	4749.11.1	94
95	453.88.8	931.66.6	1409.44.4	1887.22.2	2365.00.0	2842.77.7	3320.55.5	3798.33.3	4276.11.1	4753.88.8	95
96	458.66.6	936.44.4	1414.22.2	1892.00.0	2369.77.7	2847.55.5	3325.33.3	3803.11.1	4280.88.8	4758.66.6	96
97	463.44.4	941.22.2	1419.00.0	1896.77.7	2374.55.5	2852.33.3	3330.11.1	3807.88.8	4285.66.6	4763.44.4	97
98	468.22.2	946.00.0	1423.77.7	1901.55.5	2379.33.3	2857.11.1	3334.88.8	3812.66.6	4290.44.4	4768.22.2	98
99	473.00.0	950.77.7	1428.55.5	1906.33.3	2384.11.1	2861.88.8	3339.66.6	3817.44.4	4295.22.2	4773.00.0	99

S.	10	11	12	13	14	15	16	17	18	19	S.
d.	$ cts.	$ cts.	$ cts.	$ cts.	$ cts.	$ cts.	$ cts.	$ cts.	$ cts.	$ cts.	d.
0	2.38.8	2.62.7	2.86.6	3.10.5	3.34.4	3.58.3	3.82.2	4.06.1	4.30.0	4.53.8	0
1	2.40.8	2.64.7	2.88.6	3.12.5	3.36.4	3.60.3	3.84.2	4.08.1	4.31.9	4.55.8	1
2	2.42.8	2.66.7	2.90.6	3.14.5	3.38.4	3.62.3	3.86.2	4.10.0	4.33.9	4.57.8	2
3	2.44.8	2.68.7	2.92.6	3.16.5	3.40.4	3.64.3	3.88.1	4.12.0	4.35.9	4.59.8	3
4	2.46.8	2.70.7	2.94.6	3.18.5	3.42.4	3.66.2	3.90.1	4.14.0	4.37.9	4.61.8	4
5	2.48.8	2.72.7	2.96.6	3.20.5	3.44.3	3.68.2	3.92.1	4.16.0	4.39.9	4.63.8	5
6	2.50.8	2.74.7	2.98.6	3.22.4	3.46.3	3.70.2	3.94.1	4.18.0	4.41.9	4.65.8	6
7	2.52.8	2.76.7	3.00.5	3.24.4	3.48.3	3.72.2	3.96.1	4.20.0	4.43.9	4.67.8	7
8	2.54.8	2.78.6	3.02.5	3.26.4	3.50.3	3.74.2	3.98.1	4.22.0	4.45.9	4.69.8	8
9	2.56.8	2.80.6	3.04.5	3.28.4	3.52.3	3.76.2	4.00.1	4.24.0	4.47.9	4.71.7	9
10	2.58.7	2.82.6	3.06.5	3.30.4	3.54.3	3.78.2	4.02.1	4.26.0	4.49.9	4.73.7	10
11	2.60.7	2.84.6	3.08.5	3.32.4	3.56.3	3.80.2	4.04.1	4.28.0	4.51.8	4.75.7	11

$4.78¾⁄₁₀₀ TO THE POUND.

7⅝ %

PREMIUM.

STERLING INTO DOLLARS AND CENTS, AND VICE-VERSA.

£	0	100	200	300	400	500	600	700	800	900	£
£	$ cts.	$ cts.	$ cts.	$ cts.	$ cts.	$ cts.	$ cts.	$ cts.	$ cts.	$ cts.	£
0	.....	478.33.3	956.66.6	1435.00.0	1913.33.3	2391.66.6	2870.00.0	3348.33.3	3826.66.6	4305.00.0	0
1	4.78.3	483.11.6	961.45.0	1439.78.3	1918.11.6	2396.45.0	2874.78.3	3353.11.6	3831.45.0	4309.78.3	1
2	9.56.6	487.90.0	966.23.3	1444.56.6	1922.90.0	2401.23.3	2879.56.6	3357.90.0	3836.23.3	4314.56.6	2
3	14.35.0	492.68.3	971.01.6	1449.35.0	1927.68.3	2406.01.6	2884.35.0	3362.68.3	3841.01.6	4319.35.0	3
4	19.13.3	497.46.6	975.80.0	1454.13.3	1932.46.6	2410.80.0	2889.13.3	3367.46.6	3845.80.0	4324.13.3	4
5	23.91.6	502.25.0	980.58.3	1458.91.6	1937.25.0	2415.58.3	2893.91.6	3372.25.0	3850.58.3	4328.91.6	5
6	28.70.0	507.03.3	985.36.6	1463.70.0	1942.03.3	2420.36.6	2898.70.0	3377.03.3	3855.36.6	4333.70.0	6
7	33.48.3	511.81.6	990.15.0	1468.48.3	1946.81.6	2425.15.0	2903.48.3	3381.81.6	3860.15.0	4338.48.3	7
8	38.26.6	516.60.0	994.93.3	1473.26.6	1951.60.0	2429.93.3	2908.26.6	3386.60.0	3865.93.3	4343.26.6	8
9	43.05.0	521.38.3	999.71.6	1478.05.0	1956.38.3	2434.71.6	2913.05.0	3391.38.3	3869.71.6	4348.05.0	9
10	47.83.3	526.16.6	1004.50.0	1482.83.3	1961.16.6	2439.50.0	2917.83.3	3396.16.6	3874.50.0	4352.83.3	10
11	52.61.6	530.95.0	1009.28.3	1487.61.6	1965.95.0	2444.28.3	2922.61.6	3400.95.0	3879.28.3	4357.61.6	11
12	57.40.0	535.73.3	1014.06.6	1492.40.0	1970.73.3	2449.06.6	2927.40.0	3405.73.3	3884.06.6	4362.40.0	12
13	62.18.3	540.51.6	1018.85.0	1497.18.3	1975.51.6	2453.85.0	2932.18.3	3410.51.6	3888.85.0	4367.18.3	13
14	66.96.6	545.30.0	1023.63.3	1501.96.6	1980.30.0	2458.63.3	2936.96.6	3415.30.0	3893.63.3	4371.96.6	14
15	71.75.0	550.08.3	1028.41.6	1506.75.0	1985.08.3	2463.41.6	2941.75.0	3420.08.3	3898.41.6	4376.75.0	15
16	76.53.3	554.86.6	1033.20.0	1511.53.3	1989.86.6	2468.20.0	2946.53.3	3424.86.6	3903.20.0	4381.53.3	16
17	81.31.6	559.65.0	1037.98.3	1516.31.6	1994.65.0	2472.98.3	2951.31.6	3429.65.0	3907.98.3	4386.31.6	17
18	86.10.0	564.43.3	1042.76.6	1521.10.0	1999.43.3	2477.76.6	2956.10.0	3434.43.3	3912.76.6	4391.10.0	18
19	90.88.3	569.21.6	1047.55.0	1525.88.3	2004.21.6	2482.55.0	2960.88.3	3439.21.6	3917.55.0	4395.88.3	19
20	95.66.6	574.00.0	1052.33.3	1530.66.6	2009.00.0	2487.33.3	2965.66.6	3444.00.0	3922.33.3	4400.66.6	20
21	100.45.0	578.78.3	1057.11.6	1535.45.0	2013.78.3	2492.11.6	2970.45.0	3448.78.3	3927.11.6	4405.45.0	21
22	105.23.3	583.56.6	1061.90.0	1540.23.3	2018.56.6	2496.90.0	2975.23.3	3453.56.6	3931.90.0	4410.23.3	22
23	110.01.6	588.35.0	1066.68.3	1545.01.6	2023.35.0	2501.68.3	2980.01.6	3458.35.0	3936.68.3	4415.01.6	23
24	114.80.0	593.13.3	1071.46.6	1549.80.0	2028.13.3	2506.46.6	2984.80.0	3463.13.3	3941.46.6	4419.80.0	24
25	119.58.3	597.91.6	1076.25.0	1554.58.3	2032.91.6	2511.25.0	2989.58.3	3467.91.6	3946.25.0	4424.58.3	25
26	124.36.6	602.70.0	1081.03.3	1559.36.6	2037.70.0	2516.03.3	2994.36.6	3472.70.0	3951.03.3	4429.36.6	26
27	129.15.0	607.48.3	1085.81.6	1564.15.0	2042.48.3	2520.81.6	2999.15.0	3477.48.3	3955.81.6	4434.15.0	27
28	133.93.3	612.26.6	1090.60.0	1568.93.3	2047.26.6	2525.60.0	3003.93.3	3482.26.6	3960.60.0	4438.93.3	28
29	138.71.6	617.05.0	1095.38.3	1573.71.6	2052.05.0	2530.38.3	3008.71.6	3487.05.0	3965.38.3	4443.71.6	29
30	143.50.0	621.83.3	1100.16.6	1578.50.0	2056.83.3	2535.16.6	3013.50.0	3491.83.3	3970.16.6	4448.50.0	30
31	148.28.3	626.61.6	1104.95.0	1583.28.3	2061.61.6	2539.95.0	3018.28.3	3496.61.6	3974.95.0	4453.28.3	31
32	153.06.6	631.40.0	1109.73.3	1588.06.6	2066.40.0	2544.73.3	3023.06.6	3501.40.0	3979.73.3	4458.06.6	32
33	157.85.0	636.18.3	1114.51.6	1592.85.0	2071.18.3	2549.51.6	3027.85.0	3506.18.3	3984.51.6	4462.85.0	33
34	162.63.3	640.96.6	1119.30.0	1597.63.3	2075.96.6	2554.30.0	3032.63.3	3510.96.6	3989.30.0	4467.63.3	34
35	167.41.6	645.75.0	1124.08.3	1602.41.6	2080.75.0	2559.08.3	3037.41.6	3515.75.0	3994.08.3	4472.41.6	35
36	172.20.0	650.53.3	1128.86.6	1607.20.0	2085.53.3	2563.86.6	3042.20.0	3520.53.3	3998.86.6	4477.20.0	36
37	176.98.3	655.31.6	1133.65.0	1611.98.3	2090.31.6	2568.65.0	3046.98.3	3525.31.6	4003.65.0	4481.98.3	37
38	181.76.6	660.10.0	1138.43.3	1616.76.6	2095.10.0	2573.43.3	3051.76.6	3530.10.0	4008.43.3	4486.76.6	38
39	186.55.0	664.88.3	1143.21.6	1621.55.0	2099.88.3	2578.21.6	3056.55.0	3534.88.3	4013.21.6	4491.55.0	39
40	191.33.3	669.66.6	1148.00.0	1626.33.3	2104.66.6	2583.00.0	3061.33.3	3539.66.6	4018.00.0	4496.33.3	40
41	196.11.6	674.45.0	1152.78.3	1631.11.6	2109.45.0	2587.78.3	3066.11.6	3544.45.0	4022.78.3	4501.11.6	41
42	200.90.0	679.23.3	1157.56.6	1635.90.0	2114.23.3	2592.56.6	3070.90.0	3549.23.3	4027.56.6	4505.90.0	42
43	205.68.3	684.01.6	1162.35.0	1640.68.3	2119.01.6	2597.35.0	3075.68.3	3554.01.6	4032.35.0	4510.68.3	43
44	210.46.6	688.80.0	1167.13.3	1645.46.6	2123.80.0	2602.13.3	3080.46.6	3558.80.0	4037.13.3	4515.46.6	44
45	215.25.0	693.58.3	1171.91.6	1650.25.0	2128.58.3	2606.91.6	3085.25.0	3563.58.3	4041.91.6	4520.25.0	45
46	220.03.3	698.36.6	1176.70.0	1655.03.3	2133.36.6	2611.70.0	3090.03.3	3568.36.6	4046.70.0	4525.03.3	46
47	224.81.6	703.15.0	1181.48.3	1659.81.6	2138.15.0	2616.48.3	3094.81.6	3573.15.0	4051.48.3	4529.81.6	47
48	229.60.0	707.93.3	1186.26.6	1664.60.0	2142.93.3	2621.26.6	3099.60.0	3577.93.3	4056.26.6	4534.60.0	48
49	234.38.3	712.71.6	1191.05.0	1669.38.3	2147.71.6	2626.05.0	3104.38.3	3582.71.6	4061.05.0	4539.38.3	49

S.	0	1	2	3	4	5	6	7	8	9	S.
d.	$ cts.	$ cts.	$ cts.	$ cts.	$ cts.	$ cts.	$ cts.	$ cts.	$ cts.	$ cts.	d.
0	.....	.23.9	.47.8	.71.7	.95.6	1.19.5	1.43.5	1.67.4	1.91.3	2.15.2	0
1	.01.9	.25.9	.49.8	.73.7	.97.6	1.21.5	1.45.4	1.69.4	1.93.3	2.17.2	1
2	.03.9	.27.9	.51.8	.75.7	.99.6	1.23.5	1.47.4	1.71.4	1.95.3	2.19.2	2
3	.05.9	.29.8	.53.8	.77.7	1.01.6	1.25.5	1.49.4	1.73.3	1.97.3	2.21.2	3
4	.07.9	.31.8	.55.8	.79.7	1.03.6	1.27.5	1.51.4	1.75.3	1.99.3	2.22.2	4
5	.09.9	.33.8	.57.7	.81.7	1.05.6	1.29.5	1.53.4	1.77.3	2.01.2	2.25.2	5
6	.11.9	.35.8	.59.7	.83.7	1.07.6	1.31.5	1.55.4	1.79.3	2.03.2	2.27.2	6
7	.13.9	.37.8	.61.7	.85.6	1.09.6	1.33.5	1.57.4	1.81.3	2.05.2	2.29.1	7
8	.15.9	.39.8	.63.7	.87.6	1.11.6	1.35.5	1.59.4	1.83.3	2.07.2	2.31.1	8
9	.17.9	.41.8	.65.7	.89.6	1.13.5	1.37.5	1.61.4	1.85.3	2.09.2	2.33.1	9
10	.19.9	.43.8	.67.7	.91.6	1.15.5	1.39.4	1.63.4	1.87.3	2.11.2	2.35.1	10
11	.21.8	.45.8	.69.7	.93.6	1.17.5	1.41.4	1.65.3	1.89.3	2.13.2	2.37.1	11

4.78\frac{9}{100}$ TO THE POUND.

7$\frac{5}{8}$ %

PREMIUM.

STERLING INTO DOLLARS AND CENTS, AND VICE-VERSA.

£	0	100	200	300	400	500	600	700	800	900	£
	$ cts.	$ cts.	$ cts.	$ cts.	$ cts.	$ cts.	$ cts.	$ cts.	$ cts.	$ cts.	
50	239.16.6	717.50.0	1195.83.3	1674.16.6	2152.50.0	2630.83.3	3109.16.6	3587.50.0	4065.83.3	4544.16.6	50
51	243.95.0	722.28.3	1200.61.6	1678.95.0	2157.28.3	2635.61.6	3113.95.0	3592.28.3	4070.61.6	4548.95.0	51
52	248.73.3	727.06.6	1205.40.0	1683.73.3	2162.06.6	2640.40.0	3118.73.3	3597.06.6	4075.40.0	4553.73.3	52
53	253.51.6	731.85.0	1210.18.3	1688.51.6	2166.85.0	2645.18.3	3123.51.6	3601.85.0	4080.18.3	4558.51.6	53
54	258.30.0	736.63.3	1214.96.6	1693.30.0	2171.63.3	2649.96.6	3128.30.0	3606.63.3	4084.96.6	4563.30.0	54
55	263.08.3	741.41.6	1219.75.0	1698.08.3	2176.41.6	2654.75.0	3133.08.3	3611.41.6	4089.75.0	4568.08.3	55
56	267.86.6	746.20.0	1224.53.3	1702.86.6	2181.20.0	2659.53.3	3137.86.6	3616.20.0	4094.53.3	4572.86.6	56
57	272.65.0	750.98.3	1229.31.6	1707.65.0	2185.98.3	2664.31.6	3142.65.0	3620.98.3	4099.31.6	4577.65.0	57
58	277.43.3	755.76.6	1234.10.0	1712.43.3	2190.76.6	2669.10.0	3147.43.3	3625.76.6	4104.10.0	4582.43.3	58
59	282.21.6	760.55.0	1238.88.3	1717.21.6	2195.55.0	2673.88.3	3152.21.6	3630.55.0	4108.88.3	4587.21.6	59
60	287.00.0	765.33.3	1243.66.6	1722.00.0	2200.33.3	2678.66.6	3157.00.0	3635.33.3	4113.66.6	4592.00.0	60
61	291.78.3	770.11.6	1248.45.0	1726.78.3	2205.11.6	2683.45.0	3161.78.3	3640.11.6	4118.45.0	4596.78.3	61
62	296.56.6	774.90.0	1253.23.3	1731.56.6	2209.90.0	2688.23.3	3166.56.6	3644.90.0	4123.23.3	4601.56.6	62
63	301.35.0	779.68.3	1258.01.6	1736.35.0	2214.68.3	2693.01.6	3171.35.0	3649.68.3	4128.01.6	4606.35.0	63
64	306.13.3	784.46.6	1262.80.0	1741.13.3	2219.46.6	2697.80.0	3176.13.3	3654.46.6	4132.80.0	4611.13.3	64
65	310.91.6	789.25.0	1267.58.3	1745.91.6	2224.25.0	2702.58.3	3180.91.6	3659.25.0	4137.58.3	4615.91.6	65
66	315.70.0	794.03.3	1272.36.6	1750.70.0	2229.03.3	2707.36.6	3185.70.0	3664.03.3	4142.36.6	4620.70.0	66
67	320.48.3	798.81.6	1277.15.0	1755.48.3	2233.81.6	2712.15.0	3190.48.3	3668.81.6	4147.15.0	4625.48.3	67
68	325.26.6	803.60.0	1281.93.3	1760.26.6	2238.60.0	2716.93.3	3195.26.6	3673.60.0	4151.93.3	4630.26.6	68
69	330.05.0	808.38.3	1286.71.6	1765.05.0	2243.38.3	2721.71.6	3200.05.0	3678.38.3	4156.71.6	4635.05.0	69
70	334.83.3	813.16.6	1291.50.0	1769.83.3	2248.16.6	2726.50.0	3204.83.3	3683.16.6	4161.50.0	4639.83.3	70
71	339.61.6	817.95.0	1296.28.3	1774.61.6	2252.95.0	2731.28.3	3209.61.6	3687.95.0	4166.28.3	4644.61.6	71
72	344.40.0	822.73.3	1301.06.6	1779.40.0	2257.73.3	2736.06.6	3214.40.0	3692.73.3	4171.06.6	4649.40.0	72
73	349.18.3	827.51.6	1305.85.0	1784.18.3	2262.51.6	2740.85.0	3219.18.3	3697.51.6	4175.85.0	4654.18.3	73
74	353.96.6	832.30.0	1310.63.3	1788.96.6	2267.30.0	2745.63.3	3223.96.6	3702.30.0	4180.63.3	4658.96.6	74
75	358.75.0	837.08.3	1315.41.6	1793.75.0	2272.08.3	2750.41.6	3228.75.0	3707.08.3	4185.41.6	4663.75.0	75
76	363.53.3	841.86.6	1320.20.0	1798.53.3	2276.86.6	2755.20.0	3233.53.3	3711.86.6	4190.20.0	4668.53.3	76
77	368.31.6	846.65.0	1324.98.3	1803.31.6	2281.65.0	2759.98.3	3238.31.6	3716.65.0	4194.98.3	4673.31.6	77
78	373.10.0	851.43.3	1329.76.6	1808.10.0	2286.43.3	2764.76.6	3243.10.0	3721.43.3	4199.76.6	4678.10.0	78
79	377.88.3	856.21.6	1334.55.0	1812.88.3	2291.21.6	2769.55.0	3247.88.3	3726.21.6	4204.55.0	4682.88.3	79
80	382.66.6	861.00.0	1339.33.3	1817.66.6	2296.00.0	2774.33.3	3252.66.6	3731.00.0	4209.33.3	4687.66.6	80
81	387.45.0	865.78.3	1344.11.6	1822.45.0	2300.78.3	2779.11.6	3257.45.0	3735.78.3	4214.11.6	4692.45.0	81
82	392.23.3	870.56.6	1348.90.0	1827.23.3	2305.56.6	2783.90.0	3262.23.3	3740.56.6	4218.90.0	4697.23.3	82
83	397.01.6	875.35.0	1353.68.3	1832.01.6	2310.35.0	2788.68.3	3267.01.6	3745.35.0	4223.68.3	4702.01.6	83
84	401.80.0	880.13.3	1358.46.6	1836.80.0	2315.13.3	2793.46.6	3271.80.0	3750.13.3	4228.46.6	4706.80.0	84
85	406.58.3	884.91.6	1363.25.0	1841.58.3	2319.91.6	2798.25.0	3276.58.3	3754.91.6	4233.25.0	4711.58.3	85
86	411.36.6	889.70.0	1368.03.3	1846.36.6	2324.70.0	2803.03.3	3281.36.6	3759.70.0	4238.03.3	4716.36.6	86
87	416.15.0	894.48.3	1372.81.6	1851.15.0	2329.48.3	2807.81.6	3286.15.0	3764.48.3	4242.81.6	4721.15.0	87
88	420.93.3	899.26.6	1377.60.0	1855.93.3	2334.26.6	2812.60.0	3290.93.3	3769.26.6	4247.60.0	4725.93.3	88
89	425.71.6	904.05.0	1382.38.3	1860.71.6	2339.05.0	2817.38.3	3295.71.6	3774.05.0	4252.38.3	4730.71.6	89
90	430.50.0	908.83.3	1387.16.6	1865.50.0	2343.83.3	2822.16.6	3300.50.0	3778.83.3	4257.16.6	4735.50.0	90
91	435.28.3	913.61.6	1391.95.0	1870.28.3	2348.61.6	2826.95.0	3305.28.3	3783.61.6	4261.95.0	4740.28.3	91
92	440.06.6	918.40.0	1396.73.3	1875.06.6	2353.40.0	2831.73.3	3310.06.6	3788.40.0	4266.73.3	4745.06.6	92
93	444.85.0	923.18.3	1401.51.6	1879.85.0	2358.18.3	2836.51.6	3314.85.0	3793.18.3	4271.51.6	4749.85.0	93
94	449.63.3	927.96.6	1406.30.0	1884.63.3	2362.96.6	2841.30.0	3319.63.3	3797.96.6	4276.30.0	4754.63.3	94
95	454.41.6	932.75.0	1411.08.3	1889.41.6	2367.75.0	2846.08.3	3324.41.6	3802.75.0	4281.08.3	4759.41.6	95
96	459.20.0	937.53.3	1415.86.6	1894.20.0	2372.53.3	2850.86.6	3329.20.0	3807.53.3	4285.86.6	4764.20.0	96
97	463.98.3	942.31.6	1420.65.0	1898.98.3	2377.31.6	2855.65.0	3333.98.3	3812.31.6	4290.65.0	4768.98.3	97
98	468.76.6	947.10.0	1425.43.3	1903.76.6	2382.10.0	2860.43.3	3338.76.6	3817.10.0	4295.43.3	4773.76.6	98
99	473.55.0	951.88.3	1430.21.6	1908.55.0	2386.88.3	2865.21.6	3343.55.0	3821.88.3	4300.21.6	4778.55.0	99

S.	10	11	12	13	14	15	16	17	18	19	S.
d.	$ cts.	$ cts.	$ cts.	$ cts.	$ cts.	$ cts.	$ cts.	$ cts.	$ cts.	$ cts.	d.
0	2.30.1	2.63.0	2.87.0	3.10.9	3.34.8	3.58.7	3.82.6	4.06.5	4.30.5	4.54.4	0
1	2.41.1	2.65.0	2.88.9	3.12.9	3.36.8	3.60.7	3.84.6	4.08.6	4.32.4	4.56.4	1
2	2.43.1	2.67.0	2.90.9	3.14.8	3.38.8	3.62.7	3.86.6	4.10.5	4.34.4	4.58.3	2
3	2.45.1	2.69.0	2.92.9	3.16.8	3.40.8	3.64.7	3.88.6	4.12.5	4.36.4	4.60.3	3
4	2.47.1	2.71.0	2.94.9	3.18.8	3.42.7	3.66.7	3.90.6	4.14.5	4.38.4	4.62.3	4
5	2.49.1	2.73.0	2.96.9	3.20.8	3.44.7	3.68.7	3.92.6	4.16.5	4.40.4	4.64.3	5
6	2.51.1	2.75.0	2.98.9	3.22.8	3.46.7	3.70.6	3.94.6	4.18.5	4.42.4	4.66.3	6
7	2.53.0	2.77.0	3.00.9	3.24.8	3.48.7	3.72.6	3.96.5	4.20.5	4.44.4	4.68.3	7
8	2.55.0	2.79.0	3.02.9	3.26.8	3.50.7	3.74.6	3.98.5	4.22.5	4.46.4	4.70.3	8
9	2.57.0	2.80.9	3.04.9	3.28.8	3.52.7	3.76.6	4.00.5	4.24.5	4.48.4	4.72.3	9
10	2.59.0	2.82.9	3.06.9	3.30.8	3.54.7	3.78.6	4.02.5	4.26.4	4.50.4	4.74.3	10
11	2.61.0	2.84.9	3.08.9	3.32.8	3.56.7	3.80.6	4.04.5	4.28.4	4.52.3	4.76.3	11

$4.78 87/100 TO THE POUND.

7¾ %

PREMIUM.

STERLING INTO DOLLARS AND CENTS, AND VICE-VERSA.

£	0	100	200	300	400	500	600	700	800	900	£
	$ cts.	$ cts.	$ cts.	$ cts.	$ cts.	$ cts.	$ cts.	$ cts.	$ cts.	$ cts.	
0	......	478.88.8	957.77.7	1436.66.6	1915.55.5	2394.44.4	2873.33.3	3352.22.2	3831.11.1	4310.00.0	0
1	4.78.8	483.67.7	962.56.6	1441.45.5	1920.34.4	2399.23.3	2878.12.2	3357.01.1	3835.90.0	4314.78.8	1
2	9.57.7	488.46.6	967.35.5	1446.24.4	1925.13.3	2404.02.2	2882.91.1	3361.80.0	3840.68.8	4319.57.7	2
3	14.36.6	493.25.5	972.14.4	1451.03.3	1929.92.2	2408.81.1	2887.70.0	3366.58.8	3845.47.7	4324.36.6	3
4	19.15.5	498.04.4	976.93.3	1455.82.2	1934.71.1	2413.60.0	2892.48.8	3371.37.7	3850.26.6	4329.15.5	4
5	23.94.4	502.83.3	981.72.2	1460.61.1	1939.50.0	2418.38.8	2897.27.7	3376.16.6	3855.05.5	4333.94.4	5
6	28.73.3	507.62.2	986.51.1	1465.40.0	1944.28.8	2423.17.7	2902.06.6	3380.95.5	3859.84.4	4338.73.3	6
7	33.52.2	512.41.1	991.30.0	1470.18.8	1949.07.7	2427.96.6	2906.85.5	3385.74.4	3864.63.3	4343.52.2	7
8	38.31.1	517.20.0	996.08.8	1474.97.7	1953.86.6	2432.75.5	2911.64.4	3390.53.3	3869.42.2	4348.31.1	8
9	43.10.0	521.98.8	1000.87.7	1479.76.6	1958.65.5	2437.54.4	2916.43.3	3395.32.2	3874.21.1	4353.10.0	9
10	47.88.8	526.77.7	1005.66.6	1484.55.5	1963.44.4	2442.33.3	2921.22.2	3400.11.1	3879.00.0	4357.88.8	10
11	52.67.7	531.56.6	1010.45.5	1489.34.4	1968.23.3	2447.12.2	2926.01.1	3404.90.0	3883.78.8	4362.67.7	11
12	57.46.6	536.35.5	1015.24.4	1494.13.3	1973.02.2	2451.91.1	2930.80.0	3409.68.8	3888.57.7	4367.46.6	12
13	62.25.5	541.14.4	1020.03.3	1498.92.2	1977.81.1	2456.70.0	2935.58.8	3414.47.7	3893.36.6	4372.25.5	13
14	67.04.4	545.93.3	1024.82.2	1503.71.1	1982.60.0	2461.48.8	2940.37.7	3419.26.6	3898.15.5	4377.04.4	14
15	71.83.3	550.72.2	1029.61.1	1508.50.0	1987.38.8	2466.27.7	2945.16.6	3424.05.5	3902.94.4	4381.83.3	15
16	76.62.2	555.51.1	1034.40.0	1513.28.8	1992.17.7	2471.06.6	2949.95.5	3428.84.4	3907.73.3	4386.62.2	16
17	81.41.1	560.30.0	1039.18.8	1518.07.7	1996.96.6	2475.85.5	2954.74.4	3433.63.3	3912.52.2	4391.41.1	17
18	86.20.0	565.08.8	1043.97.7	1522.86.6	2001.75.5	2480.64.4	2959.53.3	3438.42.2	3917.31.1	4396.20.0	18
19	90.98.8	569.87.7	1048.76.6	1527.65.5	2006.54.4	2485.43.3	2964.32.2	3443.21.1	3922.10.0	4400.98.8	19
20	95.77.7	574.66.6	1053.55.5	1532.44.4	2011.33.3	2490.22.2	2969.11.1	3448.00.0	3926.88.8	4405.77.7	20
21	100.56.6	579.45.5	1058.34.4	1537.23.3	2016.12.2	2495.01.1	2973.90.0	3452.78.8	3931.67.7	4410.56.6	21
22	105.35.5	584.24.4	1063.13.3	1542.02.2	2020.91.1	2499.80.0	2978.68.8	3457.57.7	3936.46.6	4415.35.5	22
23	110.14.4	589.03.3	1067.92.2	1546.81.1	2025.70.0	2504.58.8	2983.47.7	3462.36.6	3941.25.5	4420.14.4	23
24	114.93.3	593.82.2	1072.71.1	1551.60.0	2030.48.8	2509.37.7	2988.26.6	3467.15.5	3946.04.4	4424.93.3	24
25	119.72.2	598.61.1	1077.50.0	1556.38.8	2035.27.7	2514.16.6	2993.05.5	3471.94.4	3950.83.3	4429.72.2	25
26	124.51.1	603.40.0	1082.28.8	1561.17.7	2040.06.6	2518.95.5	2997.84.4	3476.73.3	3955.62.2	4434.51.1	26
27	129.30.0	608.18.8	1087.07.7	1565.96.6	2044.85.5	2523.74.4	3002.63.3	3481.52.2	3960.41.1	4439.30.0	27
28	134.08.8	612.97.7	1091.86.6	1570.75.5	2049.64.4	2528.53.3	3007.42.2	3486.31.1	3965.20.0	4444.08.8	28
29	138.87.7	617.76.6	1096.65.5	1575.54.4	2054.43.3	2533.32.2	3012.21.1	3491.10.0	3969.98.8	4448.87.7	29
30	143.66.6	622.55.5	1101.44.4	1580.33.3	2059.22.2	2538.11.1	3017.00.0	3495.88.8	3974.77.7	4453.66.6	30
31	148.45.5	627.34.4	1106.23.3	1585.12.2	2064.01.1	2542.90.0	3021.78.8	3500.67.7	3979.56.6	4458.45.5	31
32	153.24.4	632.13.3	1111.02.2	1589.91.1	2068.80.0	2547.68.8	3026.57.7	3505.46.6	3984.35.5	4463.24.4	32
33	158.03.3	636.92.2	1115.81.1	1594.70.0	2073.58.8	2552.47.7	3031.36.6	3510.25.5	3989.14.4	4468.03.3	33
34	162.82.2	641.71.1	1120.60.0	1599.48.8	2078.37.7	2557.26.6	3036.15.5	3515.04.4	3993.93.3	4472.82.2	34
35	167.61.1	646.50.0	1125.38.8	1604.27.7	2083.16.6	2562.05.5	3040.94.4	3519.83.3	3998.72.2	4477.61.1	35
36	172.40.0	651.28.8	1130.17.7	1609.06.6	2087.95.5	2566.84.4	3045.73.3	3524.62.2	4003.51.1	4482.40.0	36
37	177.18.8	656.07.7	1134.96.6	1613.85.5	2092.74.4	2571.63.3	3050.52.2	3529.41.1	4008.30.0	4487.18.8	37
38	181.97.7	660.86.6	1139.75.5	1618.64.4	2097.53.3	2576.42.2	3055.31.1	3534.20.0	4013.08.8	4491.97.7	38
39	186.76.6	665.65.5	1144.54.4	1623.43.3	2102.32.2	2581.21.1	3060.10.0	3538.98.8	4017.87.7	4496.76.6	39
40	191.55.5	670.44.4	1149.33.3	1628.22.2	2107.11.1	2586.00.0	3064.88.8	3543.77.7	4022.66.6	4501.55.5	40
41	196.34.4	675.23.3	1154.12.2	1633.01.1	2111.90.0	2590.78.8	3069.67.7	3548.56.6	4027.45.5	4506.34.4	41
42	201.13.3	680.02.2	1158.91.1	1637.80.0	2116.68.8	2595.57.7	3074.46.6	3553.35.5	4032.24.4	4511.13.3	42
43	205.92.2	684.81.1	1163.70.0	1642.58.8	2121.47.7	2600.36.6	3079.25.5	3558.14.4	4037.03.3	4515.92.2	43
44	210.71.1	689.60.0	1168.48.8	1647.37.7	2126.26.6	2605.15.5	3084.04.4	3562.93.3	4041.82.2	4520.71.1	44
45	215.50.0	694.38.8	1173.27.7	1652.16.6	2131.05.5	2609.94.4	3088.83.3	3567.72.2	4046.61.1	4525.50.0	45
46	220.28.8	699.17.7	1178.06.6	1656.95.5	2135.84.4	2614.73.3	3093.62.2	3572.51.1	4051.40.0	4530.28.8	46
47	225.07.7	703.96.6	1182.85.5	1661.74.4	2140.63.3	2619.52.2	3098.41.1	3577.30.0	4056.18.8	4535.07.7	47
48	229.86.6	708.75.5	1187.64.4	1666.53.3	2145.42.2	2624.31.1	3103.20.0	3582.08.8	4060.97.7	4539.86.6	48
49	234.65.5	713.54.4	1192.43.3	1671.32.2	2150.21.1	2629.10.0	3107.98.8	3586.87.7	4065.76.6	4544.65.5	49

S.	0	1	2	3	4	5	6	7	8	9	S.
d.	$ cts.	$ cts.	$ cts.	$ cts.	$ cts.	$ cts.	$ cts.	$ cts.	$ cts.	$ cts.	d.
0	......	.23.9	.47.8	.71.8	.95.7	1.19.7	1.43.6	1.67.6	1.91.5	2.15.5	0
1	.01.9	.25.9	.49.8	.73.8	.97.7	1.21.7	1.45.6	1.69.6	1.93.5	2.17.4	1
2	.03.9	.27.9	.51.8	.75.8	.99.7	1.23.7	1.47.6	1.71.6	1.95.5	2.19.4	2
3	.05.9	.29.9	.53.8	.77.8	1.01.7	1.25.7	1.49.6	1.73.6	1.97.5	2.21.4	3
4	.07.9	.31.9	.55.8	.79.8	1.03.7	1.27.7	1.51.6	1.75.6	1.99.5	2.23.4	4
5	.09.9	.33.9	.57.8	.81.8	1.05.7	1.29.7	1.53.6	1.77.6	2.01.5	2.25.4	5
6	.11.9	.35.9	.59.8	.83.8	1.07.7	1.31.7	1.55.6	1.79.6	2.03.5	2.27.4	6
7	.13.9	.37.9	.61.8	.85.8	1.09.7	1.33.7	1.57.6	1.81.6	2.05.5	2.29.4	7
8	.15.9	.39.9	.63.8	.87.8	1.11.7	1.35.7	1.59.6	1.83.6	2.07.5	2.31.4	8
9	.17.9	.41.9	.65.8	.89.8	1.13.7	1.37.7	1.61.6	1.85.6	2.09.5	2.33.4	9
10	.19.9	.43.9	.67.8	.91.8	1.15.7	1.39.7	1.63.6	1.87.6	2.11.5	2.35.4	10
11	.21.9	.45.9	.69.8	.93.8	1.17.7	1.41.7	1.65.6	1.89.6	2.13.5	2.37.4	11

(21)

$4.78⁷⁸⁄₁₀₀ TO THE POUND.

7¾%

PREMIUM.

STERLING INTO DOLLARS AND CENTS, AND VICE-VERSA.

£	0	100	200	300	400	500	600	700	800	900	£
	$ cts.	$ cts.	$ cts.	$ cts.	$ cts.	$ cts.	$ cts.	$ cts.	$ cts.	$ cts.	
50	239.44.4	718.33.3	1197.22.2	1676.11.1	2155.00.0	2633.88.8	3112.77.7	3591.66.6	4070.55.5	4549.44.4	50
51	244.23.3	723.12.2	1202.01.1	1680.90.0	2159.78.8	2638.67.7	3117.56.6	3596.45.5	4075.34.4	4554.23.3	51
52	249.02.2	727.91.1	1206.80.0	1685.68.8	2164.57.7	2643.46.6	3122.35.5	3601.24.4	4080.13.3	4559.02.2	52
53	253.81.1	732.70.0	1211.58.8	1690.47.7	2169.36.6	2648.25.5	3127.14.4	3606.03.3	4084.92.2	4563.81.1	53
54	258.60.0	737.48.8	1216.37.7	1695.26.6	2174.15.5	2653.04.4	3131.93.3	3610.82.2	4089.71.1	4568.60.0	54
55	263.38.8	742.27.7	1221.16.6	1700.05.5	2178.94.4	2657.83.3	3136.72.2	3615.61.1	4094.50.0	4573.38.8	55
56	268.17.7	747.06.6	1225.95.5	1704.84.4	2183.73.3	2662.62.2	3141.51.1	3620.40.0	4099.28.8	4578.17.7	56
57	272.96.6	751.85.5	1230.74.4	1709.63.3	2188.52.2	2667.41.1	3146.30.0	3625.18.8	4104.07.7	4582.96.6	57
58	277.75.5	756.64.4	1235.53.3	1714.42.2	2193.31.1	2672.20.0	3151.08.8	3629.97.7	4108.86.6	4587.75.5	58
59	282.54.4	761.43.3	1240.32.2	1719.21.1	2198.10.0	2676.98.8	3155.87.7	3634.76.6	4113.65.5	4592.54.4	59
60	287.33.3	766.22.2	1245.11.1	1724.00.0	2202.88.8	2681.77.7	3160.66.6	3639.55.5	4118.44.4	4597.33.3	60
61	292.12.2	771.01.1	1249.90.0	1728.78.8	2207.67.7	2686.56.6	3165.45.5	3644.34.4	4123.23.3	4602.12.2	61
62	296.91.1	775.80.0	1254.68.8	1733.57.7	2212.46.6	2691.35.5	3170.24.4	3649.13.3	4128.02.2	4606.91.1	62
63	301.70.0	780.58.8	1259.47.7	1738.36.6	2217.25.5	2696.14.4	3175.03.3	3653.92.2	4132.81.1	4611.70.0	63
64	306.48.8	785.37.7	1264.26.6	1743.15.5	2222.04.4	2700.93.3	3179.82.2	3658.71.1	4137.60.0	4616.48.8	64
65	311.27.7	790.16.6	1269.05.5	1747.94.4	2226.83.3	2705.72.2	3184.61.1	3663.50.0	4142.38.8	4621.27.7	65
66	316.06.6	794.95.5	1273.84.4	1752.73.3	2231.62.2	2710.51.1	3189.40.0	3668.28.8	4147.17.7	4626.06.6	66
67	320.85.5	799.74.4	1278.63.3	1757.52.2	2236.41.1	2715.30.0	3194.18.8	3673.07.7	4151.96.6	4630.85.5	67
68	325.64.4	804.53.3	1283.42.2	1762.31.1	2241.20.0	2720.08.8	3198.97.7	3677.86.6	4156.75.5	4635.64.4	68
69	330.43.3	809.32.2	1288.21.1	1767.10.0	2245.98.8	2724.87.7	3203.76.6	3682.65.5	4161.54.4	4640.43.3	69
70	335.22.2	814.11.1	1293.00.0	1771.88.8	2250.77.7	2729.66.6	3208.55.5	3687.44.4	4166.33.3	4645.22.2	70
71	340.01.1	818.90.0	1297.78.8	1776.67.7	2255.56.6	2734.45.5	3213.34.4	3692.23.3	4171.12.2	4650.01.1	71
72	344.80.0	823.68.8	1302.57.7	1781.46.6	2260.35.5	2739.24.4	3218.13.3	3697.02.2	4175.91.1	4654.80.0	72
73	349.58.8	828.47.7	1307.36.6	1786.25.5	2265.14.4	2744.03.3	3222.92.2	3701.81.1	4180.70.0	4659.58.8	73
74	354.37.7	833.26.6	1312.15.5	1791.04.4	2269.93.3	2748.82.2	3227.71.1	3706.60.0	4185.48.8	4664.37.7	74
75	359.16.6	838.05.5	1316.94.4	1795.83.3	2274.72.2	2753.61.1	3232.50.0	3711.38.8	4190.27.7	4669.16.6	75
76	363.95.5	842.84.4	1321.73.3	1800.62.2	2279.51.1	2758.40.0	3237.28.8	3716.17.7	4195.06.6	4673.95.5	76
77	368.74.4	847.63.3	1326.52.2	1805.41.1	2284.30.0	2763.18.8	3242.07.7	3720.96.6	4199.85.5	4678.74.4	77
78	373.53.3	852.42.2	1331.31.1	1810.20.0	2289.08.8	2767.97.7	3246.86.6	3725.75.5	4204.64.4	4683.53.3	78
79	378.32.2	857.21.1	1336.10.0	1814.98.8	2293.87.7	2772.76.6	3251.65.5	3730.54.4	4209.43.3	4688.32.2	79
80	383.11.1	862.00.0	1340.88.8	1819.77.7	2298.66.6	2777.55.5	3256.44.4	3735.33.3	4214.22.2	4693.11.1	80
81	387.90.0	866.78.8	1345.67.7	1824.56.6	2303.45.5	2782.34.4	3261.23.3	3740.12.2	4219.01.1	4697.90.0	81
82	392.68.8	871.57.7	1350.46.6	1829.35.5	2308.24.4	2787.13.3	3266.02.2	3744.91.1	4223.80.0	4702.68.8	82
83	397.47.7	876.36.6	1355.25.5	1834.14.4	2313.03.3	2791.92.2	3270.81.1	3749.70.0	4228.58.8	4707.47.7	83
84	402.26.6	881.15.5	1360.04.4	1838.93.3	2317.82.2	2796.71.1	3275.60.0	3754.48.8	4233.37.7	4712.26.6	84
85	407.05.5	885.94.4	1364.83.3	1843.72.2	2322.61.1	2801.50.0	3280.38.8	3759.27.7	4238.16.6	4717.05.5	85
86	411.84.4	890.73.3	1369.62.2	1848.51.1	2327.40.0	2806.28.8	3285.17.7	3764.06.6	4242.95.5	4721.84.4	86
87	416.63.3	895.52.2	1374.41.1	1853.30.0	2332.18.8	2811.07.7	3289.96.6	3768.85.5	4247.74.4	4726.63.3	87
88	421.42.2	900.31.1	1379.20.0	1858.08.8	2336.97.7	2815.86.6	3294.75.5	3773.64.4	4252.53.3	4731.42.2	88
89	426.21.1	905.10.0	1383.98.8	1862.87.7	2341.76.6	2820.65.5	3299.54.4	3778.43.3	4257.32.2	4736.21.1	89
90	431.00.0	909.88.8	1388.77.7	1867.66.6	2346.55.5	2825.44.4	3304.33.3	3783.22.2	4262.11.1	4741.00.0	90
91	435.78.8	914.67.7	1393.56.6	1872.45.5	2351.34.4	2830.23.3	3309.12.2	3788.01.1	4266.90.0	4745.78.8	91
92	440.57.7	919.46.6	1398.35.5	1877.24.4	2356.13.3	2835.02.2	3313.91.1	3792.80.0	4271.68.8	4750.57.7	92
93	445.36.6	924.25.5	1403.14.4	1882.03.3	2360.92.2	2839.81.1	3318.70.0	3797.58.8	4276.47.7	4755.36.6	93
94	450.15.5	929.04.4	1407.93.3	1886.82.2	2365.71.1	2844.60.0	3323.48.8	3802.37.7	4281.26.6	4760.15.5	94
95	454.94.4	933.83.3	1412.72.2	1891.61.1	2370.50.0	2849.38.8	3328.27.7	3807.16.6	4286.05.5	4764.94.4	95
96	459.73.3	938.62.2	1417.51.1	1896.40.0	2375.28.8	2854.17.7	3333.06.6	3811.95.5	4290.84.4	4769.73.3	96
97	464.52.2	943.41.1	1422.30.0	1901.18.8	2380.07.7	2858.96.6	3337.85.5	3816.74.4	4295.63.3	4774.52.2	97
98	469.31.1	948.20.0	1427.08.8	1905.97.7	2384.86.6	2863.75.5	3342.64.4	3821.53.3	4300.42.2	4779.31.1	98
99	474.10.0	952.98.8	1431.87.7	1910.76.6	2389.65.5	2868.54.4	3347.43.3	3826.32.2	4305.21.1	4784.10.0	99

S.	10	11	12	13	14	15	16	17	18	19	S.
d.	$ cts.	$ cts.	$ cts.	$ cts.	$ cts.	$ cts.	$ cts.	$ cts.	$ cts.	$ cts.	d.
0	2.39.4	2.63.3	2.87.3	3.11.2	3.35.2	3.59.1	3.83.1	4.07.0	4.31.0	4.54.9	0
1	2.41.4	2.65.3	2.89.3	3.13.2	3.37.2	3.61.1	3.85.1	4.09.0	4.32.9	4.56.9	1
2	2.43.4	2.67.3	2.91.3	3.15.2	3.39.2	3.63.1	3.87.1	4.11.0	4.34.9	4.58.9	2
3	2.45.4	2.69.3	2.93.3	3.17.2	3.41.2	3.65.1	3.89.1	4.13.0	4.36.9	4.60.9	3
4	2.47.4	2.71.3	2.95.3	3.19.2	3.43.2	3.67.1	3.91.1	4.15.0	4.38.9	4.62.9	4
5	2.49.4	2.73.3	2.97.3	3.21.2	3.45.2	3.69.1	3.93.1	4.17.0	4.40.9	4.64.9	5
6	2.51.4	2.75.3	2.99.3	3.23.2	3.47.2	3.71.1	3.95.1	4.19.0	4.42.9	4.66.9	6
7	2.53.4	2.77.3	3.01.3	3.25.2	3.49.2	3.73.1	3.97.1	4.21.0	4.44.9	4.68.9	7
8	2.55.4	2.79.3	3.03.3	3.27.2	3.51.2	3.75.1	3.99.1	4.23.0	4.46.9	4.70.9	8
9	2.57.4	2.81.3	3.05.3	3.29.2	3.53.2	3.77.1	4.01.1	4.25.0	4.48.9	4.72.9	9
10	2.59.4	2.83.3	3.07.3	3.31.2	3.55.2	3.79.1	4.03.1	4.27.0	4.50.9	4.74.9	10
11	2.61.4	2.85.3	3.09.3	3.33.2	3.57.2	3.81.1	4.05.1	4.29.0	4.52.9	4.76.9	11

$4.79,⁴⁵⁄₁₀₀ TO THE POUND

7 7/8 %

PREMIUM.

STERLING INTO DOLLARS AND CENTS, AND VICE-VERSA.

£	0	100	200	300	400	500	600	700	800	900	£
	$ cts.	$ cts.	$ cts.	$ cts.	$ cts.	$ cts.	$ cts.	$ cts.	$ cts.	$ cts.	
0	.....	479.44.4	958.88.8	1438.33.3	1917.77.7	2397.22.2	2876.66.6	3356.11.1	3835.55.5	4315.00.0	0
1	4.79.4	484 23.8	963.68.3	1443.12.7	1922.57.2	2402.01.6	2881.46.1	3360.90.5	3840.35.0	4319.79.4	1
2	9.58.8	489.03.3	968.47.7	1447.92.2	1927.36.6	2406.81.1	2886.25.5	3365.70.0	3845.14.4	4324.58.8	2
3	14.38.3	493.82.7	973.27.2	1452.71.6	1932.16.1	2411.60 5	2891.05.0	3370.49.4	3849.93.8	4329.38.3	3
4	19.17.7	498.62.2	978.06.6	1457.51.1	1936.95.5	2416.40.0	2895.84.4	3375.28.8	3854.73.3	4334.17.7	4
5	23.97.2	503.41.6	982.86.1	1462.30.5	1941.75.0	2421.19.4	2900.63.8	3380 08.3	3859.52.7	4338.97.2	5
6	28.76.6	508.21.1	987 65.5	1467.10.0	1946.54.4	2425.98.8	2905.43.3	3384.87.7	3864.32.2	4343.76.6	6
7	33.56.1	513.00.5	992.45.0	1471.89.4	1951.33 8	2430.78.3	2910.22 7	3389.67.2	3869.11.6	4348.56.1	7
8	38.35.5	517.80.0	997.24.4	1476.68.8	1956.13.3	2435.57.7	2915.02.2	3394.46.6	3873.91.1	4353.35.5	8
9	43.15.0	522.59.4	1002.03.8	1481.48.3	1960.92 7	2440.37.2	2919.81.6	3399.26.1	3878.70.5	4358.15.0	9
10	47.94.4	527.38.8	1006.83.3	1486.27.7	1965.72.2	2445.16.6	2924.61.1	3404.05.5	3883.50.0	4362.94.4	10
11	52.73.8	532.18.3	1011.62.7	1491.07.2	1970.51.6	2449.96.1	2929.40.5	3408.85.0	3888.29.4	4367.73 8	11
12	57.53.3	536.97.7	1016.42.2	1495.86.6	1975.31.1	2454.75.5	2934.20.0	3413.64.4	3893.08.8	4372.53.3	12
13	62.32.7	541.77.2	1021.21.6	1500.66.1	1980 10.5	2459.55.0	2938.99.4	3418.43.8	3897.88 3	4377.32.7	13
14	67.12.2	546.56.6	1026.01.1	1505.45.5	1984.90.0	2464.34.4	2943.78.8	3423.23.3	3902.67.7	4382.12.2	14
15	71.91.6	551.36.1	1030.80 5	1510.25.0	1989.69.4	2469 13.8	2948.58.3	3428.02.7	3907.47.2	4386.91.6	15
16	76.71.1	556.15.5	1035.60 0	1515.04.4	1994.48 8	2473.93.3	2953.37.7	3432.82.2	3912 26.6	4391.71.1	16
17	81.50.5	560.95.0	1040.39.4	1519.83 8	1999.28.3	2478.72.7	2958.17.2	3437.61.6	3917.06.1	4396.50.5	17
18	86.30.0	565.74.4	1045.18.8	1524.63 3	2004 07.7	2483.52.2	2962.96.6	3442.41.1	3921.85.5	4401.30.0	18
19	91.09.4	570.53.8	1049.98.3	1529.42.7	2008.87.2	2488.31.6	2967.76.1	3447.20.5	3926 65 0	4406.09.4	19
20	95.88.8	575.33.3	1054.77.7	1534 22 2	2013.66.6	2493 11.1	2972.55.5	3452.00 0	3931.44.4	4410.88.8	20
21	100.68.3	580.12.7	1059.57.2	1539.01.6	2018.46.1	2497 90.5	2977.35.0	3456.79.4	3936.23 8	4415.68.3	21
22	105.47.7	584.92.2	1064.36.6	1543.81.1	2023.25.5	2502.70.0	2982 14.4	3461.58.8	3941.03.3	4420.47.7	22
23	110.27.2	589.71.6	1069.16.1	1548 60 5	2028.05 0	2507.49.4	2986.93.8	3466.38.3	3945.82.7	4425.27.2	23
24	115.06.6	594 51.1	1073.95.5	1553.40.0	2032 84.4	2512.28.8	2991.73.3	3471.17.7	3950.62.2	4430.06.6	24
25	119.86.1	599 30.5	1078.75.0	1558.19.4	2037.63 8	2517.08.3	2996.52.7	3475.97.2	3955.41.6	4434.86.1	25
26	124.65.5	604.10.0	1083.54.4	1562.98.8	2042 43.3	2521.87.7	3001.32.2	3480.76.6	3960.21.1	4439 65.5	26
27	129.45.0	608 89.4	1088.33.8	1567.78.3	2047 22.7	2526.67.2	3006.11.6	3485.56.1	3965.00.5	4444.45.0	27
28	134.24.4	613.68.8	1093.13.3	1572 57.7	2052.02.2	2531.46.6	3010.91.1	3490.35.5	3969.80.0	4449.24.4	28
29	139.03.8	618.48.3	1097.92.7	1577.37.2	2056.81.6	2536.26.1	3015.70.5	3495.15.0	3974.59.4	4454.03.8	29
30	143.83.3	623.27.7	1102.72.2	1582.16.6	2061.61.1	2541.05.5	3020.50.0	3499.94.4	3979.38.8	4458.83.3	30
31	148.62.7	628.07.2	1107.51.6	1586.96 1	2066.40.5	2545.85.0	3025.29.4	3504.73.8	3984.18.3	4463.62.7	31
32	153.42.2	632.86.6	1112.31.1	1591.75.5	2071.20.0	2550.64.4	3030.08.8	3509.53.3	3988.97.7	4468.42.2	32
33	158.21.6	637.66.1	1117.10.5	1596.55.0	2075.99.4	2555.43.8	3034.88.3	3514.32.7	3993.77.2	4473.21.6	33
34	163.01.1	642.45.5	1121.90.0	1601.34 4	2080.78 8	2560.23.3	3039.67.7	3519.12.2	3998.56.6	4478.01.1	34
35	167.80.5	647.25.0	1126.69.4	1606.13.8	2085.58.3	2565.02.7	3044.47.2	3523 91.6	4003.36.1	4482.80.5	35
36	172.60.0	652.04.4	1131.48.8	1610.93.3	2090.37.7	2569.82.2	3049.26.6	3528.71.1	4008.15.5	4487.60.0	36
37	177.39.4	656.83.8	1136.28.3	1615.72.7	2095.17.2	2574.61.6	3054.06.1	3533.50.5	4012.95.0	4192.39 4	37
38	182 18.8	661.63.3	1141.07.7	1620.52.2	2099.96 6	2579.41.1	3058.85.5	3538.30.0	4017.74.4	4497.18.8	38
39	186.98.3	666.42.7	1145.87.2	1625.31.6	2104.76.1	2584.20.5	3063.65.0	3543.09.4	4022.53.8	4501.98.3	39
40	191.77.7	671.22.2	1150.66.6	1630.11.1	2109.55.5	2589.00.0	3068.44.4	3547.88.8	4027.33.3	4506.77 7	40
41	196.57.2	676.01.6	1155.46.1	1634.90.5	2114.35.0	2593.79.4	3073.23.8	3552.68.3	4032.12.7	4511.57.2	41
42	201.36.6	680.81.1	1160.25.5	1639.70.0	2119.14.4	2598.58.8	3078.03.3	3557.47.7	4036.92.2	4516.36.6	42
43	206.16.1	685.60.5	1165.05.0	1644.49.4	2123 93.8	2603.38.3	3082.82.7	3562.27.2	4041.71.6	4521.16.1	43
44	210.95.5	690.40.0	1169.84.4	1649.28.8	2128.73.3	2608.17.7	3087.62.2	3567.06.6	4046.51.1	4525.95.5	44
45	215.75.0	695.19.4	1174.63.8	1654.08.3	2133 52.7	2612.97.2	3092.41.6	3571.86.1	4051.30.5	4530.75.0	45
46	220.54.4	699.98.8	1179.43.3	1658.87.7	2138.32.2	2617.76.6	3097.21.1	3576.65.5	4056.10.0	4535.54.4	46
47	225.33.8	704.78.3	1184.22.7	1663.67.2	2143.11.6	2622.56.1	3102.00.5	3581.45.0	4060.89.4	4540.33.8	47
48	230.13.3	709.57.7	1189.02.2	1668.46.6	2147.91.1	2627.35.5	3106.80.0	3586.24.4	4065.68.8	4545.13.3	48
49	234 92.7	714.37.2	1193.81.6	1673.26.1	2152.70.5	2632.15.0	3111.59.4	3591.03.8	4070.48.3	4549.92.7	49

S.	0	1	2	3	4	5	6	7	8	9	S.
d.	$ cts.	$ cts.	$ cts.	$ cts.	$ cts.	$ cts.	$ cts.	$ cts.	$ cts.	$ cts.	d.
0	.....	.23.9	.47.9	.71.9	.95.8	1.19.6	1.43.8	1.67.8	1.91.7	2.15.7	0
1	.01.9	.25.9	.49.9	.73.9	.97.8	1.21.8	1.45.8	1.69.8	1.93.7	2.17.7	1
2	.03.9	.27.9	.51.9	.75.9	.99.8	1.23.8	1.47.8	1.71.8	1.95.7	2 19.7	2
3	.05.9	.29.9	.53.9	.77.9	1.01.8	1.25.8	1.49.8	1.73.8	1.97.7	2.21.7	3
4	.07.9	.31.9	.55.9	.79.9	1.03.8	1.27.8	1.51.8	1.75.8	1.99.7	2.23.7	4
5	.09.9	.33.9	.57.9	.81.9	1.05.8	1.29.8	1.53.8	1.77.8	2.01.7	2.25.7	5
6	.11.9	.35.9	.59.9	.83.9	1.07.8	1.31.8	1.55 8	1.79.8	2.03.7	2.27.7	6
7	.13.9	.37.9	.61.9	.85.9	1.09.8	1.33 8	1.57.8	1.81.8	2.05.7	2.29.7	7
8	.15.9	.39.9	.63.9	.87.9	1.11.8	1.35.8	1.59.8	1.83 8	2.07.7	2.31.7	8
9	.17.9	.41.9	.65.9	.89.9	1.13.8	1.37.8	1.61.8	1.85.8	2.09.7	2.33.7	9
10	.19.9	.43.9	.67.9	.91.9	1.15.8	1.39.8	1.63.8	1.87.8	2.11.7	2.35.7	10
11	.21.9	.45.9	.69.9	.93.9	1.17.8	1.41.8	1.65.8	1.89.8	2.13.7	2.37.7	11

$4.79¹⁄₁₀₀ TO THE POUND.

7⅞%

PREMIUM.

STERLING INTO DOLLARS AND CENTS, AND VICE-VERSA.

£	0	100	200	300	400	500	600	700	800	900	£
	$ cts.	$ cts.	$ cts.	$ cts.	$ cts.	$ cts.	$ cts.	$ cts.	$ cts.	$ cts.	
50	239.72.2	719.16.6	1198.61.1	1678.05.5	2157.50.0	2636.94.4	3116.38.8	3595.83.3	4075.27.7	4554.72.2	50
51	244.51.6	723.96.1	1203.40.5	1682.85.0	2162.29.4	2641.73.8	3121.18.3	3600.62.7	4080.07.2	4559.51.6	51
52	249.31.1	728.75.5	1208.20.0	1687.64.4	2167.08.8	2646.53.3	3125.97.7	3605.42.2	4084.86.6	4564.31.1	52
53	254.10.5	733.55.0	1212.99.4	1692.43.8	2171.88.3	2651.32.7	3130.77.2	3610.21.6	4089.66.1	4569.10.5	53
54	258.90.0	738.34.4	1217.78.8	1697.23.3	2176.67.7	2656.12.2	3135.56.6	3615.01.1	4094.45.5	4573.90.0	54
55	263.69.4	743.13.8	1222.58.2	1702.02.7	2181.47.2	2660.91.6	3140.36.1	3619.80.5	4099.25.0	4578.69.4	55
56	268.48.8	747.93.3	1227.37.7	1706.82.2	2186.26.6	2665.71.1	3145.15.5	3624.60.0	4104.04.4	4583.48.8	56
57	273.28.3	752.72.7	1232.17.2	1711.61.6	2191.06.1	2670.50.5	3149.95.0	3629.39.4	4108.83.8	4588.28.3	57
58	278.07.7	757.52.2	1236.96.6	1716.41.1	2195.85.5	2675.30.0	3154.74.4	3634.18.8	4113.63.3	4593.07.7	58
59	282.87.2	762.31.6	1241.76.1	1721.20.5	2200.65.0	2680.09.4	3159.53.8	3638.98.3	4118.42.7	4597.87.2	59
60	287.66.6	767.11.1	1246.55.5	1726.00.0	2205.44.4	2684.88.8	3164.33.3	3643.77.7	4123.22.2	4602.66.6	60
61	292.46.1	771.90.5	1251.35.0	1730.79.4	2210.23.8	2689.68.3	3169.12.7	3648.57.2	4128.01.6	4607.46.1	61
62	297.25.5	776.70.0	1256.14.4	1735.58.8	2215.03.3	2694.47.7	3173.92.2	3653.36.6	4132.81.1	4612.25.5	62
63	302.05.0	781.49.4	1260.93.8	1740.38.3	2219.82.7	2699.27.2	3178.71.6	3658.16.1	4137.60.5	4617.05.0	63
64	306.84.4	786.28.8	1265.73.3	1745.17.7	2224.62.2	2704.06.6	3183.51.1	3662.95.5	4142.40.0	4621.84.4	64
65	311.63.8	791.08.3	1270.52.7	1749.97.2	2229.41.6	2708.86.1	3188.30.5	3667.75.0	4147.19.4	4626.63.8	65
66	316.43.3	795.87.7	1275.32.2	1754.76.6	2234.21.1	2713.65.5	3193.10.0	3672.54.4	4151.98.8	4631.43.3	66
67	321.22.7	800.67.2	1280.11.6	1759.56.1	2239.00.5	2718.45.0	3197.89.4	3677.33.8	4156.78.3	4636.22.7	67
68	326.02.2	805.46.6	1284.91.1	1764.35.5	2243.80.0	2723.24.4	3202.68.8	3682.13.3	4161.57.7	4641.02.2	68
69	330.81.6	810.26.1	1289.70.5	1769.15.0	2248.59.4	2728.03.8	3207.48.3	3686.92.7	4166.37.2	4645.81.6	69
70	335.61.1	815.05.5	1294.50.0	1773.94.4	2253.38.8	2732.83.3	3212.27.7	3691.72.2	4171.16.6	4650.61.1	70
71	340.40.5	819.85.0	1299.29.4	1778.73.8	2258.18.3	2737.62.7	3217.07.2	3696.51.6	4175.96.1	4655.40.5	71
72	345.20.0	824.64.4	1304.08.8	1783.53.3	2262.97.7	2742.42.2	3221.86.6	3701.31.1	4180.75.5	4660.20.0	72
73	349.99.4	829.43.8	1308.88.3	1788.32.7	2267.77.2	2747.21.6	3226.66.1	3706.10.5	4185.55.0	4664.99.4	73
74	354.78.8	834.23.3	1313.67.7	1793.12.2	2272.56.6	2752.01.1	3231.45.5	3710.90.0	4190.34.4	4669.78.8	74
75	359.58.3	839.02.7	1318.47.2	1797.91.6	2277.36.1	2756.80.5	3236.25.0	3715.69.4	4195.13.8	4674.58.3	75
76	364.37.7	843.82.2	1323.26.6	1802.71.1	2282.15.5	2761.60.0	3241.04.4	3720.48.8	4199.93.3	4679.37.7	76
77	369.17.2	848.61.6	1328.06.1	1807.50.5	2286.95.0	2766.39.4	3245.83.8	3725.28.3	4204.72.7	4684.17.2	77
78	373.96.6	853.41.1	1332.85.5	1812.30.0	2291.74.4	2771.18.8	3250.63.3	3730.07.7	4209.52.2	4688.96.6	78
79	378.76.1	858.20.5	1337.65.0	1817.09.4	2296.53.8	2775.98.3	3255.42.7	3734.87.2	4214.31.6	4693.76.1	79
80	383.55.5	863.00.0	1342.44.4	1821.88.8	2301.33.3	2780.77.7	3260.22.2	3739.66.6	4219.11.1	4698.55.5	80
81	388.35.0	867.79.4	1347.23.8	1826.68.3	2306.12.7	2785.57.2	3265.01.6	3744.46.1	4223.90.5	4703.35.0	81
82	393.14.4	872.58.8	1352.03.3	1831.47.7	2310.92.2	2790.36.6	3269.81.1	3749.25.5	4228.70.0	4708.14.4	82
83	397.93.8	877.38.3	1356.82.7	1836.27.2	2315.71.6	2795.16.1	3274.60.5	3754.05.0	4233.49.4	4712.93.8	83
84	402.73.3	882.17.7	1361.62.2	1841.06.6	2320.51.1	2799.95.5	3279.40.0	3758.84.4	4238.28.8	4717.73.3	84
85	407.52.7	886.97.2	1366.41.6	1845.86.1	2325.30.5	2804.75.0	3284.19.4	3763.63.8	4243.08.3	4722.52.7	85
86	412.32.2	891.76.6	1371.21.1	1850.65.5	2330.10.0	2809.54.4	3288.98.8	3768.43.3	4247.87.7	4727.32.2	86
87	417.11.6	896.56.1	1376.00.5	1855.45.0	2334.89.4	2814.33.8	3293.78.3	3773.22.7	4252.67.2	4732.11.6	87
88	421.91.1	901.35.5	1380.80.0	1860.24.4	2339.68.8	2819.13.3	3298.57.7	3778.02.2	4257.46.6	4736.91.1	88
89	426.70.5	906.15.0	1385.59.4	1865.03.8	2344.48.3	2823.92.7	3303.37.2	3782.81.6	4262.26.1	4741.70.5	89
90	431.50.0	910.94.4	1390.38.8	1869.83.3	2349.27.7	2828.72.2	3308.16.6	3787.61.1	4267.05.5	4746.50.0	90
91	436.29.4	915.73.8	1395.18.3	1874.62.7	2354.07.2	2833.51.6	3312.96.1	3792.40.5	4271.85.0	4751.29.4	91
92	441.08.8	920.53.3	1399.97.7	1879.42.2	2358.86.6	2838.31.1	3317.75.5	3797.20.0	4276.64.4	4756.08.8	92
93	445.88.3	925.32.7	1404.77.2	1884.21.6	2363.66.1	2843.10.5	3322.55.0	3801.99.4	4281.43.8	4760.88.3	93
94	450.67.7	930.12.2	1409.56.6	1889.01.1	2368.45.5	2847.90.0	3327.34.4	3806.78.8	4286.23.3	4765.67.7	94
95	455.47.2	934.91.6	1414.36.1	1893.80.5	2373.25.0	2852.69.4	3332.13.8	3811.58.3	4291.02.7	4770.47.2	95
96	460.26.6	939.71.1	1419.15.5	1898.60.0	2378.04.4	2857.48.8	3336.93.3	3816.37.7	4295.82.2	4775.26.6	96
97	465.06.1	944.50.5	1423.95.0	1903.39.4	2382.83.8	2862.28.3	3341.72.7	3821.17.2	4300.61.6	4780.06.1	97
98	469.85.5	949.30.0	1428.74.4	1908.18.8	2387.63.3	2867.07.7	3346.52.2	3825.96.6	4305.41.1	4784.85.5	98
99	474.65.0	954.09.4	1433.53.8	1912.98.3	2392.42.7	2871.87.2	3351.31.6	3830.76.1	4310.20.5	4789.65.0	99

S.	10	11	12	13	14	15	16	17	18	19	S.
d.	$ cts.	$ cts.	$ cts.	$ cts.	$ cts.	$ cts.	$ cts.	$ cts.	$ cts.	$ cts.	d.
0	2.39.7	2.63.6	2.87.6	3.11.6	3.35.6	3.59.5	3.83.5	4.07.5	4.31.5	4.55.4	0
1	2.41.7	2.65.6	2.89.6	3.13.6	3.37.6	3.61.5	3.85.5	4.09.5	4.33.5	4.57.4	1
2	2.43.7	2.67.6	2.91.6	3.15.6	3.39.6	3.63.5	3.87.5	4.11.5	4.35.5	4.59.4	2
3	2.45.7	2.69.6	2.93.6	3.17.6	3.41.6	3.65.5	3.89.5	4.13.5	4.37.5	4.61.4	3
4	2.47.7	2.71.6	2.95.6	3.19.6	3.43.6	3.67.5	3.91.5	4.15.5	4.39.5	4.63.4	4
5	2.49.7	2.73.6	2.97.6	3.21.6	3.45.6	3.69.5	3.93.5	4.17.5	4.41.5	4.65.4	5
6	2.51.7	2.75.6	2.99.6	3.23.6	3.47.6	3.71.5	3.95.5	4.19.5	4.43.5	4.67.4	6
7	2.53.7	2.77.6	3.01.6	3.25.6	3.49.6	3.73.5	3.97.5	4.21.5	4.45.5	4.69.4	7
8	2.55.7	2.79.6	3.03.6	3.27.6	3.51.6	3.75.5	3.99.5	4.23.5	4.47.5	4.71.4	8
9	2.57.7	2.81.6	3.05.6	3.29.6	3.53.6	3.77.5	4.01.5	4.25.5	4.49.5	4.73.4	9
10	2.59.7	2.83.6	3.07.6	3.31.6	3.55.6	3.79.5	4.03.5	4.27.5	4.51.5	4.75.4	10
11	2.61.7	2.85.6	3.09.6	3.33.6	3.57.6	3.81.5	4.05.5	4.29.5	4.53.5	4.77.4	11

$4.80 1/2/3 TO THE POUND.

8 %

PREMIUM.

STERLING INTO DOLLARS AND CENTS, AND VICE-VERSA.

£	0	100	200	300	400	500	600	700	800	900	£
	$ cts.	$ cts.	$ cts.	$ cts.	$ cts.	$ cts.	$ cts.	$ cts.	$ cts.	$ cts.	
0	......	480.00.0	960.00.0	1440.00.0	1920.00.0	2400.00.0	2880.00.0	3360.00.0	3840.00.0	4320.00.0	0
1	4.80.0	484 80.0	964.80.0	1444.80.0	1924.80.0	2404.80 0	2884.80.0	3364.80.0	3844.80.0	4324 80.0	1
2	9.60.0	489.60.0	969.60.0	1449.60.0	1929.60.0	2409.60.0	2889.60.0	3369.60.0	3849.60.0	4329.60.0	2
3	14.40.0	494.40.0	974.40.0	1454.40.0	1934.40 0	2414.40.0	2894.40.0	3374.40.0	3854.40.0	4334 40.0	3
4	19.20.0	499.20.0	979.20.0	1459.20.0	1939.20.0	2419.20.0	2899.20.0	3379.20.0	3859.20.0	4339.20.0	4
5	24.00.0	504.00.0	984.00.0	1464.00.0	1944.00.0	2424 00.0	2904.00.0	3384.00.0	3864.00.0	4344.00.0	5
6	28.80.0	508.80.0	988.80.0	1468.80.0	1948.80.0	2428.80.0	2908.80.0	3388.80.0	3868.80.0	4348.80.0	6
7	33.60.0	513.60.0	993 60.0	1473.60.0	1953.60.0	2433.60.0	2913.60.0	3393.60.0	3873.60.0	4353 60.0	7
8	38.40.0	518.40.0	998.40.0	1478.40.0	1958.40.0	2438.40.0	2918.40.0	3398.40.0	3878.40.0	4358 40.0	8
9	43.20.0	523.20.0	1003.20.0	1483.20.0	1963.20.0	2443.20.0	2923.20.0	3403.20.0	3883.20.0	4363.20.0	9
10	48.00.0	528.00.0	1008.00.0	1488.00.0	1968.00.0	2448.00.0	2928.00.0	3408.00.0	3888.00.0	4368.00 0	10
11	52.80.0	532.80.0	1012.80.0	1492.80.0	1972.80.0	2452 80.0	2932.80.0	3412.80.0	3892.80.0	4372 80.0	11
12	57.60.0	537.60.0	1017.60.0	1497.60.0	1977.60.0	2457.60.0	2937.60.0	3417.60.0	3897.60.0	4377.60.0	12
13	62.40.0	542 40.0	1022.40.0	1502.40.0	1982.40.0	2462.40.0	2942 40.0	3422.40.0	3902.40.0	4382.40.0	13
14	67.20.0	547.20.0	1027.20.0	1507.20.0	1987.20.0	2467.20.0	2947.20.0	3427.20.0	3907.20.0	4387.20.0	14
15	72.00.0	552.00.0	1032.00.0	1512.00.0	1992.00.0	2472.00.0	2952.00.0	3432.00.0	3912.00.0	4392.00.0	15
16	76.80.0	556.80.0	1036.80.0	1516.80.0	1996.80.0	2476.80.0	2956.80.0	3436.80.0	3916.80.0	4396.80.0	16
17	81.60.0	561.60 0	1041.60.0	1521.60.0	2001.60 0	2481.60.0	2961.60.0	3441.60.0	3921.60.0	4401.60 0	17
18	86.40.0	566.40.0	1046.40 0	1526.40.0	2006.40.0	2486.40.0	2966.40.0	3446.40.0	3926.40.0	4406.40.0	18
19	91.20.0	571.20.0	1051.20.0	1531.20.0	2011.20.0	2491.20 0	2971.20.0	3451.20.0	3931.20.0	4411.20.0	19
20	96.00.0	576 00.0	1056.00.0	1536.00.0	2016.00.0	2496.00.0	2976.00.0	3456.00.0	3936.00.0	4416.00.0	20
21	100.80.0	580.80.0	1060.80.0	1540.80.0	2020.80.0	2500.80.0	2980.80.0	3460.80.0	3940.80.0	4420.80.0	21
22	105.60.0	585.60.0	1065.60.0	1545.60.0	2025.60.0	2505.60.0	2985.60.0	3465.60.0	3945.60.0	4425.60 0	22
23	110.40.0	590.40.0	1070.40.0	1550.40.0	2030.40.0	2510.40.0	2990.40.0	3470.40.0	3950.40.0	4430.40.0	23
24	115.20.0	595.20.0	1075.20.0	1555.20.0	2035.20.0	2515.20.0	2995.20.0	3475.20.0	3955.20.0	4435.20.0	24
25	120.00.0	600.00.0	1080.00.0	1560.00.0	2040.00.0	2520.00.0	3000.00.0	3480.00.0	3960.00.0	4440.00.0	25
26	124.80.0	604.80.0	1084.80.0	1564.80.0	2044.80.0	2524.80.0	3004.80.0	3484.80.0	3964.80.0	4444.80.0	26
27	129.60.0	609.60.0	1089.60.0	1569.60.0	2049.60.0	2529.60.0	3009.60.0	3489.60.0	3969.60.0	4449.60.0	27
28	134.40.0	614.40.0	1094.40.0	1574.40.0	2054 40.0	2534.40.0	3014.40.0	3494.40.0	3974.40.0	4454.40.0	28
29	139.20.0	619.20.0	1099.20.0	1579.20.0	2059.20.0	2539.20.0	3019.20.0	3499.20.0	3979.20.0	4459.20.0	29
30	144.00.0	624.00.0	1104.00.0	1584.00.0	2064.00.0	2544.00.0	3024.00.0	3504.00.0	3984.00.0	4464.00.0	30
31	148.80.0	628.80.0	1108.80.0	1588.80.0	2068.80.0	2548.80.0	3028.80.0	3508.80.0	3988.80.0	4468.80.0	31
32	153.60.0	633.60.0	1113.60.0	1593.60.0	2073.60.0	2553.60.0	3033.60.0	3513.60.0	3993.60.0	4473.60.0	32
33	158.40.0	638.40.0	1118.40.0	1598.40.0	2078.40.0	2558.40.0	3038.40.0	3518.40.0	3998.40.0	4478.40.0	33
34	163.20.0	643.20.0	1123.20.0	1603.20.0	2083.20.0	2563.20.0	3043.20.0	3523.20.0	4003.20.0	4483.20.0	34
35	168.00.0	648.00.0	1128.00.0	1608.00.0	2088.00.0	2568.00.0	3048.00.0	3528.00.0	4008.00.0	4488.00.0	35
36	172.80.0	652.80.0	1132.80.0	1612.80.0	2092.80.0	2572.80.0	3052.80.0	3532.80.0	4012.80.0	4492.80.0	36
37	177.60.0	657.60.0	1137.60.0	1617.60.0	2097.60.0	2577.60.0	3057.60.0	3537.60.0	4017.60.0	4497.60.0	37
38	182.40.0	662.40.0	1142.40.0	1622.40.0	2102.40.0	2582.40.0	3062.40.0	3542.40.0	4022.40.0	4502.40.0	38
39	187.20.0	667.20.0	1147.20.0	1627.20.0	2107.20.0	2587.20.0	3067.20.0	3547.20.0	4027.20.0	4507.20.0	39
40	192.00.0	672.00.0	1152.00.0	1632.00.0	2112.00.0	2592.00.0	3072.00.0	3552.00.0	4032.00.0	4512.00.0	40
41	196.80.0	676.80.0	1156.80.0	1636.80.0	2116.80.0	2596.80.0	3076.80.0	3556.80.0	4036.80.0	4516.80.0	41
42	201.60.0	681.60.0	1161.60.0	1641.60.0	2121.60.0	2601.60.0	3081.60.0	3561.60.0	4041.60.0	4521.60.0	42
43	206.40.0	686.40.0	1166.40.0	1646.40.0	2126.40.0	2606.40.0	3086.40.0	3566.40.0	4046.40.0	4526.40.0	43
44	211.20.0	691.20.0	1171.20.0	1651.20.0	2131.20.0	2611.20.0	3091.20.0	3571.20.0	4051.20.0	4531.20.0	44
45	216.00.0	696.00.0	1176.00.0	1656.00.0	2136.00.0	2616.00.0	3096.00.0	3576.00.0	4056.00.0	4536.00.0	45
46	220.80.0	700.80.0	1180.80.0	1660.80.0	2140.80.0	2620.80.0	3100.80.0	3580.80.0	4060.80.0	4540.80.0	46
47	225.60.0	705.60.0	1185.60.0	1665.60.0	2145.60 0	2625.60.0	3105.60.0	3585.60.0	4065.60.0	4545.60.0	47
48	230.40.0	710.40.0	1190.40.0	1670.40.0	2150.40.0	2630.40.0	3110.40.0	3590.40.0	4070.40.0	4550.40.0	48
49	235.20.0	715.20.0	1195.20.0	1675.20.0	2155.20.0	2635.20.0	3115.20.0	3595.20.0	4075.20.0	4555.20.0	49

S.	0	1	2	3	4	5	6	7	8	9	S.
d.	$ cts.	$ cts.	$ cts.	$ cts.	$ cts.	$ cts.	$ cts.	$ cts.	$ cts.	$ cts.	d.
0	......	.24.0	.48.0	.72.0	.96.0	1.20.0	1.44.0	1.68.0	1.92.0	2.16.0	0
1	.02.0	.26.0	.50.0	.74.0	.98.0	1.22.0	1.46.0	1.70.0	1.94.0	2.18.0	1
2	.04.0	.28.0	.52.0	.76.0	1.00.0	1.24.0	1.48.0	1.72.0	1.96.0	2.20.0	2
3	.06.0	.30.0	.54.0	.78.0	1.02.0	1.26.0	1.50.0	1.74.0	1.98.0	2.22.0	3
4	.08.0	.32.0	.56.0	.80.0	1.04.0	1.28.0	1.52.0	1.76.0	2.00.0	2.24.0	4
5	.10.0	.34.0	.58.0	.82.0	1.06.0	1.30.0	1.54.0	1.78.0	2.02.0	2.26.0	5
6	.12.0	.36.0	.60.0	.84.0	1.08.0	1.32.0	1.56.0	1.80.0	2.04.0	2.28.0	6
7	.14.0	.38.0	.62.0	.86.0	1.10.0	1.34 0	1.58.0	1.82.0	2.06.0	2.30.0	7
8	.16.0	.40.0	.64.0	.88.0	1.12.0	1.36.0	1.60.0	1.84.0	2.08.0	2.32.0	8
9	.18.0	.42.0	.66.0	.90.0	1.14.0	1.38.0	1.62.0	1.86.0	2.10.0	2.34.0	9
10	.20.0	.44.0	.68.0	.92.0	1.16.0	1.40.0	1.64.0	1.88.0	2.12.0	2.36.0	10
11	.22.0	.46.0	.70.0	.94.0	1.18.0	1.42.0	1.66.0	1.90.0	2.14.0	2.38.0	11

$4.80 8/100 TO THE POUND.

8%

PREMIUM.

STERLING INTO DOLLARS AND CENTS, AND VICE-VERSA.

£	0	100	200	300	400	500	600	700	800	900	£
£	$ cts.	$ cts.	$ cts.	$ cts.	$ cts.	$ cts.	$ cts.	$ cts.	$ cts.	$ cts.	£
50	240.00.0	720.00.0	1200.00.0	1680.00.0	2160.00.0	2640.00.0	3120.00.0	3600.00.0	4080.00.0	4560.00.0	50
51	244.80.0	724.80.0	1204.80.0	1684.80.0	2164.80.0	2644.80.0	3124.80.0	3604.80.0	4084.80.0	4564.80.0	51
52	249.60.0	729.60.0	1209.60.0	1689.60.0	2169.60.0	2649.60.0	3129.60.0	3609.60.0	4089.60.0	4569.60.0	52
53	254.40.0	734.40.0	1214.40.0	1694.40.0	2174.40.0	2654.40.0	3134.40.0	3614.40.0	4094.40.0	4574.40.0	53
54	259.20.0	739.20.0	1219.20.0	1699.20.0	2179.20.0	2659.20.0	3139.20.0	3619.20.0	4099.20.0	4579.20.0	54
55	264.00.0	744.00.0	1224.00.0	1704.00.0	2184.00.0	2664.00.0	3144.00.0	3624.00.0	4104.00.0	4584.00.0	55
56	268.80.0	748.80.0	1228.80.0	1708.80.0	2188.80.0	2668.80.0	3148.80.0	3628.80.0	4108.80.0	4588.80.0	56
57	273.60.0	753.60.0	1233.60.0	1713.60.0	2193.60.0	2673.60.0	3153.60.0	3633.60.0	4113.60.0	4593.60.0	57
58	278.40.0	758.40.0	1238.40.0	1718.40.0	2198.40.0	2678.40.0	3158.40.0	3638.40.0	4118.40.0	4598.40.0	58
59	283.20.0	763.20.0	1243.20.0	1723.20.0	2203.20.0	2683.20.0	3163.20.0	3643.20.0	4123.20.0	4603.20.0	59
60	288.00.0	768.00.0	1248.00.0	1728.00.0	2208.00.0	2688.00.0	3168.00.0	3618.00.0	4128.00.0	4608.00.0	60
61	292.80.0	772.80.0	1252.80.0	1732.80.0	2212.80.0	2692.80.0	3172.80.0	3652.80.0	4132.80.0	4612.80.0	61
62	297.60.0	777.60.0	1257.60.0	1737.60.0	2217.60.0	2697.60.0	3177.60.0	3657.60.0	4137.60.0	4617.60.0	62
63	302.40.0	782.40.0	1262.40.0	1742.40.0	2222.40.0	2702.40.0	3182.40.0	3662.40.0	4142.40.0	4622.40.0	63
64	307.20.0	787.20.0	1267.20.0	1747.20.0	2227.20.0	2707.20.0	3187.20.0	3667.20.0	4147.20.0	4627.20.0	64
65	312.00.0	792.00.0	1272.00.0	1752.00.0	2232.00.0	2712.00.0	3192.00.0	3672.00.0	4152.00.0	4632.00.0	65
66	316.80.0	796.80.0	1276.80.0	1756.80.0	2236.80.0	2716.80.0	3196.80.0	3676.80.0	4156.80.0	4636.80.0	66
67	321.60.0	801.60.0	1281.60.0	1761.60.0	2241.60.0	2721.60.0	3201.60.0	3681.60.0	4161.60.0	4641.60.0	67
68	326.40.0	806.40.0	1286.40.0	1766.40.0	2246.40.0	2726.40.0	3206.40.0	3686.40.0	4166.40.0	4646.40.0	68
69	331.20.0	811.20.0	1291.20.0	1771.20.0	2251.20.0	2731.20.0	3211.20.0	3691.20.0	4171.20.0	4651.20.0	69
70	336.00.0	816.00.0	1296.00.0	1776.00.0	2256.00.0	2736.00.0	3216.00.0	3696.00.0	4176.00.0	4656.00.0	70
71	340.80.0	820.80.0	1300.80.0	1780.80.0	2260.80.0	2740.80.0	3220.80.0	3700.80.0	4180.80.0	4660.80.0	71
72	345.60.0	825.60.0	1305.60.0	1785.60.0	2265.60.0	2745.60.0	3225.60.0	3705.60.0	4185.60.0	4665.60.0	72
73	350.40.0	830.40.0	1310.40.0	1790.40.0	2270.40.0	2750.40.0	3230.40.0	3710.40.0	4190.40.0	4670.40.0	73
74	355.20.0	835.20.0	1315.20.0	1795.20.0	2275.20.0	2755.20.0	3235.20.0	3715.20.0	4195.20.0	4675.20.0	74
75	360.00.0	840.00.0	1320.00.0	1800.00.0	2280.00.0	2760.00.0	3240.00.0	3720.00.0	4200.00.0	4680.00.0	75
76	364.80.0	844.80.0	1324.80.0	1804.80.0	2284.80.0	2764.80.0	3244.80.0	3724.80.0	4204.80.0	4684.80.0	76
77	369.60.0	849.60.0	1329.60.0	1809.60.0	2289.60.0	2769.60.0	3249.60.0	3729.60.0	4209.60.0	4689.60.0	77
78	374.40.0	854.40.0	1334.40.0	1814.40.0	2294.40.0	2774.40.0	3254.40.0	3734.40.0	4214.40.0	4694.40.0	78
79	379.20.0	859.20.0	1339.20.0	1819.20.0	2299.20.0	2779.20.0	3259.20.0	3739.20.0	4219.20.0	4699.20.0	79
80	384.00.0	864.00.0	1344.00.0	1824.00.0	2304.00.0	2784.00.0	3264.00.0	3744.00.0	4224.00.0	4704.00.0	80
81	388.80.0	868.80.0	1348.80.0	1828.80.0	2308.80.0	2788.80.0	3268.80.0	3748.80.0	4228.80.0	4708.80.0	81
82	393.60.0	873.60.0	1353.60.0	1833.60.0	2313.60.0	2793.60.0	3273.60.0	3753.60.0	4233.60.0	4713.60.0	82
83	398.40.0	878.40.0	1358.40.0	1838.40.0	2318.40.0	2798.40.0	3278.40.0	3758.40.0	4238.40.0	4718.40.0	83
84	403.20.0	883.20.0	1363.20.0	1843.20.0	2323.20.0	2803.20.0	3283.20.0	3763.20.0	4243.20.0	4723.20.0	84
85	408.00.0	888.00.0	1368.00.0	1848.00.0	2328.00.0	2808.00.0	3288.00.0	3768.00.0	4248.00.0	4728.00.0	85
86	412.80.0	892.80.0	1372.80.0	1852.80.0	2332.80.0	2812.80.0	3292.80.0	3772.80.0	4252.80.0	4732.80.0	86
87	417.60.0	897.60.0	1377.60.0	1857.60.0	2337.60.0	2817.60.0	3297.60.0	3777.60.0	4257.60.0	4737.60.0	87
88	422.40.0	902.40.0	1382.40.0	1862.40.0	2342.40.0	2822.40.0	3302.40.0	3782.40.0	4262.40.0	4742.40.0	88
89	427.20.0	907.20.0	1387.20.0	1867.20.0	2347.20.0	2827.20.0	3307.20.0	3787.20.0	4267.20.0	4747.20.0	89
90	432.00.0	912.00.0	1392.00.0	1872.00.0	2352.00.0	2832.00.0	3312.00.0	3792.00.0	4272.00.0	4752.00.0	90
91	436.80.0	916.80.0	1396.80.0	1876.80.0	2356.80.0	2836.80.0	3316.80.0	3796.80.0	4276.80.0	4756.80.0	91
92	441.60.0	921.60.0	1401.60.0	1881.60.0	2361.60.0	2841.60.0	3321.60.0	3801.60.0	4281.60.0	4761.60.0	92
93	446.40.0	926.40.0	1406.40.0	1886.40.0	2366.40.0	2846.40.0	3326.40.0	3806.40.0	4286.40.0	4766.40.0	93
94	451.20.0	931.20.0	1411.20.0	1891.20.0	2371.20.0	2851.20.0	3331.20.0	3811.20.0	4291.20.0	4771.20.0	94
95	456.00.0	936.00.0	1416.00.0	1896.00.0	2376.00.0	2856.00.0	3336.00.0	3816.00.0	4296.00.0	4776.00.0	95
96	460.80.0	940.80.0	1420.80.0	1900.80.0	2380.80.0	2860.80.0	3340.80.0	3820.80.0	4300.80.0	4780.80.0	96
97	465.60.0	945.60.0	1425.60.0	1905.60.0	2385.60.0	2865.60.0	3345.60.0	3825.60.0	4305.60.0	4785.60.0	97
98	470.40.0	950.40.0	1430.40.0	1910.40.0	2390.40.0	2870.40.0	3350.40.0	3830.40.0	4310.40.0	4790.40.0	98
99	475.20.0	955.20.0	1435.20.0	1915.20.0	2395.20.0	2875.20.0	3355.20.0	3835.20.0	4315.20.0	4795.20.0	99

S.	10	11	12	13	14	15	16	17	18	19	S.
d.	$ cts.	$ cts.	$ cts.	$ cts.	$ cts.	$ cts.	$ cts.	$ cts.	$ cts.	$ cts.	d.
0	2.40.0	2.64.0	2.88.0	3.12.0	3.36.0	3.60.0	3.84.0	4.08.0	4.32.0	4.56.0	0
1	2.42.0	2.66.0	2.90.0	3.14.0	3.38.0	3.62.0	3.86.0	4.10.0	4.34.0	4.58.0	1
2	2.44.0	2.68.0	2.92.0	3.16.0	3.40.0	3.64.0	3.88.0	4.12.0	4.36.0	4.60.0	2
3	2.46.0	2.70.0	2.94.0	3.18.0	3.42.0	3.66.0	3.90.0	4.14.0	4.38.0	4.62.0	3
4	2.48.0	2.72.0	2.96.0	3.20.0	3.44.0	3.68.0	3.92.0	4.16.0	4.40.0	4.64.0	4
5	2.50.0	2.74.0	2.98.0	3.22.0	3.46.0	3.70.0	3.94.0	4.18.0	4.42.0	4.66.0	5
6	2.52.0	2.76.0	3.00.0	3.24.0	3.48.0	3.72.0	3.96.0	4.20.0	4.44.0	4.68.0	6
7	2.54.0	2.78.0	3.02.0	3.26.0	3.50.0	3.74.0	3.98.0	4.22.0	4.46.0	4.70.0	7
8	2.56.0	2.80.0	3.04.0	3.28.0	3.52.0	3.76.0	4.00.0	4.24.0	4.48.0	4.72.0	8
9	2.58.0	2.82.0	3.06.0	3.30.0	3.54.0	3.78.0	4.02.0	4.26.0	4.50.0	4.74.0	9
10	2.60.0	2.84.0	3.08.0	3.32.0	3.56.0	3.80.0	4.04.0	4.28.0	4.52.0	4.76.0	10
11	2.62.0	2.86.0	3.10.0	3.34.0	3.58.0	3.82.0	4.06.0	4.30.0	4.54.0	4.78.0	11

$4.80 1/10 TO THE POUND.

8⅛ %

PREMIUM.

STERLING INTO DOLLARS AND CENTS, AND VICE-VERSA.

£	0	100	200	300	400	500	600	700	800	900	£
	$ cts.	$ cts.	$ cts.	$ cts.	$ cts.	$ cts.	$ cts.	$ cts.	$ cts.	$ cts.	
0	......	480.55.5	961.11.1	1441.66.6	1922.22.2	2402.77.7	2883.33.3	3363.88.8	3844.44.4	4325.00.0	0
1	4.80.5	485.36.1	965.91.6	1446.47.2	1927.02.7	2407.58.3	2888.13.8	3368.69.4	3849.25.0	4329.80.5	1
2	9.61.1	490.16.6	970.72.2	1451.27.7	1931.83.3	2412.38.8	2892.94.4	3373.50.0	3854.05.5	4334.61.1	2
3	14.41.6	494.97.2	975.52.7	1456.08.3	1936.63.8	2417.19.4	2897.74.9	3378.30.5	3858.86.1	4339.41.6	3
4	19.22.2	499.77.7	980.33.3	1460.88.8	1941.44.4	2422.00.0	2902.55.5	3383.11.1	3863.66.6	4344.22.2	4
5	24.02.7	504.58.3	985.13.8	1465.69.4	1946.25.0	2426.80.5	2907.36.1	3387.91.6	3868.47.2	4349.02.7	5
6	28.83.3	509.38.8	989.94.4	1470.50.0	1951.05.5	2131.61.1	2912.16.6	3392.72.2	3873.27.7	4353.83.3	6
7	33.63.8	514.19.4	994.75.0	1475.30.5	1955.86.1	2436.41.6	2916.97.2	3397.52.7	3878.08.3	4358.63.8	7
8	38.44.4	519.00.0	999.55.5	1480.11.1	1960.66.6	2441.22.2	2921.77.7	3402.33.3	3882.88.8	4363.44.4	8
9	43.25.0	523.80.5	1004.36.1	1484.91.6	1965.47.2	2446.02.7	2926.58.3	3407.13.8	3887.69.4	4368.25.0	9
10	48.05.5	528.61.1	1009.16.6	1489.72.2	1970.27.7	2450.83.3	2931.38.8	3411.94.4	3892.50.0	4373.05.5	10
11	52.86.1	533.41.6	1013.97.2	1494.52.7	1975.08.3	2455.63.8	2936.19.4	3416.75.0	3897.30.5	4377.86.1	11
12	57.66.6	538.22.2	1018.77.7	1499.33.3	1979.88.8	2460.44.4	2941.00.0	3421.55.5	3902.11.1	4382.66.6	12
13	62.47.2	543.02.7	1023.58.3	1504.13.8	1984.69.4	2465.25.0	2945.80.5	3426.36.1	3906.91.6	4387.47.2	13
14	67.27.7	547.83.3	1028.38.8	1508.94.4	1989.50.0	2470.05.5	2950.61.1	3431.16.6	3911.72.2	4392.27.7	14
15	72.08.3	552.63.8	1033.19.4	1513.75.0	1994.30.5	2474.86.1	2955.41.6	3435.97.2	3916.52.7	4397.08.3	15
16	76.88.8	557.44.4	1038.00.0	1518.55.5	1999.11.1	2479.66.6	2960.22.2	3440.77.7	3921.33.3	4401.88.8	16
17	81.69.4	562.25.0	1042.80.5	1523.36.1	2003.91.6	2484.47.2	2965.02.7	3445.58.3	3926.13.8	4406.69.4	17
18	86.50.0	567.05.5	1047.61.1	1528.16.6	2008.72.2	2489.27.7	2969.83.3	3450.38.8	3930.91.4	4411.50.0	18
19	91.30.5	571.86.1	1052.41.6	1532.97.2	2013.52.7	2494.08.3	2974.63.8	3455.19.4	3935.75.0	4416.30.5	19
20	96.11.1	576.66.6	1057.22.2	1537.77.7	2018.33.3	2498.88.8	2979.44.4	3460.00.0	3940.55.5	4421.11.1	20
21	100.91.6	581.47.2	1062.02.7	1542.58.3	2023.13.8	2503.69.4	2984.25.0	3464.80.5	3945.36.1	4425.91.6	21
22	105.72.2	586.27.7	1066.83.3	1547.38.8	2027.94.4	2508.50.0	2989.05.5	3469.61.1	3950.16.6	4430.72.2	22
23	110.52.7	591.08.3	1071.63.8	1552.19.4	2032.75.0	2513.30.5	2993.86.1	3474.41.6	3954.97.2	4435.52.7	23
24	115.33.3	595.88.8	1076.44.4	1557.00.0	2037.55.5	2518.11.1	2998.66.6	3479.22.2	3959.77.7	4440.33.3	24
25	120.13.8	600.69.4	1081.25.0	1561.80.5	2042.36.1	2522.91.6	3003.47.2	3484.02.7	3964.58.3	4445.13.8	25
26	124.94.4	605.50.0	1086.05.5	1566.61.1	2047.16.6	2527.72.2	3008.27.7	3488.83.3	3969.38.8	4449.94.4	26
27	129.75.0	610.30.5	1090.86.1	1571.41.6	2051.97.2	2532.52.7	3013.08.3	3493.63.8	3974.19.4	4454.75.0	27
28	134.55.5	615.11.1	1095.66.6	1576.22.2	2056.77.7	2537.33.3	3017.88.8	3498.44.4	3979.00.0	4459.55.5	28
29	139.36.1	619.91.6	1100.47.2	1581.02.7	2061.58.3	2542.13.8	3022.69.4	3503.25.0	3983.80.5	4464.36.1	29
30	144.16.6	624.72.2	1105.27.7	1585.83.3	2066.38.8	2546.94.4	3027.50.0	3508.05.5	3988.61.1	4469.16.6	30
31	148.97.2	629.52.7	1110.08.3	1590.63.8	2071.19.4	2551.75.0	3032.30.5	3512.86.1	3993.41.6	4473.97.2	31
32	153.77.7	634.33.3	1114.88.8	1595.44.4	2076.00.0	2556.55.5	3037.11.1	3517.66.6	3998.22.2	4478.77.7	32
33	158.58.3	639.13.8	1119.69.4	1600.25.0	2080.80.5	2561.36.1	3041.91.6	3522.47.2	4003.02.7	4483.58.3	33
34	163.38.8	643.94.4	1124.50.0	1605.05.5	2085.61.1	2566.16.6	3046.72.2	3527.27.7	4007.83.3	4488.38.8	34
35	168.19.4	648.75.0	1129.30.5	1609.86.1	2090.41.6	2570.97.2	3051.52.7	3532.08.3	4012.63.8	4493.19.4	35
36	173.00.0	653.55.5	1134.11.1	1614.66.6	2095.22.2	2575.77.7	3056.33.3	3536.88.8	4017.44.4	4498.00.0	36
37	177.80.5	658.36.1	1138.91.6	1619.47.2	2100.02.7	2580.58.3	3061.13.8	3541.69.4	4022.25.0	4502.80.5	37
38	182.61.1	663.16.6	1143.72.2	1624.27.7	2104.83.3	2585.38.8	3065.94.4	3546.50.0	4027.05.5	4507.61.1	38
39	187.41.6	667.97.2	1148.52.7	1629.08.3	2109.63.8	2590.19.4	3070.75.0	3551.30.5	4031.86.1	4512.41.6	39
40	192.22.2	672.77.7	1153.33.3	1633.88.8	2114.44.4	2595.00.0	3075.55.5	3556.11.1	4036.66.6	4517.22.2	40
41	197.02.7	677.58.3	1158.13.8	1638.69.4	2119.25.0	2599.80.5	3080.36.1	3560.91.6	4041.47.2	4522.02.7	41
42	201.83.3	682.38.8	1162.94.4	1643.50.0	2124.05.5	2604.61.1	3085.16.6	3565.72.2	4046.27.7	4526.83.3	42
43	206.63.8	687.19.4	1167.75.0	1648.30.5	2128.86.1	2609.41.6	3089.97.2	3570.52.7	4051.08.3	4531.63.8	43
44	211.44.4	692.00.0	1172.55.5	1653.11.1	2133.66.6	2614.22.2	3094.77.7	3575.33.3	4055.88.8	4536.44.4	44
45	216.25.0	696.80.5	1177.36.1	1657.91.6	2138.47.2	2619.02.7	3099.58.3	3580.13.8	4060.69.4	4541.25.0	45
46	221.05.5	701.61.1	1182.16.6	1662.72.2	2143.27.7	2623.83.3	3104.38.8	3584.94.4	4065.50.0	4546.05.5	46
47	225.86.1	706.41.6	1186.97.2	1667.52.7	2148.08.3	2628.63.8	3109.19.4	3589.75.0	4070.30.5	4550.86.1	47
48	230.66.6	711.22.2	1191.77.7	1672.33.3	2152.88.8	2633.44.4	3114.00.0	3594.55.5	4075.11.1	4555.66.6	48
49	235.47.2	716.02.7	1196.58.3	1677.13.8	2157.69.4	2638.25.0	3118.80.5	3599.36.1	4079.91.6	4560.47.2	49

S.	0	1	2	3	4	5	6	7	8	9	S.
d.	$ cts.	$ cts.	$ cts.	$ cts.	$ cts.	$ cts.	$ cts.	$ cts.	$ cts.	$ cts.	d.
0	......	.24.0	.48.0	.72.0	.96.1	1.20.1	1.44.1	1.68.1	1.92.2	2.16.2	0
1	.02.0	.26.0	.50.0	.74.0	.98.1	1.22.1	1.46.1	1.70.1	1.94.2	2.18.2	1
2	.04.0	.28.0	.52.0	.76.0	1.00.1	1.24.1	1.48.1	1.72.1	1.96.2	2.20.2	2
3	.06.0	.30.0	.54.0	.78.0	1.02.1	1.26.1	1.50.1	1.74.1	1.98.2	2.22.2	3
4	.08.0	.32.0	.56.0	.80.0	1.04.1	1.28.1	1.52.1	1.76.1	2.00.2	2.24.2	4
5	.10.0	.34.0	.58.0	.82.0	1.06.1	1.30.1	1.54.1	1.78.1	2.02.2	2.26.2	5
6	.12.0	.36.0	.60.0	.84.0	1.08.1	1.32.1	1.56.1	1.80.1	2.04.2	2.28.2	6
7	.14.0	.38.0	.62.0	.86.0	1.10.1	1.34.1	1.58.1	1.82.1	2.06.2	2.30.2	7
8	.16.0	.40.0	.64.0	.88.0	1.12.1	1.36.1	1.60.1	1.84.1	2.08.2	2.32.2	8
9	.18.0	.42.0	.66.0	.90.0	1.14.1	1.38.1	1.62.1	1.86.1	2.10.2	2.34.2	9
10	.20.0	.44.0	.68.0	.92.0	1.16.1	1.40.1	1.64.1	1.88.1	2.12.2	2.36.2	10
11	.22.0	.46.0	.70.0	.94.0	1.18.1	1.42.1	1.66.1	1.90.1	2.14.2	2.38.2	11

$4.80 1/100 TO THE POUND.

8 1/8 %

PREMIUM.

STERLING INTO DOLLARS AND CENTS, AND VICE-VERSA.

£	0	100	200	300	400	500	600	700	800	900	£
	$ cts.	$ cts.	$ cts.	$ cts.	$ cts.	$ cts.	$ cts.	$ cts.	$ cts.	$ cts.	
50	240.27.7	720.83.3	1201.38.8	1681.94.4	2162.50.0	2643.05.5	3123.61.1	3604.16.6	4084.72.2	4565.27.7	50
51	245.08.3	725.63.8	1206.19.4	1686.75.0	2167.30.5	2647.86.1	3128.41.6	3608.97.2	4089.52.7	4570.08.3	51
52	249.88.8	730.44.4	1211.00.0	1691.55.5	2172.11.1	2652.66.6	3133.22.2	3613.77.7	4094.33.3	4574.88.8	52
53	254.69.4	735.25.0	1215.80.5	1696.36.1	2176.91.6	2657.47.2	3138.02.7	3618.58.3	4099.13.8	4579.69.4	53
54	259.50.0	740.05.5	1220.61.1	1701.16.6	2181.72.2	2662.27.7	3142.83.3	3623.38.8	4103.94.4	4584.50.0	54
55	264.30.5	744.86.1	1225.41.6	1705.97.2	2186.52.7	2667.08.3	3147.63.8	3628.19.4	4108.75.0	4589.30.5	55
56	269.11.1	749.66.6	1230.22.2	1710.77.7	2191.33.3	2671.88.8	3152.44.4	3633.00.0	4113.55.5	4594.11.1	56
57	273.91.6	754.47.2	1235.02.7	1715.58.3	2196.13.8	2676.69.4	3157.25.0	3637.80.5	4118.36.1	4598.91.6	57
58	278.72.2	759.27.7	1239.83.3	1720.38.8	2200.94.4	2681.50.0	3162.05.5	3642.61.1	4123.16.6	4603.72.2	58
59	283.52.7	764.08.3	1244.63.8	1725.19.4	2205.75.0	2686.30.5	3166.86.1	3647.41.6	4127.97.2	4608.52.7	59
60	288.33.3	768.88.8	1249.44.4	1730.00.0	2210.55.5	2691.11.1	3171.66.6	3652.22.2	4132.77.7	4613.33.3	60
61	293.13.8	773.69.4	1254.25.0	1734.80.5	2215.36.1	2695.91.6	3176.47.2	3657.02.7	4137.58.3	4618.13.8	61
62	297.94.4	778.50.0	1259.05.5	1739.61.1	2220.16.6	2700.72.2	3181.27.7	3661.83.3	4142.38.8	4622.94.4	62
63	302.75.0	783.30.5	1263.86.1	1744.41.6	2224.97.2	2705.52.7	3186.08.3	3666.63.8	4147.19.4	4627.75.0	63
64	307.55.5	788.11.1	1268.66.6	1749.22.2	2229.77.7	2710.33.3	3190.88.8	3671.44.4	4152.00.0	4632.55.5	64
65	312.36.1	792.91.6	1273.47.2	1754.02.7	2234.58.3	2715.13.8	3195.69.4	3676.25.0	4156.80.5	4637.36.1	65
66	317.16.6	797.72.2	1278.27.7	1758.83.3	2239.38.8	2719.94.4	3200.50.0	3681.05.5	4161.61.1	4642.16.6	66
67	321.97.2	802.52.7	1283.08.3	1763.63.8	2244.19.4	2724.75.0	3205.30.5	3685.86.1	4166.41.6	4646.97.2	67
68	326.77.7	807.33.3	1287.88.8	1768.44.4	2249.00.0	2729.55.5	3210.11.1	3690.66.6	4171.22.2	4651.77.7	68
69	331.58.3	812.13.8	1292.69.4	1773.25.0	2253.80.5	2734.36.1	3214.91.6	3695.47.2	4176.02.7	4656.58.3	69
70	336.38.8	816.94.4	1297.50.0	1778.05.5	2258.61.1	2739.16.6	3219.72.2	3700.27.7	4180.83.3	4661.38.8	70
71	341.19.4	821.75.0	1302.30.5	1782.86.1	2263.41.6	2743.97.2	3224.52.7	3705.08.3	4185.63.8	4666.19.4	71
72	346.00.0	826.55.5	1307.11.1	1787.66.6	2268.22.2	2748.77.7	3229.33.3	3709.88.8	4190.44.4	4671.00.0	72
73	350.80.5	831.36.1	1311.91.6	1792.47.2	2273.02.7	2753.58.3	3234.13.8	3714.69.4	4195.25.0	4675.80.5	73
74	355.61.1	836.16.6	1316.72.2	1797.27.7	2277.83.3	2758.38.8	3238.94.4	3719.50.0	4200.05.5	4680.61.1	74
75	360.41.6	840.97.2	1321.52.7	1802.08.3	2282.63.8	2763.19.4	3243.75.0	3724.30.5	4204.86.1	4685.41.6	75
76	365.22.2	845.77.7	1326.33.3	1806.88.8	2287.44.4	2768.00.0	3248.55.5	3729.11.1	4209.66.6	4690.22.2	76
77	370.02.7	850.58.3	1331.13.8	1811.69.4	2292.25.0	2772.80.5	3253.36.1	3733.91.6	4214.47.2	4695.02.7	77
78	374.83.3	855.38.8	1335.94.4	1816.50.0	2297.05.5	2777.61.1	3258.16.6	3738.72.2	4219.27.7	4699.83.3	78
79	379.63.8	860.19.4	1340.75.0	1821.30.5	2301.86.1	2782.41.6	3262.97.2	3743.52.7	4224.08.3	4704.63.8	79
80	384.44.4	865.00.0	1345.55.5	1826.11.1	2306.66.6	2787.22.2	3267.77.7	3748.33.3	4228.88.8	4709.44.4	80
81	389.25.0	869.80.5	1350.36.1	1830.91.6	2311.47.2	2792.02.7	3272.58.3	3753.13.8	4233.69.4	4714.25.0	81
82	394.05.5	874.61.1	1355.16.6	1835.72.2	2316.27.7	2796.83.3	3277.38.8	3757.94.4	4238.50.0	4719.05.5	82
83	398.86.1	879.41.6	1359.97.2	1840.52.7	2321.08.3	2801.63.8	3282.19.4	3762.75.0	4243.30.5	4723.86.1	83
84	403.66.6	884.22.2	1364.77.7	1845.33.3	2325.88.8	2806.44.4	3287.00.0	3767.55.5	4248.11.1	4728.66.6	84
85	408.47.2	889.02.7	1369.58.3	1850.13.8	2330.69.4	2811.25.0	3291.80.5	3772.36.1	4252.91.6	4733.47.2	85
86	413.27.7	893.83.3	1374.38.8	1854.94.4	2335.50.0	2816.05.5	3296.61.1	3777.16.6	4257.72.2	4738.27.7	86
87	418.08.3	898.63.8	1379.19.4	1859.75.0	2340.30.5	2820.86.1	3301.41.6	3781.97.2	4262.52.7	4743.08.3	87
88	422.88.8	903.44.4	1384.00.0	1864.55.5	2345.11.1	2825.66.6	3306.22.2	3786.77.7	4267.33.3	4747.88.8	88
89	427.69.4	908.25.0	1388.80.5	1869.36.1	2349.91.6	2830.47.2	3311.02.7	3791.58.3	4272.13.8	4752.69.4	89
90	432.50.0	913.05.5	1393.61.1	1874.16.6	2354.72.2	2835.27.7	3315.83.3	3796.38.8	4276.94.4	4757.50.0	90
91	437.30.5	917.86.1	1398.41.6	1878.97.2	2359.52.7	2840.08.3	3320.63.8	3801.19.4	4281.75.0	4762.30.5	91
92	442.11.1	922.66.6	1403.22.2	1883.77.7	2364.33.3	2844.88.8	3325.44.4	3806.00.0	4286.55.5	4767.11.1	92
93	446.91.6	927.47.2	1408.02.7	1888.58.3	2369.13.8	2849.69.4	3330.25.0	3810.80.5	4291.36.1	4771.91.6	93
94	451.72.2	932.27.7	1412.83.3	1893.38.8	2373.94.4	2854.50.0	3335.05.5	3815.61.1	4296.16.6	4776.72.2	94
95	456.52.7	937.08.3	1417.63.8	1898.19.4	2378.75.0	2859.30.5	3339.86.1	3820.41.6	4300.97.2	4781.52.7	95
96	461.33.3	941.88.8	1422.44.4	1903.00.0	2383.55.5	2864.11.1	3344.66.6	3825.22.2	4305.77.7	4786.33.3	96
97	466.13.8	946.69.4	1427.25.0	1907.80.5	2388.36.1	2868.91.6	3349.47.2	3830.02.7	4310.58.3	4791.13.8	97
98	470.94.4	951.50.0	1432.05.5	1912.61.1	2393.16.6	2873.72.2	3354.27.7	3834.83.3	4315.38.8	4795.94.4	98
99	475.75.0	956.30.5	1436.86.1	1917.41.6	2397.97.2	2878.52.7	3359.08.3	3839.63.8	4320.19.4	4800.75.0	99

S.	10	11	12	13	14	15	16	17	18	19	S.
d.	$ cts.	$ cts.	$ cts.	$ cts.	$ cts.	$ cts.	$ cts.	$ cts.	$ cts.	$ cts.	d.
0	2.40.2	2.64.3	2.88.3	3.12.3	3.36.3	3.60.4	3.84.4	4.08.4	4.32.5	4.56.5	0
1	2.42.2	2.66.3	2.90.3	3.14.3	3.38.3	3.62.4	3.86.4	4.10.4	4.34.5	4.58.5	1
2	2.44.2	2.68.3	2.92.3	3.16.3	3.40.3	3.64.4	3.88.4	4.12.4	4.36.5	4.60.5	2
3	2.46.2	2.70.3	2.94.3	3.18.3	3.42.3	3.66.4	3.90.4	4.14.4	4.38.5	4.62.5	3
4	2.48.2	2.72.3	2.96.3	3.20.3	3.44.3	3.68.4	3.92.4	4.16.4	4.40.5	4.64.5	4
5	2.50.2	2.74.3	2.98.3	3.22.3	3.46.3	3.70.4	3.94.4	4.18.4	4.42.5	4.66.5	5
6	2.52.2	2.76.3	3.00.3	3.24.3	3.48.3	3.72.4	3.96.4	4.20.4	4.44.5	4.68.5	6
7	2.54.2	2.78.3	3.02.3	3.26.3	3.50.3	3.74.4	3.98.4	4.22.4	4.46.5	4.70.5	7
8	2.56.2	2.80.3	3.04.3	3.28.3	3.52.3	3.76.4	4.00.4	4.24.4	4.48.5	4.72.5	8
9	2.58.2	2.82.3	3.06.3	3.30.3	3.54.3	3.78.4	4.02.4	4.26.4	4.50.5	4.74.5	9
10	2.60.2	2.84.3	3.08.3	3.32.3	3.56.3	3.80.4	4.04.4	4.28.4	4.52.5	4.76.5	10
11	2.62.2	2.86.3	3.10.3	3.34.3	3.58.3	3.82.4	4.06.4	4.30.4	4.54.5	4.78.5	11

$4.81 11/100 TO THE POUND.

8 ¼ %

PREMIUM.

STERLING INTO DOLLARS AND CENTS, AND VICE-VERSA.

£	0	100	200	300	400	500	600	700	800	900	£
	$ cts.	$ cts.	$ cts.	$ cts.	$ cts.	$ cts.	$ cts.	$ cts.	$ cts.	$ cts.	
0	......	481.11.1	962.22.2	1443.33.3	1924.44.4	2405.55 5	2886.66.6	3367.77.7	3848.88.8	4330.00.0	0
1	4.81.1	485.92.2	967.03.3	1448.14.4	1929.25.5	2410.36.6	2891.47.7	3372.58.8	3853.70.0	4334.81.1	1
2	9.62.2	490.73.3	971.84.4	1452.95.5	1934.06.6	2415.17.7	2896.28 8	3377.40 0	3858.51.1	4339.62.2	2
3	14.43.3	495.54.4	976.65.5	1457.76.6	1938.87.7	2419.98.8	2901.10.0	3382.21.1	3863.32.2	4344.43.3	3
4	19.24.4	500.35.5	981.46.6	1462.57.7	1943.68 8	2424.80.0	2905.91.1	3387 02.2	3868.13.3	4349.24.4	4
5	24.05.5	505.16.6	986.27.7	1467.38.8	1948.50.0	2429.61.1	2910.72.2	3391.83.3	3872.94.4	4354.05.5	5
6	28.86.6	509.97.7	991.08.8	1472.20.0	1953.31.1	2434.42.2	2915.53.3	3396 64.4	3877.75.5	4358.86.6	6
7	33.67.7	514.78.8	995.90.0	1477.01.1	1958.12.2	2439.23.3	2920.34.4	3401.45.5	3882.56.6	4363.67.7	7
8	38.48.8	519.60.0	1000.71.1	1481.82.2	1962.93.3	2444.04.4	2925.15.5	3406.26.6	3887.37.7	4368.48.8	8
9	43.30.0	524.41.1	1005.52.2	1486.63.3	1967.74.4	2448.85.5	2929.96.6	3411.07.7	3892.18.8	4373.30.0	9
10	48.11.1	529.22.2	1010.33.3	1491.44.4	1972.55.5	2453.66.6	2934.77.7	3415.88.8	3897.00.0	4378.11.1	10
11	52.92.2	534.03.3	1015.14.4	1496.25.5	1977.36.6	2458.47.7	2939.58.8	3420.70 0	3901.81.1	4382.92.2	11
12	57.73.3	538.84.4	1019.95.5	1501.06.6	1982.17.7	2463.28.8	2944.40.0	3425.51.1	3906.62.2	4387.73.3	12
13	62.54.4	543.65.5	1024.76.6	1505.87.7	1986.98.8	2468.10.0	2949.21.1	3430.32.2	3911.43.3	4392.54.4	13
14	67.35.5	548.46.6	1029.57.7	1510.68.8	1991.80.0	2472.91.1	2954.02.2	3435.13.3	3916.24.4	4397.35.5	14
15	72.16.6	553.27.7	1034.38.8	1515.50.0	1996.61.1	2477.72.2	2958.83.3	3439.94.4	3921.05.5	4402.16.6	15
16	76.97.7	558.08.8	1039.20.0	1520.31.1	2001.42.2	2482.53.3	2963.64.4	3444.75.5	3925.86.6	4406.97.7	16
17	81.78.8	562.90.0	1044.01.1	1525.12.2	2006.23.3	2487.34.4	2968.45.5	3449.56.6	3930.67.7	4411.78.8	17
18	86.60.0	567.71.1	1048.82.2	1529.93.3	2011.04.4	2492.15.5	2973.26.6	3454.37.7	3935.48.8	4416.60.0	18
19	91.41.1	572.52.2	1053.63.3	1534.74.4	2015.85.5	2496.96.6	2978.07.7	3459.18.8	3940.30.0	4421.41.1	19
20	96.22.2	577.33.3	1058.44.4	1539.55.5	2020.66.6	2501.77.7	2982.88.8	3464.00.0	3945.11.1	4426.22.2	20
21	101.03.3	582.14.4	1063.25.5	1544.36.6	2025.47.7	2506.58.8	2987.70.0	3468.81.1	3949.92.2	4431.03.3	21
22	105.84.4	586.95.5	1068.06.6	1549.17.7	2030.28.8	2511.40.0	2992.51 1	3473.62.2	3954.73.3	4435.84.4	22
23	110.65.5	591.76.6	1072.87.7	1553.98.8	2035.10.0	2516.21.1	2997.32.2	3478.43.3	3959.54.4	4440.65.5	23
24	115.46.6	596.57.7	1077.68.8	1558.80.0	2039.91.1	2521.02.2	3002.13.3	3483.24.4	3964.35.5	4445.46.6	24
25	120.27.7	601.38.8	1082.50.0	1563.61.1	2044.72.2	2525.83.3	3006.94.4	3488.05.5	3969 16.6	4450.27.7	25
26	125.08.8	606.20.0	1087.31.1	1568.42.2	2049.53.3	2530.64.4	3011.75.5	3492.86.6	3973 97.7	4455.08.8	26
27	129.90.0	611.01.1	1092.12.2	1573.23.3	2054.34.4	2535.45.5	3016.56.6	3497.67.7	3978.78.8	4459.90.0	27
28	134.71.1	615.82.2	1096.93.3	1578.04.4	2059.15.5	2540.26.6	3021.37.7	3502.48.8	3983 60 0	4464.71.1	28
29	139.52.2	620.63.3	1101.74.4	1582.85.5	2063.96.6	2545.07.7	3026.18.8	3507.30.0	3988.41.1	4469.52.2	29
30	144.33.3	625.44.4	1106.55.5	1587.66.6	2068.77.7	2549.88.8	3031.00.0	3512.11.1	3993 22.2	4474.33.3	30
31	149.14.4	630.25.5	1111.36.6	1592.47.7	2073.58.8	2554.70.0	3035.81.1	3516.92.2	3998 03.3	4479 14.4	31
32	153.95.5	635.06.6	1116.17.7	1597.28.8	2078.40.0	2559.51.1	3040.62.2	3521.73.3	4002.84 4	4483.95.5	32
33	158.76.6	639.87.7	1120.98.8	1602.10.0	2083.21.1	2564.32.2	3045.43.3	3526.54.4	4007.65.5	4488.76.6	33
34	163.57.7	644.68.8	1125.80.0	1606.91.1	2088.02.2	2569.13.3	3050.24.4	3531.35.5	4012.46.6	4493.57.7	34
35	168.38.8	649.50.0	1130.61.1	1611.72.2	2092.83.3	2573.94.4	3055.05.5	3536.16.6	4017.27.7	4498.38.8	35
36	173.20.0	654.31.1	1135.42.2	1616.53.3	2097.64.4	2578.75.5	3059.86.6	3540.97.7	4022.08.8	4503.20.0	36
37	178.01.1	659.12.2	1140.23.3	1621.34.4	2102.45.5	2583.56.6	3064.67.7	3545.78.8	4026.90.0	4508.01.1	37
38	182.82.2	663.93.3	1145.04.4	1626.15.5	2107.26.6	2588.37.7	3069.48.8	3550.60.0	4031.71.1	4512 82.2	38
39	187.63.3	668.74.4	1149.85.5	1630.96.6	2112.07.7	2593.18.8	3074.30.0	3555.41.1	4036.52.2	4517.63.3	39
40	192.44.4	673.55.5	1154.66.6	1635.77.7	2116.88.8	2598.00.0	3079.11.1	3560.23.2	4041.33.3	4522 44.4	40
41	197.25.5	678.36.6	1159.47.7	1640.58.8	2121.70.0	2602.81.1	3083.92.2	3565.03.3	4046.14.4	4527.25.5	41
42	202.06.6	683.17.7	1164.28.8	1645.40.0	2126.51.1	2607.62.2	3088.73.3	3569.84.4	4050 95.5	4532.06.6	42
43	206.87.7	687.98.8	1169.10.0	1650.21.1	2131.32.2	2612.43.3	3093.54.4	3574.65.5	4055.76.6	4536.87.7	43
44	211.68.8	692.80.0	1173.91.1	1655.02.2	2136.13.3	2617.24.4	3098.35.5	3579.46.6	4060.57.7	4541.68.8	44
45	216.50.0	697.61.1	1178.72.2	1659.83.3	2140.94.4	2622.05.5	3103.16.6	3584.27.7	4065.38.8	4546.50.0	45
46	221.31.1	702.42.2	1183.53.3	1664.64.4	2145.75.5	2626.86.6	3107.97.7	3589.08.8	4070.20 0	4551.31.1	46
47	226.12.2	707.23.3	1188.34.4	1669.45.5	2150.56.6	2631.67.7	3112.78.8	3593.90.0	4075.01.1	4556.12 2	47
48	230.93.3	712.04.4	1193.15.5	1674.26.6	2155.37.7	2636.48.8	3117.60.0	3598.71.1	4079.82.2	4560.93.3	48
49	235.74.4	716.85.5	1197.96.6	1679.07.7	2160.18.8	2641.30.0	3122.41.1	3603.52.2	4084.63.3	4565.74.4	49

s.	0	1	2	3	4	5	6	7	8	9	s.
d.	$ cts.	$ cts.	$ cts.	$ cts.	$ cts.	$ cts.	$ cts.	$ cts.	$ cts.	$ cts.	d.
0	......	.24.0	.48.1	.72.1	.96.2	1.20.2	1.44.2	1.68.3	1.92.4	2.16.5	0
1	.02.0	.26.0	.50.1	.74.1	.98.2	1.22.2	1.46.3	1.70.3	1.94.4	2.18.5	1
2	.04.0	.28.0	.52.1	.76.1	1.00.2	1.24.2	1.48.3	1.72.3	1.96.4	2.20.5	2
3	.06.0	.30.0	.54.1	.78.1	1.02.2	1.26.2	1.50.3	1.74.3	1.98.4	2.22.5	3
4	.08.0	.32.0	.56.1	.80.1	1.04.2	1.28.2	1.52.3	1.76.3	2.00.4	2.24.5	4
5	.10.0	.34.0	.58.1	.82.1	1.06.2	1.30.2	1.54.3	1.78.3	2.02.4	2.26.5	5
6	.12.0	.36.0	.60.1	.84.1	1.08.2	1.32.2	1.56.3	1.80.3	2.04.4	2.28.5	6
7	.14.0	.38.0	.62.1	.86.1	1.10.2	1.34.2	1.58.3	1.82.3	2.06.4	2.30.5	7
8	.16.0	.40.0	.64.1	.88.1	1.12.2	1.36.2	1.60.3	1.84 3	2.08.4	2.32.5	8
9	.18.0	.42.0	.66.1	.90.1	1.14.2	1.38.2	1.62.3	1.86.3	2.10.4	2.34.5	9
10	.20.0	.44.0	.68.1	.92.1	1.16.2	1.40.2	1.64.3	1.88.3	2.12.4	2.36.5	10
11	.22.0	.46.0	.70.1	.94.1	1.18.2	1.42.2	1.66.3	1.90.3	2.14.4	2.38.5	11

$4.81⁷⁄₁₀₀ TO THE POUND.

8¼%

PREMIUM.

STERLING INTO DOLLARS AND CENTS, AND VICE-VERSA.

£	0	100	200	300	400	500	600	700	800	900	£
	$ cts.	$ cts.	$ cts.	$ cts.	$ cts.	$ cts.	$ cts.	$ cts.	$ cts.	$ cts.	
50	240.55.5	721 66.6	1202.77.7	1683.88.8	2165.00.0	2646.11.1	3127.22.2	3608.33.3	4089.44.4	4570.55.5	50
51	245.36.6	726.47.7	1207.58.8	1688.70.0	2169.81.1	2650.92 2	3132.03.3	3613.14.4	4094.25.5	4575.36.6	51
52	250.17.7	731.28.8	1212.40.0	1693.51.1	2174.62.2	2655.73.3	3136.84.4	3617.95.5	4099.06.6	4580.17.7	52
53	254.98.8	736.10.0	1217.21.1	1698.32.2	2179.43.3	2660.54.4	3141.65.5	3622.76.6	4103.87.7	4584.98.8	53
54	259.80.0	740.91.1	1222.02.2	1703.13.3	2184.24.4	2665.35.5	3146.46.6	3627.57.7	4108.68.8	4589.80.0	54
55	264.61.1	745.72.2	1226.83.3	1707.94.4	2189.05.5	2670.16.6	3151.27.7	3632 38.8	4113.50.0	4594.61.1	55
56	269.42.2	750.53.3	1231.64.4	1712.75 5	2193.86.6	2674.97.7	3156.08.8	3637.20.0	4118.31.1	4599.42.2	56
57	274.23.3	755.34.4	1236.45.5	1717.56.6	2198.67.7	2679.78.8	3160.90.0	3642.01.1	4123.12.2	4604.23.3	57
58	279.04.4	760.15.5	1241 26.6	1722.37.7	2203.48.8	2684.60.0	3165.71.1	3646 82.2	4127.93.3	4609.04.4	58
59	283.85.5	764.96 6	1246.07.7	1727.18.8	2208.30.0	2689.41.1	3170.52.2	3651.63.3	4132.74.4	4613.85.5	59
60	288.66.6	769.77.7	1250.88.8	1732.00.0	2213.11.1	2694.22.2	3175.33.3	3656.44.4	4137.55.5	4618.66.6	60
61	293.47.7	774.58.8	1255.70.0	1736.81.1	2217.92.2	2699.03.3	3180.14.4	3661.25.5	4142.36.6	4623.47.7	61
62	298.28.8	779.40.0	1260.51.1	1741.62.2	2222.73.3	2703.84.4	3184.95.5	3666.06.6	4147.17.7	4628.28.8	62
63	303 10.0	784.21.1	1265.32.2	1746.43.3	2227.54.4	2708.65.5	3189 76.6	3670.87.7	4151.98.8	4633.10.0	63
64	307.91.1	789.02.2	1270.13.3	1751.24.4	2232.35.5	2713.46.6	3194.57.7	3675 68.8	4156.80.0	4637.91.1	64
65	312.72.2	793.83.3	1274 94.4	1756.05.5	2237.16.6	2718.27.7	3199.38.8	3680.50.0	4161.61.1	4642.72.2	65
66	317.53.3	798.64.4	1279.75 5	1760.86.6	2241.97.7	2723 08.8	3204.20.0	3685.31.1	4166.42.2	4647.53.3	66
67	322.34.4	803.45.5	1284.56.6	1765.67.7	2246.78.8	2727.90 0	3209.01.1	3690.12.2	4171.23.3	4652.34.4	67
68	327.15.5	808.26.6	1289 37.7	1770.48.8	2251.60.0	2732.71.1	3213.82.2	3694 93.3	4176.04.4	4657.15.5	68
69	331.96.6	813.07.7	1294.18.8	1775.30.0	2256.41.1	2737.52.2	3218.63.3	3699.74 4	4180.85.5	4661.96.6	69
70	336.77.7	817.88.8	1299 00.0	1780.11.1	2261.22.2	2742.33.3	3223.44.4	3704.55.5	4185.66.6	4666.77.7	70
71	341.58.8	822.70.0	1303.81.1	1784.92.2	2266.03.3	2747.14.4	3228.25.5	3709.36.6	4190.47.7	4671.58.8	71
72	346.40.0	827.51.1	1308.62.2	1789.73.3	2270.84.4	2751.95.5	3233.06.6	3714.17.7	4195.28.8	4676.40.0	72
73	351.21.1	832.32 2	1313.43 3	1794 54.4	2275.65.5	2756.76.6	3237.87.7	3718.98.8	4200.10.0	4681.21.1	73
74	356.02.2	837.13.3	1318.24.4	1799.35.5	2280.46.6	2761.57.7	3242.68 8	3723.80.0	4204.91.1	4686.02.2	74
75	360.83 3	841.94.4	1323.05.5	1804.16.6	2285.27.7	2766.38.8	3247.50.0	3728.61.1	4209 72.2	4690.83 3	75
76	365.64.4	846.75 5	1327.86.6	1808.97.7	2290.08.8	2771.20.0	3252.31.1	3733.42 2	4214.53.3	4695.64.4	76
77	370.45.5	851.56.6	1332.67.7	1813.78.8	2294.90.0	2776 01.1	3257.12.2	3738.23.3	4219.34.4	4700.45.5	77
78	375 26.6	856.37.7	1337.48.8	1818.60.0	2299.71.1	2780.82.2	3261.93.3	3743.04.4	4224.15.5	4705.26.6	78
79	380.07.7	861.18.8	1342.30.0	1823.41.1	2304.52.2	2785.63.3	3266.74.4	3747.85.5	4228.96.6	4710.07.7	79
80	384.88.8	866.00.0	1347.11.1	1828.22.2	2309.33.3	2790.44.4	3271.55.5	3752.66.6	4233.77.7	4714.88.8	80
81	389.70.0	870.81.1	1351.92.2	1833.03.3	2314.14.4	2795.25.5	3276.36.6	3757.47.7	4238.58.8	4719.70.0	81
82	394.51.1	875.62.2	1356.73.3	1837.84.4	2318.95.5	2800.06.6	3281.17.7	3762.28.8	4243.40.0	4724 51.1	82
83	399.32.2	880.43.3	1361.54.4	1842.65.5	2323.76.6	2804.87.7	3285.98.8	3767.10.0	4248.21.1	4729.32.2	83
84	404.13.3	885.24.4	1366.35.5	1847.46.6	2328.57.7	2809.68.8	3290.80.0	3771.91.1	4253.02.2	4734.13 3	84
85	408.94.4	890.05.5	1371.16.6	1852.27.7	2333.38.8	2814.50.0	3295.61.1	3776.72.2	4257.83.3	4738.94.4	85
86	413.75.5	894.86.6	1375.97.7	1857.08.8	2338.20.0	2819.31.1	3300.42.2	3781.53.3	4262.64.4	4743.75.5	86
87	418.56.6	899.67.7	1380.78.8	1861.90.0	2343.01.1	2824.12.2	3305.23.3	3786.34.4	4267.45.5	4748.56.6	87
88	423.37.7	904.48.8	1385.60.0	1866.71.1	2347.82.2	2828.93.3	3310.04.4	3791.15.5	4272.26.6	4753.37.7	88
89	428.18.8	909.30.0	1390.41.1	1871.52.2	2352.63.3	2833.74.4	3314.85.5	3795.96.6	4277.07.7	4758.18.8	89
90	433.00.0	914.11.1	1395.22.2	1876.33 3	2357.44.4	2838.55.5	3319.66.6	3800.77.7	4281.88.8	4763.00.0	90
91	437.81.1	918.92.2	1400.03.3	1881.14.4	2362.25.5	2843.36.6	3324.47.7	3805.58.8	4286.70.0	4767.81.1	91
92	442 62.2	923.73.3	1404.84.4	1885.95.5	2367.06.6	2848.17.7	3329.28.8	3810.40.0	4291.51.1	4772.62.2	92
93	447.43.3	928.54.4	1409.65.5	1890.76.6	2371.87.7	2852.98.8	3334.10.0	3815.21.1	4296.32.2	4777.43.3	93
94	452.24.4	933.35.5	1414.46.6	1895.57.7	2376.68.8	2857.80.0	3338.91.1	3820.02.2	4301.13.3	4782.24.4	94
95	457.05.5	938.16.6	1419.27.7	1900.38.8	2381.50.0	2862.61.1	3343.72.2	3824.83.3	4305.94.4	4787.05.5	95
96	461.86.6	942.97.7	1424 08.8	1905.20.0	2386.31.1	2867.42.2	3348.53.3	3829.64.4	4310.75.5	4791.86.6	96
97	466.67.7	947.78.8	1428.90.0	1910 01.1	2391.12.2	2872.23.3	3353.34.4	3834.45.5	4315.56.6	4796.67.7	97
98	471.48 8	952.60.0	1433.71.1	1914.82.2	2395.93.3	2877.04.4	3358.15.5	3839.26.6	4320.37.7	4801.48.8	98
99	476.30.0	957.41.1	1438.52.2	1919 63.3	2400.74.4	2881.85.5	3362.96.6	3844.07.7	4325.18.8	4806.30.0	99

S.	10	11	12	13	14	15	16	17	18	19	S.
d.	$ cts.	$ cts.	$ cts.	$ cts.	$ cts.	$ cts.	$ cts.	$ cts.	$ cts.	$ cts.	d.
0	2.40.5	2.64.6	2.88.6	3.12.7	3.36.7	3.60.8	3.84.8	4.08.9	4.33.0	4.57.0	0
1	2.42.5	2.66.6	2.90.6	3.14.7	3.38.7	3.62.8	3.86.8	4.10.9	4.35.0	4.59.0	1
2	2.44.5	2.68.6	2.92.6	3.16.7	3.40.7	3.64.8	3.88.8	4.12.9	4.37.0	4.61.0	2
3	2.46.5	2.70.6	2.94.6	3.18.7	3.42.7	3.66.8	3.90.8	4.14.9	4.39.0	4.63.0	3
4	2.48.5	2.72.6	2.96.6	3.20.7	3.44.7	3.68.8	3.92.8	4.16.9	4.41.0	4.65.0	4
5	2.50.5	2.74.6	2.98.6	3.22.7	3.46.7	3.70.8	3.94.8	4.18.9	4.43.0	4.67.0	5
6	2.52.5	2.76.6	3.00.6	3.24.7	3.48.7	3.72.8	3.96.8	4.20.9	4.45.0	4.69.0	6
7	2.54.5	2.78.6	3.02.6	3.26.7	3.50.7	3.74.8	3.98.8	4.22.9	4.47.0	4.71.0	7
8	2.56.5	2.80.6	3.04.6	3.28.7	3.52.7	3.76.8	4.00.8	4.24.9	4.49.0	4.73.0	8
9	2.58.5	2.82.6	3.06.6	3.30.7	3.54.7	3.78.8	4.02.8	4.26.9	4.51.0	4.75.0	9
10	2.60.5	2.84.6	3.08.6	3.32.7	3.56.7	3.80.8	4.04.8	4.28.9	4.53.0	4.77.0	10
11	2.62.5	2.86.6	3.10.6	3.34.7	3.58.7	3.82.8	4.06.8	4.30.9	4.55.0	4.79.0	11

(30)

$4.81 11/100 TO THE POUND.

8 3/8 %

PREMIUM.

STERLING INTO DOLLARS AND CENTS, AND VICE-VERSA.

£	0	100	200	300	400	500	600	700	800	900	£
0	......	481.66.6	963.33.3	1445.00.0	1926.66 6	2408 33.3	2890.00.0	3371.66.6	3853.33.3	4335.00.0	0
1	4.81.6	486.48.3	968.15.0	1449.81.6	1931.48.3	2413.15.0	2894.81.6	3376.48.3	3858.15 0	4339.81.6	1
2	9.63.3	491.30.0	972.96.6	1454.63.3	1936.30.0	2417.96.6	2899.63.3	3381.30.0	3862.96.6	4344.63.3	2
3	14.45.0	496.11.6	977.78.3	1459.45.0	1941.11.6	2422.78.3	2904.45.0	3386.11.6	3867.78.3	4349.45.0	3
4	19.26.6	500.93.3	982.60.0	1464 26 6	1945.93.3	2427.60.0	2909.26.6	3390.93.3	3872.60.0	4354 26.6	4
5	24.08.3	505.75.0	987.41.6	1469.08.3	1950.75.0	2432.41.6	2914.08.3	3395.75.0	3877.41.6	4359.08 3	5
6	28.90.0	510.56.6	992.23.3	1473.90.0	1955.56.6	2437.23.3	2918.90.0	3400.56.6	3882.23.3	4363.90.0	6
7	33.71.6	515.38.3	997.05.0	1478.71.6	1960.38.3	2442.05.0	2923.71.6	3405.38.3	3887.05 0	4368.71.6	7
8	38.53.3	520.20.0	1001.86.6	1483.53.3	1965.20.0	2446.86.6	2928.53.3	3410.20.0	3891.86 6	4373.53.3	8
9	43.35.0	525.01.6	1006.68.3	1488.35.0	1970.01.6	2451.68.3	2933.35.0	3415.01.6	3896.68.3	4378.35.0	9
10	48.16.6	529.83.3	1011.50.0	1493.16.6	1974.83.3	2456.50.0	2938.16.6	3419.83.3	3901.50.0	4383.16.6	10
11	52.98.3	534 65.0	1016.31.6	1497.98.3	1979.65.0	2461.31.6	2942.98 3	3424.65.0	3906.31.6	4387.98.3	11
12	57.80.0	539.46.6	1021.13.3	1502 80.0	1984.46.6	2466.13.3	2947.80.0	3429.46.6	3911.13.3	4392 80.0	12
13	62.61.6	544.28.3	1025.95.0	1507.61.6	1989.28.3	2470.95 0	2952.61.6	3434.28.3	3915.95.0	4397 61.6	13
14	67.43.3	549.10.0	1030.76.6	1512.43.3	1994.10.0	2475.76.6	2957.43.3	3439.10.0	3920.76.6	4402.43.3	14
15	72.25.0	553.91.6	1035.58.3	1517.25.0	1998.91.6	2480.58.3	2962.25.0	3443.91.6	3925.58.3	4407.25.0	15
16	77.06.6	558.73.3	1040.40.0	1522.06.6	2003.73.3	2485.40.0	2967.06.6	3448.73.3	3930.40.0	4412.06.6	16
17	81 88.3	563.55.0	1045 21.6	1526.88.3	2008.55.0	2490.21.6	2971.88.3	3453.55.0	3935.21.6	4416.88.3	17
18	86.70.0	568.36.6	1050.03.3	1531.70.0	2013.36.6	2495.03.3	2976.70.0	3458.36.6	3940.03.3	4421.70.0	18
19	91.51.6	573.18.3	1054.85.0	1536.51.6	2018.18.3	2499.85.0	2981.51.6	3463.18.3	3944.85.0	4426.51.6	19
20	96.33.3	578.00.0	1059.66.6	1541.33.3	2023.00.0	2504.66.6	2986.33.3	3468.00.0	3949.66.6	4431.33.3	20
21	101.15.0	582 81.6	1064.48.3	1546.15.0	2027.81.6	2509.48.3	2991.15.0	3472.81.6	3954.48.3	4436.15.0	21
22	105.96.6	587.63.3	1069.30.0	1550.96.6	2032.63.3	2514.30.0	2995.96.6	3477.63.3	3959.30.0	4440.96.6	22
23	110.78.3	592 45.0	1074.11.6	1555.78.3	2037.45.0	2519.11.6	3000.78.3	3482.45.0	3964.11.6	4445.78.3	23
24	115.60.0	597 26.6	1078.93.3	1560.60.0	2042.26.6	2523.93 3	3005.60.0	3487.26.6	3968.93.3	4450.60.0	24
25	120.41.6	602.08.3	1083.75.0	1565.41.6	2047.08.3	2528.75.0	3010 41.6	3492.08.3	3973.75.0	4455.41.6	25
26	125.23.3	606.90.0	1088.56.6	1570.23.3	2051.90.0	2533.56.6	3015.23.3	3496.90.0	3978.56.6	4460.23.3	26
27	130.05.0	611.71.6	1093.38.3	1575.05.0	2056.71.6	2538.38.3	3020.05.0	3501.71.6	3983.38.3	4465.05.0	27
28	134.86.6	616.53.3	1098.20.0	1579.86.6	2061.53.3	2543.20.0	3024.86.6	3506.53.3	3988.20.0	4469.86.6	28
29	139.68.3	621.35.0	1103.01.6	1584.68.3	2066.35.0	2548.01.6	3029.68.3	3511.35.0	3993.01.6	4474.68.3	29
30	144.50.0	626.16.6	1107.83.3	1589.50.0	2071.16.6	2552.83.3	3034 50.0	3516.16.6	3997.83.3	4479.50.0	30
31	149.31.6	630.98.3	1112.65.0	1594.31.6	2075.98.3	2557.65.0	3039.31.6	3520.98.3	4002.65.0	4484.31 6	31
32	154.13.3	635.80.0	1117.46.6	1599.13.3	2080.80 0	2562.46 6	3044 13.3	3525.80.0	4007.46.6	4489.13.3	32
33	158.95.0	640.61.6	1122.28.3	1603.95.0	2085.61.6	2567.28.3	3048.95.0	3530.61.6	4012.28.3	4493.95.0	33
34	163.76.6	645.43.3	1127.10.0	1608.76.6	2090.43.3	2572.10.0	3053.76.6	3535.43.3	4017.10.0	4498.76.6	34
35	168.58.3	650.25.0	1131.91.6	1613.58.3	2095.25.0	2576.91.6	3058.58 3	3540.25.0	4021.91.6	4503 58.3	35
36	173.40.0	655.06.6	1136.73.3	1618.40.0	2100.06.6	2581.73.3	3063 40.0	3545.06.6	4026.73.3	4508 40.0	36
37	178.21.6	659.88.3	1141.55.0	1623.21.6	2104.88 3	2586.55.0	3068.21.6	3549.88.3	4031.55.0	4513.21.6	37
38	183 03.3	664.70.0	1146.36 6	1628.03.3	2109.70.0	2591.36.6	3073.03.3	3554.70 0	4036.36.6	4518.03.3	38
39	187.85.0	669.51.6	1151.18.3	1632.85.0	2114.51.6	2596.18.3	3077.85.0	3559.51.6	4041.18.3	4522.85.0	39
40	192.66.6	674.33 3	1156.00.0	1637.66.6	2119.33.3	2601.00.0	3082.66.6	3564.33.3	4046 00 0	4527.66.6	40
41	197.48.3	679 15.0	1160.81.6	1642.48.3	2124.15.0	2605.81.6	3087.48.3	3569.15.0	4050.81.6	4532 48.3	41
42	202 30.0	683.96.6	1165.63 3	1647.30.0	2128.96.6	2610.63.3	3092.30.0	3573.96.6	4055.63.3	4537.30.0	42
43	207.11.6	688.78.3	1170.45.0	1652.11.6	2133.78.3	2615.45.0	3097.11.6	3578.78.3	4060.45 0	4542 11 6	43
44	211 93.3	693.60.0	1175.26.6	1656.93.3	2138.60.0	2620.26.6	3101.93.3	3583.60.0	4065.26 6	4546 93.3	44
45	216.75.0	698.41.6	1180.08.3	1661.75.0	2143.41.6	2625.08.3	3106.75.0	3588.41.6	4070.08.3	4551.75 0	45
46	221 56.6	703.23.3	1184.90.0	1666.56.6	2148.23.3	2629.90 0	3111.56.6	3593.23.3	4074.90.0	4556 56 6	46
47	226.38.3	708.05.0	1189.71.6	1671.38.3	2153.05.0	2634.71.6	3116.38.3	3598.05.0	4079 71.6	4561 38.3	47
48	231.20.0	712.86.6	1194.53.3	1676.20.0	2157.86.6	2639.53.3	3121.20.0	3602.86.6	4084.53.3	4566.20.0	48
49	236.01.6	717.68 3	1199.35.0	1681.01.6	2162.68.3	2644.35.0	3126.01.6	3607.68.3	4089.35.0	4571.01.6	49

S.	0	1	2	3	4	5	6	7	8	9	S.
d.	$ cts.	$ cts.	$ cts.	$ cts.	$ cts.	$ cts.	$ cts.	$ cts.	$ cts.	$ cts.	d.
0	......	.24.0	.48.1	.72.2	.96.3	1.20.4	1.44.5	1.68.5	1.92.6	2.16.7	0
1	.02.0	.26.0	.50.1	.74.2	.98.3	1.22.4	1.46.5	1.70.5	1.94.6	2.18.7	1
2	.04.0	.28.0	.52.1	.76.2	1.00.3	1.24.4	1.48.5	1.72.5	1.96.6	2.20.7	2
3	.06.0	.30.0	.54.1	.78.2	1.02.3	1.26.4	1.50.5	1.74.5	1.98.6	2.22.7	3
4	.08.0	.32.0	.56.1	.80.2	1.04.3	1.28.4	1.52.5	1.76.5	2.00.6	2.24.7	4
5	.10.0	.34.0	.58.1	.82.2	1.06.3	1.30.4	1.54.5	1.78.5	2.02.6	2.26.7	5
6	.12.0	.36.0	.60.1	.84.2	1.08.3	1.32.4	1.56.5	1.80.5	2.04.6	2.28.7	6
7	.14.0	.38.0	.62.1	.86.2	1.10.3	1.34.4	1.58.5	1.82.5	2.06.6	2.30.7	7
8	.16.0	.40.0	.64.1	.88.2	1.12.3	1.36.4	1.60.5	1.84.5	2.08.6	2.32.7	8
9	.18.0	.42.0	.66.1	.90 2	1.14.3	1.38.4	1.62.5	1.86.5	2.10.6	2.34.7	9
10	.20.0	.44.0	.68.1	.92.2	1.16.3	1.40.4	1.64.5	1.88.5	2.12.6	2.36.7	10
11	.22.0	.46.0	.70.1	.94.2	1.18.3	1.42.4	1.66.5	1.90.5	2.14.6	2.38.7	11

$4.81 1⁄100 TO THE POUND.

8 3⁄8 %

PREMIUM.

STERLING INTO DOLLARS AND CENTS, AND VICE-VERSA.

£	0	100	200	300	400	500	600	700	800	900	£
	$ cts.	$ cts.	$ cts.	$ cts.	$ cts.	$ cts.	$ cts.	$ cts.	$ cts.	$ cts.	
50	240.83.3	722.50.0	1204.16.6	1685.83.3	2167.50.0	2649.16.6	3130.83.3	3612.50.0	4094.16.6	4575.83.3	50
51	245.65.0	737.31.6	1208.98.3	1690.65.0	2172.31.6	2653.98.3	3135.65.0	3617.31.6	4098.98.3	4580.65.0	51
52	250.46.6	732.13.3	1213.80.0	1695.46.6	2177.13.3	2658.80.0	3140.46.6	3622.13.3	4103.80.0	4585.46.6	52
53	255.28.3	736.95.0	1218.61.6	1700.28.3	2181.95.0	2663.61.6	3145.28.3	3626.95.0	4108.61.6	4590.28.3	53
54	260.10.0	741.76.6	1223.43.3	1705.10.0	2186.76.6	2668.43.3	3150.10.0	3631.76.6	4113.43.3	4595.10.0	54
55	264.91.6	746.58.3	1228.25.0	1709.91.6	2191.58.3	2673.25.0	3154.91.6	3636.58.3	4118.25.0	4599.91.6	55
56	269.73.3	751.40.0	1233.06.6	1714.73.3	2196.40.0	2678.06.6	3159.73.3	3641.40.0	4123.06.6	4604.73.3	56
57	274.55.0	756.21.6	1237.88.3	1719.55.0	2201.21.6	2682.88.3	3164.55.0	3646.21.6	4127.88.3	4609.55.0	57
58	279.36.6	761.03.3	1242.70.0	1724.36.6	2206.03.3	2687.70.0	3169.36.6	3651.03.3	4132.70.0	4614.36.6	58
59	284.18.3	765.85.0	1247.51.6	1729.18.3	2210.85.0	2692.51.6	3174.18.3	3655.85.0	4137.51.6	4619.18.3	59
60	289.00.0	770.66.6	1252.33.3	1734.00.0	2215.66.6	2697.33.3	3179.00.0	3660.66.6	4142.33.3	4624.00.0	60
61	293.81.6	775.48.3	1257.15.0	1738.81.6	2220.48.3	2702.15.0	3183.81.6	3665.48.3	4147.15.0	4628.81.6	61
62	298.63.3	780.30.0	1261.96.6	1743.63.3	2225.30.0	2706.96.6	3188.63.3	3670.30.0	4151.96.6	4633.63.3	62
63	303.45.0	785.11.6	1266.78.3	1748.45.0	2230.11.6	2711.78.3	3193.45.0	3675.11.6	4156.78.3	4638.45.0	63
64	308.26.6	789.93.3	1271.60.0	1753.26.6	2234.93.3	2716.60.0	3198.26.6	3679.93.3	4161.60.0	4643.26.6	64
65	313.08.3	794.75.0	1276.41.6	1758.08.3	2239.75.0	2721.41.6	3203.08.3	3684.75.0	4166.41.6	4648.08.3	65
66	317.90.0	799.56.6	1281.23.3	1762.90.0	2244.56.6	2726.23.3	3207.90.0	3689.56.6	4171.23.3	4652.90.0	66
67	322.71.6	804.38.3	1286.05.0	1767.71.6	2249.38.3	2731.05.0	3212.71.6	3694.38.3	4176.05.0	4657.71.6	67
68	327.53.3	809.20.0	1290.86.6	1772.53.3	2254.20.0	2735.86.6	3217.53.3	3699.20.0	4180.86.6	4662.53.3	68
69	332.35.0	814.01.6	1295.68.3	1777.35.0	2259.01.6	2740.68.3	3222.35.0	3704.01.6	4185.68.3	4667.35.0	69
70	337.16.6	818.83.3	1300.50.0	1782.16.6	2263.83.3	2745.50.0	3227.16.6	3708.83.3	4190.50.0	4672.16.6	70
71	341.98.3	823.65.0	1305.31.6	1786.98.3	2268.65.0	2750.31.6	3231.98.3	3713.65.0	4195.31.6	4676.98.3	71
72	346.80.0	828.46.6	1310.13.3	1791.80.0	2273.46.6	2755.13.3	3236.80.0	3718.46.6	4200.13.3	4681.80.0	72
73	351.61.6	833.28.3	1314.95.0	1796.61.6	2278.28.3	2759.95.0	3241.61.6	3723.28.3	4204.95.0	4686.61.6	73
74	356.43.3	838.10.0	1319.76.6	1801.43.3	2283.10.0	2764.76.6	3246.43.3	3728.10.0	4209.76.6	4691.43.3	74
75	361.25.0	842.91.6	1324.58.3	1806.25.0	2287.91.6	2769.58.3	3251.25.0	3732.91.6	4214.58.3	4696.25.0	75
76	366.06.6	847.73.3	1329.40.0	1811.06.6	2292.73.3	2774.40.0	3256.06.6	3737.73.3	4219.40.0	4701.06.6	76
77	370.88.3	852.55.0	1334.21.6	1815.88.3	2297.55.0	2779.21.6	3260.88.3	3742.55.0	4224.21.6	4705.88.3	77
78	375.70.0	857.36.6	1339.03.3	1820.70.0	2302.36.6	2784.03.3	3265.70.0	3747.36.6	4229.03.3	4710.70.0	78
79	380.51.6	862.18.3	1343.85.0	1825.51.6	2307.18.3	2788.85.0	3270.51.6	3752.18.3	4233.85.0	4715.51.6	79
80	385.33.3	867.00.0	1348.66.6	1830.33.3	2312.00.0	2793.66.6	3275.33.3	3757.00.0	4238.66.6	4720.33.3	80
81	390.15.0	871.81.6	1353.48.3	1835.15.0	2316.81.6	2798.48.3	3280.15.0	3761.81.6	4243.48.3	4725.15.0	81
82	394.96.6	876.63.3	1358.30.0	1839.96.6	2321.63.3	2803.30.0	3284.96.6	3766.63.3	4248.30.0	4729.96.6	82
83	399.78.3	881.45.0	1363.11.6	1844.78.3	2326.45.0	2808.11.6	3289.78.3	3771.45.0	4253.11.6	4734.78.3	83
84	404.60.0	886.26.6	1367.93.3	1849.60.0	2331.26.6	2812.93.3	3294.60.0	3776.26.6	4257.93.3	4739.60.0	84
85	409.41.6	891.08.3	1372.75.0	1854.41.6	2336.08.3	2817.75.0	3299.41.6	3781.08.3	4262.75.0	4744.41.6	85
86	414.23.3	895.90.0	1377.56.6	1859.23.3	2340.90.0	2822.56.6	3304.23.3	3785.90.0	4267.56.6	4749.23.3	86
87	419.05.0	900.71.6	1382.38.3	1864.05.0	2345.71.6	2827.38.3	3309.05.0	3790.71.6	4272.38.3	4754.05.0	87
88	423.86.6	905.53.3	1387.20.0	1868.86.6	2350.53.3	2832.20.0	3313.86.6	3795.53.3	4277.20.0	4758.86.6	88
89	428.68.3	910.35.0	1392.01.6	1873.68.3	2355.35.0	2837.01.6	3318.68.3	3800.35.0	4282.01.6	4763.68.3	89
90	433.50.0	915.16.6	1396.83.3	1878.50.0	2360.16.6	2841.83.3	3323.50.0	3805.16.6	4286.83.3	4768.50.0	90
91	438.31.6	919.98.3	1401.65.0	1883.31.6	2364.98.3	2846.65.0	3328.31.6	3809.98.3	4291.65.0	4773.31.6	91
92	443.13.3	924.80.0	1406.46.6	1888.13.3	2369.80.0	2851.46.6	3333.13.3	3814.80.0	4296.46.6	4778.13.3	92
93	447.95.0	929.61.6	1411.28.3	1892.95.0	2374.61.6	2856.28.3	3337.95.0	3819.61.6	4301.28.3	4782.95.0	93
94	452.76.6	934.43.3	1416.10.0	1897.76.6	2379.43.3	2861.10.0	3342.76.6	3824.43.3	4306.10.0	4787.76.6	94
95	457.58.3	939.25.0	1420.91.6	1902.58.3	2384.25.0	2865.91.6	3347.58.3	3829.25.0	4310.91.6	4792.58.3	95
96	462.40.0	944.06.6	1425.73.3	1907.40.0	2389.06.6	2870.73.3	3352.40.0	3834.06.6	4315.73.3	4797.40.0	96
97	467.21.6	948.88.3	1430.55.0	1912.21.6	2393.88.3	2875.55.0	3357.21.6	3838.88.3	4320.55.0	4802.21.6	97
98	472.03.3	953.70.0	1435.36.6	1917.03.3	2398.70.0	2880.36.6	3362.03.3	3843.70.0	4325.36.6	4807.03.3	98
99	476.85.0	958.51.6	1440.18.3	1921.85.0	2403.51.6	2885.18.3	3366.85.0	3848.51.6	4330.18.3	4811.85.0	99

S.	10	11	12	13	14	15	16	17	18	19	S.
d.	$ cts.	$ cts.	$ cts.	$ cts.	$ cts.	$ cts.	$ cts.	$ cts.	$ cts.	$ cts.	d.
0	2.40.8	2.64.9	2.89.0	3.13.0	3.37.1	3.61.2	3.85.3	4.09.4	4.33.5	4.57.5	0
1	2.42.8	2.66.9	2.91.0	3.15.0	3.39.1	3.63.2	3.87.3	4.11.4	4.35.5	4.59.5	1
2	2.44.8	2.68.9	2.93.0	3.17.0	3.41.1	3.65.2	3.89.3	4.13.4	4.37.5	4.61.5	2
3	2.46.8	2.70.9	2.95.0	3.19.0	3.43.1	3.67.2	3.91.3	4.15.4	4.39.5	4.63.5	3
4	2.48.8	2.72.9	2.97.0	3.21.0	3.45.1	3.69.2	3.93.3	4.17.4	4.41.5	4.65.5	4
5	2.50.8	2.74.9	2.99.0	3.23.0	3.47.1	3.71.2	3.95.3	4.19.4	4.43.5	4.67.5	5
6	2.52.8	2.76.9	3.01.0	3.25.0	3.49.1	3.73.2	3.97.3	4.21.4	4.45.5	4.69.5	6
7	2.54.8	2.78.9	3.03.0	3.27.0	3.51.1	3.75.2	3.99.3	4.23.4	4.47.5	4.71.5	7
8	2.56.8	2.80.9	3.05.0	3.29.0	3.53.1	3.77.2	4.01.3	4.25.4	4.49.5	4.73.5	8
9	2.58.8	2.82.9	3.07.0	3.31.0	3.55.1	3.79.2	4.03.3	4.27.4	4.51.5	4.75.5	9
10	2.60.8	2.84.9	3.09.0	3.33.0	3.57.1	3.81.2	4.05.3	4.29.4	4.53.5	4.77.5	10
11	2.62.8	2.86.9	3.11.0	3.35.0	3.59.1	3.83.2	4.07.3	4.31.4	4.55.5	4.79.5	11

$4.82¹¹⁄₁₀₀ TO THE POUND.

8½%

PREMIUM.

STERLING INTO DOLLARS AND CENTS, AND VICE-VERSA.

£	0	100	200	300	400	500	600	700	800	900	£
£	$ cts.	$ cts.	$ cts.	$ cts.	$ cts.	$ cts.	$ cts.	$ cts.	$ cts.	$ cts.	£
0	.....	482.22 2	964.44.4	1446.66.6	1928.88.8	2411.11.1	2893.33.3	3375.55.5	3857.77.7	4340.00.0	0
1	4.82.2	487.04.4	969.26.6	1451.48.8	1933.71.1	2415.93.3	2898.15.5	3380.37.7	3862.60.0	4344.82.2	1
2	9.64.4	491.86.6	974.08.8	1456.31.1	1938.53.3	2420.75.5	2902.97.7	3385.20.0	3867.42.2	4349.64.4	2
3	14.46.6	496.68.8	978.91.1	1461.13.3	1943.35.5	2425.57.7	2907.80.0	3390.02.2	3872.24.4	4354.46.6	3
4	19.28.8	501.51.1	983.73.3	1465.95.5	1948.17.7	2430.40.0	2912.62.2	3394.84.4	3877.06.6	4359.28.8	4
5	24.11.1	506.33.3	988.55.5	1470.77.7	1953.00.0	2435.22.2	2917.44.4	3399.66.6	3881.88.8	4364.11.1	5
6	28.93.3	511.15.5	993.37.7	1475.60.0	1957.82.2	2440.04.4	2922.26.6	3404.48.8	3886.71.1	4368.93.3	6
7	33.75.5	515.97.7	998.20.0	1480.42.2	1962.64.4	2444.86.6	2927.08.8	3409.31.1	3891.53.3	4373.75.5	7
8	38.57.7	520.80.0	1003.02.2	1485.24.4	1967.46.6	2449.68.8	2931.91.1	3414.13.3	3896.35.5	4378.57.7	8
9	43.40.0	525.62.2	1007.84.4	1490.06.6	1972.28.8	2454.51.1	2936.73.3	3418.95.5	3901.17.7	4383.40.0	9
10	48.22.2	530.44.4	1012.66.6	1494.88.8	1977.11.1	2459.33.3	2941.55.5	3423.77.7	3906 00.0	4388.22.2	10
11	53.04.4	535.26.6	1017.48.8	1499.71.1	1981.93.3	2464.15 5	2946.37.7	3428.60.0	3910.82.2	4393.04.4	11
12	57.86.6	540.08.8	1022.31.1	1504.53.3	1986.75.5	2468.97.7	2951.20.0	3433.42.2	3915.64.4	4397.86.6	12
13	62.68.8	544.91.1	1027.13.3	1509.35.5	1991.57.7	2473.80.0	2956.02.2	3438.24.4	3920.46.6	4402.68.8	13
14	67.51.1	549.73.3	1031.95.5	1514.17.7	1996.40.0	2478.62.2	2960.84.4	3443.06.6	3925.28.8	4407.51.1	14
15	72.33.3	554.55.5	1036.77.7	1519 00.0	2001.22.2	2483.44.4	2965.66.6	3447.88.8	3930.11.1	4412.33.3	15
16	77.15.5	559.37.7	1041.60.0	1523.82 2	2006.04.4	2488.26.6	2970.48.8	3452.71.1	3934.93.3	4417.15.5	16
17	81 97.7	564.20.0	1046 42.2	1528.64.4	2010.86.6	2493.08.8	2975.31.1	3457.53.3	3939.75.5	4421.97.7	17
18	86.80.0	569.02.2	1051.24.4	1533.46.6	2015.68.8	2497.91.1	2980.13.3	3462.35.5	3944.57.7	4426.80.0	18
19	91.62.2	573.84.4	1056.06.6	1538.28.8	2020.51.1	2502.73.3	2984.95.5	3467.17.7	3949.40.0	4431.62.2	19
20	96.44.4	578.66.6	1060.88.8	1543.11.1	2025.33.3	2507.55.5	2989.77.7	3472.00.0	3954.22.2	4436.44.4	20
21	101.26.6	583.48.8	1065.71.1	1547.93.3	2030.15.5	2512.37.7	2994.60.0	3476.82.2	3959.04.4	4441.26.6	21
22	106.08.8	588.31.1	1070.53.3	1552.75.5	2034.97.7	2517.20.0	2999 42.2	3481.64.4	3963.86.6	4446.08.8	22
23	110.91.1	593.13.3	1075.35.5	1557.57.7	2039.80.0	2522 02.2	3004.24.4	3486.46.6	3968.68.8	4450.91.1	23
24	115.73.3	597.95.5	1080.17.7	1562.40.0	2044.62.2	2526.84.4	3009.06.6	3491.28.8	3973.51.1	4455.73.3	24
25	120.55.5	602.77.7	1085.00.0	1567.22.2	2049.44.4	2531.66.6	3013.88.8	3496.11.1	3978.33.3	4460.55.5	25
26	125.37.7	607.60.0	1089.82.2	1572.04.4	2054.26.6	2536.48.8	3018.71.1	3500.93.3	3983.15.5	4465.37.7	26
27	130.20.0	612.42.2	1094.64.4	1576.86.6	2059.08.8	2541.31.1	3023.53.3	3505.75.5	3987.97.7	4470.20.0	27
28	135.02.2	617.24.4	1099.46.6	1581.68.8	2063.91.1	2546 13 3	3028.35.5	3510.57.7	3992.80.0	4475.02.2	28
29	139.84.4	622.06.6	1104.28.8	1586.51.1	2068.73.3	2550.95.5	3033.17.7	3515.40.0	3997.62.2	4479.84.4	29
30	144.66.6	626.88.8	1109.11.1	1591.33.3	2073.55.5	2555.77.7	3038.00.0	3520.22.2	4002.44.4	4484.66.6	30
31	149.48.8	631.71.1	1113.93.3	1596 15.5	2078.37.7	2560.60.0	3042.82.2	3525.04.4	4007.26.6	4489.48.8	31
32	154.31.1	636.53.3	1118.75.5	1600.97.7	2083.20 0	2565.42.2	3047.64.4	3529.86.6	4012.08.8	4494.31.1	32
33	159.13.3	641.35.5	1123.57.7	1605 80.0	2088.02.2	2570.24.4	3052.46.6	3534.68.8	4016.91.1	4499.13.3	33
34	163.95.5	646.17.7	1128.40.0	1610.62.2	2092.84.4	2575.06.6	3057 28.8	3539.51.1	4021.73.3	4503.95.5	34
35	168.77.7	651.00.0	1133.22.2	1615.44.4	2097.66.6	2579.88.8	3062.11.1	3544.33.3	4026.55.5	4508.77.7	35
36	173.60.0	655.82.2	1138.04.4	1620.26.6	2102.48.8	2584.71.1	3066.93.3	3549.15.5	4031.37.7	4513.60.0	36
37	178.42.2	660.64.4	1142.86.6	1625.08.8	2107.31.1	2589.53.3	3071 75.5	3553.97.7	4036.20.0	4518.42.2	37
38	183.24.4	665.46.6	1147.68.8	1629.91.1	2112.13.3	2594.35.5	3076.57.7	3558.80.0	4041.02.2	4523.24.4	38
39	188.06.6	670.28.8	1152.51.1	1634.73.3	2116.95.5	2599.17.7	3081.40.0	3563.62.2	4045.84 4	4528.06.6	39
40	192.88.8	675.11.1	1157.33.3	1639 55.5	2121.77.7	2604.00.0	3086.22.2	3568.44.4	4050.66.6	4532.88.8	40
41	197.71.1	679.93.3	1162.15.5	1644.37 7	2126.60.0	2608.82.2	3091.04.4	3573.26.6	4055.48.8	4537.71.1	41
42	202.53.3	684.75.5	1166.97.7	1649.20.0	2131.42.2	2613.64.4	3095.86.6	3578.08.8	4060.31.1	4542 53.3	42
43	207.35.5	689.57.7	1171 80.0	1654.02.2	2136.24.4	2618.46.6	3100.68.8	3582.91.1	4065.13.3	4547.35.5	43
44	212.17.7	694.40.0	1176.62.2	1658.84.4	2141.06.6	2623.28.8	3105.51.1	3587.73.3	4069.95.5	4552.17.7	44
45	217.00.0	699.22.2	1181 44.4	1663.66 6	2145.88.8	2628.11.1	3110.33.3	3592.55.5	4074.77.7	4557.00.0	45
46	221.82.2	704.04.4	1186.26.6	1668.48.8	2150.71.1	2632.93.3	3115.15.5	3597.37.7	4079.60.0	4561.82.2	46
47	226.64.4	708.86.6	1191 08.8	1673.31.1	2155.53.3	2637.75.5	3119.97.7	3602.20.0	4084 42.2	4566.64.4	47
48	231.46.6	713.68.8	1195.91.1	1678.13.3	2160.35.5	2642 57.7	3124.80.0	3607.02.2	4089.24.4	4571.46.6	48
49	236.28.8	718.51.1	1200.73.3	1682.95.5	2165.17.7	2647.40.0	3129.62.2	3611.84.4	4094.06.6	4576.28.8	49

S.	0	1	2	3	4	5	6	7	8	9	S.
d.	$ cts.	$ cts.	$ cts.	$ cts.	$ cts.	$ cts.	$ cts.	$ cts.	$ cts.	$ cts.	d.
0	.....	.24.1	.48.2	.72.3	.96.4	1.20.5	1.44.6	1.68.7	1.92.8	2.17.0	0
1	.02.0	.26.1	.50.2	.74.3	.98.4	1.22.5	1.46.6	1.70.7	1.94.8	2.19.0	1
2	.04.0	.28.1	.52.2	.76.3	1.00.4	1.24.5	1.48.6	1.72.7	1.96.8	2.21.0	2
3	.06.0	.30.1	.54.2	.78.3	1.02.4	1.26.5	1.50.6	1.74.7	1.98.8	2.23.0	3
4	.08.0	.32.1	.56.2	.80.3	1.04.4	1.28.5	1.52.6	1.76.7	2.00.8	2.25.0	4
5	.10.0	.34.1	.58.2	.82.3	1.06.4	1.30.5	1.54 6	1.78.7	2.02.8	2.27.0	5
6	.12.0	.36.1	.60.2	.84.3	1.08.4	1.32.5	1.56.6	1.80.7	2.04.8	2.29.0	6
7	.14.0	.38.1	.62.2	.86.3	1.10.4	1.34.5	1.58.6	1.82.7	2.06.8	2.31.0	7
8	.16.0	.40.1	.64.2	.88.3	1.12.4	1.36.5	1.60.6	1.84.7	2.08.8	2.33.0	8
9	.18.0	.42.1	.66.2	.90.3	1.14.4	1.38.5	1.62.6	1.86.7	2.10.8	2.35.0	9
10	.20.0	.44.1	.68.2	.92.3	1.16.4	1.40.5	1.64.6	1.88.7	2.12.8	2.37.0	10
11	.22.0	.46.1	.70.2	.94.3	1.18.4	1.42.5	1.66.6	1.90.7	2.14.8	2.39.0	11

£	0	100	200	300	400	500	600	700	800	900	£
	$ cts.	$ cts.	$ cts.	$ cts.	$ cts.	$ cts.	$ cts.	$ cts.	$ cts.	$ cts.	
50	241.11.1	723.33.3	1205.55.5	1687.77.7	2170.00.0	2652.22.2	3134.44.4	3616.66.6	4098.88.8	4581.11.1	50
51	245.93.3	728.15.5	1210.37.7	1692.60.0	2174.82.2	2657.04.4	3139.26.6	3621.48.8	4103.71.1	4585.93.3	51
52	250.75.5	732.97.7	1215.20.0	1697.42.2	2179.64.4	2661.86.6	3144.08.8	3626.31.1	4108.53.3	4590.75.5	52
53	255.57.7	737.80.0	1220.02.2	1702.24.4	2184.46.6	2666.68.8	3148.91 1	3631.13.3	4113.35.5	4595.57.7	53
54	260.40.0	742.62.2	1224.84.4	1707.06.6	2189.28.8	2671.51.1	3153.73.5	3635 95.5	4118.17.7	4600.40.0	54
55	265.22.2	747.44.4	1229.66.6	1711.88.8	2194.11.1	2676.33.3	3158.55.5	3640.77.7	4123.00 0	4605.22.2	55
56	270.04.4	752.26.6	1234.48.8	1716.71.1	2198.93.3	2681.15.5	3163.37.7	3645.60.0	4127.82.2	4610.04.4	56
57	274.86 6	757.08.8	1239.31.1	1721.53.3	2203.75.5	2685.97.7	3168.20.0	3650.42.2	4132.64.4	4614.86.6	57
58	279.68.8	761.91.1	1244.13.3	1726.35.5	2208.57.7	2690.80 0	3173.02.2	3655.24.4	4137.46.6	4619.68.8	58
59	284.51.1	766.73.3	1248.95.5	1731.17.7	2213.40.0	2695.62.2	3177.84.4	3660.06 6	4142.28.8	4624.51.1	59
60	289.33.3	771.55.5	1253.77.7	1736.00.0	2218.22.2	2700.44.4	3182.66.6	3664.88.8	4147.11.1	4629.33.3	60
61	294.15.5	776.37.7	1258.60.0	1740 82.2	2223.04.4	2705.26.6	3187.48.8	3669.71.1	4151.93.3	4634.15.5	61
62	298.97.7	781.20.0	1263.42 2	1745.64.4	2227.86.6	2710.08.8	3192.31.1	3674.53.3	4156.75.5	4638.97.7	62
63	303.80.0	786 02.2	1268.24.4	1750 46.6	2232.68.8	2714.91.1	3197.13.3	3679.35.5	4161.57.7	4643.80.0	63
64	308.62.2	790.84.4	1273.06.6	1755.28.8	2237.51.1	2719 73.3	3201.95.5	3684.17.7	4166 40.0	4648.62.2	64
65	313.44.4	795.66.6	1277.88.8	1760.11.1	2242.33.3	2724.55.5	3206.77.7	3689.00.0	4171.22.2	4653.44.4	65
66	318.26.6	800.48.8	1282.71.1	1764.93.3	2247.15.5	2729.37.7	3211.60.0	3693.82.2	4176.04.4	4658.26.6	66
67	323.08.8	805.31.1	1287.53.3	1769.75.5	2251.97.7	2734.20.0	3216.42.2	3698.64.4	4180.86.6	4663.08.8	67
68	327.91.1	810.13.3	1292.35.5	1774.57.7	2256.80.0	2739.02.2	3221.24.4	3703.46.6	4185.68.8	4667.91.1	68
69	332.73.3	814.95.5	1297.17.7	1779.40.0	2261.62.2	2743.84.4	3226.06.6	3708.28.8	4190.51.1	4672.73.3	69
70	337 55.5	819.77.7	1302.00.0	1784.22.2	2266.44.4	2748.66.6	3230.88.8	3713.11.1	4195.33.3	4677.55.5	70
71	342.37.7	824.60.0	1306.82 2	1789.04.4	2271.26.6	2753.48.8	3235.71.1	3717.93.3	4200.15.5	4682.37.7	71
72	347.20.0	829.42.2	1311.64.4	1793.86.6	2276.08.8	2758.31.1	3240.53.3	3722.75.5	4204.97.7	4687.20.0	72
73	352.02.2	834.24.4	1316.46.6	1798.68.8	2280.91.1	2763.13.3	3245.35.5	3727.57.7	4209.80.0	4692.02.2	73
74	356.84 4	839.06.6	1321.28.8	1803.51.1	2285.73.3	2767.95.5	3250.17.7	3732.40.0	4214.62.2	4696.84.4	74
75	361.66.6	843.88.8	1326.11.1	1808.33.3	2290.55.5	2772.77.7	3255.00.0	3737.22.2	4219.44.4	4701.66 6	75
76	366.48.8	848.71.1	1330.93.3	1813.15.5	2295 37.7	2777.60.0	3259.82.2	3742.04.4	4224.26.6	4706.48.8	76
77	371.31.1	853.53.3	1335 75.5	1817.97.7	2300.20.0	2782.42.2	3264.64.4	3746.86.6	4229.08.8	4711.31.1	77
78	376.13.3	858.35.5	1340.57.7	1822.80.0	2305.02.2	2787.24.4	3269.46.6	3751.68.8	4233.91.1	4716.13.3	78
79	380.95.5	863.17.7	1345.40.0	1827.62.2	2309.84.4	2792.06.6	3274.28.8	3756.51.1	4238.73.3	4720.95.5	79
80	385.77.7	868.00.0	1350.22.2	1832.44.4	2314.66.6	2796.88.8	3279.11.1	3761.33.3	4243.55.5	4725.77.7	80
81	390.60.0	872.82.2	1355.04.4	1837.26.6	2319 48.8	2801.71.1	3283.93.3	3766.15 5	4248.37.7	4730.60.0	81
82	395.42 2	877.64.4	1359.86.6	1842.08.8	2324.31.1	2806.53.3	3288.75.5	3770.97.7	4253.20.0	4735.42.2	82
83	400.24.4	882.46.6	1364.68.8	1846.91.1	2329.13.3	2811.35.5	3293.57.7	3775.80.0	4258.02.2	4740.24.4	83
84	405.06.6	887.28.8	1369.51.1	1851.73.3	2333.95.5	2816.17.7	3298.40.0	3780.62.2	4262.84.4	4745.06.6	84
85	409.88.8	892.11.1	1374.33.3	1856.55.5	2338.77.7	2821.00.0	3303.22.2	3785.44.4	4267.66.6	4749.88.8	85
86	414.71.1	896.93.3	1379.15.5	1861.37.7	2343.60.0	2825.82.2	3308.04.4	3790.26.6	4272.48.8	4754.71.1	86
87	419.53.3	901.75.5	1383.97.7	1866.20.0	2348.42.2	2830.64.4	3312.86.6	3795.08.8	4277.31.1	4759.53.3	87
88	424.35.5	906.57 7	1388.80.0	1871.02.2	2353.24.4	2835.46.6	3317.68.8	3799.91.1	4282.13.3	4764.35.5	88
89	429.17.7	911.40.0	1393.62.2	1875.84.4	2358.06.6	2840.28.8	3322.51.1	3804.73.3	4286.95.5	4769.17.7	89
90	434.00.0	916.22.2	1398.44.4	1880.66.6	2362.88.8	2845.11.1	3327.33.3	3809.55.5	4291.77.7	4774.00.0	90
91	438.82.2	921.04.4	1403.26 6	1885.48.8	2367.71.1	2849.93.3	3332.15.5	3814 37.7	4296.60.0	4778.82.2	91
92	443.64.4	925.86.6	1408.08.8	1890.31.1	2372.53.3	2854.75.5	3336.97.7	3819.20.0	4301.42.2	4783.64.4	92
93	448.46.6	930.68.8	1412.91.1	1895.13.3	2377.35.5	2859.57.7	3341.80 0	3824.02.2	4306.24.4	4788.46.6	93
94	453.28.8	935.51.1	1417.73.3	1809.95.5	2382.17.7	2864.40.0	3346.62.2	3828.84.4	4311.06.6	4793.28.8	94
95	458.11.1	940.33.3	1422.55.5	1904.77.7	2387.00.0	2869.22.2	3351.44.4	3833.66.6	4315.88.8	4798.11.1	95
96	462.93.3	945.15.5	1427.37.7	1909.60.0	2391.82.2	2874.04.4	3356 26.6	3838.48.8	4320.71.1	4802.93.3	96
97	467.75.5	949.97.7	1432.20.0	1914.42.2	2396.64.4	2878.86.6	3361.08.8	3843.31.1	4325.53.3	4807.75 5	97
98	472.57 7	954.80.0	1437.02.2	1919.24.4	2401.46.6	2883.68.8	3365.91.1	3848.13.3	4330.35.5	4812.57.7	98
99	477.40.0	959.62.2	1441.84.4	1924.06.6	2406.28.8	2888.51.1	3370.73.3	3852.95.5	4335.17.7	4817.40.0	99

S.	10	11	12	13	14	15	16	17	18	19	S.
d.	$ cts.	$ cts.	$ cts.	$ cts.	$ cts.	$ cts.	$ cts.	$ cts.	$ cts.	$ cts.	d.
0	2.41.1	2.65.2	2.89.3	3.13.4	3.37.5	3.61.6	3.85.7	4.09.8	4.34.0	4.58.1	0
1	2.43.1	2.67.2	2.91.3	3.15.4	3.39.5	3.63.6	3.87.7	4.11.8	4.36.0	4.60.1	1
2	2.45.1	2.69.2	2.93.3	3.17.4	3.41.5	3.65.6	3.89.7	4.13.8	4.38.0	4.62.1	2
3	2.47.1	2.71.2	2.95.2	3.19.4	3.43.5	3.67.6	3.91.7	4.15.8	4.40.0	4.64.1	3
4	2.49.1	2.73.2	2.97.3	3.21.4	3.45.5	3.69.6	3.93.7	4.17.8	4.42.0	4.66 1	4
5	2.51.1	2.75.2	2.99.3	3.23.4	3.47.5	3.71.6	3.95.7	4.19.8	4.44.0	4.68.1	5
6	2.53.1	2.77.2	3.01.3	3.25.4	3.49.5	3.73.6	3.97.7	4.21.8	4.46.0	4.70.1	6
7	2.55.1	2.79.2	3.03.3	3.27.4	3.51.5	3.75.6	3.99.7	4.23.8	4.48.0	4.72.1	7
8	2.57.1	2.81.2	3.05.3	3.29.4	3.53.6	3.77.6	4.01.7	4.25.8	4.50.0	4.74.1	8
9	2.59.1	2.83.2	3.07.3	3.31.4	3.55.5	3.79.6	4.03.7	4.27.8	4.52.0	4.76.1	9
10	2.61.1	2.85.2	3.09.3	3.33.4	3.57.5	3.81.6	4.05.7	4.29.8	4.54.0	4.78.1	10
11	2.63.1	2.87.2	3.11.3	3.35.4	3.59.5	3.83.6	4.07.7	4.31.8	4.56.0	4 80.1	11

$4.82 11/100 TO THE POUND.

8 5/8 %

PREMIUM.

STERLING INTO DOLLARS AND CENTS, AND VICE-VERSA.

£	0	100	200	300	400	500	600	700	800	900	£
£	$ cts.	$ cts.	$ cts.	$ cts.	$ cts.	$ cts.	$ cts.	$ cts.	$ cts.	$ cts.	£
0	......	482.77.7	965.55.5	1448.33.3	1931.11.1	2413.88.8	2896.66.6	3379.44.4	3862.22.2	4345.00.0	0
1	4.82.7	487.60.5	970.38.3	1453.16.1	1935.93.8	2418.71.6	2901.49.4	3384 27.2	3867.05.0	4349.82.7	1
2	9.65.5	492.43.3	975.21.1	1457.98.8	1940.76.6	2423.54 4	2906.32.2	3389.10.0	3871.87.7	4354.65.5	2
3	14.48.3	497.26.1	980.03.8	1462.81.6	1945.59 4	2428.37.2	2911.15.0	3393.92.7	3876.70.5	4359.48.3	3
4	19.31.1	502.08.8	984.86.6	1467.64.4	1950.42.2	2433.20.0	2915.97.7	3398 75.5	3881.53.3	4364.31.1	4
5	24.13.8	506.91.6	989.69 4	1472.47.2	1955.25.0	2438.02.7	2920.80.5	3403.58.3	3886.36.1	4369.13.8	5
6	28.96.6	511.74.4	994.52.2	1477.30.0	1960.07.7	2442.85.5	2925.63 3	3408.41.1	3891.18.8	4373.96.6	6
7	33.79.4	516.57.2	999.35.0	1482.12.7	1964.90.5	2447.68.3	2930.46.1	3413.23 8	3896.01.6	4378.79.4	7
8	38.62.2	521.40 0	1004.17.7	1486.95.5	1969.73.3	2452 51.1	2935.28.8	3418.06.6	3900.84 4	4383.62.2	8
9	43.45.0	526.22.7	1009.00 5	1491.78.3	1974.56.1	2457.33.8	2940.11.6	3422.89.4	3905.67.2	4388.45.0	9
10	48.27.7	531.05 5	1013.83.3	1496.61.1	1979.38.8	2462.16.6	2944.94.4	3427.72.2	3910.50.0	4393.27.7	10
11	53.10.5	535 88.3	1018.66.1	1501.43.8	1984.21.6	2466.99.4	2949.77.2	3432.55.0	3915.32.7	4398.10.5	11
12	57.93.3	540.71.1	1023.48.8	1506.26.6	1989.04.4	2471.82.2	2954.60.0	3437.37.7	3920.15.5	4402.93.3	12
13	62.76.1	545.53.8	1028.31.6	1511.09.4	1993.87.2	2476.65.0	2959.42.7	3442.20.5	3924 98.3	4407.76.1	13
14	67.58.8	550.36.6	1033.14.4	1515 92.2	1998.70.0	2481.47.7	2964.25.5	3447.03.3	3929.81.1	4412.58.8	14
15	72.41.6	555.19.4	1037.97.2	1520.75.0	2003 52.7	2486.30.5	2969.08.3	3451.86.1	3934.63.8	4417.41.6	15
16	77.24.4	560.02.2	1042.80 0	1525.57.7	2008.35.5	2491.13.3	2973.91.1	3456.68 8	3939.46.6	4122.24.4	16
17	82 07.2	564.85.0	1047.62.7	1530.40.5	2013.18.3	2495.96.1	2978.73.8	3461.51.6	3944.29.4	4427.07.2	17
18	86.90.0	569.67.7	1052.45.5	1535.23.3	2018.01.1	2500.78.8	2983.56 6	3466.34.4	3949.12.2	4131.90.0	18
19	91.72.7	574.50.5	1057.28.3	1540.06.1	2022.83.8	2505.61.6	2988.39.4	3471.17.2	3953.95.0	4436.72.7	19
20	96.55.5	579 33.3	1062.11.1	1544.88.8	2027.66.6	2510.44.4	2993.22 2	3476 00.0	3958.77.7	4441.55.5	20
21	101.38.3	584.16.1	1066.93.8	1549.71.6	2032.49.4	2515.27.2	2998.05.0	3480.82.7	3963.60.5	4446.38.3	21
22	106.21.1	588.98.8	1071.76.6	1554.54.4	2037.32 2	2520.10.0	3002 87 7	3485.65.5	3968.43.3	4451.21.1	22
23	111.03.8	593.81.6	1076.59.4	1559.37.2	2042.15.0	2524.92.7	3007.70 5	3490.48.3	3973.26.1	4456.03.8	23
24	115.86.6	598.64.4	1081.42.2	1564.20.0	2046.97 7	2529.75.5	3012.53.3	3495.31.1	3978.08.8	4460.86.6	24
25	120.69.4	603.47.2	1086.25.0	1569.02.7	2051.80.5	2534.58.3	3017.36.1	3500 13.8	3982.91.6	4465.69.4	25
26	125.52.2	608.30.0	1091.07.7	1573.85.5	2056.63.3	2539.41.1	3022.18.8	3504.96.6	3987.74.4	4470.52.2	26
27	130.35.0	613.12.7	1095.90.5	1578.68.3	2061 46.1	2544.23.8	3027.01.6	3509.79.4	3992.57.2	4475.35.0	27
28	135.17.7	617.95.5	1100.73.3	1583.51.1	2066.28.8	2549.06.6	3031.84.4	3514.62.2	3997.40.0	4480.17.7	28
29	140.00.5	622.78.3	1105.56.1	1588.33 8	2071.11.6	2553.89.4	3036.67.2	3519.45.0	4002.22.7	4485.00.5	29
30	144.83.3	627.61.1	1110.38.8	1593.16.6	2075.94 4	2558.72.2	3041.50.0	3524.27.7	4007.05.5	4489.83.3	30
31	149.66.1	632.43 8	1115.21.6	1597.99.4	2080.77 2	2563.55.0	3046.32.7	3529.10.5	4011.88.3	4494.66.1	31
32	154.48.8	637.26 6	1120.04.4	1602.82.2	2085.60.0	2568.37.7	3051.15.5	3533.93.3	4016.71.1	4499.48.8	32
33	159.31.6	642.09.4	1124.87.2	1607.65.0	2090 42 7	2573.20.5	3055.98.3	3538.76.1	4021.53.8	4504.31.6	33
34	164.14.4	646.92.2	1129.70.0	1612.47.7	2095.25 5	2578.03.3	3060.81.1	3543.58.8	4026.36.6	4509.14.4	34
35	168.97.2	651.75.0	1134.52.7	1617.30 5	2100.08.3	2582.86.1	3065.63.8	3548.41.6	4031.19.4	4513.97.2	35
36	173.80.0	656.57.7	1139.35.5	1622 13.3	2104.91.1	2587.68.8	3070.46.6	3553.24.4	4036.02.2	4518.80.0	36
37	178.62.7	661.40.5	1144.18 3	1626 96.1	2109.73.8	2592.51.6	3075.29.4	3558.07.2	4040.85.0	4523.62.7	37
38	183.45.5	666 23.3	1149.01.1	1631.78.8	2114.56.6	2597.34.4	3080.12.2	3562.90.0	4045.67.7	4528.45.5	38
39	188.28.3	671.06.1	1153.83.8	1636.61.6	2119.39.4	2602.17.2	3084.95.0	3567.72.7	4050.50.5	4533.28.3	39
40	193.11.1	675.88.8	1158.66.6	1641.44.4	2124.22.2	2607.00.0	3089.77.7	3572.55.5	4055.33.3	4538.11.1	40
41	197.93.8	680 71.6	1163.49.4	1646.27.2	2129.05.0	2611.82.7	3094.60.5	3577.38.3	4060.16.1	4542.93 8	41
42	202.76.6	685.54 4	1168.32.2	1651.10.0	2133.87.7	2616.65.5	3099.43 3	3582.21.1	4064.98.8	4547.76.6	42
43	207.59.4	690.37.2	1173.15.0	1655 92.7	2138.70.5	2621.48.3	3104.26.1	3587.03.8	4069.81.6	4552.59 4	43
44	212.42.2	695.20.0	1177.97.7	1660.75.5	2143.53.3	2626.31.1	3109.08.8	3591.86.6	4074.64.4	4557.42.2	44
45	217.25 0	700.02.7	1182.80.5	1665.58.3	2148.36.1	2631.13.8	3113.91.6	3596.69.4	4079.47.2	4562.25.0	45
46	222.07.7	704.85.5	1187.63.3	1670.41.1	2153.18.8	2635.96.6	3118.74.4	3601.52.2	4084.30.0	4567.07.7	46
47	226.90.5	709.68.3	1192.46.1	1675.23.8	2158.01.6	2640.79.4	3123.57.2	3606.35.0	4089.12.7	4571.90.5	47
48	231.73.3	714.51.1	1197.28.8	1680.06.6	2162.84.4	2645.62.2	3128.40.0	3611.17.7	4093.95 5	4576.73.3	48
49	236.56.1	719.33.8	1202.11.6	1684.89.4	2167.67.2	2650.45.0	3133.22.7	3616.00.5	4098.78.3	4581.56.1	49

S.	0	1	2	3	4	5	6	7	8	9	S.
d.	$ cts.	$ cts.	$ cts.	$ cts.	$ cts.	$ cts.	$ cts.	$ cts.	$ cts.	$ cts.	d.
0	......	.24.1	.48.2	.72.4	.96,5	1.20,6	1.44.8	1.68.9	1,93,1	2,17.2	0
1	.02.0	.26.1	.50.2	.74.4	.98.5	1.22.7	1.46.8	1.70.9	1,95.1	2,19.2	1
2	.04.0	.28.1	.52.3	.76.4	1.00.5	1.24.7	1.48.8	1.73.0	1.97.1	2.21.2	2
3	.06.0	.30.1	.54.3	.78.4	1.02.5	1.26.7	1.50.8	1.75.0	1.99.1	2.23.2	3
4	.08.0	.32.1	.56.3	.80.4	1.04.6	1.28.7	1.52.8	1.77.0	2.01.1	2.25.3	4
5	.10.0	.34.1	.58.3	.82.4	1.06.6	1.30.7	1.54.8	1.79.0	2.03.1	2.27.3	5
6	.12.0	.36.2	.60.3	.84.4	1.08.6	1.32.7	1.56.9	1.81.0	2.05.1	2.29.3	6
7	.14.0	.38.2	.62.3	.86.4	1.10.6	1.34.7	1.58.9	1.83.0	2.07.1	2.31.3	7
8	.16.0	.40.2	.64.3	.88.5	1.12.6	1.36.7	1.60.9	1.85.0	2.09.2	2.33.3	8
9	.18.1	.42.2	.66.3	.90.5	1.14.6	1.38.8	1.62.9	1.87.0	2.11.2	2.35.3	9
10	.20.1	.44.2	.68.3	.92.5	1.16.6	1.40.8	1.64.9	1.89.0	2.13.2	2.37.3	10
11	.22.1	.46.2	.70.4	.94.5	1.18.6	1.42.8	1.66.9	1.91.1	2.15.2	2.39.3	11

$4.82 11/100 TO THE POUND.

8 5/8 %

PREMIUM.

STERLING INTO DOLLARS AND CENTS, AND VICE-VERSA.

£	0	100	200	300	400	500	600	700	800	900	£
£	$ cts.	$ cts.	$ cts.	$ cts.	$ cts.	$ cts.	$ cts.	$ cts.	$ cts.	$ cts.	£
50	241.38.8	724.16.6	1206 91.4	1689.72.2	2172.50.0	2655.27.7	3138.05.5	3620.83.3	4103.61.1	4586.38.8	50
51	246.21.6	728.99.4	1211.77.2	1694.55.0	2177.32.7	2660.10.5	3142.88.3	3625.66.1	4108.43.8	4591.21.6	51
52	251.04.4	733.82.2	1216.60.0	1699.37.7	2182.15.5	2664.93.3	3147.71.1	3630.48.8	4113.26.6	4596.04.4	52
53	255.87.2	738.65.0	1221.42.7	1704 20.5	2186.98.3	2669.76.1	3152 53.8	3635.31.6	4118 09.4	4600.87.2	53
54	260.70.0	743.47.7	1226.25.5	1709.03.3	2191.81.1	2674.58.8	3157.36.6	3640.14 4	4122.92.2	4605.70.0	54
55	265.52.7	748.30.5	1251.08.3	1713.86.1	2196 63.8	2679.41.6	3162.19 4	3644.97.2	4127.75 0	4610.52.7	55
56	270.35.5	753.13.3	1235.91.1	1718.68.8	2201.46 6	2684.24.4	3167.02.2	3649.80.0	4132.57.7	4615.35.5	56
57	275.18.3	757 96.1	1240.73.8	1723.51 6	2206.29.4	2689.07.2	3171.85 0	3654.62.7	4137.40.5	4620.18.3	57
58	280.01.1	762 78.8	1245.56.6	1728 34.1	2211.12.2	2693.90.0	3176 67 7	3659.45.5	4142.23.3	4625.01.1	58
59	284.83.*	767.61.6	1250.39.4	1733.17.2	2215.95.0	2698.72.7	3181.50.5	3664.28.3	4147.06.1	4629.83.8	59
60	289.66.6	772.44.4	1255.22.2	1738.00.0	2220.77.7	2703.55.5	3186.33.3	3669.11.1	4151 88.8	4634 66.6	60
61	294.49.4	777.27.2	1260.05.0	1742.82.7	2225.60.5	2708.38.3	3191.16.1	3673.93.8	4156 71.6	4639.49.4	61
62	299.32.2	782.10.0	1264.87.7	1747.65 5	2230 43 3	2713.21.1	3195.98.8	3678.76.6	4161.54.4	4644.32.2	62
63	304.15.0	786.92.7	1269.70.5	1752.48.3	2235.26.1	2718.03.8	3200.81.6	3683.59 4	4166.37.2	4649.15.0	63
64	308.97.7	791.75.5	1274.53 3	1757.31.1	2240.08.8	2722.86.6	3205.64.4	3688.42.2	4171.20.0	4653.97.7	64
65	313.80.5	796 58 3	1279.36.1	1762.13.8	2244 91.6	2727.69.4	3210.47.2	3693.25.0	4176.02 7	4658.80.5	65
66	318.63.3	801.41.1	1284.18.8	1766.96.6	2249 74.4	2732.52.2	3215.30.0	3698.07.7	4180.85.5	4663.63.3	66
67	323.46.1	806.23.8	1289.01.6	1771.79.4	2254.57.2	2737 35.0	3220.12.7	3702.90.5	4185.68.3	4668.46.1	67
68	328.28.8	811.06.6	1293.84.4	1776.62.2	2259 40.0	2742 17.7	3224.95.5	3707.73.3	4190 51.1	4673.28.8	68
69	333.11.6	815.89.4	1298.67.2	1781.45.0	2264.22.7	2747.00.5	3229.78.3	3712.56.1	4195.33.8	4678.11.6	69
70	337.94.4	820.72.2	1303.50.0	1786.27.7	2269.05 5	2751.83.3	3234.61.1	3717.38.8	4200.16.6	4682.94.4	70
71	342.77.2	825.55.0	1308.32.7	1791.10.5	2273.88 3	2756.66.1	3239.43.8	3722.21.6	4204.99.4	4687.77.2	71
72	347.60.0	830.37.7	1313.15.5	1795.93.3	2278.71.1	2761.48.8	3244.26.6	3727.04 4	4209.82.2	4692.60.0	72
73	352.42.7	835.20.5	1317.98.3	1800.76.1	2283.53.8	2766.31.6	3249.09.4	3731.87.2	4214.65.0	4697.42.7	73
74	357.25.5	840.03.3	1322.81.1	1805.58.8	2288.36.6	2771.14.4	3253.92.2	3736.70.0	4219 47.7	4702.25.5	74
75	362.08.3	844.86.1	1327.63.8	1810.41.6	2293 19.4	2775.97.2	3258.75.0	3741.52.7	4224.30.5	4707.08.3	75
76	366.91.1	849.68.8	1332.46.6	1815 24.4	2298.02.2	2780.80.0	3263.57.7	3746.35.5	4229.13.3	4711.91.1	76
77	371.73.8	854.51.6	1337.29.4	1820.07.2	2302.85.0	2785.62.7	3268.40.5	3751 18.3	4233.96.1	4716.73.8	77
78	376.56.6	859.34.4	1342.12.2	1824.90 0	2307.67.7	2790.45.5	3273 23.3	3756.01.1	4238.78.8	4721.56 6	78
79	381.39.4	864.17.2	1346.95.0	1829.72.7	2312.50.5	2795.28.3	3278.06.1	3760.83.8	4243.61.6	4726.39.4	79
80	386.22.2	869.00.0	1351.77.7	1834.55 5	2317.33.3	2800.11.1	3282.88.8	3765.66.6	4248.44 4	4731.22.2	80
81	391.05 0	873.82.7	1356.60.5	1839.38.3	2322.16.1	2804.93.8	3287.71.6	3770.49.4	4253.27.2	4736 05.0	81
82	395.87.7	878.65.5	1361.43.3	1844.21.1	2326 98.8	2809.76.6	3292.54.4	3775.32.2	4258.10.0	4740.87 7	82
83	400.70.5	883.48.3	1366.26.1	1849.03.8	2331.81.6	2814.59.4	3297.37.2	3780.15.0	4262.92.7	4745 70.5	83
84	405.53.3	888.31.1	1371.08.8	1853.86.6	2336.64.4	2819.42.2	3302.20 0	3784.97.7	4267.75.5	4750.53.3	84
85	410.36.1	893.13.8	1375.91.6	1858.69 4	2341.47.2	2824.25.0	3307.02.7	3789.80.5	4272.58.3	4755 36.1	85
86	415.18.8	897.96.6	1380.74.4	1863.52.2	2346.30.0	2829.07.7	3311 85.5	3794.63 3	4277.41.1	4760 18.8	86
87	420.01.6	902.79.4	1385.57.2	1868.35.0	2351.12.7	2833.90.5	3316.68.3	3799.46.1	4282.23.8	4765.01.6	87
88	424.84 4	907 62.2	1390.40.0	1873.17.7	2355.95 5	2838.73.3	3321.51.1	3804.28.8	4287.06.6	4769 84.4	88
89	429.67.2	912.45.0	1395.22.7	1878.00.5	2360.78 3	2843.56.1	3326.33.8	3809.11.6	4291.89.4	4774.67.2	89
90	434.50.0	917.27.7	1400.05.5	1882.83.3	2365.61.1	2848.38.8	3331.16.6	3813.94.4	4296.72.2	4779.50 0	90
91	439.32.7	922.10.5	1404.88.3	1887.66.1	2370.43.8	2853.21.6	3335.99.4	3818.77.2	4301.55.0	4784.32 7	91
92	444.15.5	926.93.3	1409.71.1	1892.48.8	2375.26.6	2858.04.4	3340.82.2	3823.60.0	4306.37.7	4789.15 5	92
93	448.98.3	931.76.1	1414.53.8	1897.31.6	2380.09 4	2862.87.2	3345.65.0	3828.42.7	4311.20.5	4793.98 3	93
94	453.81.1	936.58.8	1419.36.6	1902.14.4	2384.92.2	2867.70.0	3350.47.7	3833.25.5	4316.03.3	4798 81.1	94
95	458.63.8	941.41.6	1424.19.4	1906.97.2	2389.75.0	2872.52.7	3355.30.5	3838.08.3	4320.86.1	4803.63.8	95
96	463.46.6	946.24.4	1429.02.2	1911.80 0	2394.57.7	2877.35.5	3360.13.3	3842.91.1	4325.68.8	4808.46.6	96
97	468.29.4	951.07.2	1433.85.0	1916.62.7	2399.40.5	2882.18.3	3364.96.1	3847.73.8	4330.51.6	4813.29.4	97
98	473.12.2	955 90.0	1438.67.7	1921.45.5	2404.23.3	2887.01.1	3369.78.8	3852.56.6	4335.34.4	4818.12.2	98
99	477.95.0	960.72.7	1443.50.5	1926.28.3	2409.06.1	2891.83.8	3374.61.6	3857.39.4	4340.17 2	4822.95.0	99

S.	10	11	12	13	14	15	16	17	18	19	S.
d.	$ cts.	$ cts.	$ cts.	$ cts.	$ cts.	$ cts.	$ cts.	$ cts.	$ cts.	$ cts.	d.
0	2.41.3	2.65.5	2.89.6	3.13.8	3.37.9	3.62.0	3.86.2	4.10.3	4.34.5	4.58.6	0
1	2.43.4	2.67.5	2.91.6	3.15.8	3.39.9	3.64.1	3.88.2	4.12.3	4.36.5	4.60.6	1
2	2.45.4	2.69.5	2.93.6	3.17.8	3.41.9	3.66.1	3.90.2	4.14.3	4.38.5	4.62.6	2
3	2.47.4	2.71.5	2.95.7	3.19.8	3.43.9	3.68.1	3.92.2	4.16.4	4.40.5	4.64.6	3
4	2.49.4	2.73.5	2.97.7	3.21.8	3.45.9	3.70.1	3.94.2	4.18.4	4.42.5	4.66 6	4
5	2.51.4	2.75.5	2.99.7	3.23.8	3.48.0	3.72.1	3.96.2	4.20.4	4.44.5	4.68.6	5
6	2.53.4	2.77.5	3.01.7	3.25.8	3.50.0	3.74.1	3.98.2	4.22.4	4.46 5	4.70.7	6
7	2.55.4	2.79.6	3 03.7	3.27.8	3.52.0	3.76.1	4.00.3	4.24.4	4.48.5	4.72.7	7
8	2.57.4	2.81.6	3.05.7	3.29.8	3.54.0	3.78.1	4.02.3	4.26.4	4.50.5	4.74.7	8
9	2.59.4	2.83.6	3.07.7	3.31.9	3.56.0	3.80.1	4.04.3	4.28.4	4.52.6	4.76.7	9
10	2.61.5	2.85.6	3.09.7	3.33.9	3.58.0	3.82.2	4.06.3	4.30.4	4.54 6	4.78.7	10
11	2.63.5	2.87.6	3.11.7	3.35.9	3.60.0	3.84.2	4.08.3	4.32.4	4.56.6	4.80.7	11

$4.83 1⁄100 TO THE POUND.

8¾ %

PREMIUM.

STERLING INTO DOLLARS AND CENTS, AND VICE-VERSA.

£	0	100	200	300	400	500	600	700	800	900	£
	$ cts.	$ cts.	$ cts.	$ cts.	$ cts.	$ cts.	$ cts.	$ cts.	$ cts.	$ cts.	
0	......	483.33 3	966.66.6	1450.00.0	1933.33.3	2416.66.6	2900.00.0	3383.33.3	3866.66.6	4350.00.0	0
1	4.83.3	488.16.6	971.50.0	1454.83.3	1938.16.6	2421.50.0	2904.83.3	3388.16.6	3871.50.0	4354.83.3	1
2	9.66.6	493.00.0	976.33.3	1459.66.6	1943.00.0	2426.33.3	2909.66.6	3393.00.0	3876.33.3	4359.66.6	2
3	14.50.0	497.83.3	981.16.6	1464.50.0	1947.83.3	2431.16.6	2914.50.0	3397.83.3	3881.16.6	4364.50.0	3
4	19.33.3	502.66.6	986.00.0	1469.33.3	1952.66.6	2436.00.0	2919.33.3	3402.66.6	3886.00.0	4369.33.3	4
5	24.16.6	507.50.0	990.83.3	1474.16.6	1957.50.0	2440.83.3	2924.16.6	3407.50.0	3890.83.3	4374.16.6	5
6	29.00.0	512.33.3	995.66.6	1479.00.0	1962.33.3	2445.66.6	2929.00.0	3412.33.3	3895.66.6	4379.00.0	6
7	33.83.3	517.16.6	1000.50.0	1483.83.3	1967.16.6	2450.50.0	2933.83.3	3417.16.6	3900.50.0	4383.83.3	7
8	38.66.6	522.00.0	1005.33.3	1488.66.6	1972.00.0	2455.33.3	2938.66.6	3422.00.0	3905.33.3	4388.66.6	8
9	43.50.0	526.83.3	1010.16.6	1493.50.0	1976.83.3	2460.16.6	2943.50.0	3426.83.3	3910.16.6	4393.50.0	9
10	48.33.3	531.66 6	1015.00.0	1498.33.3	1981.66.6	2465.00.0	2948.33.3	3431.66.6	3915.00.0	4398.33.3	10
11	53.16.6	536.50.0	1019.83.3	1503.16.6	1986.50.0	2469.83.3	2953.16.6	3436.50.0	3919.83.3	4403.16.6	11
12	58.00.0	541.33.3	1024.66.6	1508.00.0	1991.33.3	2474.66.6	2958.00.0	3441.33.3	3924.66.6	4408.00.0	12
13	62.83.3	546.16.6	1029.50.0	1512.83.3	1996.16.6	2479.50.0	2962.83.3	3446.16.6	3929.50.0	4412.83.3	13
14	67.66.6	551.00.0	1034.33.3	1517.66.6	2001.00.0	2484.33.3	2967.66.6	3451.00.0	3934.33.3	4417.66.6	14
15	72.50.0	555.83.3	1039.16.6	1522.50.0	2005.83.3	2489.16.6	2972.50.0	3455.83.3	3939.16.6	4122.50.0	15
16	77.33.3	560.66.6	1044.00.0	1527.33.3	2010.66.6	2494.00.0	2977.33.3	3460.66.6	3944.00.0	4427.33.3	16
17	82.16.6	565.50.0	1048.83.3	1532.16.6	2015.50.0	2498.83.3	2982.16.6	3465.50.0	3948.83.3	4432.16 6	17
18	87.00.0	570.33.3	1053.66.6	1537.00.0	2020.33.3	2503.66.6	2987.00.0	3470.33.3	3953.66.6	4437.00.0	18
19	91.83.3	575.16.6	1058.50.0	1541.83.3	2025.16.6	2508.50.0	2991.83.3	3475.16.6	3958.50.0	4441.83.3	19
20	96.66.6	580.00.0	1063.33.3	1546.66.6	2030.00.0	2513.33.3	2996.66.6	3480.00.0	3963.33.3	4446.66.6	20
21	101.50.0	584.83.3	1068.16.6	1551.50.0	2034.83.3	2518.16.6	3001.50.0	3484.83.3	3968.16.6	4451.50.0	21
22	106.33.3	589.66.6	1073.00.0	1556.33.3	2039.66.6	2523.00.0	3006.33.3	3489.66.6	3973.00.0	4456.33.3	22
23	111.16.6	594.50.0	1077.83.3	1561.16.6	2044.50.0	2527.83.3	3011.16.6	3494.50.0	3977.83.3	4461.16.6	23
24	116.00.0	599.33.3	1082.66.6	1566.00.0	2049.33.3	2532.66.6	3016.00.0	3499.33.3	3982.66.6	4466.00.0	24
25	120.83.3	604.16.6	1087.50.0	1570.83.3	2054.16.6	2537.50.0	3020.83.3	3504.16.6	3987.50.0	4470.83.3	25
26	125.66.6	609.00.0	1092.33.3	1575.66.6	2059.00.0	2542.33.3	3025.66.6	3509.00.0	3992.33.3	4475.66.6	26
27	130.50.0	613.83.3	1097.16.6	1580.50.0	2063.83.3	2547.16.6	3030.50.0	3513.83.3	3997.16.6	4480.50.0	27
28	135.33.3	618.66.6	1102.00.0	1585.33.3	2068.66 6	2552.00.0	3035.33.3	3518.66.6	4002.00.0	4485.33.3	28
29	140.16.6	623.50.0	1106.83.3	1590.16.6	2073.50.0	2556.83.3	3040.16.6	3523.50.0	4006.83.3	4490.16.6	29
30	145.00.0	628.33.3	1111.66.6	1595.00.0	2078.33.3	2561.66.6	3045.00.0	3528.33.3	4011.66.6	4495.00.0	30
31	149.83.3	633.16.6	1116.50.0	1599.83.3	2083.16.6	2566.50.0	3049.83.3	3533.16.6	4016.50.0	4499.83.3	31
32	154.66.6	638.00.0	1121.33.3	1604.66.6	2088.00 0	2571.33.3	3054.66.6	3538.00.0	4021.33.3	4504.66.6	32
33	159.50.0	642.83.3	1126.16.6	1609.50.0	2092.83.3	2576.16.6	3059.50.0	3542.83.3	4026.16 6	4509.50.0	33
34	164.33.3	647.66.6	1131.00.0	1614.33.3	2097.66.6	2581.00.0	3064.33.3	3547.66.6	4031.00.0	4514.33.3	34
35	169.16.6	652.50.0	1135.83.3	1619.16.6	2102.50.0	2585.83.3	3069.16.6	3552.50.0	4035.83.3	4519.16.6	35
36	174.00.0	657.33.3	1140.66.6	1624 00.0	2107.33.3	2590.66.6	3074.00.0	3557.33.3	4040.66.6	4524.00.0	36
37	178.83.3	662.16.6	1145.50.0	1628.83.3	2112.16.6	2595.50.0	3078.83 3	3562.16.6	4045.50.0	4528.83.3	37
38	183.66.6	667.00.0	1150.33.3	1633.66.6	2117.00.0	2600.33.3	3083.66.6	3567.00.0	4050.33.3	4533.66.6	38
39	188.50.0	671.83.3	1155.16.6	1638.50.0	2121.83.3	2605.16.6	3088.50.0	3571.83.3	4055.16.6	4538.50.0	39
40	193.33.3	676.66.6	1160.00.0	1643.33.3	2126.66.6	2610.00.0	3093.33.3	3576.66.6	4060.00.0	4543.33.3	40
41	198.16.6	681.50.0	1164.83.3	1648.16.6	2131.50.0	2614.83.3	3098.16 6	3581.50.0	4064.83.3	4548.16.6	41
42	203 00.0	686.33.3	1169.66.6	1653.00.0	2136.33.3	2619.66.6	3103.00.0	3586.33.3	4069.66.6	4553.00.0	42
43	207.83.3	691.16.6	1174.50.0	1657 83.3	2141.16.6	2624.50.0	3107.83.3	3591.16.6	4074.50.0	4557.83.3	43
44	212.66.6	696.00.0	1179.33.3	1662.66.6	2146.00.0	2629.33.3	3112.66.6	3596.00.0	4079.33.3	4562.66.6	44
45	217.50.0	700.83.3	1184.16 6	1667.50.0	2150.83.3	2634.16.6	3117.50.0	3600.83.3	4084.16.6	4567.50.0	45
46	222.33.3	705.66.6	1189.00.0	1672 33.3	2155.66.6	2639.00.0	3122.33.3	3605.66.6	4089.00.0	4572.33.3	46
47	227.16.6	710.50.0	1193.83.3	1677.16.6	2160.50.0	2643.83.3	3127.16.6	3610.50.0	4093.83.3	4577.16.6	47
48	232.00.0	715.33.3	1198.66.6	1682.00.0	2165.33.3	2648.66.6	3132.00.0	3615.33.3	4098.66.6	4582.00.0	48
49	236.83.3	720.16.6	1203.50.0	1686.83.3	2170.16.6	2653.50.0	3136.83.3	3620.16.6	4103.50.0	4586.83.3	49

s.	0	1	2	3	4	5	6	7	8	9	s.
d.	$ cts.	$ cts.	$ cts.	$ cts.	$ cts.	$ cts.	$ cts.	$ cts.	$ cts.	$ cts.	d.
0	......	.24.1	.48.3	.72.5	.96.6	1.20.8	1.45.0	1.69.1	1.93.3	2.17.5	0
1	.02.0	.26.1	.50.3	.74.5	.98.6	1.22.8	1.47.0	1.71.1	1.95.3	2.19.5	1
2	.04.0	.28.1	.52.3	.76.5	1.00.6	1.24.8	1.49.0	1.73.1	1.97.3	2.21.5	2
3	.06.0	.30.2	.54.3	.78.5	1.02.7	1.26.8	1.51.0	1.75.2	1.99.3	2.23.5	3
4	.08.0	.32.2	.56.3	.80.5	1.04.7	1.28.8	1.53.0	1.77.2	2.01.3	2.25.5	4
5	.10.0	.34.2	.58.3	.82.5	1.06.7	1.30.8	1.55.0	1.79.2	2.03.3	2.27.5	5
6	.12.0	.36.2	.60.4	.84.5	1.08.7	1.32.9	1.57.0	1.81.2	2.05.5	2.29.5	6
7	.14.0	.38.2	.62.4	.86.5	1.10.7	1.34.9	1.59.0	1.83.2	2.07.5	2.31.5	7
8	.16.0	.40.2	.64.4	.88.5	1.12.7	1.36.9	1.61.0	1.85.2	2.09.5	2.33.5	8
9	.18.1	.42.2	.66.4	.90.6	1.14.7	1.38.9	1.63.1	1.87.2	2.11.5	2.35.6	9
10	.20.1	.44.2	.68.4	.92.6	1.16.7	1.40.9	1.65.1	1.89.2	2.13.5	2.37.6	10
11	.22.1	.46.2	.70.4	.94.6	1.18.7	1.42.9	1.67.1	1.91.2	2.15.5	2.39.6	11

$4.83¾ TO THE POUND.

8¾%

PREMIUM.

STERLING INTO DOLLARS AND CENTS, AND VICE-VERSA.

£	0	100	200	300	400	500	600	700	800	900	£
	$ cts.	$ cts.	$ cts.	$ cts.	$ cts.	$ cts.	$ cts.	$ cts.	$ cts.	$ cts.	
50	241.66.6	725.00.0	1208.33.3	1691.66.6	2175.00.0	2658.33.3	3141.66.6	3625.00.0	4108.33.3	4591.66.6	50
51	246.50.0	729.83.3	1213.16.6	1696.50.0	2179.83.3	2663.16.6	3146.50.0	3629.83.3	4113.16.6	4596.50.0	51
52	251.33.3	734.66.6	1218.00.0	1701.33.3	2184.66.6	2668.00.0	3151.33.3	3634.66.6	4118.00.0	4601.33.3	52
53	256.16.6	739.50.0	1222.83.3	1706.16.6	2189.50.0	2672.83.3	3156.16.6	3639.50.0	4122.83.3	4606.16.6	53
54	261.00.0	744.33.3	1227.66.6	1711.00.0	2194.33.3	2677.66.6	3161.00.0	3644.33.3	4127.66.6	4611.00.0	54
55	265.83.3	749.16.6	1232.50.0	1715.83.3	2199.16.6	2682.50.0	3165.83.3	3649.16.6	4132.50.0	4615.83.3	55
56	270.66.6	754.00.0	1237.33.3	1720.66.6	2204.00.0	2687.33.3	3170.66.6	3654.00.0	4137.33.3	4620.66.6	56
57	275.50.0	758.83.3	1242.16.6	1725.50.0	2208.83.3	2692.16.6	3175.50.0	3658.83.3	4142.16.6	4625.50.0	57
58	280.33.3	763.66.6	1247.00.0	1730.33.3	2213.66.6	2697.00.0	3180.33.3	3663.66.6	4147.00.0	4630.33.3	58
59	285.16.6	768.50.0	1251.83.3	1735.16.6	2218.50.0	2701.83.3	3185.16.6	3668.50.0	4151.83.3	4635.16.6	59
60	290.00.0	773.33.3	1256.66.6	1740.00.0	2223.33.3	2706.66.6	3190.00.0	3673.33.3	4156.66.6	4640.00.0	60
61	294.83.3	778.16.6	1261.50.0	1744.83.3	2228.16.6	2711.50.0	3194.83.3	3678.16.6	4161.50.0	4644.83.3	61
62	299.66.6	783.00.0	1266.33.3	1749.66.6	2233.00.0	2716.33.3	3199.66.6	3683.00.0	4166.33.3	4649.66.6	62
63	304.50.0	787.83.3	1271.16.6	1754.50.0	2237.83.3	2721.16.6	3204.50.0	3687.83.3	4171.16.6	4654.50.0	63
64	309.33.3	792.66.6	1276.00.0	1759.33.3	2242.66.6	2726.00.0	3209.33.3	3692.66.6	4176.00.0	4659.33.3	64
65	314.16.6	797.50.0	1280.83.3	1764.16.6	2247.50.0	2730.83.3	3214.16.6	3697.50.0	4180.83.3	4664.16.6	65
66	319.00.0	802.33.3	1285.66.6	1769.00.0	2252.33.3	2735.66.6	3219.00.0	3702.33.3	4185.66.6	4669.00.0	66
67	323.83.3	807.16.6	1290.50.0	1773.83.3	2257.16.6	2740.50.0	3223.83.3	3707.16.6	4190.50.0	4673.83.3	67
68	328.66.6	812.00.0	1295.33.3	1778.66.6	2262.00.0	2745.33.3	3228.66.6	3712.00.0	4195.33.3	4678.66.6	68
69	333.50.0	816.83.3	1300.16.6	1783.50.0	2266.83.3	2750.16.6	3233.50.0	3716.83.3	4200.16.6	4683.50.0	69
70	338.33.3	821.66.6	1305.00.0	1788.33.3	2271.66.6	2755.00.0	3238.33.3	3721.66.6	4205.00.0	4688.33.3	70
71	343.16.6	826.50.0	1309.83.3	1793.16.6	2276.50.0	2759.83.3	3243.16.6	3726.50.0	4209.83.3	4693.16.6	71
72	348.00.0	831.33.3	1314.66.6	1798.00.0	2281.33.3	2764.66.6	3248.00.0	3731.33.3	4214.66.6	4698.00.0	72
73	352.83.3	836.16.6	1319.50.0	1802.83.3	2286.16.6	2769.50.0	3252.83.3	3736.16.6	4219.50.0	4702.83.3	73
74	357.66.6	841.00.0	1324.33.3	1807.66.6	2291.00.0	2774.33.3	3257.66.6	3741.00.0	4224.33.3	4707.66.6	74
75	362.50.0	845.83.3	1329.16.6	1812.50.0	2295.83.3	2779.16.6	3262.50.0	3745.83.3	4229.16.6	4712.50.0	75
76	367.33.3	850.66.6	1334.00.0	1817.33.3	2300.66.6	2784.00.0	3267.33.3	3750.66.6	4234.00.0	4717.33.3	76
77	372.16.6	855.50.0	1338.83.3	1822.16.6	2305.50.0	2788.83.3	3272.16.6	3755.50.0	4238.83.3	4722.16.6	77
78	377.00.0	860.33.3	1343.66.6	1827.00.0	2310.33.3	2793.66.6	3277.00.0	3760.33.3	4243.66.6	4727.00.0	78
79	381.83.3	865.16.6	1348.50.0	1831.83.3	2315.16.6	2798.50.0	3281.83.3	3765.16.6	4248.50.0	4731.83.3	79
80	386.66.6	870.00.0	1353.33.3	1836.66.6	2320.00.0	2803.33.3	3286.66.6	3770.00.0	4253.33.3	4736.66.6	80
81	391.50.0	874.83.3	1358.16.6	1841.50.0	2324.83.3	2808.16.6	3291.50.0	3774.83.3	4258.16.6	4741.50.0	81
82	396.33.3	879.66.6	1363.00.0	1846.33.3	2329.66.6	2813.00.0	3296.33.3	3779.66.6	4263.00.0	4746.33.3	82
83	401.16.6	884.50.0	1367.83.3	1851.16.6	2334.50.0	2817.83.3	3301.16.6	3784.50.0	4267.83.3	4751.16.6	83
84	406.00.0	889.33.3	1372.66.6	1856.00.0	2339.33.3	2822.66.6	3306.00.0	3789.33.3	4272.66.6	4756.00.0	84
85	410.83.3	894.16.6	1377.50.0	1860.83.3	2344.16.6	2827.50.0	3310.83.3	3794.16.6	4277.50.0	4760.83.3	85
86	415.66.6	899.00.0	1382.33.3	1865.66.6	2349.00.0	2832.33.3	3315.66.6	3799.00.0	4282.33.3	4765.66.6	86
87	420.50.0	903.83.3	1387.16.6	1870.50.0	2353.83.3	2837.16.6	3320.50.0	3803.83.3	4287.16.6	4770.50.0	87
88	425.33.3	908.66.6	1392.00.0	1875.33.3	2358.66.6	2842.00.0	3325.33.3	3808.66.6	4292.00.0	4775.33.3	88
89	430.16.6	913.50.0	1396.83.3	1880.16.6	2363.50.0	2846.83.3	3330.16.6	3813.50.0	4296.83.3	4780.16.6	89
90	435.00.0	918.33.3	1401.66.6	1885.00.0	2368.33.3	2851.66.6	3335.00.0	3818.33.3	4301.66.6	4785.00.0	90
91	439.83.3	923.16.6	1406.50.0	1889.83.3	2373.16.6	2856.50.0	3339.83.3	3823.16.6	4306.50.0	4789.83.3	91
92	444.66.6	928.00.0	1411.33.3	1894.66.6	2378.00.0	2861.33.3	3344.66.6	3828.00.0	4311.33.3	4794.66.6	92
93	449.50.0	932.83.3	1416.16.6	1899.50.0	2382.83.3	2866.16.6	3349.50.0	3832.83.3	4316.16.6	4799.50.0	93
94	454.33.3	937.66.6	1421.00.0	1904.33.3	2387.66.6	2871.00.0	3354.33.3	3837.66.6	4321.00.0	4804.33.3	94
95	459.16.6	942.50.0	1425.83.3	1909.16.6	2392.50.0	2875.83.3	3359.16.6	3842.50.0	4325.83.3	4809.16.6	95
96	464.00.0	947.33.3	1430.66.6	1914.00.0	2397.33.3	2880.66.6	3364.00.0	3847.33.3	4330.66.6	4814.00.0	96
97	468.83.3	952.16.6	1435.50.0	1918.83.3	2402.16.6	2885.50.0	3368.83.3	3852.16.6	4335.50.0	4818.83.3	97
98	473.66.6	957.00.0	1440.33.3	1923.66.6	2407.00.0	2890.33.3	3373.66.6	3857.00.0	4340.33.3	4823.66.6	98
99	478.50.0	961.83.3	1445.16.6	1928.50.0	2411.83.3	2895.16.6	3378.50.0	3861.83.3	4345.16.6	4828.50.0	99

s.	10	11	12	13	14	15	16	17	18	19	s.
d.	$ cts.	$ cts.	$ cts.	$ cts.	$ cts.	$ cts.	$ cts.	$ cts.	$ cts.	$ cts.	d.
0	2.41.6	2.65.8	2.90.0	3.14.1	3.38.3	3.62.5	3.86.6	4.10.8	4.35.0	4.59.1	0
1	2.43.6	2.67.8	2.92.0	3.16.1	3.40.3	3.64.5	3.88.6	4.12.8	4.37.0	4.61.1	1
2	2.45.6	2.69.8	2.94.0	3.18.1	3.42.3	3.66.5	3.90.6	4.1..8	4.39.0	4.63.1	2
3	2.47.7	2.71.8	2.96.0	3.20.2	3.44.3	3.68.5	3.92.7	4.16.8	4.41.0	4.65.2	3
4	2.49.7	2.73.8	2.98.0	3.22.2	3.46.3	3.70.5	3.94.7	4.18.8	4.43.0	4.67.2	4
5	2.51.7	2.75.8	3.00.0	3.24.2	3.48.3	3.72.5	3.96.7	4.20.8	4.45.0	4.69.2	5
6	2.53.7	2.77.9	3.02.0	3.26.2	3.50.4	3.74.5	3.98.7	4.22.0	4.47.0	4.71.2	6
7	2.55.7	2.79.9	3.04.0	3.28.2	3.52.4	3.76.5	4.00.7	4.24.9	4.49.0	4.73.2	7
8	2.57.7	2.81.9	3.06.0	3.30.2	3.54.4	3.78.5	4.02.7	4.26.9	4.51.0	4.75.2	8
9	2.59.7	2.83.9	3.08.1	3.32.2	3.56.4	3.80.6	4.04.7	4.28.9	4.53.1	4.77.2	9
10	2.61.7	2.85.9	3.10.1	3.34.2	3.58.4	3.82.6	4.06.7	4.30.9	4.55.1	4.79.2	10
11	2.63.8	2.87.9	3.12.1	3.36.2	3.60.4	3.84.6	4.08.7	4.32.9	4.57.1	4.81.2	11

4.83\frac{58}{100}$ TO THE POUND.

8⅞%

PREMIUM.

STERLING INTO DOLLARS AND CENTS, AND VICE-VERSA.

£	0	100	200	300	400	500	600	700	800	900	£
	$ cts.	$ cts.	$ cts.	$ cts.	$ cts.	$ cts.	$ cts.	$ cts.	$ cts.	$ cts.	
0	......	483 88.8	967.77.7	1451.66.6	1935.55.5	2419.44.4	2903.33.3	3387.22.2	3871.11.1	4355.00.0	0
1	4.83.8	488.72.7	972.61.6	1456.50.5	1940.39.4	2424.28.3	2908.17.2	3392.06.1	3875.95.0	4359.83.8	1
2	9.67.7	493.56.6	977.45.5	1461.34.4	1945.23.3	2429.12.2	2913.01.1	3396.90.0	3880.78.8	4364.67.7	2
3	14.51.6	498.40.5	982.29.4	1466.18.3	1950.07.2	2433.96.1	2917.85.0	3401.73.8	3885.62.7	4369.51.6	3
4	19.35.5	503.24.4	987.13.3	1471.02.2	1954.91.1	2438.80.0	2922.68.8	3406.57.7	3890.46.6	4374.35.5	4
5	24.19.4	508.08.3	991.97.2	1475.86.1	1959.75.0	2443.63.8	2927.52.7	3411.41.6	3895.30.5	4379.19.4	5
6	29.03.3	512.92.2	996.81.1	1480.70.0	1964.58.8	2448.47.7	2932.36.6	3416.25.5	3900.14.4	4384.03.3	6
7	33.87.2	517.76.1	1001.65.0	1485.53.8	1969.42.7	2453.31.6	2937.20.5	3421.09.4	3904.98.3	4388.87.2	7
8	38.71.1	522.60.0	1006.48.8	1490.37.7	1974.26.6	2458.15.5	2942.04.4	3425.93.3	3909.82.2	4393.71.1	8
9	43.55.0	527.43.8	1011.32.7	1495.21.6	1979.10.5	2462.99.4	2946.88.3	3430.77.2	3914.66.1	4398.55.0	9
10	48.38.8	532.27.7	1016.16.6	1500.05.5	1983 94.4	2467.83.3	2951.72.2	3435.61.1	3919.50.0	4403.38.8	10
11	53.22.7	537.11.6	1021.00.5	1504.89.4	1988.78.3	2472.67.2	2956.56.1	3440.45.0	3924.33.8	4408.22.7	11
12	58.06.6	541.95.5	1025.84.4	1509.73.3	1993.62.2	2477.51.1	2961.40.0	3445.28.8	3929.17.7	4413.06.6	12
13	62.90.5	546.79.4	1030.68.3	1514.57.2	1998.46.1	2482.35.0	2966.23.8	3450.12.7	3934.01.6	4417.90.5	13
14	67.74.4	551.63.3	1035.52.2	1519.41.1	2003.30.0	2487.18.8	2971.07.7	3454.96.6	3938.85.5	4422.74.4	14
15	72.58.3	556.47.2	1040.36.1	1524.25.0	2008.13.8	2492.02.7	2975.91.6	3459.80.5	3943.69.4	4427.58.3	15
16	77.42.2	561.31.1	1045.20.0	1529.08.8	2012.97.7	2496.86.6	2980.75.5	3464.64.4	3948.53.3	4432.42.2	16
17	82 26.1	566.15.0	1050.03.8	1533.92.7	2017.81.6	2501.70.5	2985.59.4	3469.48.3	3953.37.2	4437.26.1	17
18	87.10.0	570.98 8	1054.87.7	1538.76.6	2022.65.5	2506.54.4	2990.43.3	3474.32.2	3958.21.1	4442.10.0	18
19	91.93.8	575.82.7	1059.71.6	1543.60.5	2027.49.4	2511.38.3	2995.27.2	3479.16.1	3963.05.0	4446.93.8	19
20	96.77.7	580 66.6	1064.55.5	1548.44.4	2032 33.3	2516.22.2	3000.11.1	3484.00.0	3967.88.8	4451.77.7	20
21	101.61.6	585.50 5	1069.39.4	1553.28.3	2037.17.2	2521.06.1	3004.95.0	3488.83.8	3972.72.7	4456.61.6	21
22	106.45.5	590.34 4	1074.23.3	1558.12.2	2042.01.1	2525.90.0	3009.78.8	3493.67.7	3977.56 6	4461.45.5	22
23	111.29.4	595.18.3	1079.07.2	1562.96.1	2046.85.0	2530.73.8	3014.62.7	3498.51.6	3982.40.5	4466.29.4	23
24	116.13.3	600.02.2	1083.91.1	1567.80.0	2051.68.8	2535.57.7	3019.46.6	3503.35.5	3987.24.4	4471.13.3	24
25	120.97.2	604.86.1	1088.75.0	1572.63.8	2056.52.7	2540.41.6	3024.30.5	3508.19.4	3992.08.3	4475.97.2	25
26	125.81.1	609.70.0	1093.58.8	1577.47.7	2061.36.6	2545.25.5	3029.14.4	3513.03.3	3996.92.2	4480.81.1	26
27	130.65.0	614.53.8	1098.42.7	1582.31.6	2066.20.5	2550.09.4	3033.98.3	3517.87.2	4001.76.1	4485.65.0	27
28	135.48.8	619.37.7	1103.26.6	1587.15.5	2071.04.4	2554.93.3	3038.82.2	3522.71.1	4006.60.0	4490.48.8	28
29	140.32.7	624.21.6	1108.10.5	1591.99.4	2075.88.3	2559.77.2	3043.66.1	3527.55.0	4011.43.8	4495.32.7	29
30	145.16.6	629.05.5	1112.94.4	1596.83.3	2080.72.2	2564.61.1	3048.50.0	3532.38.8	4016.27.7	4500.16.6	30
31	150.00.5	633.89.4	1117.78.3	1601.67.2	2085.56.1	2569.45.0	3053.33.8	3537.22.7	4021.11.6	4505.00.5	31
32	154.84.4	638.73.3	1122.62.2	1606.51.1	2090.40.0	2574.28.8	3058.17.7	3542.06.6	4025.95.5	4509.84.4	32
33	159.68.3	643.57.2	1127.46.1	1611.35.0	2095.23.8	2579.12.7	3063.01.6	3546.90.5	4030.79.4	4514.68.3	33
34	164.52.2	648.41.1	1132.30.0	1616.18.8	2100.07.7	2583.96.6	3067.85.5	3551.74.4	4035.63.3	4519.52.2	34
35	169.36.1	653.25.0	1137.13.8	1621.02.7	2104.91.6	2588.80.5	3072.69.4	3556.58.3	4040 47.2	4524.36.1	35
36	174.20.0	658.08.8	1141.97.7	1625.86.6	2109.75.5	2593.64.4	3077.53.3	3561.42.2	4045.31.1	4529.20.0	36
37	179.03.8	662.92.7	1146.81.6	1630.70.5	2114.59.4	2598.48.3	3082.37.2	3566.26.1	4050.15.0	4534.03.8	37
38	183.87.7	667.76.6	1151.65.5	1635.54.4	2119.43.3	2603.32.2	3087.21.1	3571.10.0	4054 98.8	4538.87.7	38
39	188.71.6	672.60.5	1156.49.4	1640.38.3	2124.27.2	2608.16.1	3092.05.0	3575.93.8	4059.82.7	4543.71.6	39
40	193.55.5	677.44.4	1161.33.3	1645.22.2	2129.11.1	2613.00.0	3096.88.8	3580.77.7	4064.66.6	4548.55.5	40
41	198.39.4	682.28.3	1166.17.2	1650.06.1	2133.95.0	2617.83.8	3101.72.7	3585.61.6	4069.50.5	4553.39.4	41
42	203.23.3	687.12.2	1171.01.1	1654.90.0	2138.78.8	2622.67.7	3106.56.6	3590.45.5	4074.34.4	4558.23.3	42
43	208.07.2	691.96.1	1175.85.0	1659.73.8	2143.62.7	2627.51.6	3111.40.5	3595.29.4	4079.18.3	4563.07.2	43
44	212.91.1	696.80.0	1180.68.8	1664.57.7	2148.46.6	2632.35.5	3116.24.4	3600.13.3	4084.02.2	4567.91.1	44
45	217.75.0	701.63.8	1185.52.7	1669.41.6	2153.30 5	2637.19.4	3121.08.3	3604.97.2	4088.86.1	4572.75.0	45
46	222.58.8	706.47.7	1190.36.6	1674.25.5	2158.14.4	2642.03.3	3125.92.2	3609.81.1	4093.70.0	4577.58.8	46
47	227.42.7	711.31.6	1195.20.5	1679.09.4	2162.98.3	2646.87.2	3130.76.1	3614.65.0	4098.53.8	4582.42.7	47
48	232.26.6	716.15.5	1200.04.4	1683.93.3	2167.82.2	2651.71.1	3135.60.0	3619.48.8	4103.37.7	4587.26.6	48
49	237.10.5	720.99 4	1204.88.3	1688.77.2	2172.66.1	2656.55.0	3140.43.8	3624.32.7	4108.21.6	4592.10.5	49

s.	0	1	2	3	4	5	6	7	8	9	s.
d.	$ cts.	$ cts.	$ cts.	$ cts.	$ cts.	$ cts.	$ cts.	$ cts.	$ cts.	$ cts.	d.
0	......	.24.1	.48.3	.72.5	.96.7	1.20.9	1.45.1	1.69.3	1.93.5	2.17.7	0
1	.02.0	.26.3	.50.4	.74.6	.98.7	1.22.9	1.47.1	1.71.3	1.95.5	2.19.7	1
2	.04.0	.28.3	.52.4	.76.6	1.00.8	1.25.0	1.49.1	1.73.3	1.97.5	2.21.7	2
3	.06.0	.30.3	.54.4	.78.6	1.02.8	1.27 0	1.51.2	1.75.4	1.99.5	2.23.7	3
4	.08.0	.32.3	.56.4	.80.6	1.04.8	1.29.0	1.53.2	1.77.4	2.01.6	2.25.8	4
5	.10.0	.34.3	.58.4	.82.6	1.06.8	1.31.0	1.55.2	1.79.4	2.03.6	2.27.8	5
6	.12.0	.36.3	.60.4	.84.6	1.08.8	1.33.0	1.57.2	1.81.4	2.05.6	2.29.8	6
7	.14.0	.38.3	.62.4	.86.6	1.10.8	1.35.0	1.59.2	1.83.4	2.07.6	2.31.8	7
8	.16.0	.40.3	.64.4	.88.6	1.12.8	1.37.0	1.61.2	1.85.4	2.09.6	2.33.8	8
9	.18.1	.42.4	.66.4	.90.6	1.14.8	1.39.0	1.63.2	1.87.4	2.11.6	2.35.8	9
10	.20.1	.44.4	.68.5	.92.7	1.16.8	1.41.0	1.65.2	1.89.4	2.13.6	2.37.8	10
11	.22.1	.46.4	.70.5	.94.7	1.18.9	1.43.1	1.67.2	1.91.4	2.15.6	2.39.8	11

(39)

$4.83,⁷⁄₁₀₀ TO THE POUND.

8⅞ %

PREMIUM.

STERLING INTO DOLLARS AND CENTS, AND VICE-VERSA.

£	0	100	200	300	400	500	600	700	800	900	£
	$ cts.	$ cts.	$ cts.	$ cts.	$ cts.	$ cts.	$ cts.	$ cts.	$ cts.	$ cts.	
50	241.94.4	725.83.3	1209.72.2	1693.61.1	2177.50.0	2661.38.8	3145.27.7	3629.16.6	4113.05.5	4596.94.4	50
51	246.78.3	730.67.2	1214.56.1	1698.45.0	2182.33.8	2666.22.7	3150.11.6	3634.00.5	4117.89.4	4601.78.3	51
52	251.62.2	735.51.1	1219.40.0	1703.28.8	2187.17.7	2671.06.6	3154.95.5	3638.84.4	4122.73.3	4606.62.2	52
53	256.46.1	740.35.0	1224.23.8	1708.12.7	2192.01.6	2675.90.5	3159.79.4	3643.68.3	4127.57.2	4611.46.1	53
54	261.30.0	745.18.8	1229.07.7	1712.96.6	2196.85.5	2680.74.4	3164.63.3	3648.52.2	4132.41.1	4616.30.0	54
55	266.13.8	750.02.7	1233.91.6	1717.80.5	2201.69.4	2685.58.3	3169.47.2	3653.36.1	4137.25.0	4621.13.8	55
56	270.97.7	754.86.6	1238.75.5	1722.64.4	2206.53.3	2690.42.2	3174.31.1	3658.20.0	4142.08.8	4625.97.7	56
57	275.81.6	759.70.5	1243.59.4	1727.48.3	2211.37.2	2695.26.1	3179.15.0	3663.03.8	4146.92.7	4630.81.6	57
58	280.65.5	764.54.4	1248.43.3	1732.32.2	2216.21.1	2700.10.0	3183.98.8	3667.87.7	4151.76.6	4635.65.5	58
59	285.49.4	769.38.3	1253.27.2	1737.16.1	2221.05.0	2704.93.8	3188.82.7	3672.71.6	4156.60.5	4640.49.4	59
60	290.33.3	774.22.2	1258.11.1	1742.00.0	2225.88.8	2709.77.7	3193.66.6	3677.55.5	4161.44.4	4645.33.3	60
61	295.17.2	779.06.1	1262.95.0	1746.83.8	2230.72.7	2714.61.6	3198.50.5	3682.39.4	4166.28.3	4650.17.2	61
62	300.01.1	783.90.0	1267.78.8	1751.67.7	2235.56.6	2719.45.5	3203.34.4	3687.23.3	4171.12.2	4655.01.1	62
63	304.85.0	788.73.8	1272.62.7	1756.51.6	2240.40.5	2724.29.4	3208.18.3	3692.07.2	4175.96.1	4659.85.0	63
64	309.68.8	793.57.7	1277.46.6	1761.35.5	2245.24.4	2729.13.3	3213.02.2	3696.91.1	4180.80.0	4664.68.8	64
65	314.52.7	798.41.6	1282.30.5	1766.19.4	2250.08.3	2733.97.2	3217.86.1	3701.75.0	4185.63.8	4669.52.7	65
66	319.36.6	803.25.5	1287.14.4	1771.03.3	2254.92.2	2738.81.1	3222.70.0	3706.58.8	4190.47.7	4674.36.6	66
67	324.20.5	808.09.4	1291.98.3	1775.87.2	2259.76.1	2743.65.0	3227.53.8	3711.42.7	4195.31.6	4679.20.5	67
68	329.04.4	812.93.3	1296.82.2	1780.71.1	2264.60.0	2748.48.8	3232.37.7	3716.26.6	4200.15.5	4684.04.4	68
69	333.88.3	817.77.2	1301.66.1	1785.55.0	2269.43.8	2753.32.7	3237.21.6	3721.10.5	4204.99.4	4688.86.3	69
70	338.72.2	822.61.1	1306.50.0	1790.38.8	2274.27.7	2758.16.6	3242.05.5	3725.94.4	4209.83.3	4693.72.2	70
71	343.56.1	827.45.0	1311.33.8	1795.22.7	2279.11.6	2763.00.5	3246.89.4	3730.78.3	4214.67.2	4698.56.1	71
72	348.40.0	832.28.8	1316.17.7	1800.06.6	2283.95.5	2767.84.4	3251.73.3	3735.62.2	4219.51.1	4703.40.0	72
73	353.23.8	837.12.7	1321.01.6	1804.90.5	2288.79.4	2772.68.3	3256.57.2	3740.46.1	4224.35.0	4708.23.8	73
74	358.07.7	841.96.6	1325.85.5	1809.74.4	2293.63.3	2777.52.2	3261.41.1	3745.30.0	4229.18.8	4713.07.7	74
75	362.91.6	846.80.5	1330.69.4	1814.58.3	2298.47.2	2782.36.1	3266.25.0	3750.13.8	4234.02.7	4717.91.6	75
76	367.75.5	851.64.4	1335.53.3	1819.42.2	2303.31.1	2787.20.0	3271.08.8	3754.97.7	4238.86.6	4722.75.5	76
77	372.59.4	856.48.3	1340.37.2	1824.26.1	2308.15.0	2792.03.8	3275.92.7	3759.81.6	4243.70.5	4727.59.4	77
78	377.43.3	861.32.2	1345.21.1	1829.10.0	2312.98.8	2796.87.7	3280.76.6	3764.65.5	4248.54.4	4732.43.3	78
79	382.27.2	866.16.1	1350.05.0	1833.93.8	2317.82.7	2801.71.6	3285.60.5	3769.49.4	4253.38.3	4737.27.2	79
80	387.11.1	871.00.0	1354.88.8	1838.77.7	2322.66.6	2806.55.5	3290.44.4	3774.33.3	4258.22.2	4742.11.1	80
81	391.95.0	875.83.8	1359.72.7	1843.61.6	2327.50.5	2811.39.4	3295.28.3	3779.17.2	4263.06.1	4746.95.0	81
82	396.78.8	880.67.7	1364.56.6	1848.45.5	2332.34.4	2816.23.3	3300.12.2	3784.01.1	4267.90.0	4751.78.8	82
83	401.62.7	885.51.6	1369.40.5	1853.29.4	2337.18.3	2821.07.2	3304.96.1	3788.85.0	4272.73.8	4756.62.7	83
84	406.46.6	890.35.5	1374.24.4	1858.13.3	2342.02.2	2825.91.1	3309.80.0	3793.68.8	4277.57.7	4761.46.6	84
85	411.30.5	895.19.4	1379.08.3	1862.97.2	2346.86.1	2830.75.0	3314.63.8	3798.52.7	4282.41.6	4766.30.5	85
86	416.14.4	900.03.3	1383.92.2	1867.81.1	2351.70.0	2835.58.8	3319.47.7	3803.36.6	4287.25.5	4771.14.4	86
87	420.98.3	904.87.2	1388.76.1	1872.65.0	2356.53.8	2840.42.7	3324.31.6	3808.20.5	4292.09.4	4775.98.3	87
88	425.82.2	909.71.1	1393.60.0	1877.48.8	2361.37.7	2845.26.6	3329.15.5	3813.04.4	4296.93.3	4780.82.2	88
89	430.66.1	914.55.0	1398.43.8	1882.32.7	2366.21.6	2850.10.5	3333.99.4	3817.88.3	4301.77.2	4785.66.1	89
90	435.50.0	919.38.8	1403.27.7	1887.16.6	2371.05.5	2854.94.4	3338.83.3	3822.72.2	4306.61.1	4790.50.0	90
91	440.33.8	924.22.7	1408.11.6	1892.00.5	2375.89.4	2859.78.3	3343.67.2	3827.56.1	4311.45.0	4795.33.8	91
92	445.17.7	929.06.6	1412.95.5	1896.84.4	2380.73.3	2864.62.2	3348.51.1	3832.40.0	4316.28.8	4800.17.7	92
93	450.01.6	933.90.5	1417.79.4	1901.68.3	2385.57.2	2869.46.1	3353.35.0	3837.23.8	4321.12.7	4805.01.6	93
94	454.85.5	938.74.4	1422.63.3	1906.52.2	2390.41.1	2874.30.0	3358.18.8	3842.07.7	4325.96.6	4809.85.5	94
95	459.69.4	943.58.3	1427.47.2	1911.36.1	2395.25.0	2879.13.8	3363.02.7	3846.91.6	4330.80.5	4814.69.4	95
96	464.53.3	948.42.2	1432.31.1	1916.20.0	2400.08.8	2883.97.7	3367.86.6	3851.75.5	4335.64.4	4819.53.3	96
97	469.37.2	953.26.1	1437.15.0	1921.03.8	2404.92.7	2888.81.6	3372.70.5	3856.59.4	4340.48.3	4824.37.2	97
98	474.21.1	958.10.0	1441.98.8	1925.87.7	2409.76.6	2893.65.5	3377.54.4	3861.43.3	4345.32.2	4829.21.1	98
99	479.05.0	962.93.8	1446.82.7	1930.71.6	2414.60.5	2898.49.4	3382.38.3	3866.27.2	4350.16.1	4834.05.0	99

s.	10	11	12	13	14	15	16	17	18	19	s.
d.	$ cts.	$ cts.	$ cts.	$ cts.	$ cts.	$ cts.	$ cts.	$ cts.	$ cts.	$ cts.	d.
0	2.41.9	2.66.1	2.90.3	3.14.5	3.38.7	3.62.9	3.87.1	4.11.3	4.35.5	4.59.6	0
1	2.43.9	2.68.1	2.92.3	3.16.5	3.40.7	3.64.9	3.89.1	4.13.3	4.37.5	4.61.7	1
2	2.45.9	2.70.1	2.94.3	3.18.5	3.42.7	3.66.9	3.91.1	4.15.3	4.39.5	4.63.7	2
3	2.47.9	2.72.1	2.96.3	3.20.5	3.44.7	3.68.9	3.93.1	4.17.3	4.41.5	4.65.7	3
4	2.50.0	2.74.1	2.98.3	3.22.5	3.46.7	3.70.9	3.95.1	4.19.3	4.43.5	4.67.7	4
5	2.52.0	2.76.1	3.00.3	3.24.5	3.48.7	3.72.9	3.97.1	4.21.3	4.45.6	4.69.7	5
6	2.54.0	2.78.2	3.02.4	3.26.5	3.50.7	3.74.9	3.99.1	4.23.3	4.47.5	4.71.7	6
7	2.56.0	2.80.2	3.04.4	3.28.5	3.52.8	3.76.9	4.01.1	4.25.3	4.49.5	4.73.7	7
8	2.58.0	2.82.2	3.06.4	3.30.6	3.54.8	3.79.0	4.03.2	4.27.3	4.51.5	4.75.7	8
9	2.60.0	2.84.2	3.08.4	3.32.6	3.56.8	3.81.0	4.05.2	4.29.4	4.53.6	4.77.7	9
10	2.62.0	2.86.2	3.10.4	3.34.6	3.58.8	3.83.0	4.07.2	4.31.4	4.55.6	4.79.8	10
11	2.64.0	2.88.2	3.12.4	3.36.6	3.60.8	3.85.0	4.09.2	4.33.4	4.57.6	4.81.8	11

$4.84¼ TO THE POUND.

9%

PREMIUM.

STERLING INTO DOLLARS AND CENTS, AND VICE-VERSA.

£	0	100	200	300	400	500	600	700	800	900	£
£	$ cts.	$ cts.	$ cts.	$ cts.	$ cts.	$ cts.	$ cts.	$ cts.	$ cts.	$ cts.	£
0	......	484.44.4	968.88.8	1453.33.3	1937.77.7	2422.22.2	2906.66.6	3391.11.1	3875.55.5	4360.00.0	0
1	4.84.4	489.28.8	973.73.3	1458.17.7	1942.62.2	2427.06.6	2911.51.1	3395.95.5	3880.40.0	4364.84.4	1
2	9.68.8	494.13.3	978.57.7	1463.02.2	1947.46.6	2431.91.1	2916.35.5	3400.80.0	3885.24.4	4369.68.8	2
3	14.53.3	498.97.7	983.42.2	1467.86.6	1952.31.1	2436.75.5	2921.20.0	3405.64.4	3890.08.8	4374.53.3	3
4	19.37.7	503.82.2	988.26.6	1472.71.1	1957.15.5	2441.60.0	2926.04.4	3410.48.8	3894.93.3	4379.37.7	4
5	24.22.2	508.66.6	993.11.1	1477.55.5	1962.00.0	2446.44.4	2930.88.8	3415.33.3	3899.77.7	4384.22.2	5
6	29.06.6	513.51.1	997.95.5	1482.40.0	1966.84.4	2451.28.8	2935.73.3	3420.17.7	3904.62.2	4389.06.6	6
7	33.91.1	518.35.5	1002.80.0	1487.24.4	1971.68.8	2456.13.3	2940.57.7	3425.02.2	3909.46.6	4393.91.1	7
8	38.75.5	523.20.0	1007.64.4	1492.08.8	1976.53.3	2460.97.7	2945.42.2	3429.86.6	3914.31.1	4398.75.5	8
9	43.60.0	528.04.4	1012.48.8	1496.93.3	1981.37.7	2465.82.2	2950.26.6	3434.71.1	3919.15.5	4403.60.0	9
10	48.44.4	532.88.8	1017.33.3	1501.77.7	1986.22.2	2470.66.6	2955.11.1	3439.55.5	3924.00.0	4408.44.4	10
11	53.28.8	537.73.3	1022.17.7	1506.62.2	1991.06.6	2475.51.1	2959.95.5	3444.40.0	3928.84.4	4413.28.8	11
12	58.13.3	542.57.7	1027.02.2	1511.46.6	1995.91.1	2480.35.5	2964.80.0	3449.24.4	3933.68.8	4418.13.3	12
13	62.97.7	547.42.2	1031.86.6	1516.31.1	2000.75.5	2485.20.0	2969.64.4	3454.08.8	3938.53.3	4422.97.7	13
14	67.82.2	552.26.6	1036.71.1	1521.15.5	2005.60.0	2490.04.4	2974.48.8	3458.93.3	3943.37.7	4427.82.2	14
15	72.66.6	557.11.1	1041.55.5	1526.00.0	2010.44.4	2494.88.8	2979.33.3	3463.77.7	3948.22.2	4432.66.6	15
16	77.51.1	561.95.5	1046.40.0	1530.84.4	2015.28.8	2499.73.3	2984.17.7	3468.62.2	3953.06.6	4437.51.1	16
17	82.35.5	566.80.0	1051.24.4	1535.68.8	2020.13.3	2504.57.7	2989.02.2	3473.46.6	3957.91.1	4442.35.5	17
18	87.20.0	571.64.4	1056.08.8	1540.53.3	2024.97.7	2509.42.2	2993.86.6	3478.31.1	3962.75.5	4447.20.0	18
19	92.04.4	576.48.8	1060.93.3	1545.37.7	2029.82.2	2514.26.6	2998.71.1	3483.15.5	3967.60.0	4452.04.4	19
20	96.88.8	581.33.3	1065.77.7	1550.22.2	2034.66.6	2519.11.1	3003.55.5	3488.00.0	3972.44.4	4456.88.8	20
21	101.73.3	586.17.7	1070.62.2	1555.06.6	2039.51.1	2523.95.5	3008.40.0	3492.84.4	3977.28.8	4461.73.3	21
22	106.57.7	591.02.2	1075.46.6	1559.91.1	2044.35.5	2528.80.0	3013.24.4	3497.68.8	3982.13.3	4466.57.7	22
23	111.42.2	595.86.6	1080.31.1	1564.75.5	2049.20.0	2533.64.4	3018.08.8	3502.53.3	3986.97.7	4471.42.2	23
24	116.26.6	600.71.1	1085.15.5	1569.60.0	2054.04.4	2538.48.8	3022.93.3	3507.37.7	3991.82.2	4476.26.6	24
25	121.11.1	605.55.5	1090.00.0	1574.44.4	2058.88.8	2543.33.3	3027.77.7	3512.22.2	3996.66.6	4481.11.1	25
26	125.95.5	610.40.0	1094.84.4	1579.28.8	2063.73.3	2548.17.7	3032.62.2	3517.06.6	4001.51.1	4485.95.5	26
27	130.80.0	615.24.4	1099.68.8	1584.13.3	2068.57.7	2553.02.2	3037.46.6	3521.91.1	4006.35.5	4490.80.0	27
28	135.64.4	620.08.8	1104.53.3	1588.97.7	2073.42.2	2557.86.6	3042.31.1	3526.75.5	4011.20.0	4495.64.4	28
29	140.48.8	624.93.3	1109.37.7	1593.82.2	2078.26.6	2562.71.1	3047.15.5	3531.60.0	4016.04.4	4500.48.8	29
30	145.33.3	629.77.7	1114.22.2	1598.66.6	2083.11.1	2567.55.5	3052.00.0	3536.44.4	4020.88.8	4505.33.3	30
31	150.17.7	634.62.2	1119.06.6	1603.51.1	2087.95.5	2572.40.0	3056.84.4	3541.28.8	4025.73.3	4510.17.7	31
32	155.02.2	639.46.6	1123.91.1	1608.35.5	2092.80.0	2577.24.4	3061.68.8	3546.13.3	4030.57.7	4515.02.2	32
33	159.86.6	644.31.1	1128.75.5	1613.20.0	2097.64.4	2582.08.8	3066.53.3	3550.97.7	4035.42.2	4519.86.6	33
34	164.71.1	649.15.5	1133.60.0	1618.04.4	2102.48.8	2586.93.3	3071.37.7	3555.82.2	4040.26.6	4524.71.1	34
35	169.55.5	654.00.0	1138.44.4	1622.88.8	2107.33.3	2591.77.7	3076.22.2	3560.66.6	4045.11.1	4529.55.5	35
36	174.40.0	658.84.4	1143.28.8	1627.73.3	2112.17.7	2596.62.2	3081.06.6	3565.51.1	4049.95.5	4534.40.0	36
37	179.24.4	663.68.8	1148.13.3	1632.57.7	2117.02.2	2601.46.6	3085.91.1	3570.35.5	4054.80.0	4539.24.4	37
38	184.08.8	668.53.3	1152.97.7	1637.42.2	2121.86.6	2606.31.1	3090.75.5	3575.20.0	4059.64.4	4544.08.8	38
39	188.93.3	673.37.7	1157.82.2	1642.26.6	2126.71.1	2611.15.5	3095.60.0	3580.04.4	4064.48.8	4548.93.3	39
40	193.77.7	678.22.2	1162.66.6	1647.11.1	2131.55.5	2616.00.0	3100.44.4	3584.88.8	4069.33.3	4553.77.7	40
41	198.62.2	683.06.6	1167.51.1	1651.95.5	2136.40.0	2620.84.4	3105.28.8	3589.73.3	4074.17.7	4558.62.2	41
42	203.46.6	687.91.1	1172.35.5	1656.80.0	2141.24.4	2625.68.8	3110.13.3	3594.57.7	4079.02.2	4563.46.6	42
43	208.31.1	692.75.5	1177.20.0	1661.64.4	2146.08.8	2630.53.3	3114.97.7	3599.42.2	4083.86.6	4568.31.1	43
44	213.15.5	697.60.0	1182.04.4	1666.48.8	2150.93.3	2635.37.7	3119.82.2	3604.26.6	4088.71.1	4573.15.5	44
45	218.00.0	702.44.4	1186.88.8	1671.33.3	2155.77.7	2640.22.2	3124.66.6	3609.11.1	4093.55.5	4578.00.0	45
46	222.84.4	707.28.8	1191.73.3	1676.17.7	2160.62.2	2645.06.6	3129.51.1	3613.95.5	4098.40.0	4582.84.4	46
47	227.68.8	712.13.3	1196.57.7	1681.02.2	2165.46.6	2649.91.1	3134.35.5	3618.80.0	4103.24.4	4587.68.8	47
48	232.53.3	716.97.7	1201.42.2	1685.86.6	2170.31.1	2654.75.5	3139.20.0	3623.64.4	4108.08.8	4592.53.3	48
49	237.37.7	721.82.2	1206.26.6	1690.71.1	2175.15.5	2659.60.0	3144.04.4	3628.48.8	4112.93.3	4597.37.7	49

S.	0	1	2	3	4	5	6	7	8	9	S.
d.	$ cts.	$ cts.	$ cts.	$ cts.	$ cts.	$ cts.	$ cts.	$ cts.	$ cts.	$ cts.	d.
0	......	.24.2	.48.4	.72.6	.96.8	1.21.1	1.45.3	1.69.5	1.93.7	2.18.0	0
1	.02.0	.26.2	.50.4	.74.6	.98.9	1.23.1	1.47.3	1.71.5	1.95.7	2.20.0	1
2	.04.0	.28.2	.52.4	.76.7	1.00.9	1.25.1	1.49.3	1.73.5	1.97.8	2.22.0	2
3	.06.0	.30.2	.54.5	.78.7	1.02.9	1.27.1	1.51.4	1.75.6	1.99.8	2.24.0	3
4	.08.0	.32.3	.56.5	.80.7	1.04.9	1.29.1	1.53.4	1.77.6	2.01.8	2.26.0	4
5	.10.1	.34.3	.58.5	.82.7	1.06.9	1.31.2	1.55.4	1.79.6	2.03.8	2.28.1	5
6	.12.1	.36.3	.60.5	.84.7	1.09.0	1.33.2	1.57.4	1.81.6	2.05.8	2.30.1	6
7	.14.1	.38.3	.62.5	.86.8	1.11.0	1.35.2	1.59.4	1.83.7	2.07.9	2.32.1	7
8	.16.1	.40.3	.64.6	.88.8	1.13.0	1.37.2	1.61.5	1.85.7	2.09.9	2.34.1	8
9	.18.1	.42.4	.66.6	.90.8	1.15.0	1.39.3	1.63.5	1.87.7	2.11.9	2.36.2	9
10	.20.2	.44.4	.68.6	.92.8	1.17.0	1.41.3	1.65.5	1.89.7	2.13.9	2.38.2	10
11	.22.2	.46.4	.70.6	.94.9	1.19.1	1.43.3	1.67.5	1.91.8	2.16.0	2.40.2	11

$4.84⁴⁄₁₀₀ TO THE POUND.

9%

PREMIUM.

STERLING INTO DOLLARS AND CENTS, AND VICE-VERSA.

£	0	100	200	300	400	500	600	700	800	900	£
	$ cts.	$ cts.	$ cts.	$ cts.	$ cts.	$ cts.	$ cts.	$ cts.	$ cts.	$ cts.	
50	242.22.2	726.66.6	1211.11.1	1695.55.5	2180.00.0	2664.44.4	3148.88.8	3633.33.3	4117.77.7	4602.22.2	50
51	247.06.6	731.51.1	1215.95.5	1700.40.0	2184.84.4	2669.28.8	3153.73.3	3638.17.7	4122.62.2	4607.06.6	51
52	251.91.1	736.35.5	1220.80.0	1705.24.4	2189.68.8	2674.13.3	3158.57.7	3643.02.2	4127.46.6	4611.91.1	52
53	256.75.5	741.20.0	1225.64.4	1710.08.8	2194.53.3	2678.97.7	3163.42.2	3647.86.6	4132.31.1	4616.75.5	53
54	261.60.0	746.04.4	1230.48.8	1714.93.3	2199.37.7	2683.82.2	3168.26.6	3652.71.1	4137.15.5	4621.60.0	54
55	266.44.4	750.88.8	1235.33.3	1719.77.7	2204.22.2	2688.66.6	3173.11.1	3657.55.5	4142.00.0	4626.44.4	55
56	271.28.8	755.73.3	1240.17.7	1724.62.2	2209.06.6	2693.51.1	3177.95.5	3662.40.0	4146.84.4	4631.28.8	56
57	276.13.3	760.57.7	1245.02.2	1729.46.6	2213.91.1	2698.35.5	3182.80.0	3667.24.4	4151.68.8	4636.13.3	57
58	280.97.7	765.42.2	1249.86.6	1734.31.1	2218.75.5	2703.20.0	3187.64.4	3672.08.8	4156.53.3	4640.97.7	58
59	285.82.2	770.26.6	1254.71.1	1739.15.5	2223.60.0	2708.04.4	3192.48.8	3676.93.3	4161.37.7	4645.82.2	59
60	290.66.6	775.11.1	1259.55.5	1744.00.0	2228.44.4	2712.88.8	3197.33.3	3681.77.7	4166.22.2	4650.66.6	60
61	295.51.1	779.95.5	1264.40.0	1748.84.4	2233.28.8	2717.73.3	3202.17.7	3686.62.2	4171.06.6	4655.51.1	61
62	300.35.5	784.80.0	1269.24.4	1753.68.8	2238.13.3	2722.57.7	3207.02.2	3691.46.6	4175.91.1	4660.35.5	62
63	305.20.0	789.64.4	1274.08.8	1758.53.3	2242.97.7	2727.42.2	3211.86.6	3696.31.1	4180.75.5	4665.20.0	63
64	310.04.4	794.48.8	1278.93.3	1763.37.7	2247.82.2	2732.26.6	3216.71.1	3701.15.5	4185.60.0	4670.04.4	64
65	314.88.8	799.33.3	1283.77.7	1768.22.2	2252.66.6	2737.11.1	3221.55.5	3706.00.0	4190.44.4	4674.88.8	65
66	319.73.3	804.17.7	1288.62.2	1773.06.6	2257.51.1	2741.95.5	3226.40.0	3710.84.4	4195.28.8	4679.73.3	66
67	324.57.7	809.02.2	1293.46.6	1777.91.1	2262.35.5	2746.80.0	3231.24.4	3715.68.8	4200.13.3	4684.57.7	67
68	329.42.2	813.86.6	1298.31.1	1782.75.5	2267.20.0	2751.64.4	3236.08.8	3720.53.3	4204.97.7	4689.42.2	68
69	334.26.6	818.71.1	1303.15.5	1787.60.0	2272.04.4	2756.48.8	3240.93.3	3725.37.7	4209.82.2	4694.26.6	69
70	339.11.1	823.55.5	1308.00.0	1792.44.4	2276.88.8	2761.33.3	3245.77.7	3730.22.2	4214.66.6	4699.11.1	70
71	343.95.5	828.40.0	1312.84.4	1797.28.8	2281.73.3	2766.17.7	3250.62.2	3735.06.6	4219.51.1	4703.95.5	71
72	348.80.0	833.24.4	1317.68.8	1802.13.3	2286.57.7	2771.02.2	3255.46.6	3739.91.1	4224.35.5	4708.80.0	72
73	353.64.4	838.08.8	1322.53.3	1806.97.7	2291.42.2	2775.86.6	3260.31.1	3744.75.5	4229.20.0	4713.64.4	73
74	358.48.8	842.93.3	1327.37.7	1811.82.2	2296.26.6	2780.71.1	3265.15.5	3749.60.0	4234.04.4	4718.48.8	74
75	363.33.3	847.77.7	1332.22.2	1816.66.6	2301.11.1	2785.55.5	3270.00.0	3754.44.4	4238.88.8	4723.33.3	75
76	368.17.7	852.62.2	1337.06.6	1821.51.1	2305.95.5	2790.40.0	3274.84.4	3759.28.8	4243.73.3	4728.17.7	76
77	373.02.2	857.46.6	1341.91.1	1826.35.5	2310.80.0	2795.24.4	3279.68.8	3764.13.3	4248.57.7	4733.02.2	77
78	377.86.6	862.31.1	1346.75.5	1831.20.0	2315.64.4	2800.08.8	3284.53.3	3768.97.7	4253.42.2	4737.86.6	78
79	382.71.1	867.15.5	1351.60.0	1836.04.4	2320.48.8	2804.93.3	3289.37.7	3773.82.2	4258.26.6	4742.71.1	79
80	387.55.5	872.00.0	1356.44.4	1840.88.8	2325.33.3	2809.77.7	3294.22.2	3778.66.6	4263.11.1	4747.55.5	80
81	392.40.0	876.84.4	1361.28.8	1845.73.3	2330.17.7	2814.62.2	3299.06.6	3783.51.1	4267.95.5	4752.40.0	81
82	397.24.4	881.68.8	1366.13.3	1850.57.7	2335.02.2	2819.46.6	3303.91.1	3788.35.5	4272.80.0	4757.24.4	82
83	402.08.8	886.53.3	1370.97.7	1855.42.2	2339.86.6	2824.31.1	3308.75.5	3793.20.0	4277.64.4	4762.08.8	83
84	406.93.3	891.37.7	1375.82.2	1860.26.6	2344.71.1	2829.15.5	3313.60.0	3798.04.4	4282.48.8	4766.93.3	84
85	411.77.7	896.22.2	1380.66.6	1865.11.1	2349.55.5	2834.00.0	3318.44.4	3802.88.8	4287.33.3	4771.77.7	85
86	416.62.2	901.06.6	1385.51.1	1869.95.5	2354.40.0	2838.84.4	3323.28.8	3807.73.3	4292.17.7	4776.62.2	86
87	421.46.6	905.91.1	1390.35.5	1874.80.0	2359.24.4	2843.68.8	3328.13.3	3812.57.7	4297.02.2	4781.46.6	87
88	426.31.1	910.75.5	1395.20.0	1879.64.4	2364.08.8	2848.53.3	3332.97.7	3817.42.2	4301.86.6	4786.31.1	88
89	431.15.5	915.60.0	1400.04.4	1884.48.8	2368.93.3	2853.37.7	3337.82.2	3822.26.6	4306.71.1	4791.15.5	89
90	436.00.0	920.44.4	1404.88.8	1889.33.3	2373.77.7	2858.22.2	3342.66.6	3827.11.1	4311.55.5	4796.00.0	90
91	440.84.4	925.28.8	1409.73.3	1894.17.7	2378.62.2	2863.06.6	3347.51.1	3831.95.5	4316.40.0	4800.84.4	91
92	445.68.8	930.13.3	1414.57.7	1899.02.2	2383.46.6	2867.91.1	3352.35.5	3836.80.0	4321.24.4	4805.68.8	92
93	450.53.3	934.97.7	1419.42.2	1903.86.6	2388.31.1	2872.75.5	3357.20.0	3841.64.4	4326.08.8	4810.53.3	93
94	455.37.7	939.82.2	1424.26.6	1908.71.1	2393.15.5	2877.60.0	3362.04.4	3846.48.8	4330.93.3	4815.37.7	94
95	460.22.2	944.66.6	1429.11.1	1913.55.5	2398.00.0	2882.44.4	3366.88.8	3851.33.3	4335.77.7	4820.22.2	95
96	465.06.6	949.51.1	1433.95.5	1918.40.0	2402.84.4	2887.28.8	3371.73.3	3856.17.7	4340.62.2	4825.06.6	96
97	469.91.1	954.35.5	1438.80.0	1923.24.4	2407.68.8	2892.13.3	3376.57.7	3861.02.2	4345.46.6	4829.91.1	97
98	474.75.5	959.20.0	1443.64.4	1928.08.8	2412.53.3	2896.97.7	3381.42.2	3865.86.6	4350.31.1	4834.75.5	98
99	479.60.0	964.04.4	1448.48.8	1932.93.3	2417.37.7	2901.82.2	3386.26.6	3870.71.1	4355.15.5	4839.60.0	99

S.	10	11	12	13	14	15	16	17	18	19	S.
d.	$ cts.	$ cts.	$ cts.	$ cts.	$ cts.	$ cts.	$ cts.	$ cts.	$ cts.	$ cts.	d.
0	2.42.2	2.66.4	2.90.6	3.14.8	3.39.1	3.63.3	3.87.5	4.11.7	4.36.0	4.60.2	0
1	2.44.2	2.68.4	2.92.6	3.16.9	3.41.1	3.65.3	3.89.5	4.13.7	4.38.0	4.62.2	1
2	2.46.2	2.70.4	2.94.7	3.18.9	3.43.1	3.67.3	3.91.5	4.15.8	4.40.0	4.64.2	2
3	2.48.2	2.72.5	2.96.7	3.20.9	3.45.1	3.69.3	3.93.6	4.17.8	4.42.0	4.66.2	3
4	2.50.3	2.74.5	2.98.7	3.22.9	3.47.1	3.71.4	3.95.6	4.19.8	4.44.0	4.68.3	4
5	2.52.3	2.76.5	3.00.7	3.24.9	3.49.2	3.73.4	3.97.6	4.21.8	4.46.1	4.70.3	5
6	2.54.3	2.78.5	3.02.7	3.27.0	3.51.2	3.75.4	3.99.6	4.23.8	4.48.1	4.72.3	6
7	2.56.3	2.80.5	3.04.8	3.29.0	3.53.2	3.77.4	4.01.6	4.25.9	4.50.1	4.74.3	7
8	2.58.3	2.82.6	3.06.8	3.31.0	3.55.2	3.79.4	4.03.7	4.27.9	4.52.1	4.76.3	8
9	2.60.4	2.84.6	3.08.8	3.33.0	3.57.2	3.81.5	4.05.7	4.29.9	4.54.1	4.78.4	9
10	2.62.4	2.86.0	3.10.8	3.35.0	3.59.3	3.83.5	4.07.7	4.31.9	4.56.2	4.80.4	10
11	2.64.4	2.88.6	3.12.8	3.37.1	3.61.3	3.85.5	4.09.7	4.33.9	4.58.2	4.82.4	11

$4.85 7/16 TO THE POUND.

9 1/8 %

PREMIUM.

STERLING INTO DOLLARS AND CENTS, AND VICE-VERSA.

£	0	100	200	300	400	500	600	700	800	900	£
	$ cts.	$ cts.	$ cts.	$ cts.	$ cts.	$ cts.	$ cts.	$ cts.	$ cts.	$ cts.	
0	......	485.00.0	970.00.0	1455.00.0	1940.00.0	2425.00.0	2910.00.0	3395.00.0	3880.00.0	4365.00.0	0
1	4.85.0	489.85.0	974.85.0	1459.85.0	1944.85.0	2429.85.0	2914.85.0	3399.85.0	3884.85.0	4369.85.0	1
2	9.70.0	494.70.0	979.70.0	1464.70.0	1949.70.0	2434.70.0	2919.70.0	3404.70.0	3889.70.0	4374.70.0	2
3	14.55.0	499.55.0	984.55.0	1469.55.0	1954.55.0	2439.55.0	2924.55.0	3409.55.0	3894.55.0	4379.55.0	3
4	19.40.0	504.40.0	989.40.0	1474.40.0	1959.40.0	2444.40.0	2929.40.0	3414.40.0	3899.40.0	4384.40.0	4
5	24.25.0	509.25.0	994.25.0	1479.25.0	1964.25.0	2449.25.0	2934.25.0	3419.25.0	3904.25.0	4389.25.0	5
6	29.10.0	514.10.0	999.10.0	1484.10.0	1969.10.0	2454.10.0	2939.10.0	3424.10.0	3909.10.0	4394.10.0	6
7	33.95.0	518.95.0	1003.95.0	1488.95.0	1973.95.0	2458.95.0	2943.95.0	3428.95.0	3913.95.0	4398.95.0	7
8	38.80.0	523.80.0	1008.80.0	1493.80.0	1978.80.0	2463.80.0	2948.80.0	3433.80.0	3918.80.0	4403.80.0	8
9	43.65.0	528.65.0	1013.65.0	1498.65.0	1983.65.0	2468.65.0	2953.65.0	3438.65.0	3923.65.0	4408.65.0	9
10	48.50.0	533.50.0	1018.50.0	1503.50.0	1988.50.0	2473.50.0	2958.50.0	3443.50.0	3928.50.0	4413.50.0	10
11	53.35.0	538.35.0	1023.35.0	1508.35.0	1993.35.0	2478.35.0	2963.35.0	3448.35.0	3933.35.0	4418.35.0	11
12	58.20.0	543.20.0	1028.20.0	1513.20.0	1998.20.0	2483.20.0	2968.20.0	3453.20.0	3938.20.0	4423.20.0	12
13	63.05.0	548.05.0	1033.05.0	1518.05.0	2003.05.0	2488.05.0	2973.05.0	3458.05.0	3943.05.0	4428.05.0	13
14	67.90.0	552.90.0	1037.90.0	1522.90.0	2007.90.0	2492.90.0	2977.90.0	3462.90.0	3947.90.0	4432.90.0	14
15	72.75.0	557.75.0	1042.75.0	1527.75.0	2012.75.0	2497.75.0	2982.75.0	3467.75.0	3952.75.0	4437.75.0	15
16	77.60.0	562.60.0	1047.60.0	1532.60.0	2017.60.0	2502.60.0	2987.60.0	3472.60.0	3957.60.0	4442.60.0	16
17	82.45.0	567.45.0	1052.45.0	1537.45.0	2022.45.0	2507.45.0	2992.45.0	3477.45.0	3962.45.0	4447.45.0	17
18	87.30.0	572.30.0	1057.30.0	1542.30.0	2027.30.0	2512.30.0	2997.30.0	3482.30.0	3967.30.0	4452.30.0	18
19	92.15.0	577.15.0	1062.15.0	1547.15.0	2032.15.0	2517.15.0	3002.15.0	3487.15.0	3972.15.0	4457.15.0	19
20	97.00.0	582.00.0	1067.00.0	1552.00.0	2037.00.0	2522.00.0	3007.00.0	3492.00.0	3977.00.0	4462.00.0	20
21	101.85.0	586.85.0	1071.85.0	1556.85.0	2041.85.0	2526.85.0	3011.85.0	3496.85.0	3981.85.0	4466.85.0	21
22	106.70.0	591.70.0	1076.70.0	1561.70.0	2046.70.0	2531.70.0	3016.70.0	3501.70.0	3986.70.0	4471.70.0	22
23	111.55.0	596.55.0	1081.55.0	1566.55.0	2051.55.0	2536.55.0	3021.55.0	3506.55.0	3991.55.0	4476.55.0	23
24	116.40.0	601.40.0	1086.40.0	1571.40.0	2056.40.0	2541.40.0	3026.40.0	3511.40.0	3996.40.0	4481.40.0	24
25	121.25.0	606.25.0	1091.25.0	1576.25.0	2061.25.0	2546.25.0	3031.25.0	3516.25.0	4001.25.0	4486.25.0	25
26	126.10.0	611.10.0	1096.10.0	1581.10.0	2066.10.0	2551.10.0	3036.10.0	3521.10.0	4006.10.0	4491.10.0	26
27	130.95.0	615.95.0	1100.95.0	1585.95.0	2070.95.0	2555.95.0	3040.95.0	3525.95.0	4010.95.0	4495.95.0	27
28	135.80.0	620.80.0	1105.80.0	1590.80.0	2075.80.0	2560.80.0	3045.80.0	3530.80.0	4015.80.0	4500.80.0	28
29	140.65.0	625.65.0	1110.65.0	1595.65.0	2080.65.0	2565.65.0	3050.65.0	3535.65.0	4020.65.0	4505.65.0	29
30	145.50.0	630.50.0	1115.50.0	1600.50.0	2085.50.0	2570.50.0	3055.50.0	3540.50.0	4025.50.0	4510.50.0	30
31	150.35.0	635.35.0	1120.35.0	1605.35.0	2090.35.0	2575.35.0	3060.35.0	3545.35.0	4030.35.0	4515.35.0	31
32	155.20.0	640.20.0	1125.20.0	1610.20.0	2095.20.0	2580.20.0	3065.20.0	3550.20.0	4035.20.0	4520.20.0	32
33	160.05.0	645.05.0	1130.05.0	1615.05.0	2100.05.0	2585.05.0	3070.05.0	3555.05.0	4040.05.0	4525.05.0	33
34	164.90.0	649.90.0	1134.90.0	1619.90.0	2104.90.0	2589.90.0	3074.90.0	3559.90.0	4044.90.0	4529.90.0	34
35	169.75.0	654.75.0	1139.75.0	1624.75.0	2109.75.0	2594.75.0	3079.75.0	3564.75.0	4049.75.0	4534.75.0	35
36	174.60.0	659.60.0	1144.60.0	1629.60.0	2114.60.0	2599.60.0	3084.60.0	3569.60.0	4054.60.0	4539.60.0	36
37	179.45.0	664.45.0	1149.45.0	1634.45.0	2119.45.0	2604.45.0	3089.45.0	3574.45.0	4059.45.0	4544.45.0	37
38	184.30.0	669.30.0	1154.30.0	1639.30.0	2124.30.0	2609.30.0	3094.30.0	3579.30.0	4064.30.0	4549.30.0	38
39	189.15.0	674.15.0	1159.15.0	1644.15.0	2129.15.0	2614.15.0	3099.15.0	3584.15.0	4069.15.0	4554.15.0	39
40	194.00.0	679.00.0	1164.00.0	1649.00.0	2134.00.0	2619.00.0	3104.00.0	3589.00.0	4074.00.0	4559.00.0	40
41	198.85.0	683.85.0	1168.85.0	1653.85.0	2138.85.0	2623.85.0	3108.85.0	3593.85.0	4078.85.0	4563.85.0	41
42	203.70.0	688.70.0	1173.70.0	1658.70.0	2143.70.0	2628.70.0	3113.70.0	3598.70.0	4083.70.0	4568.70.0	42
43	208.55.0	693.55.0	1178.55.0	1663.55.0	2148.55.0	2633.55.0	3118.55.0	3603.55.0	4088.55.0	4573.55.0	43
44	213.40.0	698.40.0	1183.40.0	1668.40.0	2153.40.0	2638.40.0	3123.40.0	3608.40.0	4093.40.0	4578.40.0	44
45	218.25.0	703.25.0	1188.25.0	1673.25.0	2158.25.0	2643.25.0	3128.25.0	3613.25.0	4098.25.0	4583.25.0	45
46	223.10.0	708.10.0	1193.10.0	1678.10.0	2163.10.0	2648.10.0	3133.10.0	3618.10.0	4103.10.0	4588.10.0	46
47	227.95.0	712.95.0	1197.95.0	1682.95.0	2167.95.0	2652.95.0	3137.95.0	3622.95.0	4107.95.0	4592.95.0	47
48	232.80.0	717.80.0	1202.80.0	1687.80.0	2172.80.0	2657.80.0	3142.80.0	3627.80.0	4112.80.0	4597.80.0	48
49	237.65.0	722.65.0	1207.65.0	1692.65.0	2177.65.0	2662.65.0	3147.65.0	3632.65.0	4117.65.0	4602.65.0	49

s.	0	1	2	3	4	5	6	7	8	9	s.
d.	$ cts.	$ cts.	$ cts.	$ cts.	$ cts.	$ cts.	$ cts.	$ cts.	$ cts.	$ cts.	d.
0	......	.24.2	.48.5	.72.7	.97.0	1.21.2	1.45.5	1.69.7	1.94.0	2.18.2	0
1	.02.0	.26.2	.50.5	.74.7	.99.0	1.23.2	1.47.5	1.71.7	1.96.0	2.20.2	1
2	.04.0	.28.3	.52.5	.76.8	1.01.0	1.25.3	1.49.5	1.73.8	1.98.0	2.22.3	2
3	.06.0	.30.3	.54.5	.78.8	1.03.0	1.27.3	1.51.5	1.75.8	2.00.0	2.24.3	3
4	.08.0	.32.3	.56.5	.80.8	1.05.0	1.29.3	1.53.5	1.77.8	2.02.0	2.26.3	4
5	.10.1	.34.3	.58.6	.82.8	1.07.1	1.31.3	1.55.6	1.79.8	2.04.1	2.28.3	5
6	.12.1	.36.3	.60.6	.84.8	1.09.1	1.33.3	1.57.6	1.81.8	2.06.1	2.30.3	6
7	.14.1	.38.4	.62.6	.86.9	1.11.1	1.35.4	1.59.6	1.83.9	2.08.1	2.32.4	7
8	.16.1	.40.4	.64.6	.88.9	1.13.1	1.37.4	1.61.6	1.85.9	2.10.1	2.34.4	8
9	.18.1	.42.4	.66.6	.90.9	1.15.1	1.39.4	1.63.6	1.87.9	2.12.1	2.36.4	9
10	.20.2	.44.4	.68.7	.92.9	1.17.2	1.41.4	1.65.7	1.89.9	2.14.2	2.38.4	10
11	.22.2	.46.4	.70.7	.94.9	1.19.2	1.43.4	1.67.7	1.91.9	2.16.2	2.40.4	11

$4.85 88/100 TO THE POUND.

9 1/8 %

PREMIUM.

STERLING INTO DOLLARS AND CENTS, AND VICE-VERSA.

£	0	100	200	300	400	500	600	700	800	900	£
	$ cts.	$ cts.	$ cts.	$ cts.	$ cts.	$ cts.	$ cts.	$ cts.	$ cts.	$ cts.	
50	242.50.0	727.50.0	1212.50.0	1697.50.0	2182.50.0	2667.50.0	3152.50.0	3637.50.0	4122.50.0	4607.50.0	50
51	247.35.0	732.35.0	1217.35.0	1702.35.0	2187.35.0	2672.35.0	3157.35.0	3642.35.0	4127.35.0	4612.35.0	51
52	252.20.0	737.20.0	1222.20.0	1707.20.0	2192.20.0	2677.20.0	3162.20.0	3647.20.0	4132.20.0	4617.20.0	52
53	257.05.0	742.05.0	1227.05.0	1712.05.0	2197.05.0	2682.05.0	3167.05.0	3652.05.0	4137.05.0	4622.05.0	53
54	261.90.0	746.90.0	1231.90.0	1716.90.0	2201.90.0	2686.90.0	3171.90.0	3656.90.0	4141.90.0	4626.90.0	54
55	266.75.0	751.75.0	1236.75.0	1721.75.0	2206.75.0	2691.75.0	3176.75.0	3661.75.0	4146.75.0	4631.75.0	55
56	271.60.0	756.60.0	1241.60.0	1726.60.0	2211.60.0	2696.60.0	3181.60.0	3666.60.0	4151.60.0	4636.60.0	56
57	276.45.0	761.45.0	1246.45.0	1731.45.0	2216.45.0	2701.45.0	3186.45.0	3671.45.0	4156.45.0	4641.45.0	57
58	281.30.0	766.30.0	1251.30.0	1736.30.0	2221.30.0	2706.30.0	3191.30.0	3676.30.0	4161.30.0	4646.30.0	58
59	286.15.0	771.15.0	1256.15.0	1741.15.0	2226.15.0	2711.15.0	3196.15.0	3681.15.0	4166.15.0	4651.15.0	59
60	291.00.0	776.00.0	1261.00.0	1746.00.0	2231.00.0	2716.00.0	3201.00.0	3686.00.0	4171.00.0	4656.00.0	60
61	295.85.0	780.85.0	1265.85.0	1750.85.0	2235.85.0	2720.85.0	3205.85.0	3690.85.0	4175.85.0	4660.85.0	61
62	300.70.0	785.70.0	1270.70.0	1755.70.0	2240.70.0	2725.70.0	3210.70.0	3695.70.0	4180.70.0	4665.70.0	62
63	305.55.0	790.55.0	1275.55.0	1760.55.0	2245.55.0	2730.55.0	3215.55.0	3700.55.0	4185.55.0	4670.55.0	63
64	310.40.0	795.40.0	1280.40.0	1765.40.0	2250.40.0	2735.40.0	3220.40.0	3705.40.0	4190.40.0	4675.40.0	64
65	315.25.0	800.25.0	1285.25.0	1770.25.0	2255.25.0	2740.25.0	3225.25.0	3710.25.0	4195.25.0	4680.25.0	65
66	320.10.0	805 10.0	1290.10.0	1775.10.0	2260.10.0	2745 10.0	3230.10.0	3715 10.0	4200.10.0	4685.10.0	66
67	324.95.0	809.95.0	1294.95.0	1779.95.0	2264.95.0	2749.95.0	3234.95.0	3719.95.0	4204.95.0	4689.95.0	67
68	329.80.0	814.80.0	1299.80.0	1784.80.0	2269.80.0	2754.80.0	3239.80.0	3724.80.0	4209.80.0	4694.80.0	68
69	334.65.0	819.65.0	1304.65.0	1789.65.0	2274.65.0	2759.65.0	3244.65.0	3729.65.0	4214.65.0	4699.65.0	69
70	339.50.0	824.50.0	1309.50.0	1794.50.0	2279.50.0	2764.50.0	3249.50.0	3734.50.0	4219.50.0	4704.50.0	70
71	344.35.0	829.35.0	1314.35.0	1799.35.0	2284.35.0	2769.35.0	3254.35.0	3739.35.0	4224.35.0	4709.35.0	71
72	349.20.0	834.20.0	1319.20.0	1804.20.0	2289.20.0	2774.20.0	3259.20.0	3744.20.0	4229.20.0	4714.20.0	72
73	354.05.0	839.05.0	1324.05.0	1809.05.0	2294.05.0	2779.05.0	3264.05.0	3749.05.0	4234.05.0	4719.05.0	73
74	358.90.0	843.90.0	1328.90.0	1813.90.0	2298.90.0	2783.90.0	3268.90.0	3753.90.0	4238.90.0	4723.90.0	74
75	363.75.0	848.75.0	1333.75.0	1818.75.0	2303.75.0	2788.75.0	3273.75.0	3758.75.0	4243.75.0	4728.75.0	75
76	368.60.0	853.60.0	1338.60.0	1823.60.0	2308.60.0	2793.60.0	3278.60.0	3763.60.0	4248.60.0	4733.60.0	76
77	373.45.0	858.45.0	1343.45.0	1828.45.0	2313.45.0	2798.45.0	3283.45.0	3768.45.0	4253.45.0	4738.45.0	77
78	378.30.0	863.30.0	1348.30.0	1833.30.0	2318.30.0	2803.30.0	3288.30.0	3773.30.0	4258.30.0	4743.30.0	78
79	383.15.0	868.15.0	1353.15.0	1838.15.0	2323.15.0	2808.15.0	3293.15.0	3778.15.0	4263.15.0	4748.15.0	79
80	388.00.0	873.00.0	1358.00.0	1843.00.0	2328.00.0	2813.00.0	3298.00.0	3783.00.0	4268.00.0	4753.00.0	80
81	392.85.0	877.85.0	1362.85.0	1847.85.0	2332.85.0	2817.85.0	3302.85.0	3787.85.0	4272.85.0	4757.85.0	81
82	397.70.0	882.70.0	1367.70.0	1852.70.0	2337.70.0	2822.70.0	3307.70.0	3792.70.0	4277.70.0	4762.70.0	82
83	402.55.0	887.55.0	1372.55.0	1857.55.0	2342.55.0	2827.55.0	3312.55.0	3797.55.0	4282.55.0	4767.55.0	83
84	407.40.0	892.40.0	1377.40.0	1862.40.0	2347.40.0	2832.40.0	3317.40.0	3802.40.0	4287.40.0	4772.40.0	84
85	412.25.0	897.25.0	1382 25.0	1867.25.0	2352.25.0	2837.25.0	3322.25.0	3807.25.0	4292.25.0	4777.25.0	85
86	417.10.0	902.10.0	1387.10.0	1872.10.0	2357.10.0	2842.10.0	3327.10.0	3812.10.0	4297.10.0	4782.10.0	86
87	421.95.0	906.95.0	1391.95.0	1876.95.0	2361.95.0	2846.95.0	3331.95.0	3816.95.0	4301.95.0	4786.95.0	87
88	426.80.0	911.80.0	1396.80.0	1881.80.0	2366.80.0	2851.80.0	3336.80.0	3821.80.0	4306.80.0	4791.80.0	88
89	431.65.0	916.65.0	1401.65.0	1886.65.0	2371.65.0	2856.65.0	3341.65.0	3826.65.0	4311.65.0	4796.65.0	89
90	436.50.0	921.50.0	1406.50.0	1891.50.0	2376.50.0	2861.50.0	3316.50.0	3831.50.0	4316.50.0	4801.50.0	90
91	441.35.0	926.35.0	1411.35.0	1896.35.0	2381.35.0	2866.35.0	3351.35.0	3836.35.0	4321.35.0	4806.35.0	91
92	446.20.0	931.20.0	1416.20.0	1901.20.0	2386.20.0	2871.20.0	3356.20.0	3841.20.0	4326.20.0	4811.20.0	92
93	451.05.0	936.05.0	1421.05.0	1906.05.0	2391.05.0	2876.05.0	3361.05.0	3846.05.0	4331.05.0	4816.05.0	93
94	455.90.0	940.90.0	1425.90.0	1910.90.0	2395.90.0	2880.90.0	3365.90.0	3850.90.0	4335.90.0	4820.90.0	94
95	460.75.0	945.75.0	1430.75.0	1915.75.0	2400.75.0	2885.75.0	3370.75.0	3855.75.0	4340.75.0	4825.75.0	95
96	465.60.0	950.60.0	1435.60.0	1920.60.0	2405.60.0	2890.60.0	3375.60.0	3860.60.0	4345.60.0	4830.60.0	96
97	470.45.0	955.45.0	1440.45.0	1925.45.0	2410.45.0	2895.45.0	3380.45.0	3865.45.0	4350.45.0	4835.45.0	97
98	475.30.0	960.30.0	1445.30.0	1930.30.0	2415.30.0	2900.30.0	3385.30.0	3870.30.0	4355.30.0	4840.30.0	98
99	480.15.0	965.15.0	1450.15.0	1935.15.0	2420.15.0	2905.15.0	3390.15.0	3875.15.0	4360.15.0	4845.15.0	99

s.	10	11	12	13	14	15	16	17	18	19	s.
d.	$ cts.	$ cts.	$ cts.	$ cts.	$ cts.	$ cts.	$ cts.	$ cts.	$ cts.	$ cts.	d.
0	2.42.5	2.66.7	2.91.0	3.15.2	3.39.5	3.63.7	3.88.0	4.12.2	4.36.5	4.60.7	0
1	2.44.5	2.68.7	2.93.0	3.17.2	3.41.5	3.65.7	3.90.0	4.14.2	4.38.5	4.62.7	1
2	2.46.5	2.70.8	2.95.0	3.19.3	3.43.5	3.67.8	3.92.0	4.16.3	4.40.5	4.64.8	2
3	2.48.5	2.72.8	2.97.0	3.21.3	3.45.5	3.69.8	3.94.0	4.18.3	4.42.5	4.66.8	3
4	2.50.5	2.74.8	2.99.0	3.23.3	3.47.5	3.71.8	3.96.0	4.20.3	4.44.5	4.68.8	4
5	2.52.6	2.76.8	3.01.1	3.25.3	3.49.6	3.73.8	3.98.1	4.22.3	4.46.6	4.70.8	5
6	2.54.6	2.78.8	3.03.1	3.27.3	3.51.6	3.75.8	4.00.1	4.24.3	4.48.6	4.72.8	6
7	2.56.6	2.80.9	3.05.1	3.29.4	3.53.6	3.77.9	4.02.1	4.26.4	4.50.6	4.74.8	7
8	2.58.6	2.82.9	3.07.1	3.31.4	3.55.6	3.79.9	4.04.1	4.28.4	4.52.6	4.76.9	8
9	2.60.6	2.84.9	3.09.1	3.33.4	3.57.6	3.81.9	4.06.1	4.30.4	4.54.6	4.78.9	9
10	2.62.7	2.86.9	3.11.2	3.35.4	3.59.7	3.83.9	4.08.2	4.32.4	4.56.7	4.80.9	10
11	2.64.7	2.88.9	3.13.2	3.37.4	3.61.7	3.85.9	4.10.2	4.34.4	4.58.7	4.82.9	11

$4.85 44/100 TO THE POUND.

9¼%

PREMIUM.

STERLING INTO DOLLARS AND CENTS, AND VICE-VERSA.

£	0	100	200	300	400	500	600	700	800	900	£
	$ cts.	$ cts.	$ cts.	$ cts.	$ cts.	$ cts.	$ cts.	$ cts.	$ cts.	$ cts.	
0	......	485.55.5	971.11.1	1456.66.6	1942.22.2	2427.77.7	2913.33.3	3398.88.8	3884.44.4	4370.00.0	0
1	4.85.5	490.41.1	975.96.6	1461.52.2	1947.07.7	2432.63.3	2918.18.8	3403.74.4	3889.30.0	4374.85.5	1
2	9.71.1	495.26.6	980 82.2	1466.37.7	1951.93.3	2437.48.8	2923.04.4	3408.60.0	3894.15.5	4379.71 1	2
3	14.56.6	500.12.2	985.67.7	1471.23.3	1956.78 8	2442.34.4	2927.90.0	3413.45.5	3899.01.1	4384.56.6	3
4	19.42.2	504.97.7	990.53.3	1476.08.8	1961.64.4	2447.20.0	2932.75.5	3418.31.1	3903.86.6	4389.42.2	4
5	24.27.7	509.83.3	995.38.8	1480.94.4	1966.50.0	2452.05.5	2937.61.1	3423.16.6	3908.72.2	4394.27.7	5
6	29.13.3	514 68.8	1000.24.4	1485.80.0	1971.35.5	2456.91.1	2942.46.6	3428.02.2	3913.57.7	4399.13.3	6
7	33.98.8	519.54.4	1005.10.0	1490.65.5	1976.21.1	2461.76.6	2947.32.2	3432.87.7	3918.43.3	4403.98.8	7
8	38.84.4	524.40.0	1009.95.5	1495.51.1	1981.06 6	2466.62.2	2952.17.7	3437.73.3	3923.28.8	4408.84.4	8
9	43.70.0	529.25.5	1014.81.1	1500.36.6	1985.92.2	2471.47.7	2957.03.3	3442.58'8	3928.14.4	4413.70.0	9
10	48.55 5	534.11.1	1019.66.6	1505.22.2	1990.77.7	2476.33.3	2961.88.8	3447.44.4	3933.00.0	4418.55.5	10
11	53.41.1	538.96.6	1024.52.2	1510.07.7	1995.63.3	2481.18.8	2966.74.4	3452.30.0	3937.85.5	4423.41.1	11
12	58.26.6	543.82.2	1029.37.7	1514.93.3	2000.48.8	2486.04.4	2971.60.0	3457.15.5	3942.71.1	4428.26.6	12
13	63.12.2	548.67.7	1034.23.3	1519.78.8	2005.34.4	2490.90.0	2976.45.5	3462.01.1	3947.56.6	4433.12.2	13
14	67.97.7	553.53.3	1039.08.8	1524.64.4	2010.20.0	2495.75.5	2981.31.1	3466.86.6	3952.42.2	4437.97.7	14
15	72.83.3	558.38.8	1043.94.4	1529.50.0	2015.05.5	2500.61.1	2986.16.6	3471.72.2	3957.27.7	4442.83.3	15
16	77.68.8	563.24.4	1048.80.0	1534.35.5	2019.91.1	2505.46.6	2991.02.2	3476.57.7	3962.13.3	4447.68.8	16
17	82 54.4	568.10.0	1053.65.5	1539.21.1	2024.76.6	2510.32.2	2995.87.7	3481.43.3	3966.98.8	4452.54.4	17
18	87.40.0	572 95.5	1058.51.1	1544.06.6	2029.62.2	2515.17.7	3000.73.3	3486.28.8	3971.84.4	4457.40.0	18
19	92.25.5	577.81.1	1063.36.6	1548.92.2	2034.47.7	2520.03.3	3005.58.8	3491.14.4	3976.70.0	4462.25.5	19
20	97.11.1	582.66.6	1068.22 2	1553.77.7	2039.33.3	2524.88.8	3010.44.4	3496.00.0	3981.55.5	4467.11.1	20
21	101.96.6	587.52.2	1073.07.7	1558.63.3	2044.18.8	2529.74.4	3015.30.0	3500.85.5	3986.41.1	4471.96 6	21
22	106.82.2	592.37.7	1077.93.3	1563.48.8	2049.04.4	2534.60.0	3020.15.5	3505.71.1	3991.26.6	4476.82.2	22
23	111.67.7	597.23.3	1082.78.8	1568.34.4	2053.90.0	2539.45.5	3025.01.1	3510.56.6	3996.12.2	4481.67.7	23
24	116.53.3	602.08.8	1087.64.4	1573.20.0	2058.75.5	2544.31.1	3029.86.6	3515.42.2	4000.97.7	4486.53.3	24
25	121.38.8	606.94 4	1092.50.0	1578.05.5	2063.61.1	2549.16.6	3034.72.2	3520.27.7	4005.83.3	4491.38.8	25
26	126.24.4	611.80.0	1097.35.5	1582.91.1	2068.46.6	2554.02.2	3039.57.7	3525.13.3	4010.68.8	4496.24.4	26
27	131.10.0	616 65.5	1102.21.1	1587.76.6	2073.32.2	2558.87.7	3044.43.3	3529.98.8	4015.54.4	4501.10.0	27
28	135.95.5	621.51.1	1107.06.6	1592.62.2	2078.17.7	2563.73.3	3049.28.8	3534.84.4	4020.40.0	4505.95.5	28
29	140.81.1	626.36.6	1111.92.2	1597.47.7	2083.03.3	2568.58.8	3054.14.4	3539.70.0	4025.25.5	4510.81.1	29
30	145.66.6	631.22.2	1116.77.7	1602.33.3	2087.88.8	2573.44.4	3059.00.0	3544.55.5	4030.11.1	4515.66.6	30
31	150.52.2	636.07.7	1121.63.3	1607.18.8	2092.74.4	2578.30.0	3063.85.5	3549.41.1	4034 96.6	4520.52.2	31
32	155.37.7	640.93.3	1126.48.8	1612.04.4	2097.60.0	2583.15.5	3068.71.1	3554.26.6	4039.82.2	4525.37.7	32
33	160.23.3	645.78.8	1131.34.4	1616.90.0	2102.45.5	2588.01.1	3073.56.6	3559.12.2	4044.67.7	4530.23.3	33
34	165.08.8	650.64.4	1136.20.0	1621.75.5	2107.31.1	2592.86.6	3078.42.2	3563.97.7	4049.53.3	4535.08 8	34
35	169.94.4	655.50.0	1141.05.5	1626.61.1	2112.16.6	2597.72.2	3083.27.7	3568.83.3	4054.38.8	4539.94.4	35
36	174.80.0	660.35 5	1145.91.1	1631.46.6	2117.02.2	2602.57.7	3088.13.3	3573.68.8	4059.24.4	4544.80.0	36
37	179.65.5	665.21.1	1150 76.6	1636.32.2	2121.87.7	2607.43.3	3092.98.8	3578.54.4	4064.10.0	4549.65.5	37
38	184.51.1	670.06.6	1155.62.2	1641.17.7	2126.73.3	2612.28.8	3097.84.4	3583.40.0	4068.95.5	4554.51.1	38
39	189.36.6	674.92.2	1160.47.7	1646.03.3	2131.58.8	2617.14.4	3102.70.0	3588.25.5	4073.81.1	4559.36.6	39
40	194.22.2	679.77.7	1165.33.3	1650.88.8	2136.44.4	2622.00.0	3107.55.5	3593.11.1	4078.66.6	4564.22.2	40
41	199.07.7	684.63.3	1170.18.8	1655.74.4	2141.30.0	2626.85.5	3112.41.1	3597.96.6	4083.52.2	4569.07.7	41
42	203.93.3	689.48.8	1175.04.4	1660.60.0	2146.15.5	2631.71.1	3117.26.6	3602.82.2	4088.37.7	4573.93.3	42
43	208.78.8	694.34.4	1179.90.0	1665.45.5	2151.01.1	2636.56.6	3122.12.2	3607.67.7	4093.23.3	4578.78.8	43
44	213.64.4	699.20.0	1184.75.5	1670.31.1	2155.86.6	2641.42.2	3126.97.7	3612.53.3	4098.08.8	4583.64.4	44
45	218.50.0	704.05.5	1189.61.1	1675.16.6	2160.72.2	2646.27.7	3131.83.3	3617.38.8	4102.94.4	4588.50.0	45
46	223.35.5	708.91.1	1194.46.6	1680.02.2	2165.57.7	2651.13.3	3136.68.8	3622.24.4	4107.80.0	4593.35.5	46
47	228.21.1	713.76.6	1199.32.2	1684.87.7	2170.43.3	2655.98.8	3141.54.4	3627.10.0	4112.65.5	4598.21.1	47
48	233.06.6	718 62.2	1204.17.7	1689.73.3	2175.28.8	2660.84.4	3146.40.0	3631 95.5	4117.51.1	4603.06.6	48
49	237.92.2	723.47.7	1209.03.3	1694.58.8	2180.14.4	2665.70.0	3151.25.5	3636.81.1	4122.36.6	4607.92.2	49

s.	0	1	2	3	4	5	6	7	8	9	s.
d.	$ cts.	$ cts.	$ cts.	$ cts.	$ cts.	$ cts.	$ cts.	$ cts.	$ cts.	$ cts.	d.
0	......	.24,2	.48.5	.72.8	.97.1	1.21.3	1.45.6	1.69.9	1.94.2	2.18.5	0
1	.02.0	.26.3	.50.5	.74.8	.99.1	1.23.4	1.47.6	1.71.9	1.96.2	2.20.5	1
2	.04.0	.28.3	.52.6	.76.8	1.01.1	1.25.4	1.49.7	1.73.9	1.98.2	2.22.5	2
3	.06.0	.30.3	.54.6	.78.9	1.03.1	1.27.4	1.51.7	1.76.0	2.00.2	2.24.5	3
4	.08.0	.32.3	.56.6	.80.9	1.05.2	1.29.4	1.53.7	1.78.0	2.02.3	2.26.5	4
5	.10.1	34.3	.58.6	.82.9	1.07.2	1.31.5	1.55.7	1.80.0	2.04.3	2.28.6	5
6	.12.1	.36.4	.60.6	.84.9	1.09.2	1.33.5	1.57.8	1.82.0	2.06.3	2.30.6	6
7	.14.1	.38.4	.62.7	.86.9	1.11.2	1.35.5	1.59.8	1.84.1	2.08.3	2.32.6	7
8	.16.1	.40.4	.64.7	.89.0	1.13.2	1.37.5	1.61.8	1.80.1	2.10.3	2.34.6	8
9	.18.1	.42.4	.66.7	.91.0	1.15.3	1.39.5	1.63.8	1.88.1	2.12.4	2.36.6	9
10	.20.2	.44.4	.68.7	.93.0	1.17.3	1.41.6	1.65.8	1.90.1	2.14.4	2.38.7	10
11	.22.2	.46.4	.70.7	.95.0	1.19.3	1.43.6	1.67.9	1.92.1	2.16.4	2.40.7	11

$4.85⁷⁄₁₀ TO THE POUND

9¼ %

PREMIUM.

STERLING INTO DOLLARS AND CENTS, AND VICE-VERSA.

£	0	100	200	300	400	500	600	700	800	900	£
£	$ cts.	$ cts.	$ cts.	$ cts.	$ cts.	$ cts.	$ cts.	$ cts.	$ cts.	$ cts.	£
50	242.77.7	728.33.3	1213.88.8	1699.44.4	2185.00.0	2670.55.5	3156.11.1	3641.66.6	4127.22.2	4612.77.7	50
51	247.63.3	733.18.8	1218.74.4	1704.30.0	2189.85.5	2675.41.1	3160.96.6	3646.52.2	4132.07.7	4617.63.3	51
52	252.48.8	738.04.4	1223.60.0	1709.15.5	2194.71.1	2680.26.6	3165.82.2	3651.37.7	4136.93.3	4622.48.8	52
53	257.34.4	742.90.0	1228.45.5	1714.01.1	2199.56.6	2685.12.2	3170.67.7	3656.23.3	4141.78.8	4627.34.4	53
54	262.20.0	747.75.5	1233.31.1	1718.86.6	2204.42.2	2689.97.7	3175.53.3	3661.08.8	4146.64.4	4632.20.0	54
55	267.05.5	752.61.1	1238.16.6	1723.72.2	2209.27.7	2694.83.3	3180.38.8	3665.94.4	4151.50.0	4637.05.5	55
56	271.91.1	757.46.6	1243.02.2	1728.57.7	2214.13.3	2699.68.8	3185.24.4	3670.80.0	4156.35.5	4641.91.1	56
57	276.76.6	762.32.2	1247.87.7	1733.43.3	2218.98.8	2704.54.4	3190.10.0	3675.65.5	4161.21.1	4646.76.6	57
58	281.62.2	767.17.7	1252.73.3	1738.28.8	2223.84.4	2709.40.0	3194.95.5	3680.51.1	4166.06.6	4651.62.2	58
59	286.47.7	772.03.3	1257.58.8	1743.14.4	2228.70.0	2714.25.5	3199.81.1	3685.36.6	4170.92.2	4656.47.7	59
60	291.33.3	776.88.8	1262.44.4	1748.00.0	2233.55.5	2719.11.1	3204.66.6	3690.22.2	4175.77.7	4661.33.3	60
61	296.18.8	781.74.4	1267.30.0	1752.85.5	2238.41.1	2723.96.6	3209.52.2	3695.07.7	4180.63.3	4666.18.8	61
62	301.04.4	786.60.0	1272.15.5	1757.71.1	2243.26.6	2728.82.2	3214.37.7	3699.93.3	4185.48.8	4671.04.4	62
63	305.90.0	791.45.5	1277.01.1	1762.56.6	2248.12.2	2733.67.7	3219.23.3	3704.78.8	4190.34.4	4675.90.0	63
64	310.75.5	796.31.1	1281.86.6	1767.42.2	2252.97.7	2738.53.3	3224.08.8	3709.64.4	4195.20.0	4680.75.5	64
65	315.61.1	801.16.6	1286.72.2	1772.27.7	2257.83.3	2743.38.8	3228.94.4	3714.50.0	4200.05.5	4685.61.1	65
66	320.46.6	806.02.2	1291.57.7	1777.13.3	2262.68.8	2748.24.4	3233.80.0	3719.35.5	4204.91.1	4690.46.6	66
67	325.32.2	810.87.7	1296.43.3	1781.98.8	2267.54.4	2753.10.0	3238.65.5	3724.21.1	4209.76.6	4695.32.2	67
68	330.17.7	815.73.3	1301.28.8	1786.84.4	2272.40.0	2757.95.5	3243.51.1	3729.06.6	4214.62.2	4700.17.7	68
69	335.03.3	820.58.8	1306.14.4	1791.70.0	2277.25.5	2762.81.1	3248.36.6	3733.92.2	4219.47.7	4705.03.3	69
70	339.88.8	825.44.4	1311.00.0	1796.55.5	2282.11.1	2767.66.6	3253.22.2	3738.77.7	4224.33.3	4709.88.8	70
71	344.74.4	830.30.0	1315.85.5	1801.41.1	2286.96.6	2772.52.2	3258.07.7	3743.63.3	4229.18.8	4714.74.4	71
72	349.60.0	835.15.5	1320.71.1	1806.26.6	2291.82.2	2777.37.7	3262.93.3	3748.48.8	4234.04.4	4719.60.0	72
73	354.45.5	840.01.1	1325.56.6	1811.12.2	2296.67.7	2782.23.3	3267.78.8	3753.34.4	4238.90.0	4724.45.5	73
74	359.31.1	844.86.6	1330.42.2	1815.97.7	2301.53.3	2787.08.8	3272.64.4	3758.20.0	4243.75.5	4729.31.1	74
75	364.16.6	849.72.2	1335.27.7	1820.83.3	2306.38.8	2791.94.4	3277.50.0	3763.05.5	4248.61.1	4734.16.6	75
76	369.02.2	854.57.7	1340.13.3	1825.68.8	2311.24.4	2796.80.0	3282.35.5	3767.91.1	4253.46.6	4739.02.2	76
77	373.87.7	859.43.3	1344.98.8	1830.54.4	2316.10.0	2801.65.5	3287.21.1	3772.76.6	4258.32.2	4743.87.7	77
78	378.73.3	864.28.8	1349.84.4	1835.40.0	2320.95.5	2806.51.1	3292.06.6	3777.62.2	4263.17.7	4748.73.3	78
79	383.58.8	869.14.4	1354.70.0	1840.25.5	2325.81.1	2811.36.6	3296.92.2	3782.47.7	4268.03.3	4753.58.8	79
80	388.44.4	874.00.0	1359.55.5	1845.11.1	2330.66.6	2816.22.2	3301.77.7	3787.33.3	4272.88.8	4758.44.4	80
81	393.30.0	878.85.5	1364.41.1	1849.96.6	2335.52.2	2821.07.7	3306.63.3	3792.18.8	4277.74.4	4763.30.0	81
82	398.15.5	883.71.1	1369.26.6	1854.82.2	2340.37.7	2825.93.3	3311.48.8	3797.04.4	4282.60.0	4768.15.5	82
83	403.01.1	888.56.6	1374.12.2	1859.67.7	2345.23.3	2830.78.8	3316.34.4	3801.90.0	4287.45.5	4773.01.1	83
84	407.86.6	893.42.2	1378.97.7	1864.53.3	2350.08.8	2835.64.4	3321.20.0	3806.75.5	4292.31.1	4777.86.6	84
85	412.72.2	898.27.7	1383.83.3	1869.38.8	2354.94.4	2840.50.0	3326.05.5	3811.61.1	4297.16.6	4782.72.2	85
86	417.57.7	903.13.3	1388.68.8	1874.24.4	2359.80.0	2845.35.5	3330.91.1	3816.46.6	4302.02.2	4787.57.7	86
87	422.43.3	907.98.8	1393.54.4	1879.10.0	2364.65.5	2850.21.1	3335.76.6	3821.32.2	4306.87.7	4792.43.3	87
88	427.28.8	912.84.4	1398.40.0	1883.95.5	2369.51.1	2855.06.6	3340.62.2	3826.17.7	4311.73.3	4797.28.8	88
89	432.14.4	917.70.0	1403.25.5	1888.81.1	2374.36.6	2859.92.2	3345.47.7	3831.03.3	4316.58.8	4802.14.4	89
90	437.00.0	922.55.5	1408.11.1	1893.66.6	2379.22.2	2864.77.7	3350.33.3	3835.88.8	4321.44.4	4807.00.0	90
91	441.85.5	927.41.1	1412.96.6	1898.52.2	2384.07.7	2869.63.3	3355.18.8	3840.74.4	4326.30.0	4811.85.5	91
92	446.71.1	932.26.6	1417.82.2	1903.37.7	2388.93.3	2874.48.8	3360.04.4	3845.60.0	4331.15.5	4816.71.1	92
93	451.56.6	937.12.2	1422.67.7	1908.23.3	2393.78.8	2879.34.4	3364.90.0	3850.45.5	4336.01.1	4821.56.6	93
94	456.42.2	941.97.7	1427.53.3	1913.08.8	2398.64.4	2884.20.0	3369.75.5	3855.31.1	4340.86.6	4826.42.2	94
95	461.27.7	946.83.3	1432.38.8	1917.94.4	2403.50.0	2889.05.5	3374.61.1	3860.16.6	4345.72.2	4831.27.7	95
96	466.13.3	951.68.8	1437.24.4	1922.80.0	2408.35.5	2893.91.1	3379.46.6	3865.02.2	4350.57.7	4836.13.3	96
97	470.98.8	956.54.4	1442.10.0	1927.65.5	2413.21.1	2898.76.6	3384.32.2	3869.87.7	4355.43.3	4840.98.8	97
98	475.84.4	961.40.0	1446.95.5	1932.51.1	2418.06.6	2903.62.2	3389.17.7	3874.73.3	4360.28.8	4845.84.4	98
99	480.70.0	966.25.5	1451.81.1	1937.36.6	2422.92.2	2908.47.7	3394.03.3	3879.58.8	4365.14.4	4850.70.0	99

s.	10	11	12	13	14	15	16	17	18	19	s.
d.	$ cts.	$ cts.	$ cts.	$ cts.	$ cts.	$ cts.	$ cts.	$ cts.	$ cts.	$ cts.	d.
0	2.42.7	2.67.0	2.91.3	3.15.6	3.39.8	3.64.1	3.88.4	4.12.7	4.37.0	4.61.2	0
1	2.44.8	2.69.0	2.93.3	3.17.6	3.41.9	3.66.1	3.90.4	4.14.7	4.39.0	4.63.3	1
2	2.46.8	2.71.1	2.95.3	3.19.6	3.43.9	3.68.2	3.92.4	4.16.7	4.41.0	4.65.3	2
3	2.48.8	2.73.1	2.97.4	3.21.6	3.45.9	3.70.2	3.94.5	4.18.7	4.43.0	4.67.3	3
4	2.50.8	2.75.1	2.99.4	3.23.7	3.47.9	3.72.2	3.96.5	4.20.8	4.45.0	4.69.3	4
5	2.52.8	2.77.1	3.01.4	3.25.7	3.49.9	3.74.2	3.98.5	4.22.8	4.47.1	4.71.3	5
6	2.54.9	2.79.1	3.03.4	3.27.7	3.52.0	3.76.2	4.00.5	4.24.8	4.49.1	4.73.4	6
7	2.56.9	2.81.2	3.05.4	3.29.7	3.54.0	3.78.3	4.02.5	4.26.8	4.51.1	4.75.4	7
8	2.58.9	2.83.2	3.07.5	3.31.7	3.56.0	3.80.3	4.04.6	4.28.8	4.53.1	4.77.4	8
9	2.60.9	2.85.2	3.09.5	3.33.8	3.58.0	3.82.3	4.06.6	4.30.9	4.55.1	4.79.4	9
10	2.62.9	2.87.2	3.11.5	3.35.8	3.60.0	3.84.3	4.08.6	4.32.9	4.57.2	4.81.4	10
11	2.65.0	2.89.2	3.13.6	3.37.8	3.62.1	3.86.3	4.10.6	4.34.9	4.59.2	4.83.5	11

$4.86₁₁/₁₀₀ TO THE POUND.

9 3/8 %

PREMIUM.

STERLING INTO DOLLARS AND CENTS, AND VICE-VERSA.

£	0	100	200	300	400	500	600	700	800	900	£
	$ cts.	$ cts.	$ cts.	$ cts.	$ cts.	$ cts.	$ cts.	$ cts.	$ cts.	$ cts.	
0	......	486.11.1	972.22.2	1458.33.3	1944.44.4	2430 55.5	2916.66.6	3402.77.7	3888.88.8	4375.00.0	0
1	4.86.1	490.97.2	977.08.3	1463.19.4	1949.30.5	2435.41.6	2921.52.7	3407.63.8	3893.75.0	4379.86.1	1
2	9.72.2	495.83.3	981.94 4	1468.05.5	1954.16.6	2440.27.7	2926.38.8	3412.50.0	3898.61 1	4384.72.2	2
3	14.58.3	500.69.4	986.80.5	1472.91.6	1959.02.7	2445.13.8	2931.25.0	3417.36.1	3903.47.2	4389 58.3	3
4	19.14.4	505.55.5	991.66.6	1477.77.7	1963.88.8	2450.00.0	2936.11.1	3422.22.2	3908.33.3	4394.44.4	4
5	24.30.5	510.41.6	996.52.7	1482.63.8	1968.75.0	2454.86.1	2940.97.2	3427.08.3	3913.19.4	4399.30.5	5
6	29.16.6	515.27.7	1001.38.8	1487.50.0	1973.61.1	2459.72.2	2945.83 3	3431.94.4	3918.05.5	4404.16.6	6
7	34.02.7	520.13.8	1006.25.0	1492.36.1	1978.47.2	2464.58.3	2950.69.4	3436.80.5	3922.91.6	4409.02.7	7
8	38.88.8	525 00.0	1011.11.1	1497.22.2	1983.33.3	2469.44.4	2955.55.5	3441.66.6	3927.77.7	4413.88.8	8
9	43.75.0	529.86.1	1015.97.2	1502.08.3	1988.19.4	2474.30.5	2960.41.6	3446.52.7	3932.63.8	4418.75.0	9
10	48.61.1	534.72.2	1020.83.3	1506.94.4	1993.05.5	2479.16.6	2965.27.7	3451.38.8	3937.50.0	4423.61.1	10
11	53.47.2	539.58 3	1025.69.4	1511.80.5	1997.91.6	2484.02.7	2970 13.8	3456.25.0	3942.36.1	4428.47.2	11
12	58.33.3	544.44 4	1030.55.5	1516.66.6	2002.77.7	2488.88 8	2975.00.0	3461.11.1	3947.22.2	4433.33.3	12
13	63.19.4	549.30.5	1035.41.6	1521.52.7	2007.63.8	2493.75.0	2979.86.1	3465.97.2	3952 08.3	4438.19.4	13
14	68.05.5	554.16 6	1040.27.7	1526.38.8	2012.50.0	2498.61.1	2984.72.2	3470.83.3	3956.94.4	4443.05.5	14
15	72.91.6	559.02.7	1045.13.8	1531.25.0	2017.36.1	2503.47.2	2989.58.3	3475.69.4	3961.80.5	4447.91.6	15
16	77.77.7	563.88.8	1050.00.0	1536.11.1	2022.22.2	2508.33.3	2994.44.4	3480.55.5	3966.66.6	4452.77.7	16
17	82 63.8	568.75.0	1054.86.1	1540 97.2	2027.08.3	2513.19.4	2999.30.5	3485.41.6	3971.52 7	4457.63.8	17
18	87.50.0	573.61.1	1059.72.2	1589.58.3	2031.94.4	2518.05.5	3004.16.6	3490.27.7	3976.38.8	4462.50.0	18
19	92.36.1	578.47.2	1064.58.3	1550.69.4	2036.80.5	2522 91.6	3009.02.7	3495.13.8	3981.25.0	4467.36.1	19
20	97.22.2	583.33.3	1069 44.4	1555.55.5	2041.66.6	2527.77.7	3013.88.8	3500.00.0	3986.11.1	4472.22.2	20
21	102.08.3	588.19.4	1074.30.5	1560.41.6	2046.52.7	2532.63.8	3018.75.0	3504.86.1	3990.97 2	4477.08.3	21
22	106.94.4	593.05.5	1079.16.6	1565.27.7	2051.38.8	2537.50.0	3023.61.1	3509.72.2	3995.83.3	4481.94.4	22
23	111.80.5	597.91.6	1084.02.7	1570.13.8	2056.25.0	2542.36.1	3028.47.2	3514.58.3	4000.69.4	4486.80.5	23
24	116.66.6	602.77.7	1088.88.8	1575.00.0	2061.11.1	2547.22.2	3033.33.3	3519.44.4	4005.55.5	4491.66.6	24
25	121.52.7	607.63.8	1093.75.0	1579.86.1	2065.97.2	2552.08.3	3038.19.4	3524.30.5	4010.41.6	4496.52.7	25
26	126.38.8	612.50.0	1098.61.1	1584.72.2	2070.83.3	2556.94.4	3043.05.5	3529.16.6	4015.27.7	4501.38.8	26
27	131.25.0	617.36.1	1103.47.2	1589.58.3	2075.69.4	2561.80.5	3047.91.6	3534.02.7	4020.13.8	4506.25.0	27
28	136.11.1	622.22.2	1108.33.3	1594.44.4	2080.55.5	2566.66.6	3052.77.7	3538.88.8	4025.00.0	4511.11.1	28
29	140.97.2	627.08.3	1113.19.4	1599.30.5	2085.41.6	2571.52.7	3057.63.8	3543.75.0	4029.86.1	4515.97.2	29
30	145.83.3	631.94.4	1118.05.5	1604 16.6	2090.27.7	2576.38.8	3062.50.0	3548.61.1	4034.72.2	4520.83.3	30
31	150.69.4	636.80.5	1122.91.6	1609.02.7	2095.13.8	2581.25.0	3067.36.1	3553.47.2	4039.58.3	4525.69.4	31
32	155.55.5	641.66 6	1127.77.7	1613.88.8	2100.00.0	2586.11.1	3072.22.2	3558.33.3	4044.44.4	4530.55.5	32
33	160.41.6	646.52.7	1132.63.8	1618.75.0	2104.86.1	2590.97.2	3077.08.3	3563.19.4	4049.30.5	4535.41.6	33
34	165.27.7	651.38.8	1137.50.0	1623.61.1	2109.72.2	2595.83.3	3081.94.4	3568.05.5	4054.16.6	4540.27.7	34
35	170.13.8	656.25.0	1142.36.1	1628.47.2	2114.58.3	2600.69.4	3086.80.5	3572.91.6	4059.02.7	4545.13.8	35
36	175.00.0	661.11.1	1147.22.2	1633.33.3	2119.44.4	2605.55.5	3091.66.6	3577.77.7	4063.88.8	4550.00.0	36
37	179.86.1	665.97.2	1152.08.3	1638.19.4	2124.30.5	2610.41.6	3096.52.7	3582.63.8	4068.75.0	4554.86.1	37
38	184.72.2	670.83.3	1156.94.4	1643.05.5	2129.16.6	2615.27.7	3101.38.8	3587.50.0	4073.61.1	4559.72.2	38
39	189.58.3	675.69 4	1161.80.5	1647.91.6	2134.02.7	2620.13.8	3106.25.0	3592 36.1	4078.47.2	4564.58.3	39
40	194.44 4	680.55.5	1166.66.6	1652.77.7	2138.88.8	2625.00.0	3111.11.1	3597.22.2	4083.33.3	4569.44.4	40
41	199.30.5	685.41.6	1171.52.7	1657.63 8	2143.75.0	2629.86.1	3115.97.2	3602.08.3	4088.19.4	4574.30.5	41
42	204.16.6	690.27.7	1176.38.8	1662.50.0	2148.61.1	2634.72.2	3120.83.3	3606.94 4	4093.05.5	4579.16.6	42
43	209.02.7	695.13.8	1181.25.0	1667.36.1	2153.47.2	2639.58.3	3125 69.4	3611.80.5	4097.91.6	4584.02.7	43
44	213.88.8	700.00.0	1186.11.1	1672.22 2	2158.33.3	2644.44.4	3130.55.5	3616.66.6	4102.77.7	4588.88.8	44
45	218.75.0	704.86.1	1190.97.2	1677.08.3	2163 19.4	2649 30.5	3135.41.6	3621.52.7	4107.63.8	4593.75.0	45
46	223.61.1	709.72.2	1195.83.3	1681.94.4	2168.05.5	2654.16.6	3140.27.7	3626.38.8	4112.50.0	4598.61.1	46
47	228.47.2	714.58.3	1200.69.4	1686.80.5	2172.91.6	2659.02.7	3145.13.8	3631.25.0	4117.36 1	4603.47.2	47
48	233.33.3	719.44.4	1205.55.5	1691.66.6	2177.77.7	2663.88.8	3150.00.0	3636.11.1	4122.22.2	4608.33.3	48
49	238.19.4	724.30.5	1210.41.6	1696.52.7	2182.63.8	2668.75.0	3154.86.1	3640.97.2	4127.08.3	4613.19.4	49

s.	0	1	2	3	4	5	6	7	8	9	s.
d.	$ cts.	$ cts.	$ cts.	$ cts.	$ cts.	$ cts.	$ cts.	$ cts.	$ cts.	$ cts.	d.
0	......	.24.3	.48.6	.72.9	.97.2	1.21.5	1.45.8	1.70.1	1.94.4	2.18.7	0
1	.02.0	.26.3	.50.6	.74.9	.99.2	1.23.5	1.47.8	1.72.1	1.96.4	2.20.7	1
2	.04.0	.28.3	.52.6	.76.9	1.01.2	1.25.5	1.49.8	1.74.1	1.98.4	2.22.8	2
3	.06.0	.30.3	.54.6	.78.9	1.03.2	1.27.5	1.51.9	1.76.2	2.00.5	2.24.8	3
4	.08.0	.32.3	.56.7	.81.0	1.05.3	1.29.6	1.53.9	1.78.2	2.02.5	2.26.8	4
5	.10.1	.34.4	.58.7	.83.0	1.07.3	1.31.6	1.55.9	1.80.2	2.04.5	2.28.8	5
6	.12.1	.36.4	.60.7	.85.0	1.09.3	1.33.6	1.57.9	1.82.2	2.06.5	2.30.8	6
7	.14.1	.38.4	.62.7	.87.0	1.11.3	1.35.6	1.59.9	1.84.2	2.08.5	2.32.9	7
8	.16.1	.40.4	.64.7	.89.0	1.13.3	1.37.6	1.62.0	1.86.3	2.10.6	2.34.9	8
9	.18.1	.42.4	.66.8	.91.1	1.15.4	1.39.7	1.64.0	1.88.3	2.12.6	2.36.9	9
10	.20.2	.44.5	.68.8	.93.1	1.17.4	1.41.7	1.66.0	1.90.3	2.14.6	2.38.9	10
11	.22.2	.46 5	.70.8	.95.1	1.19.4	1.43.7	1.68.0	1.92.3	2.16.6	2.40.9	11

$4.86 11/16 TO THE POUND.

9 3/8 %

PREMIUM.

STERLING INTO DOLLARS AND CENTS, AND VICE-VERSA.

£	0	100	200	300	400	500	600	700	800	900	£
	$ cts.	$ cts.	$ cts.	$ cts.	$ cts.	$ cts.	$ cts.	$ cts.	$ cts.	$ cts.	
50	243.05.5	729.16.6	1215.27.7	1701.38.8	2187.50.0	2673.61.1	3159.72.2	3645.83.3	4131.94.4	4618.05.5	50
51	247.91.6	734.02.7	1220.13.8	1706.25.0	2192.36.1	2678.47.2	3164.58.3	3650.69.4	4136.80.5	4622.91.6	51
52	252.77.7	738.88.8	1225.00.0	1711.11.1	2197.22.2	2683.33.3	3169.44.4	3655.55.5	4141.66.6	4627.77.7	52
53	257.63.8	743.75.0	1229.86.1	1715.97.2	2202.08.3	2688.19.4	3174.30.5	3660.41.6	4146.52.7	4632.63.8	53
54	262.50.0	748.61.1	1234.72.2	1720.83.3	2206.94.4	2693.05.5	3179.16.6	3665.27.7	4151.38.8	4637.50.0	54
55	267.36.1	753.47.2	1239.58.3	1725.69.4	2211.80.5	2697.91.6	3184.02.7	3670.13.8	4156.25.0	4642.36.1	55
56	272.22.2	758.33.3	1244.44.4	1730.55.5	2216.66.6	2702.77.7	3188.88.8	3675.00.0	4161.11.1	4647.22.2	56
57	277.08.3	763.19.4	1249.30.5	1735.41.6	2221.52.7	2707.63.8	3193.75.0	3679.86.1	4165.97.2	4652.08.3	57
58	281.94.4	768.05.5	1254.16.6	1740.27.7	2226.38.8	2712.50.0	3198.61.1	3684.72.2	4170.83.3	4656.94.4	58
59	286.80.5	772.91.6	1259.02.7	1745.13.8	2231.25.0	2717.36.1	3203.47.2	3689.58.3	4175.69.4	4661.80.5	59
60	291.66.6	777.77.7	1263.88.8	1750.00.0	2236.11.1	2722.22.2	3208.33.3	3694.44.4	4180.55.5	4666.66.6	60
61	296.52.7	782.63.8	1268.75.0	1754.86.1	2240.97.2	2727.08.3	3213.19.4	3699.30.5	4185.41.6	4671.52.7	61
62	301.38.8	787.50.0	1273.61.1	1759.72.2	2245.83.3	2731.94.4	3218.05.5	3704.16.6	4190.27.7	4676.38.8	62
63	306.25.0	792.36.1	1278.47.2	1764.58.3	2250.69.4	2736.80.5	3222.91.6	3709.02.7	4195.13.8	4681.25.0	63
64	311.11.1	797.22.2	1283.33.3	1769.44.4	2255.55.5	2741.66.6	3227.77.7	3713.88.8	4200.00.0	4686.11.1	64
65	315.97.2	802.08.3	1288.19.4	1774.30.5	2260.41.6	2746.52.7	3232.63.8	3718.75.0	4204.86.1	4690.97.2	65
66	320.83.3	806.94.4	1293.05.5	1779.16.6	2265.27.7	2751.38.8	3237.50.0	3723.61.1	4209.72.2	4695.83.3	66
67	325.69.4	811.80.5	1297.91.6	1784.02.7	2270.13.8	2756.25.0	3242.36.1	3728.47.2	4214.58.3	4700.69.4	67
68	330.55.5	816.66.6	1302.77.7	1788.88.8	2275.00.0	2761.11.1	3247.22.2	3733.33.3	4219.44.4	4705.55.5	68
69	335.41.6	821.52.7	1307.63.8	1793.75.0	2279.86.1	2765.97.2	3252.08.3	3738.19.4	4224.30.5	4710.41.6	69
70	340.27.7	826.38.8	1312.50.0	1798.61.1	2284.72.2	2770.83.3	3256.94.4	3743.05.5	4229.16.6	4715.27.7	70
71	345.13.8	831.25.0	1317.36.1	1803.47.2	2289.58.3	2775.69.4	3261.80.5	3747.91.6	4234.02.7	4720.13.8	71
72	350.00.0	836.11.1	1322.22.2	1808.33.3	2294.44.4	2780.55.5	3266.66.6	3752.77.7	4238.88.8	4725.00.0	72
73	354.86.1	840.97.2	1327.08.3	1813.19.4	2299.30.5	2785.41.6	3271.52.7	3757.63.8	4243.75.0	4729.86.1	73
74	359.72.2	845.83.3	1331.94.4	1818.05.5	2304.16.6	2790.27.7	3276.38.8	3762.50.0	4248.61.1	4734.72.2	74
75	364.58.3	850.69.4	1336.80.5	1822.91.6	2309.02.7	2795.13.8	3281.25.0	3767.36.1	4253.47.2	4739.58.3	75
76	369.44.4	855.55.5	1341.66.6	1827.77.7	2313.88.8	2800.00.0	3286.11.1	3772.22.2	4258.33.3	4744.44.4	76
77	374.30.5	860.41.6	1346.52.7	1832.63.8	2318.75.0	2804.86.1	3290.97.2	3777.08.3	4263.19.4	4749.30.5	77
78	379.16.6	865.27.7	1351.38.8	1837.50.0	2323.61.1	2809.72.2	3295.83.3	3781.94.4	4268.05.5	4754.16.6	78
79	384.02.7	870.13.8	1356.25.0	1842.36.1	2328.47.2	2814.58.3	3300.69.4	3786.80.5	4272.91.6	4759.02.7	79
80	388.88.8	875.00.0	1361.11.1	1847.22.2	2333.33.3	2819.44.4	3305.55.5	3791.66.6	4277.77.7	4763.88.8	80
81	393.75.0	879.86.1	1365.97.2	1852.08.3	2338.19.4	2824.30.5	3310.41.6	3796.52.7	4282.63.8	4768.75.0	81
82	398.61.1	884.72.2	1370.83.3	1856.94.4	2343.05.5	2829.16.6	3315.27.7	3801.38.8	4287.50.0	4773.61.1	82
83	403.47.2	889.58.3	1375.69.4	1861.80.5	2347.91.6	2834.02.7	3320.13.8	3806.25.0	4292.36.1	4778.47.2	83
84	408.33.3	894.44.4	1380.55.5	1866.66.6	2352.77.7	2838.88.8	3325.00.0	3811.11.1	4297.22.2	4783.33.3	84
85	413.19.4	899.30.5	1385.41.6	1871.52.7	2357.63.8	2843.75.0	3329.86.1	3815.97.2	4302.08.3	4788.19.4	85
86	418.05.5	904.16.6	1390.27.7	1876.38.8	2362.50.0	2848.61.1	3334.72.2	3820.83.3	4306.94.4	4793.05.5	86
87	422.91.6	909.02.7	1395.13.8	1881.25.0	2367.36.1	2853.47.2	3339.58.3	3825.69.4	4311.80.5	4797.91.6	87
88	427.77.7	913.88.8	1400.00.0	1886.11.1	2372.22.2	2858.33.3	3344.44.4	3830.55.5	4316.66.6	4802.77.7	88
89	432.63.8	918.75.0	1404.86.1	1890.97.2	2377.08.3	2863.19.4	3349.30.5	3835.41.6	4321.52.7	4807.63.8	89
90	437.50.0	923.61.1	1409.72.2	1895.83.3	2381.94.4	2868.05.5	3354.16.6	3840.27.7	4326.38.8	4812.50.0	90
91	442.36.1	928.47.2	1414.58.3	1900.69.4	2386.80.5	2872.91.6	3359.02.7	3845.13.8	4331.25.0	4817.36.1	91
92	447.22.2	933.33.3	1419.44.4	1905.55.5	2391.66.6	2877.77.7	3363.88.8	3850.00.0	4336.11.1	4822.22.2	92
93	452.08.3	938.19.4	1424.30.5	1910.41.6	2396.52.7	2882.63.8	3368.75.0	3854.86.1	4340.97.2	4827.08.3	93
94	456.94.4	943.05.5	1429.16.6	1915.27.7	2401.38.8	2887.50.0	3373.61.1	3859.72.2	4345.83.3	4831.94.4	94
95	461.80.5	947.91.6	1434.02.7	1920.13.8	2406.25.0	2892.36.1	3378.47.2	3864.58.3	4350.69.4	4836.80.5	95
96	466.66.6	952.77.7	1438.88.8	1925.00.0	2411.11.1	2897.22.2	3383.33.3	3869.44.4	4355.55.5	4841.66.6	96
97	471.52.7	957.63.8	1443.75.0	1929.86.1	2415.97.2	2902.08.3	3388.19.4	3874.30.5	4360.41.6	4846.52.7	97
98	476.38.8	962.50.0	1448.61.1	1934.72.2	2420.83.3	2906.94.4	3393.05.5	3879.16.6	4365.27.7	4851.38.8	98
99	481.25.0	967.36.1	1453.47.2	1939.58.3	2425.69.4	2911.80.5	3397.91.6	3884.02.7	4370.13.8	4856.25.0	99

S.	10	11	12	13	14	15	16	17	18	19	S.
d.	$ cts.	$ cts.	$ cts.	$ cts.	$ cts.	$ cts.	$ cts.	$ cts.	$ cts.	$ cts.	d.
0	2.43.0	2.67.3	2.91.6	3.15.9	3.40.2	3.64.5	3.88.8	4.13.1	4.37.5	4.61.8	0
1	2.45.0	2.69.3	2.93.6	3.18.0	3.42.3	3.66.6	3.90.9	4.15.2	4.39.5	4.63.8	1
2	2.47.1	2.71.4	2.95.7	3.20.0	3.44.3	3.68.6	3.92.9	4.17.2	4.41.5	4.65.8	2
3	2.49.1	2.73.4	2.97.7	3.22.0	3.46.3	3.70.6	3.94.9	4.19.2	4.43.5	4.67.8	3
4	2.51.1	2.75.4	2.99.7	3.24.0	3.48.3	3.72.6	3.96.9	4.21.2	4.45.5	4.69.8	4
5	2.53.1	2.77.4	3.01.7	3.26.0	3.50.3	3.74.6	3.98.9	4.23.3	4.47.6	4.71.9	5
6	2.55.1	2.79.4	3.03.7	3.28.1	3.52.4	3.76.7	4.01.0	4.25.3	4.49.6	4.73.9	6
7	2.57.2	2.81.5	3.05.8	3.30.1	3.54.4	3.78.7	4.03.0	4.27.3	4.51.6	4.75.9	7
8	2.59.2	2.83.5	3.07.8	3.32.1	3.56.4	3.80.7	4.05.0	4.29.3	4.53.6	4.77.9	8
9	2.61.2	2.85.5	3.09.8	3.34.1	3.58.4	3.82.7	4.07.0	4.31.3	4.55.6	4.79.9	9
10	2.63.2	2.87.5	3.11.8	3.36.1	3.60.4	3.84.7	4.09.0	4.33.4	4.57.7	4.82.0	10
11	2.65.2	2.89.5	3.13.8	3.38.2	3.62.5	3.86.8	4.11.1	4.35.4	4.59.7	4.84.0	11

$4.86⅔ TO THE POUND.

9½%

PREMIUM.

STERLING INTO DOLLARS AND CENTS, AND VICE-VERSA.

£	0	100	200	300	400	500	600	700	800	900	£
	$ cts.	$ cts.	s cts.	$ cts.	$ cts.	$ cts.	$ cts.	$ cts.	$ cts.	$ cts.	
0	......	486.66.6	973.33.3	1460.00.0	1946.66.6	2433.33.3	2920.00.0	3406.66.6	3893.33.3	4380.00.0	0
1	4.86.6	491.53.3	978.20.0	1464.86.6	1951.53.3	2438.20.0	2924.86.6	3411.53.3	3898.20.0	4384.86.6	1
2	9.73.3	496.40.0	983.06.6	1469.73.3	1956.40.0	2443.06.6	2929.73.3	3416.40.0	3903.06.6	4389.73.3	2
3	14.60.0	501.26.6	987.93.3	1474.60.0	1961.26.6	2447.93.3	2934.60.0	3421.26.6	3907.93.3	4394.60.0	3
4	19.46.6	506.13.3	992.80.0	1479.46.6	1966.13.3	2452.80.0	2939.46.6	3426.13.3	3912.80.0	4399.46.6	4
5	24.33.3	511.00.0	997.66.6	1484.33.3	1971.00.0	2457.66.6	2944.33.3	3431.00.0	3917.66.6	4404.33.3	5
6	29.20.0	515.86.6	1002.53.3	1489.20.0	1975.86.6	2462.53.3	2949.20.0	3435.86.6	3922.53.3	4409.20.0	6
7	34.06.6	520.73.3	1007.40.0	1494.06.6	1980.73.3	2467.40.0	2954.06.6	3440.73.3	3927.40.0	4414.06.6	7
8	38.93.3	525.60.0	1012.26.6	1498.93.3	1985.60.0	2472.26.6	2958.93.3	3445.60.0	3932.26.6	4418.93.3	8
9	43.80.0	530.46.6	1017.13.3	1503.80.0	1990.46.6	2477.13.3	2963.80.0	3450.46.6	3937.13.3	4423.80.0	9
10	48.66.6	535.33.3	1022.00.0	1508.66.6	1995.33.3	2482.00.0	2968.66.6	3455.33.3	3942.00.0	4428.66.6	10
11	53.53.3	540.20.0	1026.66.6	1513.53.3	2000.20.0	2486.86.6	2973.53.3	3460.20.0	3946.86.6	4433.53.3	11
12	58.40.0	545.06.6	1031.73.3	1518.40.0	2005.06.6	2491.73.3	2978.40.0	3165.06.6	3951.73.3	4438.40.0	12
13	63.26.6	549.93.3	1036.60.0	1523.26.6	2009.93.3	2496.60.0	2983.26.6	3469.93.3	3956.60.0	4443.26.6	13
14	68.13.3	554.80.0	1041.46.6	1528.13.3	2014.80.0	2501.46.6	2988.13.3	3474.80.0	3961.46.6	4448.13.3	14
15	73.00.0	559.66.6	1046.33.3	1533.00.0	2019.66.6	2506.33.3	2993.00.0	3479.66.6	3966.33.3	4453.00.0	15
16	77.86.6	564.53.3	1051.20.0	1537.86.6	2024.53.3	2511.20.0	2997.86.6	3484.53.3	3971.20.0	4457.86.6	16
17	82.73.3	569.40.0	1056.06.6	1542.73.3	2029.40.0	2516.06.6	3002.73.3	3489.40.0	3976.06.6	4462.73.3	17
18	87.60.0	574.26.6	1060.93.3	1547.60.0	2034.26.6	2520.93.3	3007.60.0	3494.26.6	3980.93.3	4467.60.0	18
19	92.46.6	579.13.3	1065.80.0	1552.46.6	2039.13.3	2525.80.0	3012.46.6	3499.13.3	3985.80.0	4472.46.6	19
20	97.33.3	584.00.0	1070.66.6	1557.33.3	2044.00.0	2530.66.6	3017.33.3	3504.00.0	3990.66.6	4477.33.3	20
21	102.20.0	588.86.6	1075.53.3	1562.20.0	2048.86.6	2535.53.3	3022.20.0	3508.86.6	3995.53.3	4482.20.0	21
22	107.06.6	593.73.3	1080.40.0	1567.06.6	2053.73.3	2540.40.0	3027.06.6	3513.73.3	4000.40.0	4487.06.6	22
23	111.93.3	598.60.0	1085.26.6	1571.93.3	2058.60.0	2545.26.6	3031.93.3	3518.60.0	4005.26.6	4491.93.3	23
24	116.80.0	603.46.6	1090.13.3	1576.80.0	2063.46.6	2550.13.3	3036.80.0	3523.46.6	4010.13.3	4496.80.0	24
25	121.66.6	608.33.3	1095.00.0	1581.66.6	2068.33.3	2555.00.0	3041.66.6	3528.33.3	4015.00.0	4501.66.6	25
26	126.53.3	613.20.0	1099.86.6	1586.53.3	2073.20.0	2559.86.6	3046.53.3	3533.20.0	4019.86.6	4506.53.3	26
27	131.40.0	618.06.6	1104.73.3	1591.40.0	2078.06.6	2564.73.3	3051.40.0	3538.06.6	4024.73.3	4511.40.0	27
28	136.26.6	622.93.3	1109.60.0	1596.26.6	2082.93.3	2569.60.0	3056.26.6	3542.93.3	4029.60.0	4516.26.6	28
29	141.13.3	627.80.0	1114.46.6	1601.13.3	2087.80.0	2574.46.6	3061.13.3	3547.80.0	4034.46.6	4521.13.3	29
30	146.00.0	632.66.6	1119.33.3	1606.00.0	2092.66.6	2579.33.3	3066.00.0	3552.66.6	4039.33.3	4526.00.0	30
31	150.86.6	637.53.3	1124.20.0	1610.86.6	2097.53.3	2584.20.0	3070.86.6	3557.53.3	4044.20.0	4530.86.6	31
32	155.73.3	642.40.0	1129.06.6	1615.73.3	2102.40.0	2589.06.6	3075.73.3	3562.40.0	4049.06.6	4535.73.3	32
33	160.60.0	647.26.6	1133.93.3	1620.60.0	2107.26.6	2593.93.3	3080.60.0	3567.26.6	4053.93.3	4540.60.0	33
34	165.46.6	652.13.3	1138.80.0	1625.46.6	2112.13.3	2598.80.0	3085.46.6	3572.13.3	4058.80.0	4545.46.6	34
35	170.33.3	657.00.0	1143.66.6	1630.33.3	2117.00.0	2603.66.6	3090.33.3	3577.00.0	4063.66.6	4550.33.3	35
36	175.20.0	661.86.6	1148.53.3	1635.20.0	2121.86.6	2608.53.3	3095.20.0	3581.86.6	4068.53.3	4555.20.0	36
37	180.06.6	666.73.3	1153.40.0	1640.06.6	2126.73.3	2613.40.0	3100.06.6	3586.73.3	4073.40.0	4560.06.6	37
38	184.93.3	671.60.0	1158.26.6	1644.93.3	2131.60.0	2618.26.6	3104.93.3	3591.60.0	4078.26.6	4564.93.3	38
39	189.80.0	676.46.6	1163.13.3	1649.80.0	2136.46.6	2623.13.3	3109.80.0	3596.46.6	4083.13.3	4569.80.0	39
40	194.66.6	681.33.3	1168.00.0	1654.66.6	2141.33.3	2628.00.0	3114.66.6	3601.33.3	4088.00.0	4574.66.6	40
41	199.53.3	686.20.0	1172.86.6	1659.53.3	2146.20.0	2632.86.6	3119.53.3	3606.20.0	4092.86.6	4579.53.3	41
42	204.40.0	691.06.6	1177.73.3	1664.40.0	2151.06.6	2637.73.3	3124.40.0	3611.06.6	4097.73.3	4584.40.0	42
43	209.26.6	695.93.3	1182.60.0	1669.26.6	2155.93.3	2642.60.0	3129.26.6	3615.93.3	4102.60.0	4589.26.6	43
44	214.13.3	700.80.0	1187.46.6	1674.13.3	2160.80.0	2647.46.6	3134.13.3	3620.80.0	4107.46.6	4594.13.3	44
45	219.00.0	705.66.6	1192.33.3	1679.00.0	2165.66.6	2652.33.3	3139.00.0	3625.66.6	4112.33.3	4599.00.0	45
46	223.86.6	710.53.3	1197.20.0	1683.86.6	2170.53.3	2657.20.0	3143.86.6	3630.53.3	4117.20.0	4603.86.6	46
47	228.73.3	715.40.0	1202.06.6	1688.73.3	2175.40.0	2662.06.6	3148.73.3	3635.40.0	4122.06.6	4608.73.3	47
48	233.60.0	720.26.6	1206.93.3	1693.60.0	2180.26.6	2666.93.3	3153.60.0	3640.26.6	4126.93.3	4613.60.0	48
49	238.46.6	725.13.3	1211.80.0	1698.46.6	2185.13.3	2671.80.0	3158.46.6	3645.13.3	4131.80.0	4618.46.6	49

S.	0.	1.	2.	3.	4.	5.	6.	7.	8.	9.	S.
d.	$ cts.	$ cts.	$ cts.	$ cts.	$ cts.	$ cts.	$ cts.	$ cts.	$ cts.	$ cts.	d.
0	......	.24.3	.48.6	.73.0	.97.3	1.21.6	1.46.0	1.70.3	1.94.6	2.19.0	0
1	.02.0	.26.3	.50.7	.75.0	.99.3	1.23.7	1.48.0	1.72.3	1.96.7	2.21.0	1
2	.04.0	.28.4	.52.7	.77.0	1.01.4	1.25.7	1.50.0	1.74.4	1.98.7	2.23 1	2
3	.06.0	.30.4	.54.7	.79.1	1.03.4	1.27.7	1.52.0	1.76.4	2.00.7	2.26 1	3
4	.08.1	.32.4	.56.8	.81.1	1.05.4	1.29.8	1.54.1	1.78.4	2.02.8	2.27.1	4
5	.10.1	.34.5	.58.8	.83.1	1.07.5	1.31.8	1.56.1	1.80.5	2.04.8	2.29.2	5
6	.12.1	.36.5	.60.8	.85.1	1.09.5	1.33.8	1.58.1	1.82.5	2.06.8	2.31.2	6
7	.14.2	.38.5	.62.9	.87.2	1.11.5	1.35.9	1.60.2	1.84.5	2.08.9	2.33 2	7
8	.16.2	.40.6	.64.9	.89.2	1.13.6	1.37.9	1.62.2	1.86.6	2.10.9	2.35.3	8
9	.18.2	.42.6	.66.9	.91.2	1.15.6	1.39.9	1.64.2	1.88.6	2.12.9	2.37.3	9
10	.20.3	.44.6	.69.0	.93.3	1.17.6	1.42.0	1.66.3	1.90.6	2.15.0	2.39 3	10
11	.22.3	.46.7	.71.0	.95.3	1.19.7	1.44.0	1.68.3	1.92.7	2.17.0	2.41.4	11

$4.86⅞ TO THE POUND.

9½%

PREMIUM.

STERLING INTO DOLLARS AND CENTS, AND VICE-VERSA.

£	0	100	200	300	400	500	600	700	800	900	£
	$ cts.	$ cts.	$ cts.	$ cts.	$ cts.	$ cts.	$ cts.	$ cts.	$ cts.	$ cts.	
50	243.33.3	730.00.0	1216.66.6	1703.33.3	2190.00.0	2676.66.6	3163.33.3	3650.00.0	4136.66.6	4623.33.3	50
51	248.20.0	734.86.6	1221.53.3	1708.20.0	2194.86.6	2681.53.3	3168.20.0	3654.86.6	4141.53.3	4628.20.0	51
52	253.06.6	739.73.3	1226.40.0	1713.06.6	2199.73.3	2686.40.0	3173.06.6	3659.73.3	4146.40.0	4633.06.6	52
53	257.93.3	744.60.0	1231.26.6	1717.93.3	2204.60.0	2691.26.6	3177.93.3	3664.60.0	4151.26.6	4637.93.3	53
54	262.80.0	749.46.6	1236.13.3	1722.80.0	2209.46.6	2696.13.3	3182.80.0	3669.46.6	4156.13.3	4642.80.0	54
55	267.66.6	754.33.3	1241.00.0	1727.66.6	2214.33.3	2701.00.0	3187.66.6	3674.33.3	4161.00.0	4647.66.6	55
56	272.53.3	759.20.0	1245.86.6	1732.53.3	2219.20.0	2705.86.6	3192.53.3	3679.20.0	4165.86.6	4652.53.3	56
57	277.40.0	764.06.6	1250.73.3	1737.40.0	2224.06.6	2710.73.3	3197.40.0	3684.06.6	4170.73.3	4657.40.0	57
58	282.26.6	768.93.3	1255.60.0	1742.26.6	2228.93.3	2715.60.0	3202.26.6	3688.93.3	4175.60.0	4662.26.6	58
59	287.13.3	773.80.0	1260.46.6	1747.13.3	2233.80.0	2720.46.6	3207.13.3	3693.80.0	4180.46.6	4667.13.3	59
60	292.00.0	778.66.6	1265.33.3	1752.00.0	2238.66.6	2725.33.3	3212.00.0	3698.66.6	4185.33.3	4672.00.0	60
61	296.86.6	783.53.3	1270.20.0	1756.86.6	2243.53.3	2730.20.0	3216.86.6	3703.53.3	4190.20.0	4676.86.6	61
62	301.73.3	788.40.0	1275.06.6	1761.73.3	2248.40.0	2735.06.6	3221.73.3	3708.40.0	4195.06.6	4681.73.3	62
63	306.60.0	793.26.6	1279.93.3	1766.60.0	2253.26.6	2739.93.3	3226.60.0	3713.26.6	4199.93.3	4686.60.0	63
64	311.46.6	798.13.3	1284.80.0	1771.46.6	2258.13.3	2744.80.0	3231.46.6	3718.13.3	4204.80.0	4691.46.6	64
65	316.33.3	803.00.0	1289.66.6	1776.33.3	2263.00.0	2749.66.6	3236.33.3	3723.00.0	4209.66.6	4696.33.3	65
66	321.20.0	807.86.6	1294.53.3	1781.20.0	2267.86.6	2754.53.3	3241.20.0	3727.86.6	4214.53.3	4701.20.0	66
67	326.06.6	812.73.3	1299.40.0	1786.06.6	2272.73.3	2759.40.0	3246.06.6	3732.73.3	4219.40.0	4706.06.6	67
68	330.93.3	817.60.0	1304.26.6	1790.93.3	2277.60.0	2764.26.6	3250.93.3	3737.60.0	4224.26.6	4710.93.3	68
69	335.80.0	822.46.6	1309.13.3	1795.80.0	2282.46.6	2769.13.3	3255.80.0	3742.46.6	4229.13.3	4715.80.0	69
70	340.66.6	827.33.3	1314.00.0	1800.66.6	2287.33.3	2774.00.0	3260.66.6	3747.33.3	4234.00.0	4720.66.6	70
71	345.53.3	832.20.0	1318.86.6	1805.53.3	2292.20.0	2778.86.6	3265.53.3	3752.20.0	4238.86.6	4725.53.3	71
72	350.40.0	837.06.6	1323.73.3	1810.40.0	2297.06.6	2783.73.3	3270.40.0	3757.06.6	4243.73.3	4730.40.0	72
73	355.26.6	841.93.3	1328.60.0	1815.26.6	2301.93.3	2788.60.0	3275.26.6	3761.93.3	4248.60.0	4735.26.6	73
74	360.13.3	846.80.0	1333.46.6	1820.13.3	2306.80.0	2793.46.6	3280.13.3	3766.80.0	4253.46.6	4740.13.3	74
75	365.00.0	851.66.6	1338.33.3	1825.00.0	2311.66.6	2798.33.3	3285.00.0	3771.66.6	4258.33.3	4745.00.0	75
76	369.86.6	856.53.3	1343.20.0	1829.86.6	2316.53.3	2803.20.0	3289.86.6	3776.53.3	4263.20.0	4749.86.6	76
77	374.73.3	861.40.0	1348.06.6	1834.73.3	2321.40.0	2808.06.6	3294.73.3	3781.40.0	4268.06.6	4754.73.3	77
78	379.60.0	866.26.6	1352.93.3	1839.60.0	2326.26.6	2812.93.3	3299.60.0	3786.26.6	4272.93.3	4759.60.0	78
79	384.46.6	871.13.3	1357.80.0	1844.46.6	2331.13.3	2817.80.0	3304.46.6	3791.13.3	4277.80.0	4764.46.6	79
80	389.33.3	876.00.0	1362.66.6	1849.33.3	2336.00.0	2822.66.6	3309.33.3	3796.00.0	4282.66.6	4769.33.3	80
81	394.20.0	880.86.6	1367.53.3	1854.20.0	2340.86.6	2827.53.3	3314.20.0	3800.86.6	4287.53.3	4774.20.0	81
82	399.06.6	885.73.3	1372.40.0	1859.06.6	2345.73.3	2832.40.0	3319.06.6	3805.73.3	4292.40.0	4779.06.6	82
83	403.93.3	890.60.0	1377.26.6	1863.93.3	2350.60.0	2837.26.6	3323.93.3	3810.60.0	4297.26.6	4783.93.3	83
84	408.80.0	895.46.6	1382.13.3	1868.80.0	2355.46.6	2842.13.3	3328.80.0	3815.46.6	4302.13.3	4788.80.0	84
85	413.66.6	900.33.3	1387.00.0	1873.66.6	2360.33.3	2847.00.0	3333.66.6	3820.33.3	4307.00.0	4793.66.6	85
86	418.53.3	905.20.0	1391.86.6	1878.53.3	2365.20.0	2851.86.6	3338.53.3	3825.20.0	4311.86.6	4798.53.3	86
87	423.40.0	910.06.6	1396.73.3	1883.40.0	2370.06.6	2856.73.3	3343.40.0	3830.06.6	4316.73.3	4803.40.0	87
88	428.26.6	914.93.3	1401.60.0	1888.26.6	2374.93.3	2861.60.0	3348.26.6	3834.93.3	4321.60.0	4808.26.6	88
89	433.13.3	919.80.0	1406.46.6	1893.13.3	2379.80.0	2866.46.6	3353.13.3	3839.80.0	4326.46.6	4813.13.3	89
90	438.00.0	924.66.6	1411.33.3	1898.00.0	2384.66.6	2871.33.3	3358.00.0	3844.66.6	4331.33.3	4818.00.0	90
91	442.86.6	929.53.3	1416.20.0	1902.86.6	2389.53.3	2876.20.0	3362.86.6	3849.53.3	4336.20.0	4822.86.6	91
92	447.73.3	934.40.0	1421.06.6	1907.73.3	2394.40.0	2881.06.6	3367.73.3	3854.40.0	4341.06.6	4827.73.3	92
93	452.60.0	939.26.6	1425.93.3	1912.60.0	2399.26.6	2885.93.3	3372.60.0	3859.26.6	4345.93.3	4832.60.0	93
94	457.46.6	944.13.3	1430.80.0	1917.46.6	2404.13.3	2890.80.0	3377.46.6	3864.13.3	4350.80.0	4837.46.6	94
95	462.33.3	949.00.0	1435.66.6	1922.33.3	2409.00.0	2895.66.6	3382.33.3	3869.00.0	4355.66.6	4842.33.3	95
96	467.20.0	953.86.6	1440.53.3	1927.20.0	2413.86.6	2900.53.3	3387.20.0	3873.86.6	4360.53.3	4847.20.0	96
97	472.06.6	958.73.3	1445.40.0	1932.06.6	2418.73.3	2905.40.0	3392.06.6	3878.73.3	4365.40.0	4852.06.6	97
98	476.93.3	963.60.0	1450.26.6	1936.93.3	2423.60.0	2910.26.6	3396.93.3	3883.60.0	4370.26.6	4856.93.3	98
99	481.80.0	968.46.6	1455.13.3	1941.80.0	2428.46.6	2915.13.3	3401.80.0	3888.46.6	4375.13.3	4861.80.0	99

S.	10	11	12	13	14	15	16	17	18	19	S.
d.	$ cts.	$ cts.	$ cts.	$ cts.	$ cts.	$ cts.	$ cts.	$ cts.	$ cts.	$ cts.	d.
0	2.43.3	2.67.6	2.92.0	3.16.3	3.40.6	3.65.0	3.89.3	4.13.6	4.38.0	4.62.3	0
1	2.45.3	2.69.7	2.94.0	3.18.3	3.42.7	3.67.0	3.91.3	4.15.7	4.40.0	4.64.3	1
2	2.47.4	2.71.7	2.96.0	3.20.4	3.44.7	3.69.0	3.93.4	4.17.7	4.42.0	4.66.4	2
3	2.49.4	2.73.7	2.98.1	3.22.4	3.46.7	3.71.1	3.95.4	4.19.7	4.44.1	4.68.4	3
4	2.51.4	2.75.8	3.00.1	3.24.4	3.48.8	3.73.1	3.97.4	4.21.8	4.46.1	4.70.4	4
5	2.53.5	2.77.8	3.02.1	3.26.5	3.50.8	3.75.1	3.99.5	4.23.8	4.48.1	4.72.5	5
6	2.55.5	2.79.8	3.04.2	3.28.5	3.52.8	3.77.2	4.01.5	4.25.8	4.50.2	4.74.5	6
7	2.57.5	2.81.9	3.06.2	3.30.5	3.54.9	3.79.2	4.03.5	4.27.9	4.52.2	4.76.5	7
8	2.59.6	2.83.9	3.08.2	3.32.6	3.56.9	3.81.2	4.05.6	4.29.9	4.54.2	4.78.6	8
9	2.61.6	2.85.9	3.10.3	3.34.6	3.58.9	3.83.3	4.07.6	4.31.9	4.56.3	4.80.6	9
10	2.63.6	2.88.0	3.12.9	3.36.6	3.61.0	3.85.3	4.09.6	4.34.0	4.58.3	4.82.6	10
11	2.65.7	2.90.0	3.14.3	3.38.7	3.63.0	3.87.3	4.11.7	4.36.0	4.60.3	4.84.7	11

$4.87 11/100 TO THE POUND.

9 5/8 %

PREMIUM.

STERLING INTO DOLLARS AND CENTS, AND VICE-VERSA.

£	0	100	200	300	400	500	600	700	800	900	£
	$ cts.	$ cts.	$ cts.	$ cts.	$ cts.	$ cts.	$ cts.	$ cts.	$ cts.	$ cts.	
0	......	487.22.2	974.44.4	1461.66.6	1948.88.8	2436.11.1	2923.33.3	3410.55.5	3897.77.7	4385.00.0	0
1	4.87.2	492.09.4	979.31.6	1466.53.8	1953.76.1	2440.98.3	2928.20.5	3415.42.7	3902.65.0	4389.87.2	1
2	9.74.4	496.96.6	984.18.8	1471.41.1	1958.63.3	2445.85.5	2933.07.7	3420.30.0	3907.52.2	4394.74.4	2
3	14.61.6	501.83.8	989.06.1	1476.28.3	1963.50.5	2450.72.7	2937.95.0	3425.17.2	3912.39.4	4399.61.6	3
4	19.48.8	506.71.1	993.93.3	1481.15.5	1968.37.7	2455.60.0	2942.82.2	3430.04.4	3917.26.6	4404.48.8	4
5	24.36.1	511.58.3	998.80.5	1486.02.7	1973.25.0	2460.47.2	2947.69.4	3434.91.6	3922.13.8	4409.36.1	5
6	29.23.3	516.45.5	1003.67.7	1490.90.0	1978.12.2	2465.34.4	2952.56.6	3439.78.8	3927.01.1	4414.23.3	6
7	34.10.5	521.32.7	1008.55.0	1495.77.2	1982.99.4	2470.21.6	2957.43.8	3444.66.1	3931.88.3	4419.10.5	7
8	38.97.7	526.20.0	1013.42.2	1500.64.4	1987.86.6	2475.08.8	2962.31.1	3449.53.3	3936.75.5	4423.97.7	8
9	43.85.0	531.07.2	1018.29.4	1505.51.6	1992.73.8	2479.96.1	2967.18.3	3454.40.5	3941.62.7	4428.85.0	9
10	48.72.2	535.94.4	1023.16.6	1510.38.8	1997.61.1	2484.83.3	2972.05.5	3459.27.7	3946.50.0	4433.72.2	10
11	53.59.4	540.81.6	1028.03.8	1515.26.1	2002.48.3	2489.70.5	2976.92.7	3464.15.0	3951.37.2	4438.59.4	11
12	58.46.6	545.68.8	1032.91.1	1520.13.3	2007.35.5	2494.57.7	2981.80.0	3469.02.2	3956.24.4	4443.46.6	12
13	63.33.8	550.56.1	1037.78.3	1525.00.5	2012.22.7	2499.45.0	2986.67.2	3473.89.4	3961.11.6	4448.33.8	13
14	68.21.1	555.43.3	1042.65.5	1529.87.7	2017.10.0	2504.32.2	2991.54.4	3478.76.6	3965.98.8	4453.21.1	14
15	73.08.3	560.30.5	1047.52.7	1534.75.0	2021.97.2	2509.19.4	2996.41.6	3483.63.8	3970.86.1	4458.08.3	15
16	77.95.5	565.17.7	1052.40.0	1539.62.2	2026.84.4	2514.06.6	3001.28.8	3488.51.1	3975.73.3	4462.95.5	16
17	82.82.7	570.05.0	1057.27.2	1544.49.4	2031.71.6	2518.93.8	3006.16.1	3493.38.3	3980.60.5	4467.82.7	17
18	87.70.0	574.92.2	1062.14.4	1549.36.6	2036.58.8	2523.81.1	3011.03.3	3498.25.5	3985.47.7	4472.70.0	18
19	92.57.2	579.79.4	1067.01.6	1554.23.8	2041.46.1	2528.68.3	3015.90.5	3503.12.7	3990.35.0	4477.57.2	19
20	97.44.4	584.66.6	1071.88.8	1559.11.1	2046.33.3	2533.55.5	3020.77.7	3508.00.0	3995.22.2	4482.44.4	20
21	102.31.6	589.53.8	1076.76.1	1563.98.3	2051.20.5	2538.42.7	3025.65.0	3512.87.2	4000.09.4	4487.31.6	21
22	107.18.8	594.41.1	1081.63.3	1568.85.5	2056.07.7	2543.30.0	3030.52.2	3517.74.4	4004.96.6	4492.18.8	22
23	112.06.1	599.28.3	1086.50.5	1573.72.7	2060.95.0	2548.17.2	3035.39.4	3522.61.6	4009.83.8	4497.06.1	23
24	116.93.3	604.15.5	1091.37.7	1578.60.0	2065.82.2	2553.04.4	3040.26.6	3527.48.8	4014.71.1	4501.93.3	24
25	121.80.5	609.02.7	1096.25.0	1583.47.2	2070.69.4	2557.91.6	3045.13.8	3532.36.1	4019.58.3	4506.80.5	25
26	126.67.7	613.90.0	1101.12.2	1588.34.4	2075.56.6	2562.78.8	3050.01.1	3537.23.3	4024.45.5	4511.67.7	26
27	131.55.0	618.77.2	1105.99.4	1593.21.6	2080.43.8	2567.66.1	3054.88.3	3542.10.5	4029.32.7	4516.55.0	27
28	136.42.2	623.64.4	1110.86.6	1598.08.8	2085.31.1	2572.53.3	3059.75.5	3546.97.7	4034.20.0	4521.42.2	28
29	141.29.4	628.51.6	1115.73.8	1602.96.1	2090.18.3	2577.40.5	3064.62.7	3551.85.0	4039.07.2	4526.29.4	29
30	146.16.6	633.38.8	1120.61.1	1607.83.3	2095.05.5	2582.27.7	3069.50.0	3556.72.2	4043.94.4	4531.16.6	30
31	151.03.8	638.26.1	1125.48.3	1612.70.5	2099.92.7	2587.15.0	3074.37.2	3561.59.4	4048.81.6	4536.03.8	31
32	155.91.1	643.13.3	1130.35.5	1617.57.7	2104.80.0	2592.02.2	3079.24.4	3566.46.6	4053.68.8	4540.91.1	32
33	160.78.3	648.00.5	1135.22.7	1622.45.0	2109.67.2	2596.89.4	3084.11.6	3571.33.8	4058.56.1	4545.78.3	33
34	165.65.5	652.87.7	1140.10.0	1627.32.2	2114.54.4	2601.76.6	3088.98.8	3576.21.1	4063.43.3	4550.65.5	34
35	170.52.7	657.75.0	1144.97.2	1632.19.4	2119.41.6	2606.63.8	3093.86.1	3581.08.3	4068.30.5	4555.52.7	35
36	175.40.0	662.62.2	1149.84.4	1637.06.6	2124.28.8	2611.51.1	3098.73.3	3585.95.5	4073.17.7	4560.40.0	36
37	180.27.2	667.49.4	1154.71.6	1641.93.8	2129.16.1	2616.38.3	3103.60.5	3590.82.7	4078.05.0	4565.27.2	37
38	185.14.4	672.36.6	1159.58.8	1646.81.1	2134.03.3	2621.25.5	3108.47.7	3595.70.0	4082.92.2	4570.14.4	38
39	190.01.6	677.23.8	1164.46.1	1651.68.3	2138.90.5	2626.12.7	3113.35.0	3600.57.2	4087.79.4	4575.01.6	39
40	194.88.8	682.11.1	1169.33.3	1656.55.5	2143.77.7	2631.00.0	3118.22.2	3605.44.4	4092.66.6	4579.88.8	40
41	199.76.1	686.98.3	1174.20.5	1661.42.7	2148.65.0	2635.87.2	3123.09.4	3610.31.6	4097.53.8	4584.76.1	41
42	204.63.3	691.85.5	1179.07.7	1666.30.0	2153.52.2	2640.74.4	3127.96.6	3615.18.8	4102.41.1	4589.63.3	42
43	209.50.5	696.72.7	1183.95.0	1671.17.2	2158.39.4	2645.61.6	3132.83.8	3620.06.1	4107.28.3	4594.50.5	43
44	214.37.7	701.60.0	1188.82.2	1676.04.4	2163.26.6	2650.48.8	3137.71.1	3624.93.3	4112.15.5	4599.37.7	44
45	219.25.0	706.47.2	1193.69.4	1680.91.6	2168.13.8	2655.36.1	3142.58.3	3629.80.5	4117.02.7	4604.25.0	45
46	224.12.2	711.34.4	1198.56.6	1685.78.8	2173.01.1	2660.23.3	3147.45.5	3634.67.7	4121.90.0	4609.12.2	46
47	228.99.4	716.21.6	1203.43.8	1690.66.1	2177.88.3	2665.10.5	3152.32.7	3639.55.0	4126.77.2	4613.99.4	47
48	233.86.6	721.08.8	1208.31.1	1695.53.3	2182.75.5	2669.97.7	3157.20.0	3644.42.2	4131.64.4	4618.86.6	48
49	238.73.8	725.96.1	1213.18.3	1700.40.5	2187.62.7	2674.85.0	3162.07.2	3649.29.4	4136.51.6	4623.73.8	49

S.	0	1	2	3	4	5	6	7	8	9	S.
d.	$ cts.	$ cts.	$ cts.	$ cts.	$ cts.	$ cts.	$ cts.	$ cts.	$ cts.	$ cts.	d.
0	......	.24.3	.48.7	.73.0	.97.4	1.21.8	1.46.1	1.70.5	1.94.8	2.19.2	0
1	.02.0	.26.3	.50.7	.75.1	.99.4	1.23.8	1.48.1	1.72.5	1.96.9	2.21.2	1
2	.04.0	.28.4	.52.7	.77.1	1.01.5	1.26.8	1.50.2	1.74.5	1.98.9	2.23.3	2
3	.06.0	.30.4	.54.8	.79.1	1.03.6	1.27.8	1.52.2	1.76.6	2.00.9	2.25.3	3
4	.08.1	.32.4	.56.8	.81.2	1.05.6	1.29.9	1.54.2	1.78.6	2.03.0	2.27.3	4
5	.10.1	.34.5	.58.8	.83.2	1.07.5	1.31.9	1.56.3	1.80.6	2.05.0	2.29.4	5
6	.12.1	.36.5	.60.9	.85.2	1.09.6	1.33.9	1.58.3	1.82.7	2.07.0	2.31.4	6
7	.14.2	.38.5	.62.9	.87.2	1.11.6	1.36.0	1.60.3	1.84.7	2.09.1	2.33.4	7
8	.16.2	.40.6	.64.9	.89.3	1.13.6	1.38.0	1.62.4	1.86.7	2.11.1	2.35.4	8
9	.18.2	.42.6	.66.9	.91.3	1.15.7	1.40.0	1.64.4	1.88.7	2.13.1	2.37.5	9
10	.20.3	.44.6	.69.0	.93.3	1.17.7	1.42.1	1.66.4	1.90.8	2.15.1	2.30.5	10
11	.22.3	.46.6	.71.0	.95.4	1.19.7	1.44.1	1.68.4	1.92.8	2.17.2	2.41.5	11

$4.87 13/100 TO THE POUND.

9 5/8 %

PREMIUM.

STERLING INTO DOLLARS AND CENTS, AND VICE-VERSA.

£	0	100	200	300	400	500	600	700	800	900	£
	$ cts.	$ cts.	$ cts.	$ cts.	$ cts.	$ cts.	$ cts.	$ cts.	$ cts.	$ cts.	
50	243,61.1	730,83.3	1218,05.5	1705,27.7	2192,50.0	2679 72.2	3166,94.4	3654,16.6	4141,38.8	4628,61.1	50
51	248,48.3	735,70.5	1222,92.7	1710,15 0	2197,37.2	2684,59.4	3171,81.6	3659,03.8	4146,26.1	4633,48.3	51
52	253,35,5	740,57.7	1227,80.0	1715,02,2	2202,24,4	2689,46,6	3176,68,8	3663,91,1	4151,13 3	4638,35.5	52
53	258,22.7	745,45.0	1232,67.2	1719,89,4	2207,11.6	2694,33.8	3181,56,1	3668,78,3	4156,00.5	4643,22.7	53
54	263,10.0	750,32.2	1237,54,4	1724,76,6	2211,98 8	2699 21.1	3186,43.3	3673,65,5	4160,87.7	4648,10.0	54
55	267,97.2	755,19.4	1242,41.6	1729 63,8	2216,86,1	2704,08,3	3191,30.5	3678,52,7	4165,75.0	4652,97,2	55
56	272,84.4	760,06.6	1247,28,8	1734,51,1	2221,73.3	2708,95.5	3196,17.7	3683,40.0	4170,62,2	4657,84.4	56
57	277,71.6	764,93.8	1252,16,1	1739,38,3	2226,60,5	2713,82.7	3201,05,0	3688,27,2	4175,49.4	4662,71.6	57
58	282,58,8	769,81.1	1257,03,3	1744,25,5	2231,47,7	2718,70 0	3205,92,2	3693,14,4	4180,36,6	4667,58 8	58
59	287,46.1	774,68.3	1261,90,5	1749,12,7	2236,35,0	2723,57.2	3210,79,4	3698,01.6	4185,23.8	4672,46.1	59
60	292,33.3	779,55,5	1266,77.7	1754,00.0	2241,22.2	2728,44.4	3215,66,6	3702,88.8	4190,11.1	4677,33.3	60
61	297,20.5	784,42.7	1271,65.0	1758,87,2	2246,09.4	2733 31,6	3220,53 8	3707,76 1	4194,98.3	4682,20.5	61
62	302,07.7	789,30.0	1276,52,2	1763,74.4	2250,96,6	2738,18,8	3225,41,1	3712,63,3	4199,85,5	4687,07.7	62
63	306,95.0	794,17.2	1281,39,4	1768,61,6	2255,83,8	2743,06,1	3230,28,3	3717,50,5	4204,72,7	4691,95,0	63
64	311,82.2	799,04.4	1286 26,6	1773,48,8	2260,71,1	2747 93,3	3235,15,5	3722,37,7	4209,60 0	4696 82,2	64
65	316,69.4	803,91.6	1291,13,8	1778 36,1	2265,58,3	2752,80 5	3240,02.7	3727,25.0	4214,47,2	4701,69,4	65
66	321,56,6	808,78.8	1296,01,1	1783,23,3	2270,45,5	2757,67,7	3244,90.0	3732,12,2	4219,34,4	4706,56 6	66
67	326,43,8	813,66,1	1300,88 3	1788,10 5	2275,32,7	2762 55,0	3249,77,2	3736,99,4	4224,21.6	4711,43,8	67
68	331,31,1	818,53,3	1305,75,5	1792 97,7	2280,20,0	2767,42,2	3254,64 4	3741,86,6	4229,08,8	4716,31,1	68
69	336,18,3	823,40.5	1310,62,7	1797,85,0	2285,07,2	2772,29,4	3259,51,6	3746,73,8	4233,96,1	4721,18.3	69
70	341,05.5	828,27.7	1315,50.0	1802,72,2	2289,94,4	2777,16,6	3264,38,8	3751,61,1	4238,83,3	4726,05,5	70
71	345,92,7	833,15.0	1320,37,2	1807,59,4	2294,81,6	2782,03,8	3269,26,1	3756,48 3	4243,70.5	4730,92,7	71
72	350,80,0	838,02,2	1325,24,4	1812,46,6	2299 68,8	2786,91,1	3274,13,3	3761 35 5	4248,57,7	4735,80,0	72
73	355,67,2	842,89 4	1330,11,6	1817,33,8	2304,56,1	2791,78,3	3279,00,5	3766,22 7	4253,45.0	4740,67,2	73
74	360,54,4	847,76,6	1334,98,8	1822,21,1	2309,43,3	2796,65,5	3283,87,7	3771,10,0	4258,32,2	4745,54,4	74
75	365,41,6	852,63 8	1339 86,1	1827,08,3	2314,30,5	2801,52,7	3288,75,0	3775,97,2	4263,19,4	4750,41,6	75
76	370,28,8	857,51,1	1344,73,3	1831,95 5	2319,17,7	2806,40,0	3293,62,2	3780,84,4	4268,06,6	4755,28,8	76
77	375 16,1	862,38,3	1349,60,5	1836,82,7	2324,05,0	2811,27,2	3298,49,4	3785,71,6	4272,93,8	4760,16,1	77
78	380,03,3	867,25,5	1354,47,7	1841 70,0	2328,92,2	2816,14,4	3303,36,6	3790 58 8	4277,81,1	4765,03,3	78
79	384,90.5	872,12.7	1359,35.0	1846,57,2	2333,79,4	2821,01,6	3308,23,8	3795,46 1	4282,68,3	4769,90 5	79
80	389,77.7	877,00.0	1364,22,2	1851,44,4	2338,66,6	2825,88,8	3313,11,1	3800,33,3	4287,55,5	4774,77,7	80
81	394 65.0	881 87.2	1369,09,4	1856,31,6	2343,53,8	2830,76,1	3317,98,3	3805,20,5	4292,42,7	4779,65,0	81
82	399,52,2	886 74.4	1373,96,6	1861,18,8	2348,41,1	2835,63,3	3322,85 5	3810,07,7	4297,30,0	4784,52,2	82
83	404 39.4	891,61,6	1378,83,8	1866,06,1	2353,28 3	2840,50,5	3327,72,7	3814,95 0	4302,17,2	4789,39 4	83
84	409,26,6	896,48 8	1383,71,1	1870,93,3	2358,15,5	2845,37,7	3332,60,0	3819,82 2	4307,04,4	4794,26,6	84
85	414 13.8	901,36,1	1388,58,3	1875,80,5	2363,02,7	2850,25,0	3337,47,2	3824,69,4	4311,91,6	4799,13,8	85
86	419,01.1	906 23.3	1393 45,5	1880,67,7	2367,90,0	2855,12,2	3342,34,4	3829,56,6	4316,78,8	4804,01,1	86
87	423,88,3	911,10.5	1398,32,7	1885,55,0	2372,77,2	2859 99,4	3347,21,6	3834,43,8	4321,66,1	4808,88,3	87
88	428,75 5	915,97.7	1403,20,0	1890,42,2	2377,64,4	2864,86,6	3352,08,8	3839,31,1	4326,53,3	4813,75,5	88
89	433,62,7	920,85.0	1408,07,2	1895,29,4	2382,51,6	2869,73,8	3356,96,1	3844,18,3	4331,40,5	4818,62,7	89
90	438,50,0	925,72,2	1412,94,4	1900,16,6	2387,38,8	2874,61,1	3361,83,3	3849,05,5	4336,27,7	4823,50,0	90
91	443,37,2	930,59,4	1417,81,6	1905,03,8	2392,26,1	2879,48,3	3366,70,5	3853,92,7	4341,15,0	4828,37,2	91
92	448,24,4	935,46,6	1422,68,8	1909,91,1	2397,13,3	2884 35,5	3371,57,7	3858,80,0	4346,02,2	4833,24,4	92
93	453,11,6	940,33,8	1427,56,1	1914,78,3	2402,00.5	2889,22,7	3376 45,0	3863,67,2	4350,89,4	4838,11,6	93
94	457,98,8	945,21,1	1432,43,3	1919,65,5	2406,87,7	2894,10,0	3381,32,2	3868,54,4	4355,76,6	4842,98,8	94
95	462,86,1	950,08,3	1437,30,5	1924,52 7	2411,75,0	2898,97,2	3386,19,4	3873,41,6	4360,63,8	4847,86,1	95
96	467,73,3	954,95,5	1442,17,7	1929,40,0	2416,62,2	2903,84,4	3391,06,6	3878,28,8	4365,51 1	4852,73,3	96
97	472,60 5	959,82,7	1447,05,0	1934,27,2	2421 49,4	2908,71,6	3395,93,8	3883,16,1	4370,38,3	4857,60,5	97
98	477,47,7	964,70,0	1451,92,2	1939,14,4	2426,36,6	2013,58,8	3400,81,1	3888,03,3	4375,25,5	4862,47,7	98
99	482,35,0	969,57,2	1456,79,4	1944,01,6	2431,23,8	2918,46,1	3405,68,3	3892,90,5	4380,12,7	4867,35,0	99

s.	10	11	12	13	14	15	16	17	18	19	s.
d.	$ cts.	$ cts.	$ cts.	$ cts.	$ cts.	$ cts.	$ cts.	$ cts.	$ cts.	$ cts.	d.
0	2.43.6	2.67.9	2.92.3	3.16.6	3.41.0	3.65.4	3.89.7	4.14.1	4.38.5	4.62.8	0
1	2.45.6	2.70.0	2.94.3	3.18.7	3.43.0	3.67.4	3.91.8	4.16.1	4.40.5	4.64.8	1
2	2.47.6	2.72.0	2.96.3	3.20.7	3.45.1	3.69.4	3.93.8	4.18.1	4.42.5	4.66.9	2
3	2.49.7	2.74.0	2.98.4	3.22.7	3.47.1	3.71.5	3.95.8	4.20.2	4.44.5	4.68.9	3
4	2.51.7	2.76.0	3.00.4	3.24.8	3.49.1	3.73.5	3.97.9	4.22.2	4.46.6	4.70.9	4
5	2.53.7	2.78.1	3.02.4	3.26.8	3.51.2	3.75.5	3.99.9	4.24.2	4.48.6	4.73.0	5
6	2.55.7	2.80.1	3.04 5	3.28.8	3.53.2	3.77.5	4.01.9	4.26.3	4.50,6	4.75.0	6
7	2.57.8	2.82.1	3.06,5	3.30.8	3.55,2	3.79.6	4.03,9	4.28,3	4.52.7	4.77.0	7
8	2.59.8	2.84.2	3.08.5	3.32.9	3.57.2	3.81.6	4.06,0	4.30,3	4.54.7	4.79.1	8
9	2.61.8	2.86.2	3.10.6	3.34.9	3.59.3	3.83.6	4.08.0	4.32,4	4.56,7	4.81.1	9
10	2.63.9	2.88.2	3.12.6	3.36,9	3.61,3	3.85.7	4.10,0	4.34.4	4.58.8	4.83,1	10
11	2.65.9	2.90.3	3.14.6	3.39.0	3.63.3	3.87.7	4.12.1	4.36.4	4.60.8	4.85.1	11

$4.87 11/100 TO THE POUND.

9 3/4 %

PREMIUM.

STERLING INTO DOLLARS AND CENTS, AND VICE-VERSA.

£	0	100	200	300	400	500	600	700	800	900	£
£	$ cts.	$ cts.	$ cts.	$ cts.	$ cts.	$ cts.	$ cts.	$ cts.	$ cts.	$ cts.	£
0	.....	487.77.7	975.55.5	1463.33.3	1951.11.1	2438.88.8	2926.66.6	3414.44.4	3902.22.2	4390.00.0	0
1	4.87.7	492.65.5	980.43.3	1468.21.1	1955.98.8	2443.76.6	2931.54.4	3419.32.2	3907.10.0	4394.87.7	1
2	9.75.5	497.53.3	985.31.1	1473.08.8	1960.86.6	2448.64.4	2936.42.2	3424.20.0	3911.97.7	4399.75.5	2
3	14.63.3	502.41.1	990.18.8	1477.96.6	1965.74.4	2453.52.2	2941.30.0	3429.07.7	3916.85.5	4404.63.3	3
4	19.51.1	507.28.8	995.06.6	1482.84.4	1970.62.2	2458.40.0	2946.17.7	3433.95.5	3921.73.3	4409.51.1	4
5	24.38.8	512.16.6	909.94.4	1487.72.2	1975.50.0	2463.27.7	2951.05.5	3438.83.3	3926.61.1	4414.38.8	5
6	29.26.6	517.04.4	1004.82.2	1492.60.0	1980.37.7	2468.15.5	2955.93.3	3443.71.1	3931.48 8	4419.26.6	6
7	34.14.4	521.92.2	1009.70.0	1497.47.7	1985.25.5	2473.03.3	2960.81.1	3448.58.8	3936.36.6	4424.14.4	7
8	39.02.2	526.80.0	1014.57.7	1502.35.5	1990.13.3	2477.91.1	2965.68.8	3453.46.6	3941.24.4	4429.02.2	8
9	43.90.0	531.67.7	1019.45.5	1507.23.3	1995.01.1	2482.78.8	2970.56.6	3458.34.4	3946.12.2	4433.90.0	9
10	48.77.7	536.55.5	1024.33.3	1512.11.1	1999.88.8	2487.66.6	2975.44.4	3463.22.2	3951.00.0	4438.77.7	10
11	53.65.5	541.43.3	1029.21.1	1516.98.8	2004.76 6	2492.54.4	2980.32.2	3468.10.0	3955.87.7	4443.65.5	11
12	58.53.3	546.31.1	1034.08.8	1521.86.6	2009.64 4	2497.42.2	2985.20.0	3472.97.7	3960.75.5	4448.53.3	12
13	63.41.1	551.18.8	1038.96.6	1526.74.4	2014.52.2	2502.30.0	2990.07.7	3477.85.5	3965.63.3	4453.41.1	13
14	68.28.8	556.06.6	1043.84.4	1531.62.2	2019.40.0	2507.17.7	2994.95.5	3482.73.3	3970.51.1	4458.28.8	14
15	73.16.6	560.94.4	1048.72.2	1536.50.0	2024.27.7	2512.05.5	2999.83.3	3487.61.1	3975.38.8	4463.16.6	15
16	78.04.4	565.82 2	1053.60.0	1541.37.7	2029.15.5	2516.93.3	3004.71.1	3492.48.8	3980.26.6	4468.04.4	16
17	82 92.2	570.70.0	1058.47.7	1546.25.5	2034.03.3	2521.81.1	3009.58.8	3497.36.6	3985.14.4	4472.92.2	17
18	87.80.0	575.57.7	1063.35.5	1551.13.3	2038 91.1	2526.68.8	3014.46.6	3502.24.4	3990.02.2	4477.80.0	18
19	92.67.7	580.45.5	1068.23.3	1556.01.1	2043.78.8	2531.56.6	3019.34.4	3507.12.2	3994.90.0	4482.67.7	19
20	97.55.5	585.33.3	1073.11.1	1560.88.8	2048.66.6	2536.44.4	3024.22.2	3512.00.0	3999.77.7	4487.55.5	20
21	102.43.3	590.21.1	1077.98.8	1565.76.6	2053.54.4	2541.32.2	3029.10.0	3516.87.7	4004.65.5	4492.43.3	21
22	107.31.1	595.08.8	1082.86.6	1570.64.4	2058.42.2	2546.20.0	3033.97.7	3521.75.5	4009.53.3	4497.31.1	22
23	112.18.8	599.96.6	1087.74.4	1575.52.2	2063.30.0	2551.07.7	3038.85.5	3526.63.3	4014.41.1	4502.18.8	23
24	117.06.6	604.84.4	1092.62.2	1580.40.0	2068.17.7	2555.95.5	3043.73.3	3531.51.1	4019.28.8	4507.06.6	24
25	121.94.4	609 72 2	1097.50.0	1585.27.7	2073.05.5	2560.83.3	3048.61.1	3536.38.8	4024.16.6	4511.94.4	25
26	126.82.2	614 60.0	1102.37.7	1590.15.5	2077.93.3	2565.71.1	3053.48.8	3541.26.6	4029 04.4	4516.82.2	26
27	131.70.0	619.47.7	1107 25 5	1595.03.3	2082.81.1	2570.58.8	3058.36.6	3546.14.4	4033.92.2	4521.70.0	27
28	136.57.7	624.35.5	1112 13.3	1599.91.1	2087.68.8	2575.46.6	3063.24.4	3551.02.2	4038.80.0	4526.57.7	28
29	141.45.5	629.23.3	1117.01.1	1604.78.8	2092.56.6	2580 34.4	3068.12.2	3555 90.0	4043.67.7	4531.45.5	29
30	146.33.3	634.11.1	1121.88.8	1609.66.6	2097.44.4	2585.22.2	3073.00.0	3560.77.7	4048.55.5	4536.33.3	30
31	151.21.1	638.98.8	1126.76.6	1614.54.4	2102.32.2	2590.10.0	3077.87.7	3565.65.5	4053.43.3	4541.21.1	31
32	156.08.8	643 86.6	1131.64.4	1619.42.2	2107.20.0	2594.97.7	3082.75.5	3570.53.3	4058.31.1	4546.08.8	32
33	160.96.6	648 74.4	1136.52.2	1624.30.0	2112.07.7	2599.85.5	3087.63.3	3575.41.1	4063.18.8	4550.96.6	33
34	165.84.4	653.62.2	1141.40.0	1629.17.7	2116.95.5	2604.73.3	3092.51.1	3580.28.8	4068.06.6	4555.84.4	34
35	170.72.2	658.50.0	1146.27.7	1634.05.5	2121.83.3	2609.61.1	3097.38.8	3585.16.6	4072.94.4	4560.72.2	35
36	175.60.0	663.37.7	1151.15.5	1638.93.3	2126.71.1	2614.48.8	3102.26.6	3590.04.4	4077.82.2	4565.60.0	36
37	180.47.7	668.25.5	1156 03.3	1643.81.1	2131.58.8	2619.36.6	3107.14.4	3594.92.2	4082.70.0	4570.47.7	37
38	185.35.5	673.13.3	1160.91.1	1648.68.8	2136.46.6	2624.24.4	3112.02.2	3599.80.0	4087.57.7	4575.35.5	38
39	190.23.3	678.01.1	1165.78.8	1653.56.6	2141.34.4	2629.12.2	3116.90.0	3604.67.7	4092.45.5	4580.23.3	39
40	195.11.1	682.88 8	1170.66.6	1658.44.4	2146.22.2	2634.00.0	3121.77.7	3609.55.5	4097.33.3	4585.11.1	40
41	199.98.8	687.76.6	1175.54.4	1663.32.2	2151.10.0	2638.87.7	3126.65.5	3614.43.3	4102.21.1	4589.98.8	41
42	204.86.6	692.64.4	1180.42.2	1668.20.0	2155.97.7	2643.75.5	3131.53.3	3619.31.1	4107.08.8	4594.86.6	42
43	209.74.4	697.52.2	1185.30.0	1673.07.7	2160.85.5	2648.63.3	3136.41.1	3624.18.8	4111 96.6	4599.74.4	43
44	214.62.2	702.40.0	1190.17.7	1677.95.5	2165.73.3	2653.51.1	3141.28.8	3629.06.6	4116.84.4	4604.62.2	44
45	219.50.0	707.27.7	1195.05.5	1682.83.3	2170.61.1	2658.38.8	3146.16.6	3633.94.4	4121.72.2	4609.50.0	45
46	224.37.7	712.15.5	1199.93.3	1687.71.1	2175.48.8	2663.26.6	3151.04.4	3638.82.2	4126.60.0	4614 37.7	46
47	229 25 5	717.03.3	1204.81.1	1692.58.8	2180.36.6	2668.14.4	3155.92.2	3643.70.0	4131 47.7	4619.25.5	47
48	234.13.3	721.91 1	1209.68.8	1697.46.6	2185.24.4	2673.02.2	3160.80.0	3648.57.7	4136.35.5	4624.13.3	48
49	239.01.1	726.78.8	1214.56.6	1702.34.4	2190.12.2	2677.90.0	3165.67.7	3653.45.5	4141.23.3	4629.01.1	49

S.	0	1	2	3	4	5	6	7	8	9	S.
d.	$ cts.	$ cts.	$ cts.	$ cts.	$ cts.	$ cts.	$ cts.	$ cts.	$ cts.	$ cts.	d.
0	.....	.24.3	.48.7	.73.1	.97.5	1.21.9	1.46.3	1.70.7	1.95.1	2.19.5	0
1	.02.0	.26.4	.50.8	.75.2	.99.5	1.23.9	1.48.3	1.72.7	1.97.1	2.21.5	1
2	.04.0	.28.4	.52.8	.77.2	1.01.6	1.26.0	1.50.4	1.74.7	1.99.1	2.23.5	2
3	.06.1	.30.4	.54.8	.79.2	1.03.6	1.28.0	1.52.4	1.76.8	2.01.2	2.25.6	3
4	.08.1	.32.5	.56.9	.81.3	1.05.6	1.30.0	1.54.4	1.78.8	2.03.2	2.27.6	4
5	.10.1	.34.5	.58.9	.83.3	1.07.7	1.32.1	1.56.5	1.80.8	2.05.2	2.29.6	5
6	.12.2	.36.5	.60.9	.85.3	1.09.7	1.34.1	1.58.5	1.82.9	2.07.3	2.31.7	6
7	.14.2	.38.6	.63.0	.87.4	1.11.7	1.36.1	1.60.5	1.84.9	2.09.3	2.33.7	7
8	.16.2	.40.6	.65.0	.89.4	1.13.8	1.38.2	1.62.6	1.86.9	2.11.3	2.35.7	8
9	.18.3	.42.6	.67.0	.91.4	1.15.8	1.40.2	1.64.6	1.89.0	2.13.4	2.37.8	9
10	.20.3	.44.7	.69.1	.93 5	1.17.8	1.42.2	1.66.6	1.91.0	2.15.4	2.39.8	10
11	.22.3	.46.7	.71.1	.95.5	1.19.9	1.44.3	1.68.7	1.93.0	2.17.4	2.41.8	11

$4.87 1/100 TO THE POUND.

9¾ %

PREMIUM.

STERLING INTO DOLLARS AND CENTS, AND VICE-VERSA.

£	0	100	200	300	400	500	600	700	800	900	£
£	$ cts.	$ cts.	$ cts.	$ cts.	$ cts.	$ cts.	$ cts.	$ cts.	$ cts.	$ cts.	£
50	243.88.8	731.66.6	1219.44.4	1707.22.2	2195.00.0	2682.77.7	3170.55.5	3658.33.3	4146.11.1	4633.88.8	50
51	248.76.6	736.54.4	1224.32.2	1712.10.0	2199.87.7	2687.65.5	3175.43.3	3663.21.1	4150.98.8	4638.76.6	51
52	253.64.4	741.42.2	1229.20.0	1716.97.7	2204.75.5	2692.53.3	3180.31.1	3668.08.8	4155.86.6	4643.64.4	52
53	258.52.2	746.30.0	1234.07.7	1721.85.5	2209.63.3	2697.41.1	3185.18.8	3672.96.6	4160.74.4	4648.52.2	53
54	263.40.0	751.17.7	1238.95.5	1726.73.3	2214.51.1	2702.28.8	3190.06.6	3677.84.4	4165.62.2	4653.40.0	54
55	268.27.7	756.05.5	1243.83.3	1731.61.1	2219.38.8	2707.16.6	3194.94.4	3682.72.2	4170.50.0	4658.27.7	55
56	273.15.5	760.93.3	1248.71.1	1736.48.8	2224.26.6	2712.04.4	3199.82.2	3687.60.0	4175.37.7	4663.15.5	56
57	278.03.3	765.81.1	1253.58.8	1741.36.6	2229.14.4	2716.92.2	3204.76.0	3692.47.7	4180.25.5	4668.03.3	57
58	282.91.1	770.68.8	1258.46.6	1746.24.4	2234.02.2	2721.80.0	3209.57.7	3697.35.5	4185.13.3	4672.91.1	58
59	287.78.8	775.56.6	1263.34.4	1751.12.2	2238.90.0	2726.67.7	3214.45.5	3702.23.3	4190.01.1	4677.78.8	59
60	292.66.6	780.44.4	1268.22.2	1756.00.0	2243.77.7	2731.55.5	3219.33.3	3707.11.1	4194.88.8	4682.66.6	60
61	297.54.4	785.32.2	1273.10.0	1760.87.7	2248.65.5	2736.43.3	3224.21.1	3711.98.8	4199.76.6	4687.54.4	61
62	302.42.2	790.20.0	1277.97.7	1765.75.5	2253.53.3	2741.31.1	3229.08.8	3716.86.6	4204.64.4	4692.42.2	62
63	307.30.0	795.07.7	1282.85.5	1770.63.3	2258.41.1	2746.18.8	3233.96.6	3721.74.4	4209.52.2	4697.30.0	63
64	312.17.7	799.95.5	1287.73.3	1775.51.1	2263.28.8	2751.06.6	3238.84.4	3726.62.2	4214.40.0	4702.17.7	64
65	317.05.5	804.83.3	1292.61.1	1780.38.8	2268.16.6	2755.94.4	3243.72.2	3731.50.0	4219.27.7	4707.05.5	65
66	321.93.3	809.71.1	1297.48.8	1785.26.6	2273.04.4	2760.82.2	3248.60.0	3736.37.7	4224.15.5	4711.93.3	66
67	326.81.1	814.58.8	1302.36.6	1790.14.4	2277.92.2	2765.70.0	3253.47.7	3741.25.5	4229.03.3	4716.81.1	67
68	331.68.8	819.46.6	1307.24.4	1795.02.2	2282.80.0	2770.57.7	3258.35.5	3746.13.3	4233.91.1	4721.68.8	68
69	336.56.6	824.34.4	1312.12.2	1799.90.0	2287.67.7	2775.45.5	3263.23.3	3751.01.1	4238.78.8	4726.56.6	69
70	341.44.4	829.22.2	1317.00.0	1804.77.7	2292.55.5	2780.33.3	3268.11.1	3755.88.8	4243.66.6	4731.44.4	70
71	346.32.2	834.10.0	1321.87.7	1809.65.5	2297.43.3	2785.21.1	3272.98.8	3760.76.6	4248.54.4	4736.32.2	71
72	351.20.0	838.97.7	1326.75.5	1814.53.3	2302.31.1	2790.08.8	3277.86.6	3765.64.4	4253.42.2	4741.20.0	72
73	356.07.7	843.85.5	1331.63.3	1819.41.1	2307.18.8	2794.96.6	3282.74.4	3770.52.2	4258.30.0	4746.07.7	73
74	360.95.5	848.73.3	1336.51.1	1824.28.8	2312.06.6	2799.84.4	3287.62.2	3775.40.0	4263.17.7	4750.95.5	74
75	365.83.3	853.61.1	1341.38.8	1829.16.6	2316.94.4	2804.72.2	3292.50.0	3780.27.7	4268.05.5	4755.83.3	75
76	370.71.1	858.48.8	1346.26.6	1834.04.4	2321.82.2	2809.60.0	3297.37.7	3785.15.5	4272.93.3	4760.71.1	76
77	375.58.8	863.36.6	1351.14.4	1838.92.2	2326.70.0	2814.47.7	3302.25.5	3790.03.3	4277.81.1	4765.58.8	77
78	380.46.6	868.24.4	1356.02.2	1843.80.0	2331.57.7	2819.35.5	3307.13.3	3794.91.1	4282.68.8	4770.46.6	78
79	385.34.4	873.12.2	1360.90.0	1848.67.7	2336.45.5	2824.23.3	3312.01.1	3799.78.8	4287.56.6	4775.34.4	79
80	390.22.2	878.00.0	1365.77.7	1853.55.5	2341.33.3	2829.11.1	3316.88.8	3804.66.6	4292.44.4	4780.22.2	80
81	395.10.0	882.87.7	1370.65.5	1858.43.3	2346.21.1	2833.98.8	3321.76.6	3809.54.4	4297.32.2	4785.10.0	81
82	399.97.7	887.75.5	1375.53.3	1863.31.1	2351.08.8	2838.86.6	3326.64.4	3814.42.2	4302.20.0	4789.97.7	82
83	404.85.5	892.63.3	1380.41.1	1868.18.8	2355.96.6	2843.74.4	3331.52.2	3819.30.0	4307.07.7	4794.85.5	83
84	409.73.3	897.51.1	1385.28.8	1873.06.6	2360.84.4	2848.62.2	3336.40.0	3824.17.7	4311.95.5	4799.73.3	84
85	414.61.1	902.38.8	1390.16.6	1877.94.4	2365.72.2	2853.50.0	3341.27.7	3829.05.5	4316.83.3	4804.61.1	85
86	419.48.8	907.26.6	1395.04.4	1882.82.2	2370.60.0	2858.37.7	3346.15.5	3833.93.3	4321.71.1	4809.48.8	86
87	424.36.6	912.14.4	1399.92.2	1887.70.0	2375.47.7	2863.25.5	3351.03.3	3838.81.1	4326.58.8	4814.36.6	87
88	429.24.4	917.02.2	1404.80.0	1892.57.7	2380.35.5	2868.13.3	3355.91.1	3843.68.8	4331.46.6	4819.24.4	88
89	434.12.2	921.90.0	1409.67.7	1897.45.5	2385.23.3	2873.01.1	3360.78.8	3848.56.6	4336.34.4	4824.12.2	89
90	439.00.0	926.77.7	1414.55.5	1902.33.3	2390.11.1	2877.88.8	3365.66.6	3853.44.4	4341.22.2	4829.00.0	90
91	443.87.7	931.65.5	1419.43.3	1907.21.1	2394.98.8	2882.76.6	3370.54.4	3858.32.2	4346.10.0	4833.87.7	91
92	448.75.5	936.53.3	1424.31.1	1912.08.8	2399.86.6	2887.64.4	3375.42.2	3863.20.0	4350.97.7	4838.75.5	92
93	453.63.3	941.41.1	1429.18.8	1916.96.6	2404.74.4	2892.52.2	3380.30.0	3868.07.7	4355.85.5	4843.63.3	93
94	458.51.1	946.28.8	1434.06.6	1921.84.4	2409.62.2	2897.40.0	3385.17.7	3872.95.5	4360.73.3	4848.51.1	94
95	463.38.8	951.16.6	1438.94.4	1926.72.2	2414.50.0	2902.27.7	3390.05.5	3877.83.3	4365.61.1	4853.38.8	95
96	468.26.6	956.04.4	1443.82.2	1931.60.0	2419.37.7	2907.15.5	3394.93.3	3882.71.1	4370.48.8	4858.26.6	96
97	473.14.4	960.92.2	1448.70.0	1936.47.7	2424.25.5	2912.03.3	3399.81.1	3887.58.8	4375.36.6	4863.14.4	97
98	478.02.2	965.80.0	1453.57.7	1941.35.5	2429.13.3	2916.91.1	3404.68.8	3892.46.6	4380.24.4	4868.02.2	98
99	482.90.0	970.67.7	1458.45.5	1946.23.3	2434.01.1	2921.78.8	3409.56.6	3897.34.4	4385.12.2	4872.90.0	99

S.	10	11	12	13	14	15	16	17	18	19	S.
d.	$ cts.	$ cts.	$ cts.	$ cts.	$ cts.	$ cts.	$ cts.	$ cts.	$ cts.	$ cts.	d.
0	2.43.8	2.68.2	2.92.6	3.17.0	3.41.4	3.65.8	3.90.2	4.14.6	4.39.0	4.63.3	0
1	2.45.9	2.70.3	2.94.7	3.19.0	3.43.4	3.67.8	3.92.2	4.16.6	4.41.0	4.65.4	1
2	2.47.9	2.72.3	2.96.7	3.21.1	3.45.5	3.69.9	3.94.2	4.18.6	4.43.0	4.67.4	2
3	2.50.0	2.74.3	2.98.7	3.23.1	3.47.5	3.71.9	3.96.3	4.20.7	4.45.1	4.69.4	3
4	2.52.0	2.76.4	3.00.8	3.25.1	3.49.5	3.73.9	3.98.3	4.22.7	4.47.1	4.71.5	4
5	2.54.0	2.78.4	3.02.8	3.27.2	3.51.6	3.76.0	4.00.3	4.24.7	4.49.1	4.73.5	5
6	2.56.0	2.80.4	3.04.8	3.29.2	3.53.6	3.78.0	4.02.4	4.26.8	4.51.2	4.75.5	6
7	2.58.1	2.82.5	3.06.9	3.31.2	3.55.6	3.80.0	4.04.4	4.28.8	4.53.2	4.77.6	7
8	2.60.1	2.84.5	3.08.9	3.33.3	3.57.7	3.82.1	4.06.4	4.30.8	4.55.2	4.79.6	8
9	2.62.1	2.86.5	3.10.9	3.35.3	3.59.7	3.84.1	4.08.5	4.32.9	4.57.3	4.81.6	9
10	2.64.2	2.88.6	3.13.0	3.27.3	3.61.7	3.86.1	4.10.5	4.34.9	4.59.3	4.83.7	10
11	2.66.2	2.90.6	3.15.0	3.39.4	3.63.8	3.88.2	4.12.5	4.36.9	4.61.3	4.85.7	11

$4.88¹³⁄₁₆ TO THE POUND.

9⅞ %

PREMIUM.

STERLING INTO DOLLARS AND CENTS, AND VICE-VERSA.

£	0	100	200	300	400	500	600	700	800	900	£
	$ cts.	$ cts.	$ cts.	$ cts.	$ cts.	$ cts.	$ cts.	$ cts.	$ cts.	$ cts.	
0	......	488.33.3	976.66.6	1465.00.0	1953.33.3	2441.66.6	2930.00.0	3418.33.3	3906.66.6	4395.00.0	0
1	4.88.3	493.21.6	981.55.0	1469.88.3	1958.21.6	2446.55.0	2934.88.3	3423.21.6	3911.55.0	4399 88.3	1
2	9.76.6	498.10.0	986.43.3	1474.76.6	1963.10.0	2451.43.3	2939 76.6	3428.10.0	3916.43.3	4404.76.6	2
3	14.65.0	502.98.3	991.31.6	1479.65.0	1967.98.3	2456.31.6	2944.65.0	3432.98.3	3921.31.6	4409.65.0	3
4	19.53.3	507.86.6	996.20.0	1484.53.3	1972.86.6	2461.20.0	2949.53.3	3437.86 6	3926.20.0	4414.53.3	4
5	24.41.6	512.75.0	1001 08.3	1489.41.6	1977.75.0	2466.08.3	2954.41.6	3442.75.0	3931.08.3	4419.41.6	5
6	29.30.0	517.63.3	1005.96.6	1494.30 0	1982.63.3	2470.96.6	2959.30.0	3447.63.3	3935.96.6	4424.30.0	6
7	34.18.3	522.51.6	1010.85 0	1499 18.3	1987.51 6	2475.85.0	2964.18.3	3452.51.6	3940,85.0	4429.18.3	7
8	39.06.6	527.40.0	1015.73.3	1504 06.6	1992.40.0	2480.73.3	2969.06.6	3457.40.0	3945.73.3	4434.06.6	8
9	43.95.0	532.28.3	1020.61.6	1508.95.0	1997.28.3	2485.61.6	2973.95.0	3462.28.3	3950.61.6	4438.95.0	9
10	48.83.3	537.16.6	1025 50.0	1513.83.3	2002.16.6	2490.50.0	2978.83.3	3467.16.6	3955.50.0	4443.83.3	10
11	53.71.6	542.05.0	1030.38.3	1518.71.6	2007.05.0	2495.38.3	2983.71.6	3472.05.0	3960.38.3	4448.71.6	11
12	58.60.0	546.93.3	1035.26.6	1523.60.0	2011.93.3	2500.26.6	2988.60.0	3476 93.3	3965.26.6	4453.60.0	12
13	63.48.3	551.81.6	1040.15.0	1528.48.3	2016.81.6	2505.15.0	2993.48.3	3481.81.6	3970.15.0	4458 48.3	13
14	68.36.6	556.70.0	1045.03.3	1533.36.6	2021.70.0	2510.03.3	2998.36.6	3486.70 0	3975.03.3	4463.36.6	14
15	73.25.0	561.58.3	1049 91.6	1538.25.0	2026.58.3	2514.91.6	3003.25.0	3491 58.3	3979.91.6	4468.25.0	15
16	78.13.3	566.46 6	1054.80.0	1543.13.3	2031.46.6	2519.80.0	3008.13.3	3496.46.6	3984.80.0	4473.13.3	16
17	83 01.6	571.35.0	1059.68.3	1548.01.6	2036.35.0	2524.68.3	3013.01.6	3501.35.0	3989.68.3	4478.01.6	17
18	87.90.0	576.23.3	1064.56.6	1552.90.0	2041.23.3	2529.56.6	3017.90.0	3506.23.3	3994.56.6	4482.90.0	18
19	92.78.3	581.11.6	1069.45.0	1557.78.3	2046.11.6	2534.45.0	3022.78.3	3511.11.6	3999.45.0	4487.78.3	19
20	97.66.6	586.00.0	1074.33.3	1562.66.6	2051.00.0	2539.33.3	3027.66.6	3516.00.0	4004.33.3	4492.66.6	20
21	102.55.0	590 88.3	1079.21.6	1567.55.0	2055.88.3	2544.21.6	3032.55.0	3520.88.3	4009.21.6	4497.55.0	21
22	107.43.3	595.76 6	1084.10.0	1572.43.3	2060.76.6	2549.10.0	3037.43.3	3525.76.6	4014.10.0	4502.43.3	22
23	112.31.6	600.65.0	1088.98.3	1577.31.6	2065.65.0	2553 98.3	3042.31.6	3530.65.0	4018 98.3	4507.31.6	23
24	117.20.0	605.53.3	1093.86.6	1582.20.0	2070.53.3	2558.86.6	3047.20.0	3535.53.3	4023.86.6	4512.20.0	24
25	122.08.3	610.41.6	1098.75.0	1587.08.3	2075.41.6	2563.75.0	3052.08.3	3540.41.6	4028 75.0	4517.08.3	25
26	126.96.6	615.30.0	1103.63.3	1591 96.6	2080.30.0	2568.63.3	3056.96.6	3545.30.0	4033.63.3	4521.96.6	26
27	131.85.0	620.18.3	1108.51.6	1596.85.0	2085.18.3	2573.51.6	3061.85.0	3550.18.3	4038.51.6	4526 85.0	27
28	136.73.3	625.06.6	1113.40.0	1601.73.3	2090.06.6	2578.40.0	3066.73.3	3555.06.6	4043.40.0	4531.73.3	28
29	141.61.6	629.95.0	1118.28.3	1606.61.6	2094.95.0	2583.28.3	3071.61.6	3559.95 0	4048.28.3	4536.61.6	29
30	146.50.0	634.83.3	1123.16.6	1611.50.0	2099.83.3	2588.16.6	3076.50.0	3564.83.3	4053.16.6	4541.50.0	30
31	151.38.3	639.71.6	1128.05.0	1616.38.3	2104.71.6	2593.05.0	3081.38.3	3569.71.6	4058.05.0	4546.38.3	31
32	156.26.6	644.60.0	1132.93.3	1621 26.6	2109.60.0	2597.93.3	3086.26.6	3574.60.0	4062.93.3	4551.26.6	32
33	161.15.0	649.48.3	1137.81.6	1626.15.0	2114.48.3	2602.81.6	3091.15.0	3579.48.3	4067.81.6	4556.15.0	33
34	166.03.3	654.36.6	1142.70.0	1631.03.3	2119.36.6	2607.70.0	3096.03.3	3584.36.6	4072.70.0	4561.03.3	34
35	170.91.6	659.25.0	1147.58.3	1635 91.6	2124.25 0	2612.58.3	3100 91.6	3589.25.0	4077.58.3	4565 91 6	35
36	175.80.0	664 13.3	1152.46.6	1640.80.0	2129.13.3	2617.46.6	3105.80.0	3594.13.3	4082,46.6	4570.80.0	36
37	180.68.3	669.01.6	1157.35.0	1645.68.3	2134.01.6	2622.35.0	3110.68.3	3599 01.6	4087.35.0	4575.68.3	37
38	185.56.6	673 90 0	1162.23.3	1650.56.6	2138.90.0	2627.23.3	3115.56.6	3603.90.0	4092 23.3	4580 56.6	38
39	190.45.0	678.78.3	1167.11.6	1655.45.0	2143.78.3	2632.11.6	3120 45.0	3608.78.3	4097.11.6	4585.45.0	39
40	195.33.3	683.66.6	1172.00.0	1660.33.3	2148.66.6	2637.00.0	3125.33.3	3613.66.6	4102.00.0	4590.33.3	40
41	200.21.6	688.55.0	1176.88.3	1665.21.6	2153.55.0	2641.88.3	3130.21.6	3618.55.0	4106.88.3	4595.21.6	41
42	205.10.0	693.43.3	1181.76.6	1670.10.0	2158.43.3	2646.76.6	3135.10.0	3623.43.3	4111.76.6	4600.10.0	42
43	209.98 3	698.31.6	1186.65.0	1674.98.3	2163.31.6	2651.65.0	3139.98.3	3628.31.6	4116.65.0	4604 98..3	43
44	214.86.6	703.20.0	1191.53.3	1679.86.6	2168.20.0	2656.53.3	3144.86.6	3633.20.0	4121.53.3	4609.86.6	44
45	219.75.0	708.08.3	1196.41.6	1684 75.0	2173.08.3	2661.41.6	3149.75.0	3638.08.3	4126.41.6	4614.75.0	45
46	224.63 3	712.96.6	1201.30.0	1689.63.3	2177.96 6	2666.30.0	3154.63.3	3642.96 6	4131.30.0	4619.63.3	46
47	229 51.6	717.85.0	1206.18 3	1694.51.6	2182.85.0	2671.18.3	3159.51 6	3647.85.0	4136.18.3	4624.51.6	47
48	234.40.0	722.73.3	1211.06.6	1699.40.0	2187.73.3	2676.06.6	3164.40.0	3652.73.3	4141.06.6	4629.40.0	48
49	239.28.3	727.61.6	1215.95.0	1704.28.3	2192.61.6	2680.95.0	3169.28.3	3657.61.6	4145.95.0	4634.28.3	49

S.	0	1	2	3	4	5	6	7	8	9	S.
d.	$ cts.	$ cts.	$ cts.	$ cts.	$ cts.	$ cts.	$ cts.	$ cts.	$ cts.	$ cts.	d.
0	......	.24,4	.48,8	.73.2	.97.6	1.22,0	1.46.5	1.70.9	1.95.3	2.19.7	0
1	.02.0	.26.4	.50.8	.75.2	.99.7	1.24.1	1.48.5	1.72.9	1.97.3	2.21.7	1
2	.04.0	.28.4	.52.9	.77.3	1.01.7	1.26.1	1.50.5	1.74.9	1.99.4	2.23.8	2
3	.06.1	.30.5	.54.9	.79.3	1.03.7	1.28.1	1.52.6	1.77.0	2.01.4	2.25.8	3
4	.08.1	.32.5	.56.9	.81.3	1.05.8	1.30.2	1.54.6	1.79.0	2.03.4	2.27.8	4
5	.10.1	.34.5	.59.0	.83.4	1.07.8	1.32.2	1.56.6	1.81.0	2.05.5	2.29.9	5
6	.12.2	.36.6	.61.0	.85.4	1.09.8	1.34.2	1.58.7	1.83.1	2.07.5	2.31.9	6
7	.14.2	.38.6	.63.1	.87.4	1.11.9	1.36.3	1.60.7	1.85.1	2.09.5	2.33.9	7
8	.16.2	.40.6	.65.2	.89.5	1.13.9	1.38.3	1.62.7	1.87.1	2.11.6	2.36.0	8
9	.18.3	.42.7	.67.2	.91.5	1.15.9	1.40.3	1.64.8	1.89.2	2.13.6	2.38.0	9
10	.20.3	.44.7	.69.2	.93.5	1.18.0	1.42.4	1.66.8	1.91.2	2.15.6	2.40.0	10
11	.22.3	.46.7	.71.3	.95.6	1.20.0	1.44.4	1.68.8	1.93.2	2.17.7	2.42.1	11

$4.88 88/100 TO THE POUND.

9 7/8 %

PREMIUM.

STERLING INTO DOLLARS AND CENTS, AND VICE-VERSA.

£	0	100	200	300	400	500	600	700	800	900	£
	$ cts.	$ cts.	$ cts.	$ cts.	$ cts.	$ cts.	$ cts.	$ cts.	$ cts.	$ cts.	
50	244.16.6	732.50.0	1220.83.3	1709.16.6	2197.50.0	2685.83.3	3174.16.6	3662.50.0	4150.83.3	4639.16.6	50
51	249.05.0	737.38.3	1225.71.6	1714.05.0	2202.38.3	2690.71.6	3179.05.0	3667.38.3	4155.71.6	4644.05.0	51
52	253.93.3	742.26.6	1230.60.0	1718.93.3	2207.26.6	2695.60.0	3183.93.3	3672.26.6	4160.60.0	4648.93.3	52
53	258.81.6	747.15.0	1235.48.3	1723.81.6	2212.15.0	2700.48.3	3188.81.6	3677.15.0	4165.48.3	4653.81.6	53
54	263.70.0	752.03.3	1240.36.6	1728.70.0	2217.03.3	2705.36.6	3193.70.0	3682.03.3	4170.36.6	4658.70.0	54
55	268.58.3	756.91.6	1245.25.0	1733.58.3	2221.91.6	2710.25.0	3198.58.3	3686.91.6	4175.25.0	4663.58.3	55
56	273.46.6	761.80.0	1250.13.3	1738.46.6	2226.80.0	2715.13.3	3203.46.6	3691.80.0	4180.13.3	4668.46.6	56
57	278.35.0	766.68.3	1255.01.6	1743.35.0	2231.68.3	2720.01.6	3208.35.0	3696.68.3	4185.01.6	4673.35.0	57
58	283.23.3	771.56.6	1259.90.0	1748.23.3	2236.56.6	2724.90.0	3213.23.3	3701.56.6	4189.90.0	4678.23.3	58
59	288.11.6	776.45.0	1264.78.3	1753.11.6	2241.45.0	2729.78.3	3218.11.6	3706.45.0	4194.78.3	4683.11.6	59
60	293.00.0	781.33.3	1269.66.6	1758.00.0	2246.33.3	2734.66.6	3223.00.0	3711.33.3	4199.66.6	4688.00.0	60
61	297.88.3	786.21.6	1274.55.0	1762.88.3	2251.21.6	2739.55.0	3227.88.3	3716.21.6	4204.55.0	4692.88.3	61
62	302.76.6	791.10.0	1279.43.3	1767.76.6	2256.10.0	2744.43.3	3232.76.6	3721.10.0	4209.43.3	4697.76.6	62
63	307.65.0	795.98.3	1284.31.6	1772.65.0	2260.98.3	2749.31.6	3237.65.0	3725.98.3	4214.31.6	4702.65.0	63
64	312.53.3	800.86.6	1289.20.0	1777.53.3	2265.86.6	2754.20.0	3242.53.3	3730.86.6	4219.20.0	4707.53.3	64
65	317.41.6	805.75.0	1294.08.3	1782.41.6	2270.75.0	2759.08.3	3247.41.6	3735.75.0	4224.08.3	4712.41.6	65
66	322.30.0	810.63.3	1298.96.6	1787.30.0	2275.63.3	2763.96.6	3252.30.0	3740.63.3	4228.96.6	4717.30.0	66
67	327.18.3	815.51.6	1303.85.0	1792.18.3	2280.51.6	2768.85.0	3257.18.3	3745.51.6	4233.85.0	4722.18.3	67
68	332.06.6	820.40.0	1308.73.3	1797.06.6	2285.40.0	2773.73.3	3262.06.6	3750.40.0	4238.73.3	4727.06.6	68
69	336.95.0	825.28.3	1313.61.6	1801.95.0	2290.28.3	2778.61.6	3266.95.0	3755.28.3	4243.61.6	4731.95.0	69
70	341.83.3	830.16.6	1318.50.0	1806.83.3	2295.16.6	2783.50.0	3271.83.3	3760.16.6	4248.50.0	4736.83.3	70
71	346.71.6	835.05.0	1323.38.3	1811.71.6	2300.05.0	2788.38.3	3276.71.6	3765.05.0	4253.38.3	4741.71.6	71
72	351.60.0	839.93.3	1328.26.6	1816.60.0	2304.93.3	2793.26.6	3281.60.0	3769.93.3	4258.26.6	4746.60.0	72
73	356.48.3	844.81.6	1333.15.0	1821.48.3	2309.81.6	2798.15.0	3286.48.3	3774.81.6	4263.15.0	4751.48.3	73
74	361.36.6	849.70.0	1338.03.3	1826.36.6	2314.70.0	2803.03.3	3291.36.6	3779.70.0	4268.03.3	4756.36.6	74
75	366.25.0	854.58.3	1342.91.6	1831.25.0	2319.58.3	2807.91.6	3296.25.0	3784.58.3	4272.91.6	4761.25.0	75
76	371.13.3	859.46.6	1347.80.0	1836.13.3	2324.46.6	2812.80.0	3301.13.3	3789.46.6	4277.80.0	4766.13.3	76
77	376.01.6	864.35.0	1352.68.3	1841.01.6	2329.35.0	2817.68.3	3306.01.6	3794.35.0	4282.68.3	4771.01.6	77
78	380.90.0	869.23.3	1357.56.6	1845.90.0	2334.23.3	2822.56.6	3310.90.0	3799.23.3	4287.56.6	4775.90.0	78
79	385.78.3	874.11.6	1362.45.0	1850.78.3	2339.11.6	2827.45.0	3315.78.3	3804.11.6	4292.45.0	4780.78.3	79
80	390.66.6	879.00.0	1367.33.3	1855.66.6	2344.00.0	2832.33.3	3320.66.6	3809.00.0	4297.33.3	4785.66.6	80
81	395.55.0	883.88.3	1372.21.6	1860.55.0	2348.88.3	2837.21.6	3325.55.0	3813.88.3	4302.21.6	4790.55.0	81
82	400.43.3	888.76.6	1377.10.0	1865.43.3	2353.76.6	2842.10.0	3330.43.3	3818.76.6	4307.10.0	4795.43.3	82
83	405.31.6	893.65.0	1381.98.3	1870.31.6	2358.65.0	2846.98.3	3335.31.6	3823.65.0	4311.98.3	4800.31.6	83
84	410.20.0	898.53.3	1386.86.6	1875.20.0	2363.53.3	2851.86.6	3340.20.0	3828.53.3	4316.86.6	4805.20.0	84
85	415.08.3	903.41.6	1391.75.0	1880.08.3	2368.41.6	2856.75.0	3345.08.3	3833.41.6	4321.75.0	4810.08.3	85
86	419.96.6	908.30.0	1396.63.3	1884.96.6	2373.30.0	2861.63.3	3349.96.6	3838.30.0	4326.63.3	4814.96.6	86
87	424.85.0	913.18.3	1401.51.6	1889.85.0	2378.18.3	2866.51.6	3354.85.0	3843.18.3	4331.51.6	4819.85.0	87
88	429.73.3	918.06.6	1406.40.0	1894.73.3	2383.06.6	2871.40.0	3359.73.3	3848.06.6	4336.40.0	4824.73.3	88
89	434.61.6	922.95.0	1411.28.3	1899.61.6	2387.95.0	2876.28.3	3364.61.6	3852.95.0	4341.28.3	4829.61.6	89
90	439.50.0	927.83.3	1416.16.6	1904.50.0	2392.83.3	2881.16.6	3369.50.0	3857.83.3	4346.16.6	4834.50.0	90
91	444.38.3	932.71.6	1421.05.0	1909.38.3	2397.71.6	2886.05.0	3374.38.3	3862.71.6	4351.05.0	4839.38.3	91
92	449.26.6	937.60.0	1425.93.3	1914.26.6	2402.60.0	2890.93.3	3379.26.6	3867.60.0	4355.93.3	4844.26.6	92
93	454.15.0	942.48.3	1430.81.6	1919.15.0	2407.48.3	2895.81.6	3384.15.0	3872.48.3	4360.81.6	4849.15.0	93
94	459.03.3	947.36.6	1435.70.0	1924.03.3	2412.36.6	2900.70.0	3389.03.3	3877.36.6	4365.70.0	4854.03.3	94
95	463.91.6	952.25.0	1440.58.3	1928.91.6	2417.25.0	2905.58.3	3393.91.6	3882.25.0	4370.58.3	4858.91.6	95
96	468.80.0	957.13.3	1445.46.6	1933.80.0	2422.13.3	2910.46.6	3398.80.0	3887.13.3	4375.46.6	4863.80.0	96
97	473.68.3	962.01.6	1450.35.0	1938.68.3	2427.01.6	2915.35.0	3403.68.3	3892.01.6	4380.35.0	4868.68.3	97
98	478.56.6	966.90.0	1455.23.3	1943.56.6	2431.90.0	2920.23.3	3408.56.6	3896.90.0	4385.23.3	4873.56.6	98
99	483.45.0	971.78.3	1460.11.6	1948.45.0	2436.78.3	2925.11.6	3413.45.0	3901.78.3	4390.11.6	4878.45.0	99

S.	10	11	12	13	14	15	16	17	18	19	S.
d.	$ cts.	$ cts.	$ cts.	$ cts.	$ cts.	$ cts.	$ cts.	$ cts.	$ cts.	$ cts.	d.
0	2.44.1	2.68.5	2.93.0	3.17.4	3.41.8	3.66.2	3.90.6	4.15.0	4.39.5	4.63.9	0
1	2.46.2	2.70.6	2.95.0	3.19.4	3.43.8	3.68.2	3.92.7	4.17.1	4.41.5	4.65.9	1
2	2.48.2	2.72.6	2.97.0	3.21.4	3.45.9	3.70.3	3.94.7	4.19.1	4.43.5	4.67.9	2
3	2.50.2	2.74.6	2.99.1	3.23.5	3.47.9	3.72.3	3.96.7	4.21.1	4.45.6	4.70.0	3
4	2.52.3	2.76.7	3.01.1	3.25.5	3.49.9	3.74.3	3.98.8	4.23.2	4.47.6	4.72.0	4
5	2.54.3	2.78.7	3.03.1	3.27.5	3.52.0	3.76.4	4.00.8	4.25.2	4.49.6	4.74.0	5
6	2.56.3	2.80.7	3.05.2	3.29.6	3.54.0	3.78.4	4.02.8	4.27.2	4.51.7	4.76.1	6
7	2.58.4	2.82.8	3.07.2	3.31.6	3.56.0	3.80.4	4.04.9	4.29.3	4.53.7	4.78.1	7
8	2.60.4	2.84.8	3.09.2	3.33.6	3.58.1	3.82.5	4.06.9	4.31.3	4.55.7	4.80.1	8
9	2.62.4	2.86.8	3.11.3	3.35.7	3.60.1	3.84.5	4.08.9	4.33.3	4.57.8	4.82.2	9
10	2.64.5	2.88.9	3.13.3	3.37.7	3.62.1	3.86.5	4.11.0	4.35.4	4.59.8	4.84.2	10
11	2.66.5	2.90.9	3.15.3	3.39.7	3.64.2	3.88.6	4.13.0	4.37.4	4.61.8	4.86.2	11

$4.88¹⁄₁₀ TO THE POUND.

10%

PREMIUM.

STERLING INTO DOLLARS AND CENTS, AND VICE-VERSA.

£	0	100	200	300	400	500	600	700	800	900	£
	$ cts.	$ cts.	$ cts.	$ cts.	$ cts.	$ cts.	$ cts.	$ cts.	$ cts.	$ cts.	
0	......	488.88.8	977.77.7	1466.66.6	1955.55.5	2444.44.4	2933.33.3	3422.22.2	3911.11.1	4400.00.0	0
1	4.88.8	493.77.7	982.66.6	1471.55.5	1960.44.4	2449.33.3	2938.22.2	3427.11.1	3916.00.0	4404.88.8	1
2	9.77.7	498.66.6	987.55.5	1476.44.4	1965.33.3	2454.22.2	2943.11.1	3432.00.0	3920.88.8	4409.77.7	2
3	14.66.6	503.55.5	992.44.4	1481.33.3	1970.22.2	2459.11.1	2948.00.0	3436.88.8	3925.77.7	4414.66.6	3
4	19.55.5	508.44.4	997.33.3	1486.22.2	1975.11.1	2464.00.0	2952.88.8	3441.77.7	3930.66.6	4419.55.5	4
5	24.44.4	513.33.3	1002.22.2	1491.11.1	1980.00.0	2468.88.8	2957.77.7	3446.66.6	3935.55.5	4424.44.4	5
6	29.33.3	518.22.2	1007.11.1	1496.00.0	1984.88.8	2473.7.77	2962.66.6	3451.55.5	3940.44.4	4429.33.3	6
7	34.22.2	523.11.1	1012.00.0	1500.88.8	1989.77.7	2478.66.6	2967.55.5	3456.44.4	3945.33.3	4434.22.2	7
8	39.11.1	528.00.0	1016.88.8	1505.77.7	1994.66.6	2483.55.5	2972.44.4	3461.33.3	3950.22.2	4439.11.1	8
9	44.00.0	532.88.8	1021.77.7	1510.66.6	1999.55.5	2488.44.4	2977.33.3	3466.22.2	3955.11.1	4444.00.0	9
10	48.88.8	537.77.7	1026.66.6	1515.55.5	2004.44.4	2493.33.3	2982.22.2	3471.11.1	3960.00.0	4448.88.8	10
11	53.77.7	542.66.6	1031.55.5	1520.44.4	2009.33.3	2498.22.2	2987.11.1	3476.00.0	3964.88.8	4453.77.7	11
12	58.66.6	547.55.5	1036.44.4	1525.33.3	2014.22.2	2503.11.1	2992.00.0	3480.88.8	3969.77.7	4458.66.6	12
13	63.55.5	552.44.4	1041.33.3	1530.22.2	2019.11.1	2508.00.0	2996.88.8	3485.77.7	3974.66.6	4463.55.5	13
14	68.44.4	557.33.3	1046.22.2	1535.11.1	2024.00.0	2512.88.8	3001.77.7	3490.66.6	3979.55.5	4468.44.4	14
15	73.33.3	562.22.2	1051.11.1	1540.00.0	2028.88.8	2517.77.7	3006.66.6	3495.55.5	3984.44.4	4473.33.3	15
16	78.22.2	567.11.1	1056.00.0	1544.88.8	2033.77.7	2522.66.6	3011.55.5	3500.44.4	3989.33.3	4478.22.2	16
17	83.11.1	572.00.0	1060.88.8	1549.77.7	2038.66.6	2527.55.5	3016.44.4	3505.33.3	3994.22.2	4483.11.1	17
18	88.00.0	576.88.8	1065.77.7	1554.66.6	2043.55.5	2532.44.4	3021.33.3	3510.22.2	3999.11.1	4488.00.0	18
19	92.88.8	581.77.7	1070.66.6	1559.55.5	2048.44.4	2537.33.3	3026.22.2	3515.11.1	4004.00.0	4492.88.8	19
20	97.77.7	586.66.6	1075.55.5	1564.44.4	2053.33.3	2542.22.2	3031.11.1	3520.00.0	4008.88.8	4497.77.7	20
21	102.66.6	591.55.5	1080.44.4	1569.33.3	2058.22.2	2547.11.1	3036.00.0	3524.88.8	4013.77.7	4502.66.6	21
22	107.55.5	596.44.4	1085.33.3	1574.22.2	2063.11.1	2552.00.0	3040.88.8	3529.77.7	4018.66.6	4507.55.5	22
23	112.44.4	601.33.3	1090.22.2	1579.11.1	2068.00.0	2556.88.8	3045.77.7	3534.66.6	4023.55.5	4512.44.4	23
24	117.33.3	606.22.2	1095.11.1	1584.00.0	2072.88.8	2561.77.7	3050.66.6	3539.55.5	4028.44.4	4517.33.3	24
25	122.22.2	611.11.1	1100.00.0	1588.88.8	2077.77.7	2566.66.6	3055.55.5	3544.44.4	4033.33.3	4522.22.2	25
26	127.11.1	616.00.0	1104.88.8	1593.77.7	2082.66.6	2571.55.5	3060.44.4	3549.33.3	4038.22.2	4527.11.1	26
27	132.00.0	620.88.8	1109.77.7	1598.66.6	2087.55.5	2576.44.4	3065.33.3	3554.22.2	4043.11.1	4532.00.0	27
28	136.88.8	625.77.7	1114.66.6	1603.55.5	2092.44.4	2581.33.3	3070.22.2	3559.11.1	4048.00.0	4536.88.8	28
29	141.77.7	630.66.6	1119.55.5	1608.44.4	2097.33.3	2586.22.2	3075.11.1	3564.00.0	4052.88.8	4541.77.7	29
30	146.66.6	635.55.5	1124.44.4	1613.33.3	2102.22.2	2591.11.1	3080.00.0	3568.88.8	4057.77.7	4546.66.6	30
31	151.55.5	640.44.4	1129.33.3	1618.22.2	2107.11.1	2596.00.0	3084.88.8	3573.77.7	4062.66.6	4551.55.5	31
32	156.44.4	645.33.3	1134.22.2	1623.11.1	2112.00.0	2600.88.8	3089.77.7	3578.66.6	4067.55.5	4556.44.4	32
33	161.33.3	650.22.2	1139.11.1	1628.00.0	2116.88.8	2605.77.7	3094.66.6	3583.55.5	4072.44.4	4561.33.3	33
34	166.22.2	655.11.1	1144.00.0	1632.88.8	2121.77.7	2610.66.6	3099.55.5	3588.44.4	4077.33.3	4566.22.2	34
35	171.11.1	660.00.0	1148.88.8	1637.77.7	2126.66.6	2615.55.5	3104.44.4	3593.33.3	4082.22.2	4571.11.1	35
36	176.00.0	664.88.8	1153.77.7	1642.66.6	2131.55.5	2620.44.4	3109.33.3	3598.22.2	4087.11.1	4576.00.0	36
37	180.88.8	669.77.7	1158.66.6	1647.55.5	2136.44.4	2625.33.3	3114.22.2	3603.11.1	4092.00.0	4580.88.8	37
38	185.77.7	674.66.6	1163.55.5	1652.44.4	2141.33.3	2630.22.2	3119.11.1	3608.00.0	4096.88.8	4585.77.7	38
39	190.66.6	679.55.5	1168.44.4	1657.33.3	2146.22.2	2635.11.1	3124.00.0	3612.88.8	4101.77.7	4590.66.6	39
40	195.55.5	684.44.4	1173.33.3	1662.22.2	2151.11.1	2640.00.0	3128.88.8	3617.77.7	4106.66.6	4595.55.5	40
41	200.44.4	689.33.3	1178.22.2	1667.11.1	2156.00.0	2644.88.8	3133.77.7	3622.66.6	4111.55.5	4600.44.4	41
42	205.33.3	694.22.2	1183.11.1	1672.00.0	2160.88.8	2649.77.7	3138.66.6	3627.55.5	4116.44.4	4605.33.3	42
43	210.22.2	699.11.1	1188.00.0	1676.88.8	2165.77.7	2654.66.6	3143.55.5	3632.44.4	4121.33.3	4610.22.2	43
44	215.11.1	704.00.0	1192.88.8	1681.77.7	2170.66.6	2659.55.5	3148.44.4	3637.33.3	4126.22.2	4615.11.1	44
45	220.00.0	708.88.8	1197.77.7	1686.66.6	2175.55.5	2664.44.4	3153.33.3	3642.22.2	4131.11.1	4620.00.0	45
46	224.88.8	713.77.7	1202.66.6	1691.55.5	2180.44.4	2669.33.3	3158.22.2	3647.11.1	4136.00.0	4624.88.8	46
47	229.77.7	718.66.6	1207.55.5	1696.44.4	2185.33.3	2674.22.2	3163.11.1	3652.00.0	4140.88.8	4629.77.7	47
48	234.66.6	723.55.5	1212.44.4	1701.33.3	2190.22.2	2679.11.1	3168.00.0	3656.88.8	4145.77.7	4634.66.6	48
49	239.55.5	728.44.4	1217.33.3	1706.22.2	2195.11.1	2684.00.0	3172.88.8	3661.77.7	4150.66.6	4639.55.5	49

s.	0	1	2	3	4	5	6	7	8	9	s.
d.	$ cts.	$ cts.	$ cts.	$ cts.	$ cts.	$ cts.	$ cts.	$ cts.	$ cts.	$ cts.	d.
0	......	.24.4	.48.8	.73.3	.97.7	1.22.2	1.46.6	1.71.1	1.95.5	2.20.0	0
1	.02.0	.26.4	.50.9	.75.3	.99.8	1.24.2	1.48.7	1.73.1	1.97.5	2.22.0	1
2	.04.0	.28.5	.52.9	.77.4	1.01.8	1.26.2	1.50.7	1.75.1	1.99.6	2.24.0	2
3	.06.1	.30.5	.54.9	.79.4	1.03.8	1.28.3	1.52.7	1.77.2	2.01.6	2.26.1	3
4	.08.1	.32.5	.57.0	.81.4	1.05.9	1.30.3	1.54.8	1.79.2	2.03.6	2.28.1	4
5	.10.1	.34.6	.59.0	.83.5	1.07.9	1.32.3	1.56.8	1.81.2	2.05.7	2.30.1	5
6	.12.2	.36.6	.61.0	.85.5	1.09.9	1.34.4	1.58.8	1.83.3	2.07.7	2.32.2	6
7	.14.2	.38.6	.63.1	.87.5	1.12.0	1.36.4	1.60.9	1.85.3	2.09.7	2.34.2	7
8	.16.2	.40.7	.65.1	.89.6	1.14.0	1.38.4	1.62.9	1.87.3	2.11.8	2.36.2	8
9	.18.3	.42.7	.67.1	.91.6	1.16.0	1.40.5	1.64.9	1.89.4	2.13.8	2.38.3	9
10	.20.3	.44.7	.69.2	.93.6	1.18.1	1.42.5	1.67.0	1.91.4	2.15.8	2.40.3	10
11	.22.3	.46.8	.71.2	.95.7	1.20.1	1.44.5	1.69.0	1.93.4	2.17.9	2.42.3	11

$4.88 TO THE POUND.
10%
PREMIUM.
STERLING INTO DOLLARS AND CENTS, AND VICE-VERSA.

£	0	100	200	300	400	500	600	700	800	900	£
£	d cts	$ cts	$ cts	$ cts	$ cts	$ cts	$ cts	$ cts	$ cts	$ cts	£
50	244.44.4	733.33.3	1222.22.2	1711.11.1	2200.00.0	2688.88.8	3177.77.7	3666.66.6	4155.55.5	4644.44.4	50
51	249.33.3	738.22.2	1227.11.1	1716 00.0	2204.88.8	2693.77.7	3182.66.6	3671 55.5	4160.44.4	4649.33.3	51
52	254.22.2	743.11.1	1232.00.0	1720.88.8	2209.77.7	2698.66.6	3187.55.5	3676.44.4	4165.33.3	4654.22.2	52
53	259.11.1	748.00.0	1236 88.8	1725.77.7	2214 66.6	2703.55.5	3192 44.4	3681.33.3	4170.22.2	4659.11.1	53
54	264.00.0	752.88.8	1241.77.7	1730.66.6	2219.55.5	2708.44.4	3197.33.3	3686.22.2	4175.11.1	4664.00.0	54
55	268.88.8	757.77.7	1246.66.6	1735.55.5	2224.44.4	2713.33.3	3202.22.2	3691.11.1	4180.00.0	4668.88.8	55
56	273.77.7	762.66.6	1251.55.5	1740.44.4	2229.33.3	2718.22.2	3207.11.1	3696.00.0	4184.88.8	4673.77.7	56
57	278.66.6	767.55.5	1256.44.4	1745.33.3	2234.22.2	2723.11.1	3212.00.0	3700.88.8	4189.77.7	4678.66.6	57
58	283.55.5	772.44.4	1261.33.3	1750.22.2	2239.11.1	2728.00.0	3216.88.8	3705.77.7	4194.66.6	4683.55.5	58
59	288 44.4	777.33.3	1266.22 2	1755.11.1	2244 00.0	2732.88.8	3221.77.7	3710.66.6	4199.55.5	4688.44.4	59
60	293.33.3	782.22.2	1271.11.1	1760.00.0	2248.88.8	2737.77.7	3226.66.6	3715.55.5	4204 44.4	4693.33.3	60
61	298.22.2	787.11.1	1276.00.0	1764.88.8	2253.77.7	2742.66.6	3231.55.5	3720.44.4	4209.33.3	4698.22.2	61
62	303.11.1	792.00.0	1280.88.8	1769.77.7	2258.66.6	2747.55.5	3236.44.4	3725.33.3	4214.22.2	4703.11.1	62
63	308.00.0	796.88.8	1285.77.7	1774.66.6	2263.55.5	2752.44.4	3241.33.3	3730.22.2	4219 11.1	4708.00.0	63
64	312.88.8	801.77.7	1290.66.6	1779.55.5	2268.44.4	2757.33.3	3246.22.2	3735.11.1	4224 00.0	4712.88.8	64
65	317.77.7	806.66.6	1295.55.5	1784.44.4	2273.33.3	2762.22.2	3251.11.1	3740.00.0	4228.88.8	4717.77.7	65
66	322.66.6	811.55.5	1300.44.4	1789.33.3	2278.22.2	2767.11.1	3256.00.0	3744.88.8	4233.77.7	4722.66.6	66
67	327.55.5	816.44.4	1305.33.3	1794.22.2	2283.11.1	2772.00.0	3260.88.8	3749.77.7	4238.66.6	4727.55.5	67
68	332.44.4	821.33.3	1310.22.2	1799.11.1	2288.00.0	2776.88.8	3265.77.7	3754.66.6	4243.55.5	4732.44.4	68
69	337.33.3	826.22.2	1315.11.1	1804.00.0	2292.88.8	2781.77.7	3270.66.6	3759.55.5	4248.44.4	4737.33.3	69
70	342.22.2	831.11.1	1320.00.0	1808.88.8	2297.77.7	2786.66.6	3275.55.5	3764.44.4	4253.33.3	4742.22.2	70
71	347.11.1	836.00.0	1324.88.8	1813.77.7	2302.66.6	2791.55.5	3280.44.4	3769.33.3	4258.22.2	4747.11.1	71
72	352.00.0	840.88.8	1329.77.7	1818.66.6	2307.55.5	2796.44.4	3285.33.3	3774.22.2	4263.11.1	4752.00.0	72
73	356.88.8	845.77.7	1334.66.6	1823.55.5	2312.44.4	2801.33.3	3290.22.2	3779.11.1	4268.00.0	4756.88.8	73
74	361.77.7	850.66.6	1339.55.5	1828.44.4	2317.33.3	2806.22.2	3295.11.1	3784.00 0	4272.88.8	4761.77.7	74
75	366.66.6	855.55.5	1344.44.4	1833.33.3	2322 22.2	2811.11.1	3300 00.0	3788.88.8	4277.77.7	4766.66.6	75
76	371.55.5	860.44.4	1349.33.3	1838.22.2	2327.11.1	2816.00.0	3304.88.8	3793 77.7	4282.66.6	4771.55.5	76
77	376.44.4	865.33.3	1354.22.2	1843.11.1	2332.00.0	2820.88.8	3309.77.7	3798.66.6	4287.55.5	4776.44.4	77
78	381.33.3	870.22.2	1359.11.1	1848.00.0	2336.88.8	2825.77.7	3314.66.6	3803.55.5	4292.44.4	4781.33.3	78
79	386.22.2	875.11.1	1364.00.0	1852.88.8	2341.77.7	2830.66.6	3319.55.5	3808.44.4	4297.33.3	4786.22.2	79
80	391.11.1	880.00.0	1368.88.8	1857.77.7	2346.66.6	2835.55.5	3324.44.4	3813.33.3	4302.22.2	4791.11.1	80
81	396.00.0	884.88.8	1373.77.7	1862.66.6	2351.55.5	2840.44.4	3329.33.3	3818.22.2	4307.11.1	4796.00.0	81
82	400.88 8	889.77.7	1378.66.6	1867.55.5	2356.44.4	2845.33.3	3334.22.2	3823.11.1	4312.00.0	4800.88.8	82
83	405.77.7	894.66.6	1383.55.5	1872.44.4	2361.33.3	2850.22.2	3339.11.1	3828.00.0	4316.88.8	4805.77.7	83
84	410.66.6	899.55.5	1388.44.4	1877.33.3	2366.22.2	2855.11.1	3344.00.0	3832.88.8	4321.77.7	4810.66.6	84
85	415.55.5	904.44.4	1393.33.3	1882.22.2	2371.11.1	2860.00.0	3348.88.8	3837.77.7	4326.66.6	4815.55.5	85
86	420.44.4	909.33.3	1398.22.2	1887.11.1	2376.00.0	2864.88.8	3353.77.7	3842.66.6	4331.55.5	4820.44.4	86
87	425.33.3	914.22.2	1403.11.1	1892.00.0	2380.88 8	2869.77.7	3358.66.6	3847.55.5	4336.44.4	4825.33.3	87
88	430.22.2	919.11.1	1408.00.0	1896.88.8	2385.77.7	2874.66.6	3363.55.5	3852.44.4	4341.33.3	4830.22.2	88
89	435.11.1	924.00.0	1412.88.8	1901.77.7	2390.66.6	2879.55.5	3368.44.4	3857.33.3	4346.22.2	4835.11.1	89
90	440.00.0	928.88.8	1417.77.7	1906.66.6	2395.55.5	2884.44.4	3373.33.3	3862.22.2	4351.11.1	4840.00.0	90
91	444.88.8	933.77.7	1422.66.6	1911.55.5	2400.44.4	2889.33.3	3378.22.2	3867.11.1	4356.00.0	4844.88.8	91
92	449.77.7	938.66.6	1427.55.5	1916.44.4	2405.33.3	2894.22.2	3383.11.1	3872.00.0	4360.88.8	4849.77.7	92
93	454.66.6	943.55.5	1432.44.4	1921.33.3	2410.22.2	2899.11.1	3388.00.0	3876.88.8	4365.77.7	4854.66.6	93
94	459.55.5	948.44.4	1437.33.3	1926.22.2	2415.11.1	2904.00.0	3392.88.8	3881.77.7	4370.66.6	4859.55.5	94
95	464.44.4	953.33.3	1442.22.2	1931.11.1	2420.00.0	2908.88.8	3397.77.7	3886.66.6	4375.55.5	4864.44.4	95
96	469.33.3	958 22.2	1447.11.1	1936.00.0	2424.88.8	2913.77.7	3402.66.6	3891.55.5	4380.44.4	4869.33.3	96
97	474.22.2	963.11.1	1452.00.0	1940.88.8	2429.77.7	2918.66.6	3407.55.5	3896.44.4	4385.33.3	4874.22.2	97
98	479.11.1	968.00.0	1456.88.8	1945.77.7	2434.66.6	2923.55.5	3412.44.4	3901.33.3	4390.22.2	4879.11.1	98
99	484.00.0	972.88.8	1461.77.7	1950.66.6	2439.55.5	2928.44.4	3417.33.3	3906.22.2	4395.11.1	4884.00.0	99

S.	10	11	12	13	14	15	16	17	18	19	S.
d.	cts	cts	cts	cts	cts	cts	cts	cts	cts	cts	d.
0	2.44.4	2.68.8	2.93.3	3.17.7	3.42.2	3.66.6	3.91.1	4.15.5	4.40.0	4.64.4	0
1	2.46.4	2.70.9	2.95.3	3.19.8	3.44.2	3.68.7	3.93.1	4.17.5	4.42.0	4.66.4	1
2	2.48.5	2.72.9	2.97.4	3.21.8	3.46.2	3.70.7	3.95.1	4.19.6	4.44.0	4.68.5	2
3	2.50.5	2.74.9	2.99.4	3.23.8	3.48.3	3.72.7	3.97.2	4.21.6	4.46.1	4.70.5	3
4	2.52.5	2.77.0	3.01.4	3.25.9	3.50.3	3.74.8	3.99.2	4.23.6	4.48.1	4.72.5	4
5	2.54.6	2.79.0	3.03.5	3.27.9	3.52.3	3.76.8	4.01.2	4.25.7	4.50.1	4.74.6	5
6	2.56.6	2.81.0	3.05.5	3.29.9	3.54.4	3.78.8	4.03.3	4.27.7	4.52.2	4.76.6	6
7	2.58.6	2.83.1	3.07.5	3.32.0	3.56.4	3.80.9	4.05.3	4.29.7	4.54.2	4.78.6	7
8	2.60.7	2.85.1	3.09 6	3.34.0	3.58.4	3.82.9	4.07.3	4.31.8	4.56.2	4.80.7	8
9	2.62.7	2.87.1	3.11.6	3.36.0	3.60.5	3.84.9	4.09.4	4.33.8	4.58.3	4.82.7	9
10	2.64.7	2.89.2	3.13.6	3.38.1	3.62.5	3.87.0	4.11.4	4.35.8	4.60.3	4.84.7	10
11	2.66.8	2.91.2	3.15.7	3.40.1	3.64.5	3.89.0	4.13.4	4.37.9	4.62.3	4.86.8	11

$4.89.¹⁰⁰⁄₁₀₀ TO THE POUND.

10⅛ %

PREMIUM.

STERLING INTO DOLLARS AND CENTS, AND VICE-VERSA.

£	0	100	200	300	400	500	600	700	800	900	£
£	$ cts.	$ cts.	$ cts.	$ cts.	$ cts.	$ cts.	$ cts.	$ cts.	$ cts.	$ cts.	£
0	......	489.44.4	978.88.8	1468.33.3	1957.77.7	2447.22.2	2936.66.6	3426.11.1	3915.55.5	4405.00.0	0
1	4.89.4	494.33.8	983.78.3	1473.22.7	1962.67.2	2452.11.6	2941.56.1	3431.00.5	3920.45.0	4409.89.4	1
2	9.78.8	499.23.3	988.67.7	1478.12.2	1967.56.6	2457.01.1	2946.45.5	3435.90.0	3925.34.4	4414.78.8	2
3	14.68.3	504.12.7	993.57.2	1483.01.6	1972.46.1	2461.90.5	2951.35.0	3440.79.4	3930.23.8	4419.68.3	3
4	19.57.7	509.02.2	998.46.6	1487.91.1	1977.35.5	2466.80.0	2956.24.4	3445.68.8	3935.13.3	4424.57.7	4
5	24.47.2	513.91.6	1003.36.1	1492.80.5	1982.25.0	2471.69.4	2961.13.8	3450.58.3	3940.02.7	4429.47.2	5
6	29.36.6	518.81.1	1008.25.5	1497.70.0	1987.14.4	2476.58.8	2966.03.3	3455.47.7	3944.92.2	4434.36.6	6
7	34.26.1	523.70.5	1013.15.0	1502.59.4	1992.03.8	2481.48.3	2970.92.7	3460.37.2	3949.81.6	4439.26.1	7
8	39.15.5	528.60.0	1018.04.4	1507.48.8	1996.93.3	2486.37.7	2975.82.2	3465.26.6	3954.71.1	4444.15.5	8
9	44.05.0	533.49.4	1022.93.8	1512.38.3	2001.82.7	2491.27.2	2980.71.6	3470.16.1	3959.60.5	4449.05.0	9
10	48.94.4	538.38.8	1027.83.3	1517.27.7	2006.72.2	2496.16.6	2985.61.1	3475.05.5	3964.50.0	4453.94.4	10
11	53.83.8	543.28.3	1032.72.7	1522.17.2	2011.61.6	2501.06.1	2990.50.5	3479.95.0	3969.39.4	4458.83.8	11
12	58.73.3	548.17.7	1037.62.2	1527.06.6	2016.51.1	2505.95.5	2995.40.0	3484.84.4	3974.28.8	4463.73.3	12
13	63.62.7	553.07.2	1042.51.6	1531.96.1	2021.40.5	2510.85.0	3000.29.4	3489.73.8	3979.18.3	4468.62.7	13
14	68.52.2	557.96.6	1047.41.1	1536.85.5	2026.30.0	2515.74.4	3005.18.8	3494.63.3	3984.07.7	4473.52.2	14
15	73.41.6	562.86.1	1052.30.5	1541.75.0	2031.19.4	2520.63.8	3010.08.3	3499.52.7	3988.97.2	4478.41.6	15
16	78.31.1	567.75.5	1057.20.0	1546.64.4	2036.08.8	2525.53.3	3014.97.7	3504.42.2	3993.86.6	4483.31.1	16
17	83.20.5	572.65.0	1062.09.4	1551.53.8	2040.98.3	2530.42.7	3019.87.2	3509.31.6	3998.76.1	4488.20.5	17
18	88.10.0	577.54.4	1066.98.8	1556.43.3	2045.87.7	2535.32.2	3024.76.6	3514.21.1	4003.65.5	4493.10.0	18
19	92.99.4	582.43.8	1071.88.3	1561.32.7	2050.77.2	2540.21.6	3029.66.1	3519.10.5	4008.55.0	4497.99.4	19
20	97.88.8	587.33.3	1076.77.7	1566.22.2	2055.66.6	2545.11.1	3034.55.5	3524.00.0	4013.44.4	4502.88.8	20
21	102.78.3	592.22.7	1081.67.2	1571.11.6	2060.56.1	2550.00.5	3039.45.0	3528.89.4	4018.33.8	4507.78.3	21
22	107.67.7	597.12.2	1086.56.6	1576.01.1	2065.45.5	2554.90.0	3044.34.4	3533.78.8	4023.23.3	4512.67.7	22
23	112.57.2	602.01.6	1091.46.1	1580.90.5	2070.35.0	2559.79.4	3049.23.8	3538.68.3	4028.12.7	4517.57.2	23
24	117.46.6	606.91.1	1096.35.5	1585.80.0	2075.24.4	2564.68.8	3054.13.3	3543.57.7	4033.02.2	4522.46.6	24
25	122.36.1	611.80.5	1101.25.0	1590.69.4	2080.13.8	2569.58.3	3059.02.7	3548.47.2	4037.91.6	4527.36.1	25
26	127.25.5	616.70.0	1106.14.4	1595.58.8	2085.03.3	2574.47.7	3063.92.2	3553.36.6	4042.81.1	4532.25.5	26
27	132.15.0	621.59.4	1111.03.8	1600.48.3	2089.92.7	2579.37.2	3068.81.6	3558.26.1	4047.70.5	4537.15.0	27
28	137.04.4	626.48.8	1115.93.3	1605.37.7	2094.82.2	2584.26.6	3073.71.1	3563.15.5	4052.60.0	4542.04.4	28
29	141.93.8	631.38.3	1120.82.7	1610.27.2	2099.71.6	2589.16.1	3078.60.5	3568.05.0	4057.49.4	4546.93.8	29
30	146.83.3	636.27.7	1125.72.2	1615.16.6	2104.61.1	2594.05.5	3083.50.0	3572.94.4	4062.38.8	4551.83.3	30
31	151.72.7	641.17.2	1130.61.6	1620.06.1	2109.50.5	2598.95.0	3088.39.4	3577.83.8	4067.28.3	4556.72.7	31
32	156.62.2	646.06.6	1135.51.1	1624.95.5	2114.40.0	2603.84.4	3093.28.8	3582.73.3	4072.17.7	4561.62.2	32
33	161.51.6	650.96.1	1140.40.5	1629.85.0	2119.29.4	2608.73.8	3098.18.3	3587.62.7	4077.07.2	4566.51.6	33
34	166.41.1	655.85.5	1145.30.0	1634.74.4	2124.18.8	2613.63.3	3103.07.7	3592.52.2	4081.96.6	4571.41.1	34
35	171.30.5	660.75.0	1150.19.4	1639.63.8	2129.08.3	2618.52.7	3107.97.2	3597.41.6	4086.86.1	4576.30.5	35
36	176.20.0	665.64.4	1155.08.8	1644.53.3	2133.97.7	2623.42.2	3112.86.6	3602.31.1	4091.75.5	4581.20.0	36
37	181.09.4	670.53.8	1159.98.3	1649.42.7	2138.87.2	2628.31.6	3117.76.1	3607.20.5	4096.65.0	4586.09.4	37
38	185.98.8	675.43.3	1164.87.7	1654.32.2	2143.76.6	2633.21.1	3122.65.5	3612.10.0	4101.54.4	4590.98.8	38
39	190.88.3	680.32.7	1169.77.2	1659.21.6	2148.66.1	2638.10.5	3127.55.0	3616.99.4	4106.43.8	4595.88.3	39
40	195.77.7	685.22.2	1174.66.6	1664.11.1	2153.55.5	2643.00.0	3132.44.4	3621.88.8	4111.33.3	4600.77.7	40
41	200.67.2	690.11.6	1179.56.1	1669.00.5	2158.45.0	2647.89.4	3137.33.8	3626.78.3	4116.22.7	4605.67.2	41
42	205.56.6	695.01.1	1184.45.5	1673.90.0	2163.34.4	2652.78.8	3142.23.3	3631.67.7	4121.12.2	4610.56.6	42
43	210.46.1	699.90.5	1189.35.0	1678.79.4	2168.23.8	2657.68.3	3147.12.7	3636.57.2	4126.01.6	4615.46.1	43
44	215.35.5	704.80.0	1194.24.4	1683.68.8	2173.13.3	2662.57.7	3152.02.2	3641.46.6	4130.91.1	4620.35.5	44
45	220.25.0	709.69.4	1199.13.8	1688.58.3	2178.02.7	2667.47.2	3156.91.6	3646.36.1	4135.80.5	4625.25.0	45
46	225.14.4	714.58.8	1204.03.3	1693.47.7	2182.92.2	2672.36.6	3161.81.1	3651.25.5	4140.70.0	4630.14.4	46
47	230.03.8	719.48.3	1208.92.7	1698.37.2	2187.81.6	2677.26.1	3166.70.5	3656.15.0	4145.59.4	4635.03.8	47
48	234.93.3	724.37.7	1213.82.2	1703.26.6	2192.71.1	2682.15.5	3171.60.0	3661.04.4	4150.48.8	4639.93.3	48
49	239.82.7	729.27.2	1218.71.6	1708.16.1	2197.60.5	2687.05.0	3176.49.4	3665.93.8	4155.38.3	4644.82.7	49

S.	0	1	2	3	4	5	6	7	8	9	S.
d.	$ cts.	$ cts.	$ cts.	$ cts.	$ cts.	$ cts.	$ cts.	$ cts.	$ cts.	$ cts.	d.
0	......	.24.4	.48.9	.73.4	.97.8	1.22.3	1.46.8	1.71.3	1.95.7	2,20.2	0
1	.02.0	.26.5	.50.9	.75.4	.99.9	1.24.4	1.48.8	1.73.3	1.97.8	2.22.2	1
2	.04.0	.28.5	.53.0	.77.4	1.01.9	1.26.4	1.50.9	1.75.3	1.99.8	2.24.3	2
3	.06.1	.30.5	.55.0	.79.5	1.04.0	1.28.4	1.52.9	1.77.4	2.01.8	2.26.3	3
4	.08.1	.32.6	.57.1	.81.5	1.06.0	1.30.5	1.54.9	1.79.4	2.03.9	2.28.4	4
5	.10.2	.34.6	.59.1	.83.6	1.08.0	1.32.5	1.57.0	1.81.5	2.05.9	2.30.4	5
6	.12.2	.36.7	.61.1	.85.6	1.10.1	1.34.6	1.59.0	1.83.5	2.08.0	2.32.4	6
7	.14.2	.38.7	.63.2	.87.6	1.12.1	1.36.6	1.61.1	1.85.5	2.10.0	2.34.5	7
8	.16.3	.40.7	.65.2	.89.7	1.14.2	1.38.6	1.63.1	1.87.6	2.12.0	2.36.5	8
9	.18.3	.42.8	.67.3	.91.7	1.16.2	1.40.7	1.65.1	1.89.6	2.14.1	2.38.6	9
10	.20.4	.44.8	.69.3	.93.8	1.18.2	1.42.7	1.67.2	1.91.7	2.16.1	2.40.6	10
11	.22.4	.46.9	.71.3	.95.8	1.20.3	1.44.8	1.69.2	1.93.7	2.18.2	2.42.6	11

$4.89⁴⁴⁄₁₀₀ TO THE POUND.

10⅛%

PREMIUM.

STERLING INTO DOLLARS AND CENTS, AND VICE-VERSA.

£	0	100	200	300	400	500	600	700	800	900	£
£	$ cts.	$ cts.	$ cts.	$ cts.	$ cts.	$ cts.	$ cts.	$ cts.	$ cts.	$ cts.	£
50	244.72.2	734.16.6	1223.61.1	1713.05.5	2202.50.0	2691.94.4	3181.38.8	3670.83.3	4160.27.7	4649.72.2	50
51	249.61.6	739.06.1	1228.50.5	1717.95.0	2207.39.4	2696.83.8	3186.28.3	3675.72.7	4165.17.2	4654.61.6	51
52	254.51.1	743.95.5	1233.40.0	1722.84.4	2212.28.8	2701.73.3	3191.17.7	3680.62.2	4170.06.6	4659.51.1	52
53	259.40.5	748.85.0	1238.29.4	1727.73.8	2217.18.3	2706.62.7	3196.07.2	3685.51.6	4174.96.1	4664.40.5	53
54	264.30.0	753.74.4	1243.18.8	1732.63.3	2222.07.7	2711.52.2	3200.96.6	3690.41.1	4179.85.5	4669.30.0	54
55	269.19.4	758.63.8	1248.08.3	1737.52.7	2226.97.2	2716.41.6	3205.86.1	3695.30.5	4184.75.0	4674.19.4	55
56	274.08.8	763.53.3	1252.97.7	1742.42.2	2231.86.6	2721.31.1	3210.75.5	3700.20.0	4189.64.4	4679.08.8	56
57	278.98.3	768.42.7	1257.87.2	1747.31.6	2236.76.1	2726.20.5	3215.65.0	3705.09.4	4194.53.8	4683.98.3	57
58	283.87.7	773.32.2	1262.76.6	1752.21.1	2241.65.5	2731.10.0	3220.54.4	3709.98.8	4199.43.3	4688.87.7	58
59	288.77.2	778.21.6	1267.66.1	1757.10.5	2246.55.0	2735.99.4	3225.43.8	3714.88.3	4204.32.7	4693.77.2	59
60	293.66.6	783.11.1	1272.55.5	1762.00.0	2251.44.4	2740.88.8	3230.33.3	3719.77.7	4209.22.2	4698.66.6	60
61	298.56.1	788.00.5	1277.45.0	1766.89.4	2256.33.8	2745.78.3	3235.22.7	3724.67.2	4214.11.6	4703.56.1	61
62	303.45.5	792.90.0	1282.34.4	1771.78.8	2261.23.3	2750.67.7	3240.12.2	3729.56.6	4219.01.1	4708.45.6	62
63	308.35.0	797.79.4	1287.23.8	1776.68.3	2266.12.7	2755.57.2	3245.01.6	3734.46.1	4223.90.5	4713.35.0	63
64	313.24.4	802.68.8	1292.13.3	1781.57.7	2271.02.2	2760.46.6	3249.91.1	3739.35.5	4228.80.0	4718.24.4	64
65	318.13.8	807.58.3	1297.02.7	1786.47.2	2275.91.6	2765.36.1	3254.80.5	3744.25.0	4233.69.4	4723.13.8	65
66	323.03.3	812.47.7	1301.92.2	1791.36.6	2280.81.1	2770.25.5	3259.70.0	3749.14.4	4238.58.8	4728.03.3	66
67	327.92.7	817.37.2	1306.81.6	1796.26.1	2285.70.5	2775.15.0	3264.59.4	3754.03.8	4243.48.3	4732.92.7	67
68	332.82.2	822.26.6	1311.71.1	1801.15.5	2290.60.0	2780.04.4	3269.48.8	3758.93.3	4248.37.7	4737.82.2	68
69	337.71.6	827.16.1	1316.60.5	1806.05.0	2295.49.4	2784.93.8	3274.38.3	3763.82.7	4253.27.2	4742.71.6	69
70	342.61.1	832.05.5	1321.50.0	1810.94.4	2300.38.8	2789.83.3	3279.27.7	3768.72.2	4258.16.6	4747.61.1	70
71	347.50.5	836.95.0	1326.39.4	1815.83.8	2305.28.3	2794.72.7	3284.17.2	3773.61.6	4263.06.1	4752.50.5	71
72	352.40.0	841.84.4	1331.28.8	1820.73.3	2310.17.7	2799.62.2	3289.06.6	3778.51.1	4267.95.5	4757.40.0	72
73	357.29.4	846.73.8	1336.18.3	1825.62.7	2315.07.2	2804.51.6	3293.96.1	3783.40.5	4272.85.0	4762.29.4	73
74	362.18.8	851.63.3	1341.07.7	1830.52.2	2319.96.6	2809.41.1	3298.85.5	3788.30.0	4277.74.4	4767.18.8	74
75	367.08.3	856.52.7	1345.97.2	1835.41.6	2324.86.1	2814.30.5	3303.75.0	3793.19.4	4282.63.8	4772.08.3	75
76	371.97.7	861.42.2	1350.86.6	1840.31.1	2329.75.5	2819.20.0	3308.64.4	3798.08.8	4287.53.3	4776.97.7	76
77	376.87.2	866.31.6	1355.76.1	1845.20.5	2334.65.0	2824.09.4	3313.53.8	3802.98.3	4292.42.7	4781.87.2	77
78	381.76.6	871.21.1	1360.65.5	1850.10.0	2339.54.4	2828.98.8	3318.43.3	3807.87.7	4297.32.2	4786.76.6	78
79	386.66.1	876.10.5	1365.55.0	1854.99.4	2344.43.8	2833.88.3	3323.32.7	3812.77.2	4302.21.6	4791.66.1	79
80	391.55.5	881.00.0	1370.44.4	1859.88.8	2349.33.3	2838.77.7	3328.22.2	3817.66.6	4307.11.1	4796.55.5	80
81	396.45.0	885.89.4	1375.33.8	1864.78.3	2354.22.7	2843.67.2	3333.11.6	3822.56.1	4312.00.5	4801.45.0	81
82	401.34.4	890.78.8	1380.23.3	1869.67.7	2359.12.2	2848.56.6	3338.01.1	3827.45.5	4316.90.0	4806.34.4	82
83	406.23.8	895.68.3	1385.12.7	1874.57.2	2364.01.6	2853.46.1	3342.90.5	3832.35.0	4321.79.4	4811.23.8	83
84	411.13.3	900.57.7	1390.02.2	1879.46.6	2368.91.1	2858.35.5	3347.80.0	3837.24.4	4326.68.8	4816.13.3	84
85	416.02.7	905.47.2	1394.91.6	1884.36.1	2373.86.5	2863.25.0	3352.69.4	3842.13.8	4331.58.3	4821.02.7	85
86	420.92.2	910.36.6	1399.81.1	1889.25.5	2378.70.0	2868.14.4	3357.58.8	3847.03.3	4336.47.7	4825.92.2	86
87	425.81.6	915.26.1	1404.70.5	1894.15.0	2383.59.4	2873.03.8	3362.48.3	3851.92.7	4341.37.2	4830.81.6	87
88	430.71.1	920.15.5	1409.60.0	1899.04.4	2388.48.8	2877.93.3	3367.37.7	3856.82.2	4346.26.6	4835.71.1	88
89	435.60.5	925.05.0	1414.49.4	1903.93.8	2393.38.3	2882.82.7	3372.27.2	3861.71.6	4351.16.1	4840.60.5	89
90	440.50.0	929.94.4	1419.38.8	1908.83.3	2398.27.7	2887.72.2	3377.16.6	3866.61.1	4356.05.5	4845.50.0	90
91	445.39.4	934.83.8	1424.28.3	1913.72.7	2403.17.2	2892.61.6	3382.06.1	3871.50.5	4360.95.0	4850.39.4	91
92	450.28.8	939.73.3	1429.17.7	1918.62.2	2408.06.6	2897.51.1	3386.95.5	3876.40.0	4365.84.4	4855.28.8	92
93	455.18.3	944.62.7	1434.07.2	1923.51.6	2412.96.1	2902.40.5	3391.85.0	3881.29.4	4370.73.8	4860.18.3	93
94	460.07.7	949.52.2	1438.96.6	1928.41.1	2417.85.5	2907.30.0	3396.74.4	3886.18.8	4375.63.3	4865.07.7	94
95	464.97.2	954.41.6	1443.86.1	1933.30.5	2422.75.0	2912.19.4	3401.63.8	3891.08.3	4380.52.7	4869.97.2	95
96	469.86.6	959.31.1	1448.75.5	1938.20.0	2427.64.4	2917.08.8	3406.53.3	3895.97.7	4385.42.2	4874.86.6	96
97	474.76.1	964.20.5	1453.65.0	1943.09.4	2432.53.8	2921.98.3	3411.42.7	3900.87.2	4390.31.6	4879.76.1	97
98	479.65.5	969.10.0	1458.54.4	1947.98.8	2437.43.3	2926.87.7	3416.32.2	3905.76.6	4395.21.1	4884.65.5	98
99	484.55.0	973.99.4	1463.43.8	1952.88.3	2442.32.7	2931.77.2	3421.21.6	3910.66.1	4400.10.5	4889.55.0	99

s.	10	11	12	13	14	15	16	17	18	19	s.
d.	$ cts.	$ cts.	$ cts.	$ cts.	$ cts.	$ cts.	$ cts.	$ cts.	$ cts.	$ cts.	d.
0	2.44.7	2.69.1	2.93.6	3.18.1	3.42.6	3.67.0	3.91.5	4.16.0	4.40.5	4.64.9	0
1	2.46.7	2.71.2	2.95.7	3.20.1	3.44.6	3.69.1	3.93.5	4.18.0	4.42.5	4.67.0	1
2	2.48.8	2.73.2	2.97.7	3.22.2	3.46.6	3.71.1	3.95.6	4.20.1	4.44.5	4.69.0	2
3	2.50.8	2.75.3	2.99.7	3.24.2	3.48.7	3.73.2	3.97.6	4.22.1	4.46.6	4.71.0	3
4	2.52.8	2.77.3	3.01.8	3.26.2	3.50.7	3.75.2	3.99.7	4.24.1	4.48.6	4.73.1	4
5	2.54.9	2.79.3	3.03.8	3.28.3	3.52.8	3.77.2	4.01.7	4.26.2	4.50.7	4.75.1	5
6	2.56.9	2.81.4	3.05.9	3.30.3	3.54.8	3.79.3	4.03.7	4.28.2	4.52.7	4.77.2	6
7	2.59.0	2.83.4	3.07.9	3.32.4	3.56.8	3.81.3	4.05.8	4.30.3	4.54.7	4.79.2	7
8	2.61.0	2.85.5	3.09.9	3.34.4	3.58.9	3.83.4	4.07.8	4.32.3	4.56.8	4.81.2	8
9	2.63.0	2.87.5	3.12.0	3.36.4	3.60.9	3.85.4	4.09.9	4.34.3	4.58.8	4.83.3	9
10	2.65.1	2.89.5	3.14.0	3.38.5	3.63.0	3.87.4	4.11.9	4.36.4	4.60.9	4.85.3	10
11	2.67.1	2.91.6	3.16.1	3.40.5	3.65.0	3.89.5	4.13.9	4.38.4	4.62.9	4.87.4	11

$4.90¹⁰⁰⁄₁₀₀ TO THE POUND.

10¼%

PREMIUM.

STERLING INTO DOLLARS AND CENTS, AND VICE-VERSA.

£	0	100	200	300	400	500	600	700	800	900	£
£	$ cts.	$ cts.	$ cts.	$ cts.	$ cts.	$ cts.	$ cts.	$ cts.	$ cts.	$ cts.	£
0	......	490.00.0	980.00.0	1470.00.0	1960.00.0	2450.00.0	2940.00.0	3430.00.0	3920.00.0	4410.00.0	0
1	4.90.0	494.90.0	984.90.0	1474.90.0	1964.90.0	2454.90.0	2944.90.0	3434.90 0	3924.90.0	4414.90.0	1
2	9.80.0	499.80.0	989.80.0	1479.80.0	1969.80.0	2459.80.0	2949.80.0	3439.80.0	3929 80.0	4419.80.0	2
3	14.70.0	504.70.0	994.70.0	1484.70.0	1974.70.0	2464.70.0	2954.70.0	3444.70.0	3934.70.0	4424 70.0	3
4	19.60.0	509.60.0	999.60.0	1489.60.0	1979.60.0	2469.60.0	2959.60.0	3449.60.0	3939.60.0	4429.60.0	4
5	24.50.0	514.50.0	1004.50.0	1494.50.0	1984.50.0	2474.50.0	2964.50.0	3454.50.0	3944.50.0	4434.50.0	5
6	29.40.0	519.40.0	1009.40.0	1499.40.0	1989.40.0	2479.40.0	2969.40.0	3459.40.0	3949.40.0	4439.40.0	6
7	34.30.0	524.30.0	1014.30.0	1504.30.0	1994.30.0	2484.30.0	2974.30.0	3464.30.0	3954.30.0	4444.30.0	7
8	39.20.0	529.20.0	1019.20.0	1509.20.0	1999.20.0	2489.20.0	2979.20.0	3469.20.0	3959.20.0	4449.20.0	8
9	44.10.0	534.10 0	1024.10.0	1514.10.0	2004.10.0	2494.10.0	2984.10.0	3474.10.0	3964.10.0	4454 10.0	9
10	49 00.0	539.00.0	1029.00.0	1519.00.0	2009.00.0	2499.00.0	2989.00.0	3479.00.0	3969.00.0	4459.00.0	10
11	53.90.0	543.90.0	1033.90.0	1523.90.0	2013.90.0	2503.90.0	2993 90.0	3483.90.0	3973.90.0	4463.90.0	11
12	58.80.0	548.80.0	1038.80.0	1528.80.0	2018.80.0	2508.80.0	2998.80.0	3488.80.0	3978.80.0	4468.80.0	12
13	63.70.0	553 70.0	1043.70.0	1533 70.0	2023.70.0	2513.70.0	3003.70.0	3493.70.0	3983.70.0	4473.70.0	13
14	68.60.0	558 60.0	1048.60.0	1538.60.0	2028.60.0	2518.60.0	3008.60.0	3498.60.0	3988.60.0	4478.60.0	14
15	73.50.0	563.50.0	1053.50.0	1543.50.0	2033.50.0	2523.50.0	3013.50.0	3503.50.0	3993.50.0	4483.50.0	15
16	78.40.0	568.40.0	1058.40.0	1548.40.0	2038.40.0	2528.40.0	3018.40.0	3508.40.0	3998.40.0	4488.40.0	16
17	83 30.0	573.30.0	1063.30.0	1553.30.0	2043.30.0	2533.30.0	3023.30.0	3513.30.0	4003.30.0	4493 30.0	17
18	88.20.0	578.20.0	1068.20.0	1558.20.0	2048.20.0	2538.20.0	3028.20.0	3518.20.0	4008.20.0	4498.20.0	18
19	93.10.0	583.10.0	1073.10.0	1563.10.0	2053.10.0	2543.10.0	3033.10.0	3523.10.0	4013.10.0	4503.10.0	19
20	98.00.0	588.00.0	1078.00.0	1568.00.0	2058.00.0	2548.00.0	3038.00.0	3528.00.0	4018.00.0	4508.00.0	20
21	102 90.0	592.90.0	1082.90.0	1572.90.0	2062.90.0	2552.90.0	3042.90.0	3532.90.0	4022.90.0	4512.90.0	21
22	107.80.0	597.80.0	1087.80.0	1577.80.0	2067.80.0	2557.80.0	3047.80.0	3537.80.0	4027.80.0	4517.80.0	22
23	112.70.0	602.70.0	1092.70.0	1582.70.0	2072.70 0	2562.70 0	3052.70.0	3542.70.0	4032.70.0	4522.70.0	23
24	117.60.0	607.60.0	1097.60.0	1587.60.0	2077.60.0	2567.60 0	3057.60.0	3547.60.0	4037.60.0	4527.60.0	24
25	122.50.0	612.50.0	1102.50.0	1592.50.0	2082.50 0	2572.50 0	3062.50.0	3552.50.0	4042.50.0	4532.50.0	25
26	127.40.0	617.40 0	1107.40.0	1597.40.0	2087.40.0	2577.40.0	3067.40.0	3557.40.0	4047.40.0	4537.40.0	26
27	132.30.0	622.30.0	1112.30.0	1602.30.0	2092.30.0	2582.30.0	3072.30.0	3562.30.0	4052.30.0	4542.30.0	27
28	137.20.0	627.20.0	1117.20.0	1607.20.0	2097.20.0	2587.20.0	3077.20.0	3567.20.0	4057.20.0	4547.20 0	28
29	142.10.0	632.10.0	1122.10.0	1612.10.0	2102.10.0	2592.10.0	3082 10.0	3572.10.0	4062.10.0	4552.10.0	29
30	147.00.0	637.00.0	1127.00.0	1617.00.0	2107.00.0	2597.00.0	3087.00.0	3577.00.0	4067.00.0	4557 00.0	30
31	151.90.0	641 90.0	1131.90.0	1621.90.0	2111.90.0	2601.90.0	3091.90.0	3581.90.0	4071.90.0	4561.90.0	31
32	156.80.0	646.80.0	1136 80.0	1626.80.0	2116.80.0	2606 80.0	3096.80.0	3586.80.0	4076.80.0	4566.80.0	32
33	161.70.0	651.70.0	1141.70.0	1631.70.0	2121.70.0	2611.70.0	3101.70.0	3591.70.0	4081.70.0	4571.70.0	33
34	166.60.0	656.60.0	1146.60.0	1636.60.0	2126.60.0	2616.60.0	3106.60.0	3596.60.0	4086.60.0	4576.60.0	34
35	171.50.0	661.50.0	1151.50 0	1641.50.0	2131.50.0	2621.50.0	3111.50.0	3601.50.0	4091.50.0	4581.50.0	35
36	176.40.0	666.40.0	1156.40.0	1646.40.0	2136.40.0	2626.40.0	3116.40.0	3606.40.0	4096.40 0	4586.40.0	36
37	181.30.0	671.30.0	1161.30.0	1651.30.0	2141.30.0	2631.30.0	3121.30.0	3611.30.0	4101.30.0	4591.30.0	37
38	186.20.0	676.20.0	1166.20.0	1656.20.0	2146.20.0	2636.20.0	3126.20.0	3616.20.0	4106.20.0	4596.20.0	38
39	191.10.0	681.10.0	1171.10.0	1661.10.0	2151.10.0	2641.10.0	3131.10.0	3621.10.0	4111.10.0	4601.10.0	39
40	196 00.0	686.00.0	1176.00.0	1666.00.0	2156.00.0	2646.00.0	3136.00.0	3626.00.0	4116.00.0	4606.00.0	40
41	200.90.0	690.90.0	1180.90.0	1670.90.0	2160.90.0	2650.90.0	3140.90.0	3630.90 0	4120.90.0	4610.90.0	41
42	205.80.0	695.80.0	1185.80.0	1675.80.0	2165.80.0	2655.80.0	3145.80.0	3635.80.0	4125.80.0	4615.80.0	42
43	210.70.0	700.70.0	1190.70.0	1680.70.0	2170.70.0	2660.70.0	3150.70.0	3640.70 0	4130.70.0	4620.70.0	43
44	215.60.0	705.60.0	1195.60.0	1685.60.0	2175.60.0	2665.60.0	3155.60.0	3645.60.0	4135.60.0	4625 60.0	44
45	220.50.0	710.50.0	1200.50.0	1690.50.0	2180.50.0	2670.50.0	3160.50.0	3650.50.0	4140.50.0	4630.50.0	45
46	225.40.0	715.40.0	1205.40.0	1695.40.0	2185.40.0	2675.40.0	3165.40.0	3655.40.0	4145.40.0	4635.40.0	46
47	230.30.0	720.30.0	1210.30.0	1700.30.0	2190.30.0	2680.30.0	3170.30.0	3660 30 0	4150 30.0	4640.30.0	47
48	235.20.0	725.20.0	1215.20.0	1705.20.0	2195.20.0	2685.20.0	3175.20.0	3665.20.0	4155.20.0	4645.20.0	48
49	240.10.0	730.10.0	1220.10.0	1710.10.0	2200.10.0	2690.10.0	3180.10.0	3670.10.0	4160.10 0	4650.10.0	49

S.	0	1	2	3	4	5	6	7	8	9	S.
d.	$ cts.	$ cts.	$ cts.	$ cts.	$ cts.	$ cts.	$ cts.	$ cts.	$ cts.	$ cts.	d.
0	......	.24.5	.49.0	.73.5	.98.0	1.22.5	1.47.0	1.71.5	1.96.0	2.20.5	0
1	.02.0	.26.5	.51.0	.75.5	1.00.0	1.24.5	1.49.0	1.73.5	1.98.0	2.22.5	1
2	.04.0	.28.5	.53.0	.77.5	1.02.0	1.26.5	1.51.0	1.75.5	2.00.0	2.24.5	2
3	.06.1	.30.6	.55.1	.79.6	1.04.1	1.28.6	1.53.1	1.77.6	2.02.1	2.26.6	3
4	.08.1	.32.6	.57.1	.81.6	1.06.1	1.30.6	1.55.1	1.79.6	2.04.1	2.28.6	4
5	.10.2	.34.7	.59.2	.83.7	1.08.2	1.32.7	1.57.2	1.81.7	2.06.2	2.30.7	5
6	.12.2	.36.7	.61.2	.85.7	1.10.2	1.34.7	1.59.2	1.83.7	2.08.2	2.32.7	6
7	.14.3	.38.8	.63.3	.87.8	1.12.3	1.36.8	1.61.3	1.85.8	2.10.3	2.34.8	7
8	.16.3	.40.8	.65.3	.89.8	1.14.3	1.38.8	1.63.3	1.87.8	2.12.3	2.36.8	8
9	.18.3	.42.8	.67.3	.91.8	1.16.3	1.40.8	1.65.3	1.89.8	2.14.3	2.38.8	9
10	.20.4	.44.9	.69.4	.93.9	1.18.4	1.42.9	1.67.4	1.91.9	2.16.4	2.40.9	10
11	.22.4	.46.9	.71.4	.95.9	1.20.4	1.44.9	1.69.4	1.93.9	2.18.4	2.42.9	11

$4.90 90/100 TO THE POUND.

10¼ %
PREMIUM.
STERLING INTO DOLLARS AND CENTS, AND VICE-VERSA.

£	0	100	200	300	400	500	600	700	800	900	£
£	$ cts.	$ cts.	$ cts.	$ cts.	$ cts.	$ cts.	$ cts.	$ cts.	$ cts.	$ cts.	£
50	245 00.0	735 00.0	1225.00.0	1715.00.0	2205.00.0	2695.00.0	3185.00.0	3675.00.0	4165.00.0	4655.00.0	50
51	249.90.0	739.90.0	1229.90.0	1719.90.0	2209.90.0	2699.90.0	3189.90.0	3679.90.0	4169.90.0	4659.90.0	51
52	254.80.0	744.80 0	1234 80.0	1724.80.0	2214.80.0	2704.80.0	3194.80.0	3684.80.0	4174 80.0	4664.80.0	52
53	259.70.0	749.70.0	1239.70.0	1729.70.0	2219 70.0	2709.70.0	3199.70.0	3689.70.0	4179.70 0	4669.70.0	53
54	264.60.0	754.60.0	1244.60.0	1734.60.0	2224 60.0	2714 60.0	3204.60.0	3694.60.0	4184.60.0	4674.60.0	54
55	269.50.0	759.50.0	1249.50.0	1739.50.0	2229.50.0	2719.50.0	3209.50.0	3699.50.0	4189.50.0	4679.50.0	55
56	274.40.0	764.40.0	1254.40.0	1744.40.0	2234 40.0	2724.40.0	3214.40 0	3704.40.0	4194.40.0	4684.40.0	56
57	279.30.0	769.30.0	1259 30.0	1749.30.0	2239.30.0	2729.30.0	3219.30.0	3709.30.0	4199 30.0	4689.30.0	57
58	284.20.0	774.20.0	1264 20.0	1754.20.0	2244.20.0	2734.20 0	3224.20 0	3714.20.0	4204.20.0	4694.20.0	58
59	289·10.0	779.10.0	1269.10.0	1759.10.0	2249.10 0	2739.10.0	3229.10.0	3719.10.0	4209.10.0	4699.10.0	59
60	294.00.0	784.00.0	1274.00.0	1764.00.0	2254.00.0	2744.00.0	3234.00.0	3724.00.0	4214.00.0	4704.00.0	60
61	298.90.0	788.90.0	1278.90.0	1768.90.0	2258.90.0	2748.90.0	3238.90.0	3728.90.0	4218.90.0	4708.90.0	61
62	303.80.0	793.80.0	1283.80.0	1773.80.0	2263.80.0	2753.80.0	3243.80.0	3733.80.0	4223.80.0	4713.80.0	62
63	308.70.0	798.70 0	1288.70.0	1778.70.0	2268.70.0	2758 70.0	3248.70.0	3738.70.0	4228.70.0	4718.70.0	63
64	313.60.0	803.60.0	1293 60.0	1783.60.0	2273.60.0	2763.60.0	3253.60.0	3743.60.0	4233.60.0	4723 60.0	64
65	318.50 0	808.50.0	1298.50.0	1788.50.0	2278.50.0	2768.50.0	3258.50.0	3748 50.0	4238.50.0	4728.50.0	65
66	323.40.0	813.40.0	1303 40.0	1793.40.0	2283.40.0	2773.40.0	3263.40.0	3753.40.0	4243.40 0	4733.40 0	66
67	328.30.0	818.30.0	1308.30.0	1798.30.0	2288.30.0	2778.30.0	3268.30.0	3758.30.0	4248.30.0	4738.30.0	67
68	333.20.0	823.20.0	1313 20.0	1803.20.0	2293.20.0	2783.20.0	3273.20 0	3763.20.0	4253.20.0	4743.20.0	68
69	338.10.0	828.10.0	1318.10.0	1808.10.0	2298.10.0	2788.10.0	3278.10.0	3768.10.0	4258.10.0	4748.10.0	69
70	343.00.0	833.00.0	1323.00.0	1813.00.0	2303.00.0	2793.00 0	3283.00.0	3773 00.0	4263.00.0	4753.00.0	70
71	347.90.0	837.90.0	1327 90.0	1817.90 0	2307.90.0	2797.90.0	3287.90 0	3777.90.0	4267.90.0	4757.90.0	71
72	352.80 0	842.80.0	1332.80 0	1822.80.0	2312.80.0	2802 80.0	3292.80.0	3782.80.0	4272.80.0	4762 80.0	72
73	357.70.0	847.70.0	1337.70.0	1827.70 0	2317.70 0	2807.70.0	3297.70.0	3787.70.0	4277.70.0	4767.70.0	73
74	362.60 0	852.60.0	1342.60 0	1832.60.0	2322.60.0	2812.60.0	3302.60.0	3792.60.0	4282.60.0	4772.60.0	74
75	367.50.0	857.50.0	1347.50.0	1837.50.0	2327.50.0	2817.50.0	3307.50.0	3797.50.0	4287.50 0	4777.50.0	75
76	372.40.0	862.40.0	1352.40.0	1842.40.0	2332.40.0	2822.40.0	3312.40.0	3802.40.0	4292.40.0	4782.40.0	76
77	377.30.0	867.30.0	1357.30.0	1847.30.0	2337.30.0	2827.30.0	3317.30.0	3807.30.0	4297.30.0	4787.30.0	77
78	382.20.0	872.20.0	1362.20.0	1852.20.0	2342.20.0	2832.20.0	3322.20.0	3812.20.0	4302.20.0	4792.20.0	78
79	387.10.0	877.10.0	1367.10.0	1857.10.0	2347.10.0	2837.10.0	3327.10.0	3817.10.0	4307.10.0	4797.10.0	79
80	392.00.0	882.00.0	1372.00.0	1862.00.0	2352.00.0	2842.00.0	3332.00.0	3822.00.0	4312.00.0	4802 00.0	80
81	396.90.0	886.90 0	1376.90.0	1866.90 0	2356.90.0	2846.90.0	3336.90.0	3826.90.0	4316.90.0	4806.90.0	81
82	401.80.0	891.80.0	1381.80.0	1871 80.0	2361.80.0	2851.80.0	3341.80.0	3831.80.0	4321.80.0	4811.80.0	82
83	406 70.0	896.70.0	1386.70.0	1876.70.0	2366.70.0	2856.70.0	3346.70.0	3836.70.0	4326.70.0	4816.70.0	83
84	411.60.0	901.60.0	1391.60.0	1881.60.0	2371.60.0	2861.60.0	3351.60.0	3841.60.0	4331.60.0	4821.60.0	84
85	416.50.0	906.50.0	1396.50.0	1886.50.0	2376.50.0	2866.50.0	3356.50.0	3846.50.0	4336.50.0	4826.50.0	85
86	421.40.0	911.40.0	1401.40.0	1891.40.0	2381.40.0	2871.40.0	3361.40.0	3851.40.0	4341.40.0	4831 40.0	86
87	426.30.0	916.30.0	1406.30.0	1896.30.0	2386.30.0	2876.30.0	3366.30.0	3856.30.0	4346.30.0	4836.30.0	87
88	431.20.0	921.20.0	1411.20.0	1901 20.0	2391.20.0	2881.20.0	3371.20.0	3861.20.0	4351.20.0	4841.20.0	88
89	436.10.0	926.10.0	1416.10.0	1906.10.0	2396.10.0	2886.10.0	3376.10.0	3866.10.0	4356.10.0	4846.10.0	89
90	441.00.0	931.00.0	1421.00.0	1911.00.0	2401.00.0	2891.00.0	3381.00.0	3871.00.0	4361.00.0	4851.00.0	90
91	445.90.0	935.90.0	1425.90.0	1915.90.0	2405.90.0	2895.90.0	3385.90.0	3875.90.0	4365 90.0	4855.90.0	91
92	450.80.0	940.80.0	1430.80.0	1920.80.0	2410.80.0	2900.80.0	3390.80.0	3880.80.0	4370.80.0	4860.80.0	92
93	455.70.0	945.70.0	1435.70.0	1925.70.0	2415.70.0	2905.70.0	3395.70.0	3885.70 0	4375.70.0	4865.70.0	93
94	460.60.0	950.60.0	1440.60.0	1930.60.0	2420.60.0	2910.60.0	3400.60.0	3890.60.0	4380.60.0	4870.60.0	94
95	465.50.0	955.50.0	1445.50.0	1935.50.0	2425.50.0	2915.50.0	3405.50.0	3895.50.0	4385.50.0	4875.50.0	95
96	470.40.0	960.40.0	1450.40.0	1910.40.0	2430.40.0	2920.40.0	3410.40.0	3900.40.0	4390.40.0	4880.40.0	96
97	475.30.0	965.30.0	1455.30.0	1945.30.0	2435.30.0	2925.30.0	3415.30.0	3905.30.0	4395.30.0	4885.30.0	97
98	480 20.0	970.20.0	1460.20.0	1950.20.0	2440.20.0	2930.20.0	3420.20.0	3910.20.0	4400.20.0	4890.20.0	98
99	485.10.0	975.10.0	1465.10.0	1955.10.0	2445.10.0	2935.10.0	3425.10.0	3915.10.0	4405.10 0	4895.10.0	99

S.	10	11	12	13	14	15	16	17	18	19	S.
d.	$ cts.	$ cts.	$ cts.	$ cts.	$ cts.	$ cts.	$ cts.	$ cts.	$ cts.	$ cts.	d.
0	2.45.0	2.69.5	2.94.0	3.18.5	3.43.0	3.67.5	3.92.0	4.16.5	4.41.0	4.65.5	0
1	2.47.0	2.71.5	2.96.0	3.20.5	3.45.0	3.69.5	3.94.0	4.18.5	4.43.1	4.67.5	1
2	2.49.0	2.73.5	2.98.0	3.22.5	3.47.0	3.71.5	3.96.0	4.20.5	4.45.1	4.69.5	2
3	2.51.1	2.75.6	3.00.1	3.24.6	3.49.1	3.73.6	3.98.1	4.22.6	4.47.2	4.71.6	3
4	2.53.1	2.77.6	3.02.1	3.26.6	3.51.1	3.75.6	4.00.1	4.24.6	4.49.2	4.73.6	4
5	2.55.2	2.79.7	3.04.2	3.28.7	3.53.2	3.77.7	4.02.2	4.26.7	4.51.3	4.75.7	5
6	2.57.2	2.81.7	3.06.2	3.30.7	3.55.2	3.79.7	4.04.2	4.28.7	4.53.3	4.77.7	6
7	2.59.3	2.83.8	3.08.3	3.32.8	3.57.3	3.81.8	4.06.3	4.30.8	4.55.4	4.79.8	7
8	2.61.3	2.85.8	3.10.3	3.34.8	3.59.3	3.83.8	4.08.3	4.32.8	4.57.4	4.81.8	8
9	2.63.4	2.87.9	3.12.4	3.36.9	3.61.4	3.85.9	4.10.4	4.34.9	4.59.5	4.83.9	9
10	2.65.4	2.89.9	3.14.4	3.38.9	3.63.4	3.87.9	4.12.4	4.36.9	4.61.5	4.85.9	10
11	2.67.4	2.91.9	3.16.4	3.40.9	3.65.4	3.89.9	4.14.4	4.38.9	4.63.5	4.87.9	11

$4.90 45/100 TO THE POUND.

10 3/8 %

PREMIUM.

STERLING INTO DOLLARS AND CENTS, AND VICE-VERSA.

£	0	100	200	300	400	500	600	700	800	900	£
£	$ cts.	$ cts.	$ cts.	$ cts.	$ cts.	$ cts.	$ cts.	$ cts.	$ cts.	$ cts.	£
0	......	490.55.5	981 11.1	1471.66.6	1962.22.2	2452 77.7	2943.33,3	3433.88.8	3924.44.4	4415.00,0	0
1	4.90.5	495.46,1	986.01.6	1476.57,2	1967.12.7	2457.68,3	2948.23.8	3438.79.4	3929.35.0	4419.90,5	1
2	9.81.1	500.36,6	990.92.2	1481.47.7	1972.03,3	2462.58,8	2953.14.4	3443.70.0	3934.25.5	4424.81.1	2
3	14.71.6	505.27,2	995.82.7	1486,38,3	1976.93,8	2467 49.4	2958.05,0	3448.60,5	3939.16.1	4429.71.6	3
4	19.62.2	510.17.7	1000.73,3	1491.28.8	1981.84.4	2472.40.0	2962 95.5	3453.51.1	3944.06.6	4434.62.2	4
5	24.52.7	515.08,3	1005.63.8	1496.19,4	1986.75,0	2477.30.5	2967.86,1	3458.41.6	3948.97.2	4439.52.7	5
6	29.43.3	519.98,8	1010.54.4	1501.10.0	1991.65,5	2482.21.1	2972.76,6	3463.32,2	3953.87.7	4444.43.3	6
7	34.33,8	524.89.4	1015.45.0	1506.00,5	1996.56,1	2187.11.6	2977.67.2	3468.22,7	3958.78,3	4449.33.8	7
8	39.24.4	529.80.0	1020.35,5	1510.91.1	2001.46.6	2492.02,2	2982.57.7	3473.13,3	3963.68,8	4454.24.4	8
9	44.15.0	534.70,5	1025.26,1	1515.81,6	2006.37.2	2496.92,7	2987.48.3	3478.03.8	3968.59.4	4459.15.0	9
10	49.05.5	539.61.1	1030.16,6	1520.72,2	2011.27.7	2501 83.3	2992.38.8	3482.94 4	3973.50.0	4464.05.5	10
11	53.96.1	544.51.6	1035.07 2	1525.62,7	2016.18.3	2506.73,8	2997.29.4	3487.85,0	3978.40.5	4468.96.1	11
12	58.86,6	549.42.2	1039.97.7	1530.53,3	2021.08.8	2511.64.4	3002.20.0	3492.75,5	3983.31.1	4473.86,6	12
13	63.77.2	554.32 7	1044.88,3	1535.43.8	2025.99.4	2516.55.0	3007.10.5	3497.66,1	3988.21.6	4478.77.2	13
14	68.67.7	559.23.3	1049.78.8	1540.34.4	2030.90.0	2521.45.5	3012.01.1	3502.56,6	3993.12,2	4483.67.7	14
15	73.58,3	564.13,8	1054.69.4	1545.25.0	2035.80.5	2526 36.1	3016.91.6	3507.47.2	3998.02.7	4488.58.3	15
16	78.48.8	569.04.4	1059.60.0	1550.15,5	2040.71.1	2531.26,6	3021.82.2	3512.37,7	4002.93.3	4493.48.8	16
17	83 39.4	573.95,0	1064.50,5	1555.06,1	2045.61.6	2536.17,2	3026.72.7	3517.28,3	4007.83.8	4498.39.4	17
18	88.30.0	578.85,5	1069.41.1	1559.96,6	2050.52.2	2541.07 7	3031.63,3	3522.18.8	4012.74.4	4503.30,0	18
19	93.20,5	583.76,1	1074.31,6	1564.87,2	2055.42.7	2545.98.3	3036.53,8	3527,09.4	4017.65.0	4508.20.5	19
20	98.11,1	588.66,6	1079.22.2	1569.77,7	2060.33,3	2550.88,8	3041.44.4	3532,00.0	4022.55.5	4513.11.1	20
21	103.01,6	593.57.2	1084.12.7	1574.68,3	2065.23.8	2555.79.4	3046.35.0	3536.90,5	4027.46.1	4518 01.6	21
22	107.92.2	598.47,7	1089.03,3	1579.58,8	2070.14.4	2560.70.0	3051.25,5	3541.81,1	4032.36,6	4522.92,2	22
23	112.82.7	603.38,3	1093,93.8	1584.49,4	2075.05,0	2565.60,5	3056.16,1	3546.71,6	4037.27.2	4527.82.7	23
24	117.73.3	608.28,8	1098.84.4	1589.40.0	2079.95,5	2570.51,1	3061.06,6	3551 62.2	4042.17.7	4532.73.3	24
25	122.63,8	613.19,4	1103.75,0	1594.30.5	2084.86,1	2575.41,6	3065.97.2	3556.52,7	4047.08.3	4537.63.8	25
26	127.54.4	618.10.0	1108.65,5	1599.21,1	2089.76.6	2580.32,2	3070.87.7	3561.43,3	4051.98.8	4542.54.4	26
27	132.45.0	623.00,5	1113.56,1	1604.11,6	2094.67,2	2585,22.7	3075.78,3	3566.33,8	4056.89.4	4547.45.0	27
28	137.35,5	627.91.1	1118.46,6	1609.02 2	2099.57.7	2590.13,3	3080.68 8	3571.24.4	4061.80.0	4552.35,5	28
29	142.26.1	632.81,6	1123.37,2	1613.92,7	2104.48 3	2595.03 8	3085.59.4	3576.15.0	4066.70.5	4557.26.1	29
30	147.16,6	637.72,2	1128.27.7	1618.83,3	2109.38.8	2599.94 4	3090 50.0	3581.05.5	4071.61,1	4562.16 6	30
31	152.07,2	642.62,7	1133.18,3	1623.73,8	2114.29.4	2604.85,0	3695.40.5	3585.96,1	4076.51.6	4567.07,2	31
32	156.97.7	647.53,3	1138.08,8	1628.64,4	2119.20.0	2609.75,5	3100.31,1	3590.86,6	4081.42,2	4571.97.7	32
33	161.88.3	652.43,8	1142.99.4	1633.55,0	2124.10,5	2614.66,1	3105.21,6	3595.77.2	4086.32,7	4576.88.3	33
34	166.78.8	657.34.4	1147.90.0	1638.45,5	2129.01,1	2619.56,6	3110.12,2	3600.67,7	4091.23.3	4581.78.8	34
35	171.69.4	662.25,0	1152.80,5	1643.36,1	2133.91,6	2624.47,2	3115.02,7	3605 58.3	4096.13,8	4586.69.4	35
36	176.60.0	667.15,5	1157.71,1	1648.26,6	2138.82,2	2629.37,7	3119.93,3	3610.48,8	4101.04,4	4591.60.0	36
37	181.50,5	672.06.1	1162.61,6	1653.17,2	2143.72,7	2634.28,3	3124.83,8	3615.39,4	4105.95.0	4596 50.5	37
38	186.41.1	676.96,6	1167.52,2	1658.07.7	2148.63,3	2639.18,8	3129.74.4	3620.30.0	4110.85.5	4601.41.1	38
39	191.31,6	681.87.2	1172.42,7	1662.98,3	2153.53.8	2644.09,4	3134.65,0	3625.20.5	4115.76.1	4606.31.6	39
40	196.22.2	686.77.7	1177.33,3	1667.88.8	2158.44.4	2649.00,0	3139.55,5	3630.11.1	4120.66.6	4611.22,2	40
41	201.12.7	691.68.3	1182.23 8	1672.79.4	2163.35,0	2653.90,5	3144.46,1	3635.01,6	4125.57,2	4616.12,7	41
42	206.03.3	696.58,8	1187.14,4	1677.70.0	2168.25,5	2658.81,1	3149.36,6	3639.92,2	4130.47,7	4621.03,3	42
43	210.93.8	701,49.4	1192.05,0	1682.60,5	2173.16,1	2663.71,6	3154.27,2	3644.82,7	4135.38,3	4625.93.8	43
44	215.84.4	706,10.0	1196.95,5	1687.51,1	2178.06,6	2668.62,2	3159.17,7	3649.73,3	4140.28,8	4630.84.4	44
45	220.75.0	711,30.5	1201.86 1	1692.41,6	2182.97,2	2673.52,7	3164.08,3	3654.63,8	4145.19.4	4635.75,0	45
46	225.65.5	716,21.1	1206.76,6	1697.32,2	2187.87.7	2678.43,3	3168.98,8	3659.54,4	4150.10.0	4640.65,5	46
47	230.56.1	721,11.6	1211.67,2	1702.22,7	2192.78,3	2683.33,8	3173.89,4	3664.44,8	4155.00,5	4645.56,1	47
48	235.46,6	726.02,2	1216.57,7	1707.13,3	2197.68,8	2688.24,4	3178.80,0	3669.35,5	4159 91,1	4650.46,6	48
49	240 37.2	730.92,7	1221.48,3	1712.03,8	2202.59.4	2693.15,0	3183.70,5	3674.26,1	4164.81,6	4655.37,2	49

S.	0	1	2	3	4	5	6	7	8	9	S.
d.	$ cts.	$ cts.	$ cts.	$ cts.	$ cts.	$ cts.	$ cts.	$ cts.	$ cts.	$ cts.	d.
0	......	.24.5	.49.0	.73,5	.98.1	1.22,6	1.47.1	1.71.6	1.96,2	2.20.7	0
1	.02.0	.26.5	.51.1	.75.6	1.00,1	1.24.6	1.49.2	1.73.7	1.98,2	2.22,8	1
2	.04.0	.28.6	.53.1	.77.6	1.02,2	1.26,7	1.51,2	1.75,7	2.00.3	2.24.8	2
3	.06.1	.30.6	.55.1	.79.7	1.04.2	1.28.7	1.53,3	1.77,8	2.02,3	2.26,8	3
4	.08.1	.32,7	.57.2	.81.7	1.06,2	1.30,8	1.55,3	1.79,8	2.04.4	2.28,9	4
5	.10.2	.34,7	.59,2	.83,8	1.08.3	1.32.8	1.57,3	1.81,9	2.06,4	2.30.9	5
6	.12.2	.36,7	.61,3	.85,8	1.10,3	1.34 9	1.69,4	1.83,9	2.08,4	2.33.0	6
7	.14.3	.38.8	.63,3	.87.9	1.12.4	1.36,9	1.61,4	1.86,0	2.10.5	2.35.0	7
8	.16.3	.40,8	.65,4	.89,9	1.14.4	1.38,9	1.63,5	1.88,0	2.12.5	2.37,1	8
9	.18.4	.42.9	.67.4	.91,9	1.16.5	1.41.0	1.65,5	1.90,1	2.14.6	2.39.1	9
10	.20.4	.41.9	.69.5	.94,0	1.18.5	1.43.0	1.67.6	1.92.1	2.16.6	2.41,2	10
11	.22.4	.47.0	.71.5	.96,0	1.20.6	1.45,1	1.69,6	1.94,1	2.18.7	2.43,2	11

4.90\tfrac{40}{100}$ TO THE POUND.

10$\tfrac{3}{8}$%

PREMIUM.

STERLING INTO DOLLARS AND CENTS, AND VICE-VERSA.

£	0	100	200	300	400	500	600	700	800	900	£
£	$ cts.	$ cts.	$ cts.	$ cts.	$ cts.	$ cts.	$ cts.	$ cts.	$ cts.	$ cts.	£
50	245.27.7	735.83.3	1226.38.8	1716.94.4	2207.50.0	2698.05.5	3188.61.1	3679.16.6	4169.72.2	4660.27.7	50
51	250.18.3	740.73.8	1231.29.4	1721.85.0	2212.40.5	2702.96.1	3193.51.6	3684.07.2	4174.62.7	4665.18.3	51
52	255.08.8	745.64.4	1236.20.0	1726.75.5	2217.31.1	2707.86.6	3198.42.2	3688.97.7	4179.53.3	4670.08.8	52
53	259.99.4	750.55.0	1241.10.5	1731.66.1	2222.21.6	2712.77.2	3203.32.7	3693.88.3	4184.43.8	4674.99.4	53
54	264.90.0	755.45.5	1246.01.1	1736.56.6	2227.12.2	2717.67.7	3208.23.3	3698.78.8	4189.34.4	4679.90.0	54
55	269.80.5	760.36.1	1250.91.6	1741.47.2	2232.02.7	2722.58.3	3213.13.8	3703.69.4	4194.25.0	4684.80.5	55
56	274.71.1	765.26.6	1255.82.2	1746.37.7	2236.93.3	2727.48.8	3218.04.4	3708.60.0	4199.15.5	4689.71.1	56
57	279.61.6	770.17.2	1260.72.7	1751.28.3	2241.83.8	2732.39.4	3222.95.0	3713.50.5	4204.06.1	4694.61.6	57
58	284.52.2	775.07.7	1265.63.3	1756.18.8	2246.74.4	2737.30.0	3227.85.5	3718.41.1	4208.96.6	4699.52.2	58
59	289.42.7	779.98.3	1270.53.8	1761.09.4	2251.65.0	2742.20.5	3232.76.1	3723.31.6	4213.87.2	4704.42.7	59
60	294.33.3	784.88.8	1275.44.4	1766.00.0	2256.55.5	2747.11.1	3237.66.6	3728.22.2	4218.77.7	4709.33.3	60
61	299.23.8	789.79.4	1280.35.0	1770.90.5	2261.46.1	2752.01.6	3242.57.2	3733.12.7	4223.68.3	4714.23.8	61
62	304.14.4	794.70.0	1285.25.5	1775.81.1	2266.36.6	2756.92.2	3247.47.7	3738.03.3	4228.58.8	4719.14.4	62
63	309.05.0	799.60.5	1290.16.1	1780.71.6	2271.27.2	2761.82.7	3252.38.3	3742.93.8	4233.49.4	4724.05.0	63
64	313.95.5	804.51.1	1295.06.6	1785.62.2	2276.17.7	2766.73.3	3257.28.8	3747.84.4	4238.40.0	4728.95.5	64
65	318.86.1	809.41.6	1299.97.2	1790.52.7	2281.08.3	2771.63.8	3262.19.1	3752.75.0	4243.30.5	4733.86.1	65
66	323.76.6	814.32.2	1304.87.7	1795.43.3	2285.98.8	2776.54.4	3267.10.0	3757.65.5	4248.21.1	4738.76.6	66
67	328.67.2	819.22.7	1309.78.3	1800.33.8	2290.89.4	2781.45.0	3272.00.5	3762.56.1	4253.11.6	4743.67.2	67
68	333.57.7	824.13.3	1314.68.8	1805.24.4	2295.80.0	2786.35.5	3276.91.1	3767.46.6	4258.02.2	4748.57.7	68
69	338.48.3	829.03.8	1319.59.4	1810.15.0	2300.70.5	2791.26.1	3281.81.6	3772.37.2	4262.92.7	4753.48.3	69
70	343.38.8	833.94.4	1324.50.0	1815.05.5	2305.61.1	2796.16.6	3286.72.2	3777.27.7	4267.83.3	4758.38.8	70
71	348.29.4	838.85.0	1329.40.5	1819.96.1	2310.51.6	2801.07.2	3291.62.7	3782.18.3	4272.73.8	4763.29.4	71
72	353.20.0	843.75.5	1334.31.1	1824.86.6	2315.42.2	2805.97.7	3296.53.3	3787.08.8	4277.64.4	4768.20.0	72
73	358.10.5	848.66.1	1339.21.6	1829.77.2	2320.32.7	2810.88.3	3301.43.8	3791.99.4	4282.55.0	4773.10.5	73
74	363.01.1	853.56.6	1344.12.2	1834.67.7	2325.23.3	2815.78.8	3306.34.4	3796.90.0	4287.45.5	4778.01.1	74
75	367.91.6	858.47.2	1349.02.7	1839.58.3	2330.13.8	2820.69.4	3311.25.0	3801.80.5	4292.36.1	4782.91.6	75
76	372.82.2	863.37.7	1353.93.3	1844.48.8	2335.04.4	2825.60.0	3316.15.5	3806.71.1	4297.26.6	4787.82.2	76
77	377.72.7	868.28.3	1358.83.8	1849.39.4	2339.95.0	2830.50.5	3321.06.1	3811.61.6	4302.17.2	4792.72.7	77
78	382.63.3	873.18.8	1363.74.4	1854.30.0	2344.85.5	2835.41.1	3325.96.6	3816.52.2	4307.07.7	4797.63.3	78
79	387.53.8	878.09.4	1368.65.0	1859.20.5	2349.76.1	2840.31.6	3330.87.2	3821.42.7	4311.98.3	4802.53.8	79
80	392.44.4	883.00.0	1373.55.5	1864.11.1	2354.66.6	2845.22.2	3335.77.7	3826.33.3	4316.88.8	4807.44.4	80
81	397.35.0	887.90.5	1378.46.1	1869.01.6	2359.57.2	2850.12.7	3340.68.3	3831.23.8	4321.79.4	4812.35.0	81
82	402.25.5	892.81.1	1383.36.6	1873.92.2	2364.47.7	2855.03.3	3345.58.8	3836.14.4	4326.70.0	4817.25.5	82
83	407.16.1	897.71.6	1388.27.2	1878.82.7	2369.38.3	2859.93.8	3350.49.4	3841.05.0	4331.60.5	4822.16.1	83
84	412.06.6	902.62.2	1393.17.7	1883.73.3	2374.28.8	2864.84.4	3355.40.0	3845.95.5	4336.51.1	4827.06.6	84
85	416.97.2	907.52.7	1398.08.3	1888.63.8	2379.19.4	2869.75.0	3360.30.5	3850.86.1	4341.41.6	4831.97.2	85
86	421.87.7	912.43.3	1402.98.8	1893.54.4	2384.10.0	2874.65.5	3365.21.1	3855.76.6	4346.32.2	4836.87.7	86
87	426.78.3	917.33.8	1407.89.4	1898.45.0	2389.00.5	2879.56.1	3370.11.6	3860.67.2	4351.22.7	4841.78.3	87
88	431.68.8	922.24.4	1412.80.0	1903.35.5	2393.91.1	2884.46.6	3375.02.2	3865.57.7	4356.13.3	4846.68.8	88
89	436.59.4	927.15.0	1417.70.5	1908.26.1	2398.81.6	2889.37.2	3379.92.7	3870.48.3	4361.03.8	4851.59.4	89
90	441.50.0	932.05.5	1422.61.1	1913.16.6	2403.72.2	2894.27.7	3384.83.3	3875.38.8	4365.94.4	4856.50.0	90
91	446.40.5	936.96.1	1427.51.6	1918.07.2	2408.62.7	2899.18.3	3389.73.8	3880.29.4	4370.85.0	4861.40.5	91
92	451.31.1	941.86.6	1432.42.2	1922.97.7	2413.53.3	2904.08.8	3394.64.4	3885.20.0	4375.75.5	4866.31.1	92
93	456.21.6	946.77.2	1437.32.7	1927.88.3	2418.43.8	2908.99.4	3399.55.0	3890.10.5	4380.66.1	4871.21.6	93
94	461.12.2	951.67.7	1442.23.3	1932.78.8	2423.34.4	2913.90.0	3404.45.5	3895.01.1	4385.56.6	4876.12.2	94
95	466.02.7	956.58.3	1447.13.8	1937.69.4	2428.25.0	2918.80.5	3409.36.1	3899.91.6	4390.47.2	4881.02.7	95
96	470.93.3	961.48.8	1452.04.4	1942.60.0	2433.15.5	2923.71.1	3414.26.6	3904.82.2	4395.37.7	4885.93.3	96
97	475.83.8	966.39.4	1456.95.0	1947.50.5	2438.06.1	2928.61.6	3419.17.2	3909.72.7	4400.28.3	4890.83.8	97
98	480.74.4	971.30.0	1461.85.5	1952.41.1	2442.96.6	2933.52.2	3424.07.7	3914.63.3	4405.18.8	4895.74.4	98
99	485.65.0	976.20.5	1466.76.1	1957.31.6	2447.87.2	2938.42.7	3428.98.3	3919.53.8	4410.09.4	4900.65.0	99

S.	10	11	12	13	14	15	16	17	18	19	S.
d.	$ cts.	$ cts.	$ cts.	$ cts.	$ cts.	$ cts.	$ cts.	$ cts.	$ cts.	$ cts.	d.
0	2.45.2	2.69.8	2.94.3	3.18.8	3.43.3	3.67.9	3.92.4	4.16.9	4.41.5	4.66.0	0
1	2.47.3	2.71.8	2.96.3	3.20.9	3.45.4	3.69.9	3.94.4	4.19.0	4.43.5	4.68.0	1
2	2.49.3	2.73.8	2.98.4	3.22.9	3.47.4	3.72.0	3.96.5	4.21.0	4.45.5	4.70.1	2
3	2.51.4	2.75.9	3.00.4	3.25.0	3.49.5	3.74.0	3.98.5	4.23.1	4.47.6	4.72.1	3
4	2.53.4	2.77.9	3.02.5	3.27.0	3.51.5	3.76.0	4.00.6	4.25.1	4.49.6	4.74.2	4
5	2.55.5	2.80.0	3.04.5	3.29.0	3.53.6	3.78.1	4.02.6	4.27.2	4.51.7	4.76.2	5
6	2.57.5	2.82.0	3.06.6	3.31.1	3.55.6	3.80.1	4.04.7	4.29.2	4.53.7	4.78.2	6
7	2.59.6	2.84.1	3.08.6	3.33.1	3.57.7	3.82.2	4.06.7	4.31.2	4.55.8	4.80.3	7
8	2.61.6	2.86.1	3.10.6	3.35.2	3.59.7	3.84.2	4.08.8	4.33.3	4.57.8	4.82.3	8
9	2.63.6	2.88.2	3.12.7	3.37.2	3.61.7	3.86.3	4.10.8	4.35.3	4.59.9	4.84.4	9
10	2.65.7	2.90.2	3.14.7	3.39.3	3.63.8	3.88.3	4.12.8	4.37.4	4.61.9	4.86.4	10
11	2.67.7	2.92.2	3.16.8	3.41.3	3.65.8	3.90.4	4.14.9	4.39.4	4.63.9	4.88.5	11

$4.91₁₀₀ TO THE POUND.

10½ %

PREMIUM.

STERLING INTO DOLLARS AND CENTS, AND VICE-VERSA.

£	0	100	200	300	400	500	600	700	800	900	£
£	$ cts.	$ cts.	$ cts.	$ cts.	$ cts.	$ cts.	$ cts.	$ cts.	$ cts.	$ cts.	£
0	......	491.11.1	982.22.2	1473.33.3	1964.44.4	2455.55.5	2946.66.6	3437.77.7	3928.88.8	4420.00.0	0
1	4.91.1	496.02.2	987.13.3	1478.24.4	1969.35.5	2460.46.6	2951.57.7	3442.68.8	3933.80.0	4424.91.1	1
2	9.82.2	500.93.3	992.04.4	1483.15.5	1974.26.6	2465.37.7	2956.48.8	3447.60.0	3938.71.1	4429.82.2	2
3	14.73.3	505.84.4	996.95.5	1488.06.6	1979.17.7	2470.28.8	2961.40.0	3452.51.1	3943.62.2	4434.73.3	3
4	19.64.4	510.75.5	1001.86.6	1492.97.7	1984.08.8	2475.20.0	2966.31.1	3457.42.2	3948.53.3	4439.64.4	4
5	24.55.5	515.66.6	1006.77.7	1497.88.8	1989.00.0	2480.11.1	2971.22.2	3462.33.3	3953.44.4	4444.55.5	5
6	29.46.6	520.57.7	1011.68.8	1502.80.0	1993.91.1	2485.02.2	2976.13.3	3467.24.4	3958.35.5	4449.46.6	6
7	34.37.7	525.48.8	1016.60.0	1507.71.1	1998.82.2	2489.93.3	2981.04.4	3472.15.5	3963.26.6	4454.37.7	7
8	39.28.8	530.40.0	1021.51.1	1512.62.2	2003.73.3	2494.84.4	2985.95.5	3477.06.6	3968.17.7	4459.28.8	8
9	44.20.0	535.31.1	1026.42.2	1517.53.3	2008.64.4	2499.75.5	2990.86.6	3481.97.7	3973.08.8	4464.20.0	9
10	49.11.1	540.22.2	1031.33.3	1522.44.4	2013.55.5	2504.66.6	2995.77.7	3486.88.8	3978.00.0	4469.11.1	10
11	54.02.2	545.13.3	1036.24.4	1527.35.5	2018.46.6	2509.57.7	3000.68.8	3491.80.0	3982.91.1	4474.02.2	11
12	58.93.3	550.04.4	1041.15.5	1532.26.6	2023.37.7	2514.48.8	3005.60.0	3496.71.1	3987.82.2	4478.93.3	12
13	63.84.4	554.95.5	1046.06.6	1537.17.7	2028.28.8	2519.40.0	3010.51.1	3501.62.2	3992.73.3	4483.84.4	13
14	68.75.5	559.86.6	1050.97.7	1542.08.8	2033.20.0	2524.31.1	3015.42.2	3506.53.3	3997.64.4	4488.75.5	14
15	73.66.6	564.77.7	1055.88.8	1547.00.0	2038.11.1	2529.22.2	3020.33.3	3511.44.4	4002.55.5	4493.66.6	15
16	78.57.7	569.68.8	1060.80.0	1551.91.1	2043.02.2	2534.13.3	3025.24.4	3516.35.5	4007.46.6	4498.57.7	16
17	83.48.8	574.60.0	1065.71.1	1556.82.2	2047.93.3	2539.04.4	3030.15.5	3521.26.6	4012.37.7	4503.48.8	17
18	88.40.0	579.51.1	1070.62.2	1561.73.3	2052.84.4	2543.95.5	3035.06.6	3526.17.7	4017.28.8	4508.40.0	18
19	93.31.1	584.42.2	1075.53.3	1566.64.4	2057.75.5	2548.86.6	3039.97.7	3531.08.8	4022.20.0	4513.31.1	19
20	98.22.2	589.33.3	1080.44.4	1571.55.5	2062.66.6	2553.77.7	3044.88.8	3536.00.0	4027.11.1	4518.22.2	20
21	103.13.3	594.24.4	1085.35.5	1576.46.6	2067.57.7	2558.68.8	3049.80.0	3540.91.1	4032.02.2	4523.13.3	21
22	108.04.4	599.15.5	1090.26.6	1581.37.7	2072.48.8	2563.60.0	3054.71.1	3545.82.2	4036.93.3	4528.04.4	22
23	112.95.5	604.06.6	1095.17.7	1586.28.8	2077.40.0	2568.51.1	3059.62.2	3550.73.3	4041.84.4	4532.95.5	23
24	117.86.6	608.97.7	1100.08.8	1591.20.0	2082.31.1	2573.42.2	3064.53.3	3555.64.4	4046.75.5	4537.86.6	24
25	122.77.7	613.88.8	1105.00.0	1596.11.1	2087.22.2	2578.33.3	3069.44.4	3560.55.5	4051.66.6	4542.77.7	25
26	127.68.8	618.80.0	1109.91.1	1601.02.2	2092.13.3	2583.24.4	3074.35.5	3565.46.6	4056.57.7	4547.68.8	26
27	132.60.0	623.71.1	1114.82.2	1605.93.3	2097.04.4	2588.15.5	3079.26.6	3570.37.7	4061.48.8	4552.60.0	27
28	137.51.1	628.62.2	1119.73.3	1610.84.4	2101.95.5	2593.06.6	3084.17.7	3575.28.8	4066.40.0	4557.51.1	28
29	142.42.2	633.53.3	1124.64.4	1615.75.5	2106.86.6	2597.97.7	3089.08.8	3580.20.0	4071.31.1	4562.42.2	29
30	147.33.3	638.44.4	1129.55.5	1620.66.6	2111.77.7	2602.88.8	3094.00.0	3585.11.1	4076.22.2	4567.33.3	30
31	152.24.4	643.35.5	1134.46.6	1625.57.7	2116.68.8	2607.80.0	3098.91.1	3590.02.2	4081.13.3	4572.24.4	31
32	157.15.5	648.26.6	1139.37.7	1630.48.8	2121.60.0	2612.71.1	3103.82.2	3594.93.3	4086.04.4	4577.15.5	32
33	162.06.6	653.17.7	1144.28.8	1635.40.0	2126.51.1	2617.62.2	3108.73.3	3599.84.4	4090.95.5	4582.06.6	33
34	166.97.7	658.08.8	1149.20.0	1640.31.1	2131.42.2	2622.53.3	3113.64.4	3604.75.5	4095.86.6	4586.97.7	34
35	171.88.8	663.00.0	1154.11.1	1645.22.2	2136.33.3	2627.44.4	3118.55.5	3609.66.6	4100.77.7	4591.88.8	35
36	176.80.0	667.91.1	1159.02.2	1650.13.3	2141.24.4	2632.35.5	3123.46.6	3614.57.7	4105.68.8	4596.80.0	36
37	181.71.1	672.82.2	1163.93.3	1655.04.4	2146.15.5	2637.26.6	3128.37.7	3619.48.8	4110.60.0	4601.71.1	37
38	186.62.2	677.73.3	1168.84.4	1659.95.5	2151.06.6	2642.17.7	3133.28.8	3624.40.0	4115.51.1	4606.62.2	38
39	191.53.3	682.64.4	1173.75.5	1664.86.6	2155.97.7	2647.08.8	3138.20.0	3629.31.1	4120.42.2	4611.53.3	39
40	196.44.4	687.55.5	1178.66.6	1669.77.7	2160.88.8	2652.00.0	3143.11.1	3634.22.2	4125.33.3	4616.44.4	40
41	201.35.5	692.46.6	1183.57.7	1674.68.8	2165.80.0	2656.91.1	3148.02.2	3639.13.3	4130.24.4	4621.35.5	41
42	206.26.6	697.37.7	1188.48.8	1679.60.0	2170.71.1	2661.82.2	3152.93.3	3644.04.4	4135.15.5	4626.26.6	42
43	211.17.7	702.28.8	1193.40.0	1684.51.1	2175.62.2	2666.73.3	3157.84.4	3648.95.5	4140.06.6	4631.17.7	43
44	216.08.8	707.20.0	1198.31.1	1689.42.2	2180.53.3	2671.64.4	3162.75.5	3653.86.6	4144.97.7	4636.08.8	44
45	221.00.0	712.11.1	1203.22.2	1694.33.3	2185.44.4	2676.55.5	3167.66.6	3658.77.7	4149.88.8	4641.00.0	45
46	225.91.1	717.02.2	1208.13.3	1699.24.4	2190.35.5	2681.46.6	3172.57.7	3663.68.8	4154.80.0	4645.91.1	46
47	230.82.2	721.93.3	1213.04.4	1704.15.5	2195.26.6	2686.37.7	3177.48.8	3668.60.0	4159.71.1	4650.82.2	47
48	235.73.3	726.84.4	1217.95.5	1709.06.6	2200.17.7	2691.28.8	3182.40.0	3673.51.1	4164.62.2	4655.73.3	48
49	240.64.4	731.75.5	1222.86.6	1713.97.7	2205.08.8	2696.20.0	3187.31.1	3678.42.2	4169.53.3	4660.64.4	49

S.	0	1	2	3	4	5	6	7	8	9	S.
d.	$ cts.	$ cts.	$ cts.	$ cts.	$ cts.	$ cts.	$ cts.	$ cts.	$ cts.	$ cts.	d.
0	......	.24.5	.49.1	.73.6	.98.2	1.22.7	1.47.3	1.71.8	1.96.4	2.21.0	0
1	.02.0	.26.6	.51.1	.75.7	1.00.2	1.24.8	1.49.3	1.73.9	1.98.4	2.23.0	1
2	.04.0	.28.6	.53.2	.77.7	1.02.3	1.26.8	1.51.4	1.75.9	2.00.5	2.25.0	2
3	.06.1	.30.6	.55.2	.79.8	1.04.3	1.28.9	1.53.4	1.78.0	2.02.5	2.27.1	3
4	.08.1	.32.7	.57.2	.81.8	1.06.4	1.30.9	1.55.5	1.80.0	2.04.6	2.29.1	4
5	.10.2	.34.7	.59.3	.83.8	1.08.4	1.33.0	1.57.5	1.82.1	2.06.6	2.31.2	5
6	.12.2	.36.8	.61.3	.85.9	1.10.4	1.35.0	1.59.6	1.84.1	2.08.7	2.33.2	6
7	.14.3	.38.8	.63.4	.87.9	1.12.5	1.37.0	1.61.6	1.86.2	2.10.7	2.35.3	7
8	.16.3	.40.9	.65.4	.90.0	1.14.5	1.39.1	1.63.6	1.88.2	2.12.8	2.37.3	8
9	.18.4	.42.9	.67.5	.92.0	1.16.6	1.41.1	1.65.7	1.90.2	2.14.8	2.39.4	9
10	.20.4	.45.0	.69.5	.94.1	1.18.6	1.43.2	1.67.7	1.92.3	2.16.8	2.41.4	10
11	.22.4	.47.0	.71.6	.96.1	1.20.7	1.45.2	1.69.8	1.94.3	2.18.9	2.43.4	11

$4.91¹⁄₁₀₀ TO THE POUND.

10½ %

PREMIUM.

STERLING INTO DOLLARS AND CENTS, AND VICE-VERSA.

£	0	100	200	300	400	500	600	700	800	900	£
	$ cts.	$ cts.	$ cts.	$ cts.	$ cts.	$ cts.	$ cts.	$ cts.	$ cts.	$ cts.	
50	245.55.5	736.66.6	1227.77.7	1718.88.8	2210.00.0	2701.11.1	3192.22.2	3683.33.3	4174.44.4	4665.55.5	50
51	250.46.6	741.57.7	1232.68.8	1723.80.0	2214.91.1	2706.02.2	3197.13.3	3688.24.4	4179.35.5	4670.46.6	51
52	255.37.7	746.48.8	1237.60.0	1728.71.1	2219.82.2	2710.93.3	3202.04.4	3693.15.5	4184.26.6	4675.37.7	52
53	260.28.8	751.40.0	1242.51.1	1733.62.2	2224.73.3	2715.84.4	3206.95.5	3698.06.6	4189.17.7	4680.28.8	53
54	265.20.0	756.31.1	1247.42.2	1738.53.3	2229.64.4	2720.75.5	3211.86.6	3702.97.7	4194.08.8	4685.20.0	54
55	270.11.1	761.22.2	1252.33.3	1743.44.4	2234.55.5	2725.66.6	3216.77.7	3707.88.8	4199.00.0	4690.11.1	55
56	275.02.2	766.13.3	1257.24.4	1748.35.5	2239.46.6	2730.57.7	3221.68.8	3712.80.0	4203.91.1	4695.02.2	56
57	279.93.3	771.04.4	1262.15.5	1753.26.6	2244.37.7	2735.48.8	3226.60.0	3717.71.1	4208.82.2	4699.93.3	57
58	284.84.4	775.95.5	1267.06.6	1758.17.7	2249.28.8	2740.40.0	3231.51.1	3722.62.2	4213.73.3	4704.84.4	58
59	289.75.5	780.86.6	1271.97.7	1763.08.8	2254.20.0	2745.31.1	3236.42.2	3727.53.3	4218.64.4	4709.75.5	59
60	294.66.6	785.77.7	1276.88.8	1768.00.0	2259.11.1	2750.22.2	3241.33.3	3732.44.4	4223.55.5	4714.66.6	60
61	299.57.7	790.68.8	1281.80.0	1772.91.1	2264.02.2	2755.13.3	3246.24.4	3737.35.5	4228.46.6	4719.57.7	61
62	304.48.8	795.60.0	1286.71.1	1777.82.2	2268.93.3	2760.04.4	3251.15.5	3742.26.6	4233.37.7	4724.48.8	62
63	309.40.0	800.51.1	1291.62.2	1782.73.3	2273.84.4	2764.95.5	3256.06.6	3747.17.7	4238.28.8	4729.40.0	63
64	314.31.1	805.42.2	1296.53.3	1787.64.4	2278.75.5	2769.86.6	3260.97.7	3752.08.8	4243.20.0	4734.31.1	64
65	319.22.2	810.33.3	1301.44.4	1792.55.5	2283.66.6	2774.77.7	3265.88.8	3757.00.0	4248.11.1	4739.22.2	65
66	324.13.3	815.24.4	1306.35.5	1797.46.6	2288.57.7	2779.68.8	3270.80.0	3761.91.1	4253.02.2	4744.13.3	66
67	329.04.4	820.15.5	1311.26.6	1802.37.7	2293.48.8	2784.60.0	3275.71.1	3766.82.2	4257.93.3	4749.04.4	67
68	333.95.5	825.06.6	1316.17.7	1807.28.8	2298.40.0	2789.51.1	3280.62.2	3771.73.3	4262.84.4	4753.95.5	68
69	338.86.6	829.97.7	1321.08.8	1812.20.0	2303.31.1	2794.42.2	3285.53.3	3776.64.4	4267.75.5	4758.86.6	69
70	343.77.7	834.88.8	1326.00.0	1817.11.1	2308.22.2	2799.33.3	3290.44.4	3781.55.5	4272.66.6	4763.77.7	70
71	348.68.8	839.80.0	1330.91.1	1822.02.2	2313.13.3	2804.24.4	3295.35.5	3786.46.6	4277.57.7	4768.68.8	71
72	353.60.0	844.71.1	1335.82.2	1826.93.3	2318.04.4	2809.15.5	3300.26.6	3791.37.7	4282.48.8	4773.60.0	72
73	358.51.1	849.62.2	1340.73.3	1831.84.4	2322.95.5	2814.06.6	3305.17.7	3796.28.8	4287.40.0	4778.51.1	73
74	363.42.2	854.53.3	1345.64.4	1836.75.5	2327.86.6	2818.97.7	3310.08.8	3801.20.0	4292.31.1	4783.42.2	74
75	368.33.3	859.44.4	1350.55.5	1841.66.6	2332.77.7	2823.88.8	3315.00.0	3806.11.1	4297.22.2	4788.33.3	75
76	373.24.4	864.35.5	1355.46.6	1846.57.7	2337.68.8	2828.80.0	3319.91.1	3811.02.2	4302.13.3	4793.24.4	76
77	378.15.5	869.26.6	1360.37.7	1851.48.8	2342.60.0	2833.71.1	3324.82.2	3815.93.3	4307.04.4	4798.15.5	77
78	383.06.6	874.17.7	1365.28.8	1856.40.0	2347.51.1	2838.62.2	3329.73.3	3820.84.4	4311.95.5	4803.06.6	78
79	387.97.7	879.08.8	1370.20.0	1861.31.1	2352.42.2	2843.53.3	3334.64.4	3825.75.5	4316.86.6	4807.97.7	79
80	392.88.8	884.00.0	1375.11.1	1866.22.2	2357.33.3	2848.44.4	3339.55.5	3830.66.6	4321.77.7	4812.88.8	80
81	397.80.0	888.91.1	1380.02.2	1871.13.3	2362.24.4	2853.35.5	3344.46.6	3835.57.7	4326.68.8	4817.80.0	81
82	402.71.1	893.82.2	1384.93.3	1876.04.4	2367.15.5	2858.26.6	3349.37.7	3840.48.8	4331.60.0	4822.71.1	82
83	407.62.2	898.73.3	1389.84.4	1880.95.5	2372.06.6	2863.17.7	3354.28.8	3845.40.0	4336.51.1	4827.62.2	83
84	412.53.3	903.64.4	1394.75.5	1885.86.6	2376.97.7	2868.08.8	3359.20.0	3850.31.1	4341.42.2	4832.53.3	84
85	417.44.4	908.55.5	1399.66.6	1890.77.7	2381.88.8	2873.00.0	3364.11.1	3855.22.2	4346.33.3	4837.44.4	85
86	422.35.5	913.46.6	1404.57.7	1895.68.8	2386.80.0	2877.91.1	3369.02.2	3860.13.3	4351.24.4	4842.35.5	86
87	427.26.6	918.37.7	1409.48.8	1900.60.0	2391.71.1	2882.82.2	3373.93.3	3865.04.4	4356.15.5	4847.26.6	87
88	432.17.7	923.28.8	1414.40.0	1905.51.1	2396.62.2	2887.73.3	3378.84.4	3869.95.5	4361.06.6	4852.17.7	88
89	437.08.8	928.20.0	1419.31.1	1910.42.2	2401.53.3	2892.64.4	3383.75.5	3874.86.6	4365.97.7	4857.08.8	89
90	442.00.0	933.11.1	1424.22.2	1915.33.3	2406.44.4	2897.55.5	3388.66.6	3879.77.7	4370.88.8	4862.00.0	90
91	446.91.1	938.02.2	1429.13.3	1920.24.4	2411.35.5	2902.46.6	3393.57.7	3884.68.8	4375.80.0	4866.91.1	91
92	451.82.2	942.93.3	1434.04.4	1925.15.5	2416.26.6	2907.37.7	3398.48.8	3889.60.0	4380.71.1	4871.82.2	92
93	456.73.3	947.84.4	1438.95.5	1930.06.6	2421.17.7	2912.28.8	3403.40.0	3894.51.1	4385.62.2	4876.73.3	93
94	461.64.4	952.75.5	1443.86.6	1934.97.7	2426.08.8	2917.20.0	3408.31.1	3899.42.2	4390.53.3	4881.64.4	94
95	466.55.5	957.66.6	1448.77.7	1939.88.8	2431.00.0	2922.11.1	3413.22.2	3904.33.3	4395.44.4	4886.55.5	95
96	471.46.6	962.57.7	1453.68.8	1944.80.0	2435.91.1	2927.02.2	3418.13.3	3909.24.4	4400.35.5	4891.46.6	96
97	476.37.7	967.48.8	1458.60.0	1949.71.1	2440.82.2	2931.93.3	3423.04.4	3914.15.5	4405.26.6	4896.37.7	97
98	481.28.8	972.40.0	1463.51.1	1954.62.2	2445.73.3	2936.84.4	3427.95.5	3919.06.6	4410.17.7	4901.28.8	98
99	486.20.0	977.31.1	1468.42.2	1959.53.3	2450.64.4	2941.75.5	3432.86.6	3923.97.7	4415.08.8	4906.20.0	99

s.	10	11	12	13	14	15	16	17	18	19	s.
d.	$ cts.	$ cts.	$ cts.	$ cts.	$ cts.	$ cts.	$ cts.	$ cts.	$ cts.	$ cts.	d.
0	2.45.5	2.70.1	2.94.6	3.19.2	3.43.7	3.68.3	3.92.8	4.17.4	4.42.0	4.66.5	0
1	2.47.6	2.72.1	2.96.7	3.21.2	3.45.8	3.70.3	3.94.9	4.19.4	4.44.0	4.68.6	1
2	2.49.6	2.74.2	2.98.7	3.23.3	3.47.8	3.72.4	3.96.9	4.21.5	4.46.0	4.70.6	2
3	2.51.6	2.76.2	3.00.8	3.25.3	3.49.9	3.74.4	3.99.0	4.23.5	4.48.1	4.72.6	3
4	2.53.7	2.78.2	3.02.6	3.27.4	3.51.9	3.76.5	4.01.0	4.25.6	4.50.1	4.74.7	4
5	2.55.7	2.80.3	3.04.8	3.29.4	3.54.0	3.78.5	4.03.1	4.27.6	4.52.2	4.76.7	5
6	2.57.8	2.82.3	3.06.9	3.31.4	3.56.0	3.80.6	4.05.1	4.29.7	4.54.2	4.78.8	6
7	2.59.8	2.84.4	3.08.9	3.33.5	3.58.0	3.82.6	4.07.2	4.31.7	4.56.3	4.80.8	7
8	2.61.9	2.86.4	3.11.0	3.35.5	3.60.1	3.84.6	4.09.2	4.33.8	4.58.3	4.82.9	8
9	2.63.9	2.88.5	3.13.0	3.37.6	3.62.1	3.86.7	4.11.2	4.35.8	4.60.4	4.84.9	9
10	2.66.0	2.90.5	3.15.1	3.39.6	3.64.2	3.88.7	4.13.3	4.37.8	4.62.4	4.87.0	10
11	2.68.0	2.92.6	3.17.1	3.41.7	3.66.2	3.90.8	4.15.3	4.39.9	4.64.4	4.89.0	11

$4.91₁⁰⁄₁₀₀ TO THE POUND.

10⅝%

PREMIUM.

STERLING INTO DOLLARS AND CENTS, AND VICE-VERSA.

£	0	100	200	300	400	500	600	700	800	900	£
	$ cts.	$ cts.	$ cts.	$ cts.	$ cts.	$ cts.	$ cts.	$ cts.	$ cts.	$ cts.	
0	......	491.66.6	983.33.3	1475.00.0	1966.66.6	2458.33.3	2950.00.0	3441.66.6	3933.33.3	4425.00.0	0
1	4.91.6	496.58.3	988.25.0	1479.91.6	1971.58.3	2463.25.0	2954.91.6	3446.58.3	3938.25.0	4429.91.6	1
2	9.83.3	501.50.0	993.16.6	1484.83.3	1976.50.0	2468.16.6	2959.83.3	3451.50.0	3943.16.6	4434.83.3	2
3	14.75.0	506.41.6	998.08.3	1489.75.0	1981.41.6	2473.08.3	2964.75.0	3456.41.6	3948.08.3	4439.75.0	3
4	19.66.6	511.33.3	1003.00.0	1494.66.6	1986.33.3	2478.00.0	2969.66.6	3461.33.3	3953.00.0	4444.66.6	4
5	24.58.3	516.25.0	1007.91.6	1499.58.3	1991.25.0	2482.91.6	2974.58.3	3466.25.0	3957.91.6	4449.58.3	5
6	29.50.0	521.16.6	1012.83.3	1504.50.0	1996.16.6	2487.83.3	2979.50.0	3471.16.6	3962.83.3	4454.50.0	6
7	34.41.6	526.08.3	1017.75.0	1509.41.6	2001.08.3	2492.75.0	2984.41.6	3476.08.3	3967.75.0	4459.41.6	7
8	39.33.3	531.00.0	1022.66.6	1514.33.3	2006.00.0	2497.66.6	2989.33.3	3481.00.0	3972.66.6	4464.33.3	8
9	44.25.0	535.91.6	1027.58.3	1519.25.0	2010.91.6	2502.58.3	2994.25.0	3485.91.6	3977.58.3	4469.25.0	9
10	49.16.6	540.83.3	1032.50.0	1524.16.6	2015.83.3	2507.50.0	2999.16.6	3490.83.3	3982.50.0	4474.16.6	10
11	54.08.3	545.75.0	1037.41.6	1529.08.3	2020.75.0	2512.41.6	3004.08.3	3495.75.0	3987.41.6	4479.08.3	11
12	59.00.0	550.66.6	1042.33.3	1534.00.0	2025.66.6	2517.33.3	3009.00.0	3500.66.6	3992.33.3	4484.00.0	12
13	63.91.6	555.58.3	1047.25.0	1538.91.6	2030.58.3	2522.25.0	3013.91.6	3505.58.3	3997.25.0	4488.91.6	13
14	68.83.3	560.50.0	1052.16.6	1543.83.3	2035.50.0	2527.16.6	3018.83.3	3510.50.0	4002.16.6	4493.83.3	14
15	73.75.0	565.41.6	1057.08.3	1548.75.0	2040.41.6	2532.08.3	3023.75.0	3515.41.6	4007.08.3	4498.75.0	15
16	78.66.6	570.33.3	1062.00.0	1553.66.6	2045.33.3	2537.00.0	3028.66.6	3520.33.3	4012.00.0	4503.66.6	16
17	83.58.3	575.25.0	1066.91.6	1558.58.3	2050.25.0	2541.91.6	3033.58.3	3525.25.0	4016.91.6	4508.58.3	17
18	88.50.0	580.16.6	1071.83.3	1563.50.0	2055.16.6	2546.83.3	3038.50.0	3530.16.6	4021.83.3	4513.50.0	18
19	93.41.6	585.08.3	1076.75.0	1568.41.6	2060.08.3	2551.75.0	3043.41.6	3535.08.3	4026.75.0	4518.41.6	19
20	98.33.3	590.00.0	1081.66.6	1573.33.3	2065.00.0	2556.66.6	3048.33.3	3540.00.0	4031.66.6	4523.33.3	20
21	103.25.0	594.91.6	1086.58.3	1578.25.0	2069.91.6	2561.58.3	3053.25.0	3544.91.6	4036.58.3	4528.25.0	21
22	108.16.6	599.83.3	1091.50.0	1583.16.6	2074.83.3	2566.50.0	3058.16.6	3549.83.3	4041.50.0	4533.16.6	22
23	113.08.3	604.75.0	1096.41.6	1588.08.3	2079.75.0	2571.41.6	3063.08.3	3554.75.0	4046.41.6	4538.08.3	23
24	118.00.0	609.66.6	1101.33.3	1593.00.0	2084.66.6	2576.33.3	3068.00.0	3559.66.6	4051.33.3	4543.00.0	24
25	122.91.6	614.58.3	1106.25.0	1597.91.6	2089.58.3	2581.25.0	3072.91.6	3564.58.3	4056.25.0	4547.91.6	25
26	127.83.3	619.50.0	1111.16.6	1602.83.3	2094.50.0	2586.16.6	3077.83.3	3569.50.0	4061.16.6	4552.83.3	26
27	132.75.0	624.41.6	1116.08.3	1607.75.0	2099.41.6	2591.08.3	3082.75.0	3574.41.6	4066.08.3	4557.75.0	27
28	137.66.6	629.33.3	1121.00.0	1612.66.6	2104.33.3	2596.00.0	3087.66.6	3579.33.3	4071.00.0	4562.66.6	28
29	142.58.3	634.25.0	1125.91.6	1617.58.3	2109.25.0	2600.91.6	3092.58.3	3584.25.0	4075.91.6	4567.58.3	29
30	147.50.0	639.16.6	1130.83.3	1622.50.0	2114.16.6	2605.83.3	3097.50.0	3589.16.6	4080.83.3	4572.50.0	30
31	152.41.6	644.08.3	1135.75.0	1627.41.6	2119.08.3	2610.75.0	3102.41.6	3594.08.3	4085.75.0	4577.41.6	31
32	157.33.3	649.00.0	1140.66.6	1632.33.3	2124.00.0	2615.66.6	3107.33.3	3599.00.0	4090.66.6	4582.33.3	32
33	162.25.0	653.91.6	1145.58.3	1637.25.0	2128.91.6	2620.58.3	3112.25.0	3603.91.6	4095.58.3	4587.25.0	33
34	167.16.6	658.83.3	1150.50.0	1642.16.6	2133.83.3	2625.50.0	3117.16.6	3608.83.3	4100.50.0	4592.16.6	34
35	172.08.3	663.75.0	1155.41.6	1647.08.3	2138.75.0	2630.41.6	3122.08.3	3613.75.0	4105.41.6	4597.08.3	35
36	177.00.0	668.66.6	1160.33.3	1652.00.0	2143.66.6	2635.33.3	3127.00.0	3618.66.6	4110.33.3	4602.00.0	36
37	181.91.6	673.58.3	1165.25.0	1656.91.6	2148.58.3	2640.25.0	3131.91.6	3623.58.3	4115.25.0	4606.91.6	37
38	186.83.3	678.50.0	1170.16.6	1661.83.3	2153.50.0	2645.16.6	3136.83.3	3628.50.0	4120.16.6	4611.83.3	38
39	191.75.0	683.41.6	1175.08.3	1666.75.0	2158.41.6	2650.08.3	3141.75.0	3633.41.6	4125.08.3	4616.75.0	39
40	196.60.6	688.33.3	1180.00.0	1671.66.6	2163.33.3	2655.00.0	3146.66.6	3638.33.3	4130.00.0	4621.66.6	40
41	201.58.3	693.25.0	1184.91.6	1676.58.3	2168.25.0	2659.91.6	3151.58.3	3643.25.0	4134.91.6	4626.58.3	41
42	206.50.0	698.16.6	1189.83.3	1681.50.0	2173.16.6	2664.83.3	3156.50.0	3648.16.6	4139.83.3	4631.50.0	42
43	211.41.6	703.08.3	1194.75.0	1686.41.6	2178.08.3	2669.75.0	3161.41.6	3653.08.3	4144.75.0	4636.41.6	43
44	216.33.3	708.00.0	1199.66.6	1691.33.3	2183.00.0	2674.66.6	3166.33.3	3658.00.0	4149.66.6	4641.33.3	44
45	221.25.0	712.91.6	1204.58.3	1696.25.0	2187.91.6	2679.58.3	3171.25.0	3662.91.6	4154.58.3	4646.25.0	45
46	226.16.6	717.83.3	1209.50.0	1701.16.6	2192.83.3	2684.50.0	3176.16.6	3667.83.3	4159.50.0	4651.16.6	46
47	231.08.3	722.75.0	1214.41.6	1706.08.3	2197.75.0	2689.41.6	3181.08.3	3672.75.0	4164.41.6	4656.08.3	47
48	236.00.0	727.66.6	1219.33.3	1711.00.0	2202.66.6	2694.33.3	3186.00.0	3677.66.6	4169.33.3	4661.00.0	48
49	240.91.6	732.58.3	1224.25.0	1715.91.6	2207.58.3	2699.25.0	3190.91.6	3682.58.3	4174.25.0	4665.91.6	49

s.	0	1	2	3	4	5	6	7	8	9	s.
d.	$ cts.	$ cts.	$ cts.	$ cts.	$ cts.	$ cts.	$ cts.	$ cts.	$ cts.	$ cts.	d.
0	......	.24.5	.49.1	.73.7	.98.3	1.22.9	1.47.5	1.72.0	1.96.6	2.21.2	0
1	.02.0	.26.6	.51.2	.75.8	1.00.3	1.24.9	1.49.5	1.74.1	1.98.7	2.23.3	1
2	.04.0	.28.6	.53.2	.77.8	1.02.4	1.27.0	1.51.5	1.76.1	2.00.7	2.25.3	2
3	.06.1	.30.7	.55.3	.79.8	1.04.4	1.29.0	1.53.6	1.78.2	2.02.8	2.27.3	3
4	.08.1	.32.7	.57.3	.81.9	1.06.5	1.31.0	1.55.6	1.80.2	2.04.8	2.29.4	4
5	.10.2	.34.8	.59.3	.83.9	1.08.5	1.33.1	1.57.7	1.82.3	2.06.8	2.31.4	5
6	.12.2	.36.8	.61.4	.86.0	1.10.6	1.35.1	1.59.7	1.84.3	2.08.9	2.33.5	6
7	.14.3	.38.9	.63.4	.88.0	1.12.6	1.37.2	1.61.8	1.86.4	2.10.9	2.35.5	7
8	.16.3	.40.9	.65.5	.90.1	1.14.6	1.39.2	1.63.8	1.88.4	2.13.0	2.37.6	8
9	.18.4	.42.9	.67.5	.92.1	1.16.7	1.41.3	1.65.9	1.90.4	2.15.0	2.39.6	9
10	.20.4	.45.0	.69.6	.94.2	1.18.7	1.43.3	1.67.9	1.92.5	2.17.1	2.41.7	10
11	.22.4	.47.0	.71.6	.96.2	1.20.8	1.45.4	1.69.9	1.94.5	2.19.1	2.43.7	11

$4.91 84/100 TO THE POUND.

10 5/8 %

PREMIUM.

STERLING INTO DOLLARS AND CENTS, AND VICE-VERSA.

£	0	100	200	300	400	500	600	700	800	900	£
£	$ cts.	$ cts.	$ cts.	$ cts.	$ cts.	$ cts.	$ cts.	$ cts.	$ cts.	$ cts.	£
50	245.83.3	737.50.0	1229.16.6	1720.83.3	2212.50.0	2704.16.6	3195.83.3	3687.50.0	4179.16.6	4670.83.3	50
51	250.75.0	742.41.6	1234 08.3	1725.75.0	2217.41.6	2709 08.3	3200.75.0	3692.41 6	4184.08.3	4675.75.0	51
52	255.66.6	747.33.3	1239.00.0	1730.66.6	2222 33.3	2714 00.0	3205.66 6	3697.33.3	4189.00.0	4680.66.6	52
53	260.58.3	752.25 0	1243.91.6	1735.58.3	2227.25.0	2718.91.6	3210.58.3	3702.25.0	4193.91.6	4685.58.3	53
54	265.50 0	757.16.6	1248.83.3	1740 50.0	2232.16.6	2723.83.3	3215.50 0	3707.16.6	4198.83.3	4690.50.0	54
55	270.41.6	762.08.3	1253.75.0	1745.41.6	2237.08.3	2728.75.0	3220.41.6	3712.08.3	4203.75.0	4695.41.6	55
56	275.33.3	767.00.0	1258.66.6	1750.33.3	2242.00.0	2733.66.6	3225.33.3	3717.00.0	4208.66.6	4700.33.3	56
57	280.25.0	771.91 6	1263 58.3	1755.25.0	2246 91.6	2738 58 3	3230.25.0	3721.91.6	4213.58.3	4705.25.0	57
58	285.16.6	776.83 3	1268.50.0	1760.16.6	2251.83.3	2743 50.0	3235.16.6	3726.83.3	4218.50.0	4710.16.6	58
59	290.08 3	781.75.0	1273.41 6	1765 08.3	2256.75.0	2748.41.6	3240.08.3	3731.75.0	4223.41.6	4715.08.3	59
60	295.00.0	786.66.6	1278.33 3	1770.00.0	2261.66.6	2753.33.3	3245.00.0	3736.66.6	4228.33 3	4720.00.0	60
61	299.91.6	791.58.3	1283.25 0	1774 91.6	2266.58.3	2758.25.0	3249.91.6	3741.58.3	4233.25.0	4724.91.6	61
62	304.83.3	796.50.0	1288.16 6	1779.83.3	2271.50.0	2763.16.6	3254.83.3	3746.50.0	4238.16.6	4729.83.3	62
63	309.75.0	801.41.6	1293.08.3	1784.75.0	2276.41.6	2768.08.3	3259.75.0	3751.41.6	4243.08.3	4734.75.0	63
64	314.66.6	806.33.3	1298.00.0	1789.66.6	2281.33.3	2773.00.0	3264.66.6	3756.33.3	4248 00 0	4739.66.6	64
65	319.58.3	811.25.0	1302.91.6	1794.58.3	2286.25.0	2777.91.6	3269.58.3	3761.25.0	4252.91.6	4744.58.3	65
66	324.50.0	816.16.6	1307.83.3	1799.50.0	2291.16.6	2782.83.3	3274.50.0	3766.16.6	4257.83.3	4749.50.0	66
67	329.41.6	821.08.3	1312.75.0	1804.41.6	2296.08.3	2787.75.0	3279.41.6	3771 08.3	4262.75.0	4754.41.6	67
68	334.33.3	826.00.0	1317.66.6	1809.33.3	2301.00.0	2792.66.6	3284.33.3	3776.00.0	4267.66.6	4759.33.3	68
69	339.25.0	830.91.6	1322.58.3	1814.25.0	2305.91.6	2797.58.3	3289.25.0	3780.91.6	4272.58.3	4764.25.0	69
70	344.16.6	835.83.3	1327.50.0	1819.16.6	2310.83.3	2802.50.0	3294.16.6	3785.83.3	4277.50.0	4769.16.6	70
71	349.08.3	840.75.0	1332.41.6	1824.08.3	2315.75.0	2807.41.6	3299.08.3	3790.75.0	4282.41.6	4774.08.3	71
72	354.00.0	845.66.6	1337.33.3	1829.00.0	2320.66.6	2812.33.3	3304.00.0	3795.66.6	4287.33.3	4779.00.0	72
73	358.91.6	850.58.3	1342.25 0	1833.91.6	2325.58.3	2817.25.0	3308.91.6	3800 58.3	4292 25.0	4783.91.6	73
74	363.83 3	855.50.0	1347.16.6	1838.83.3	2330.50.0	2822.16.6	3313.83.3	3805.50.0	4297 16.6	4788 83.3	74
75	368.75.0	860.41.6	1352.08.3	1843.75.0	2335.41.6	2827.08.3	3318.75.0	3810.41.6	4302.08.3	4793.75.0	75
76	373 66.6	865.33.3	1357 00.0	1848.66.6	2340.33.3	2832.00.0	3323.66.6	3815.33.3	4307.00.0	4798.66.6	76
77	378.58.3	870.25.0	1361.91.6	1853.58.3	2345 25.0	2836.91.6	3328.58.3	3820.25.0	4311.91.6	4803.58.3	77
78	383.50.0	875.16.6	1366.83.3	1858.50.0	2350.16.6	2841.83.3	3333.50.0	3825 16.6	4316.83.3	4808.50.0	78
79	388.41.6	880.08.3	1371.75.0	1863.41.6	2355.08.3	2846.75 0	3338.41.6	3830.08.3	4321.75.0	4813.41.6	79
80	393.33.3	885.00.0	1376.66.6	1868.33.3	2360.00.0	2851.66.6	3343.33.3	3835.00.0	4326.66.6	4818.33.3	80
81	398.25.0	889.91.6	1381.58.3	1873.25.0	2364.91.6	2856.58.3	3348.25.0	3839.91.6	4331.58.3	4823.25.0	81
82	403.16.6	894.83.3	1386.50.0	1878.16.6	2369.83.3	2861.50.0	3353.16.6	3844.83.3	4336.50.0	4828.16.6	82
83	408.08.3	899.75.0	1391.41.6	1883.08.3	2374.75.0	2866.41.6	3358.08.3	3849.75.0	4341.41.6	4833.08.3	83
84	413.00.0	904.66.6	1396.33.3	1888.00.0	2379.66.6	2871.33.3	3363.00.0	3854.66.6	4346.33.3	4838.00.0	84
85	417 91.6	909.58.3	1401.25.0	1892.91.6	2384.58.3	2876.25.0	3367.91.6	3859.58.3	4351.25.0	4842.91.6	85
86	422.83.3	914.50.0	1406.16.6	1897.83.3	2389 50.0	2881.16.6	3372.83.3	3864.50.0	4356.16.6	4847.83.3	86
87	427.75.0	919.41.6	1411.08.3	1902.75.0	2394.41.6	2886.08.3	3377.75.0	3869.41.6	4361.08.3	4852.75.0	87
88	432.66.6	924.33.3	1416.00.0	1907.66.6	2399.33.3	2891.00.0	3382.66 6	3874.33.3	4366.00.0	4857.66.6	88
89	437.58.3	929 25.0	1420.91.6	1912.58.3	2404.25.0	2895.91.6	3387.58.3	3879.25.0	4370.91.6	4862.58.3	89
90	442.50.0	934.16.6	1425.83.3	1917.50.0	2409.16.6	2900.83 3	3392.50.0	3884.16.6	4375 83.3	4867.50.0	90
91	447.41.6	939.08.3	1430.75.0	1922.41.6	2414.08.3	2905.75.0	3397.41.6	3889.08 3	4380.75.0	4872.41.6	91
92	452.33.3	944.00.0	1435.66.6	1927.33.3	2419.00.0	2910.66.6	3402.33.3	3894.00.0	4385.66.6	4877.33.3	92
93	457.25.0	948.91.6	1440.58.3	1932.25.0	2423.91.6	2915.58.3	3407.25.0	3898.91.6	4390.58.3	4882.25.0	93
94	462.16.6	953.83.3	1445.50.0	1937.16.6	2428.83.3	2920.50.0	3412.16.6	3903.83.3	4395.50.0	4887.16.6	94
95	467.08.3	958.75.0	1450.41.6	1942.08.3	2433.75.0	2925.41.6	3417.08.3	3908.75.0	4400.41.6	4892.08.3	95
96	472.00.0	963.66.6	1455.33.3	1947.00.0	2438.66.6	2930.33.3	3422.00.0	3913.66.6	4405.33.3	4897.00.0	96
97	476.91.6	968.58.3	1460.25.0	1951.91.6	2443.58.3	2935.25.0	3426.91.6	3918.58.3	4410.25.0	4901.91.6	97
98	481.83.3	973.50.0	1465.16.6	1956.83.3	2448.50.0	2940.16.6	3431.83.3	3923.50.0	4415.16.6	4906.83.3	98
99	486.75.0	978.41.6	1470.08.3	1961.75 0	2453.41.6	2945.08.3	3436.75.0	3928.41.6	4420.08.3	4911.75.0	99

S.	10	11	12	13	14	15	16	17	18	19	S.
d.	$ cts.	$ cts.	$ cts.	$ cts.	$ cts.	$ cts.	$ cts.	$ cts.	$ cts.	$ cts.	d.
0	2.45.8	2.70.4	2.95.0	3.19.5	3.44.1	3.68.7	3.93.3	4.17.9	4.42.5	4.67.0	0
1	2.47.8	2.72.4	2.97.0	3.21.6	3.46.2	3.70.8	3.95.3	4.19.9	4.44.5	4.69.1	1
2	2.49.9	2.74.5	2.99.0	3.23.6	3.48.2	3.72.8	3.97.4	4.22.0	4.46.5	4.71.1	2
3	2.51.9	2.76.5	3.01.1	3.25.7	3.40.3	3.74.8	3.99.4	4.24.0	4.48.6	4.73.2	3
4	2.54.0	2.78.5	3.03.1	3.27.7	3.52.3	3.76.9	4.01.5	4.26.0	4.50.6	4.75.2	4
5	2.56.0	2.80.6	3.05.2	3.29.8	3.54.3	3.78.9	4.03.5	4.28.1	4.52.7	4.77.3	5
6	2.58.1	2.82.6	3.07.2	3.31.8	3.56.4	3.81.0	4.05.6	4.30.1	4.54.7	4.79.3	6
7	2.60.1	2.84.7	3.09.3	3.33.9	3.58.4	3.83.0	4.07.6	4.32.2	4.56.8	4.81.4	7
8	2.62.1	2.86.7	3.11.3	3.35.9	3.60.5	3.85.1	4.09.6	4.34.2	4.58.8	4.83.4	8
9	2.64.2	2.88.8	3.13.4	3.37.9	3.62.5	3.87.1	4.11.7	4.36.3	4.60.9	4.85.4	9
10	2.66.2	2.90.8	3.15.4	3.40 0	3.64.6	3.89 2	4.13.7	4.38.3	4.62.9	4.87.5	10
11	2.68.3	2.92.9	3.17.4	3.42.0	3.66.6	3.91.2	4.15.8	4.40.4	4.64.9	4.89.5	11

$\$4.92\frac{22}{100}$ TO THE POUND.

$10\frac{3}{4}\%$

PREMIUM.

STERLING INTO DOLLARS AND CENTS, AND VICE-VERSA.

£	0	100	200	300	400	500	600	700	800	900	£
£	$ cts.	$ cts.	$ cts.	$ cts.	$ cts.	$ cts.	$ cts.	$ cts.	$ cts.	$ cts.	£
0	......	492.22.2	984.44.4	1476.66.6	1968.88.8	2461.11.1	2953.33.3	3445.55.5	3937.77.7	4430.00.0	0
1	4.92.2	497.14.4	989.36.6	1481.58.8	1973.81.1	2466.03.3	2958.25.5	3450.47.7	3942.70.0	4434.92.2	1
2	9.84.4	502.06.6	994.28.8	1486.51.1	1978.73.3	2470.95.5	2963.17.7	3455.40.0	3947.62.2	4439.81.4	2
3	14.76.6	506.98.8	999.21.1	1491.43.3	1983.65.5	2475.87.7	2968.10.0	3460.32.2	3952.54.4	4444.76.6	3
4	19.68.8	511.91.1	1004.13.3	1496.35.5	1988.57.7	2480.80.0	2973.02.2	3465.24.4	3957.46.6	4449.68.8	4
5	24.61.1	516.83.3	1009.05.5	1501.27.7	1993.50.0	2485.72.2	2977.94.4	3470.16.6	3962.38.8	4454.61.1	5
6	29.53.3	521.75.5	1013.97.7	1506.20.0	1998.42.2	2490.64.4	2982.86.6	3475.08.8	3967.31.1	4459.53.3	6
7	34.45.5	526.67.7	1018.90.0	1511.12.2	2003.34.4	2495.56.6	2987.78.8	3480.01.1	3972.23.3	4464.45.5	7
8	39.37.7	531.60.0	1023.82.2	1516.04.4	2008.26.6	2500.48.8	2992.71.1	3484.93.3	3977.15.5	4469.37.7	8
9	44.30.0	536.52.2	1028.74.4	1520.96.6	2013.18.8	2505.41.1	2997.63.3	3489.85.5	3982.07.7	4474.30.0	9
10	49.22.2	541.44.4	1033.66.6	1525.88.8	2018.11.1	2510.33.3	3002.55.5	3494.77.7	3987.00.0	4479.22.2	10
11	54.14.4	546.36.6	1038.58.8	1530.81.1	2023.03.3	2515.25.5	3007.47.7	3499.70.0	3991.92.2	4484.14.4	11
12	59.06.6	551.28.8	1043.51.1	1535.73.3	2027.95.5	2520.17.7	3012.40.0	3504.62.2	3996.84.4	4489.06.6	12
13	63.98.8	556.21.1	1048.43.3	1540.65.5	2032.87.7	2525.10.0	3017.32.2	3509.54.4	4001.76.6	4493.98.8	13
14	68.91.1	561.13.3	1053.35.5	1545.57.7	2037.80.0	2530.02.2	3022.24.4	3514.46.6	4006.68.8	4498.91.1	14
15	73.83.3	566.05.5	1058.27.7	1550.50.0	2042.72.2	2534.94.4	3027.16.6	3519.38.8	4011.61.1	4503.83.3	15
16	78.75.5	570.97.7	1063.20.0	1555.42.2	2047.64.4	2539.86.6	3032.08.8	3524.31.1	4016.53.3	4508.75.5	16
17	83.67.7	575.90.0	1068.12.2	1560.34.4	2052.56.6	2544.78.8	3037.01.1	3529.23.3	4021.45.5	4513.67.7	17
18	88.60.0	580.82.2	1073.04.4	1565.26.6	2057.48.8	2549.71.1	3041.93.3	3534.15.5	4026.37.7	4518.60.0	18
19	93.52.2	585.74.4	1077.96.6	1570.18.8	2062.41.1	2554.63.3	3046.85.5	3539.07.7	4031.30.0	4523.52.2	19
20	98.44.4	590.66.6	1082.88.8	1575.11.1	2067.33.3	2559.55.5	3051.77.7	3544.00.0	4036.22.2	4528.44.4	20
21	103.36.6	595.58.8	1087.81.1	1580.03.3	2072.25.5	2564.47.7	3056.70.0	3548.92.2	4041.14.4	4533.36.6	21
22	108.28.8	600.51.1	1092.73.3	1584.95.5	2077.17.7	2569.40.0	3061.62.2	3553.84.4	4046.06.6	4538.28.8	22
23	113.21.1	605.43.3	1097.65.5	1589.87.7	2082.10.0	2574.32.2	3066.54.4	3558.76.6	4050.98.8	4543.21.1	23
24	118.13.3	610.35.5	1102.57.7	1594.80.0	2087.02.2	2579.24.4	3071.46.6	3563.68.8	4055.91.1	4548.13.3	24
25	123.05.5	615.27.7	1107.50.0	1599.72.2	2091.94.4	2584.16.6	3076.38.8	3568.61.1	4060.83.3	4553.05.5	25
26	127.97.7	620.20.0	1112.42.2	1604.64.4	2096.86.6	2589.08.8	3081.31.1	3573.53.3	4065.75.5	4557.97.7	26
27	132.90.0	625.12.2	1117.34.4	1609.56.6	2101.78.8	2594.01.1	3086.23.3	3578.45.5	4070.67.7	4562.90.0	27
28	137.82.2	630.04.4	1122.26.6	1614.48.8	2106.71.1	2598.93.3	3091.15.5	3583.37.7	4075.60.0	4567.82.2	28
29	142.74.4	634.96.6	1127.18.8	1619.41.1	2111.63.3	2603.85.5	3096.07.7	3588.30.0	4080.52.2	4572.74.4	29
30	147.66.6	639.88.8	1132.11.1	1624.33.3	2116.55.5	2608.77.7	3101.60.0	3593.22.2	4085.44.4	4577.66.6	30
31	152.58.8	644.81.1	1137.03.3	1629.25.5	2121.47.7	2613.70.0	3105.92.2	3598.14.4	4090.36.6	4582.58.8	31
32	157.51.1	649.73.3	1141.95.5	1634.17.7	2126.40.0	2618.62.2	3110.84.4	3603.06.6	4095.28.8	4587.51.1	32
33	162.43.3	654.65.5	1146.87.7	1639.10.0	2131.32.2	2623.54.4	3115.76.6	3607.98.8	4100.21.1	4592.43.3	33
34	167.35.5	659.57.7	1151.80.0	1644.02.2	2136.24.4	2628.46.6	3120.68.8	3612.91.1	4105.13.3	4597.35.5	34
35	172.27.7	664.50.0	1156.72.2	1648.94.4	2141.16.6	2633.38.8	3125.61.1	3617.83.3	4110.05.5	4602.27.7	35
36	177.20.0	669.42.2	1161.64.4	1653.86.6	2146.08.8	2638.31.1	3130.53.3	3622.75.5	4114.97.7	4607.20.0	36
37	182.12.2	674.34.4	1166.56.6	1658.78.8	2151.01.1	2643.23.3	3135.45.5	3627.67.7	4119.90.0	4612.12.2	37
38	187.04.4	679.26.6	1171.48.8	1663.71.1	2155.93.3	2648.15.5	3140.37.7	3632.60.0	4124.82.2	4617.04.4	38
39	191.96.6	684.18.8	1176.41.1	1668.63.3	2160.85.5	2653.07.7	3145.30.0	3637.52.2	4129.74.4	4621.96.6	39
40	196.88.8	689.11.1	1181.33.3	1673.55.5	2165.77.7	2658.00.0	3150.22.2	3642.44.4	4134.66.6	4626.88.8	40
41	201.81.1	694.03.3	1186.25.5	1678.47.7	2170.70.0	2662.92.2	3155.14.4	3647.36.6	4139.58.8	4631.81.1	41
42	206.73.3	698.95.5	1191.17.7	1683.40.0	2175.62.2	2667.84.4	3160.06.6	3652.28.8	4144.51.1	4636.73.3	42
43	211.65.5	703.87.7	1196.10.0	1688.32.2	2180.54.4	2672.76.6	3164.98.8	3657.21.1	4149.43.3	4641.65.5	43
44	216.57.7	708.80.0	1201.02.2	1693.24.4	2185.46.6	2677.68.8	3169.91.1	3662.13.3	4154.35.5	4646.57.7	44
45	221.50.0	713.72.2	1205.94.4	1698.16.6	2190.38.8	2682.61.1	3174.83.3	3667.05.5	4159.27.7	4651.50.0	45
46	226.42.2	718.64.4	1210.86.6	1703.08.8	2195.31.1	2687.53.3	3179.75.5	3671.97.7	4164.20.0	4656.42.2	46
47	231.34.4	723.56.6	1215.78.8	1708.01.1	2200.23.3	2692.45.5	3184.67.7	3676.90.0	4169.12.2	4661.34.4	47
48	236.26.6	728.48.8	1220.71.1	1712.93.3	2205.15.5	2697.37.7	3189.60.0	3681.82.2	4174.04.4	4666.26.6	48
49	241.18.8	733.41.1	1225.63.3	1717.85.5	2210.07.7	2702.30.0	3194.52.2	3686.74.4	4178.96.6	4671.18.8	49

S.	0	1	2	3	4	5	6	7	8	9	S.
d.	$ cts.	$ cts.	$ cts.	$ cts.	$ cts.	$ cts.	$ cts.	$ cts.	$ cts.	$ cts.	d.
0	......	.24.6	.49.2	.73.8	.98.4	1.23.0	1.47.6	1.72.2	1.96.8	2.21.5	0
1	.02.0	.26.6	.51.2	.75.8	1.00.5	1.25.1	1.49.7	1.74.3	1.98.9	2.23.5	1
2	.04.1	.28.7	.53.3	.77.9	1.02.5	1.27.1	1.51.7	1.76.3	2.00.9	2.25.6	2
3	.06.1	.30.7	.55.3	.79.9	1.04.6	1.29.2	1.53.8	1.78.4	2.03.0	2.27.6	3
4	.08.2	.32.8	.57.4	.82.0	1.06.6	1.31.2	1.55.8	1.80.4	2.05.0	2.29.7	4
5	.10.2	.34.8	.59.4	.84.0	1.08.7	1.33.3	1.57.9	1.82.5	2.07.1	2.31.7	5
6	.12.3	.36.9	.61.5	.86.1	1.10.7	1.35.3	1.59.9	1.84.5	2.09.2	2.33.8	6
7	.14.3	.38.9	.63.5	.88.1	1.12.8	1.37.4	1.62.0	1.86.6	2.11.2	2.35.8	7
8	.16.4	.41.0	.65.6	.90.2	1.14.8	1.39.4	1.64.0	1.88.6	2.13.3	2.37.9	8
9	.18.4	.43.0	.67.6	.92.2	1.16.9	1.41.5	1.66.1	1.90.7	2.15.3	2.39.9	9
10	.20.5	.45.1	.69.7	.94.3	1.18.9	1.43.5	1.68.1	1.92.7	2.17.4	2.42.0	10
11	.22.5	.47.1	.71.7	.96.3	1.21.0	1.45.6	1.70.2	1.94.8	2.19.4	2.44.0	11

$4.92⁷⁄₁₀₀ TO THE POUND.

10 ¾ %

PREMIUM.

STERLING INTO DOLLARS AND CENTS, AND VICE-VERSA.

£	0	100	200	300	400	500	600	700	800	900	£
£	$ cts.	$ cts.	$ cts.	$ cts.	$ cts.	$ cts.	$ cts.	$ cts.	$ cts.	$ cts.	£
50	246.11.1	738,33.3	1230.55,5	1722.77.7	2315.00,0	2707.22,2	3199 44,4	3691.66.6	4183,88,8	4676.11.1	50
51	251.03,3	743,25.5	1235.47.7	1727.70.0	2219.92.2	2712.14,4	3204,36 6	3696.58.8	4188.81.1	4681.03,3	51
52	255.95,5	748.17.7	1240.40 0	1732.62.2	2224.84,4	2717 06.6	3209.28,8	3701.51.1	4193.73,3	4685 95.5	52
53	260.87.7	753.10.0	1245 32.2	1737.54.4	2229 76.6	2721.98.8	3214.21,1	3706.43.3	4198.65.5	4690,87.7	53
54	265.80,0	758.02,2	1250.24.4	1742.46.6	2234 68.8	2726.91,1	3219 13,3	3711.35.5	4203.57.7	4695.80,0	54
55	270.72.2	762.94,4	1255.16 6	1747.38,8	2239.61.1	2731.83.3	3224.05.5	3716.27.7	4208.50 0	4700.72.2	55
56	275.64,4	767.86.6	1260.08 8	1752.31.1	2244 53,3	2736.75.5	3228.97.7	3721.20.0	4213.42.2	4705 64 4	56
57	280.56.6	772.78,8	1265 01.1	1757.23,3	2249.45.5	2741.67.7	3233.90.0	3726.12.2	4218.34,4	4710.56.6	57
58	285.48,8	777.71.1	1269.93.3	1762.15.5	2254.37.7	2746.60.0	3238.82 2	3731.04.4	4223.26.6	4715.48.8	58
59	290.41,1	782.63.3	1274 85.5	1767.07.7	2259.30.0	2751.52 2	3243.74.4	3735.96.6	4228.18,8	4720.41.1	59
60	295.33.3	787,55.5	1279 77.7	1772.00 0	2264.22,2	2756.44.4	3248.66 6	3740.88.8	4233.11.1	4725.33,3	60
61	300.25.5	792.47.7	1284.70.0	1776 92,2	2269.14,4	2761.36.6	3253.58.8	3745.81.1	4238.03.3	4730.25.5	61
62	305.17.7	797.40 0	1289 62,2	1781.84,4	2274.06.6	2766 28.8	3258.51 1	3750.73.3	4242 95.5	4735.17.7	62
63	310.10,0	802.32,2	1294.54,4	1786.76,6	2278.98.8	2771.21,1	3263 43,3	3755.65,5	4247.87,7	4740.10,0	63
64	315.02,2	807.24,4	1299.46 6	1791.68,8	2283.91,1	2776.13,3	3268.35.5	3760,57.7	4252.80,0	4745.02,2	64
65	319.94,4	812.16 6	1304.38,8	1796.61,1	2288.83,3	2781.05,5	3273.27.7	3765.50,0	4257.72,2	4749.94,4	65
66	324.86,6	817.08,8	1309.31,1	1801.53,3	2293.75,5	2785.97,7	3278.20,0	3770.42,2	4262.64,4	4754.86,6	66
67	329.78,8	822.01,1	1314.23,3	1806.45,5	2298.67,7	2790 90,0	3283.12,2	3775.34,4	4267.56,6	4759.78.8	67
68	334.71,1	826.93,3	1319.15,5	1811.37,7	2303.60,0	2795.82,2	3288.04,4	3780.26,6	4272.48,8	4764.71.1	68
69	339.63.3	831.85,5	1324.07,7	1816.30,0	2308.52,2	2800.74,4	3292.96,6	3785.18.8	4277.41,1	4769.63.3	69
70	344.55,5	836.77,7	1329.00,0	1821.22 2	2313.44,4	2805.66,6	3297.88,8	3790.11.1	4282.33,3	4774.55,5	70
71	349.47.7	841.70,0	1333 92.2	1826.14,4	2318.36,6	2810.58,8	3302.81,1	3795.03,3	4287.25,5	4779.47.7	71
72	354.40,0	846.62 2	1338.84,4	1831.06,6	2323.28,8	2815.51,1	3307.73,3	3799 95,5	4292.17,7	4784.40,0	72
73	359.32,2	851.54,4	1343.76,6	1835.98,8	2328.21,1	2820.43,3	3312.65,5	3804 87.7	4297.10,0	4789.32,2	73
74	364.24,4	856.46,6	1348.68,8	1840.91,1	2333.13,3	2825.35,5	3317.57,7	3809.80,0	4302.02 2	4794.24.4	74
75	369.16,6	861 38.8	1353.61,1	1845.83,3	2338.05,5	2830.27,7	3322.50,0	3814.72,2	4306.94,4	4799.16,6	75
76	374.08,8	866.31.1	1358.53,3	1850.75,5	2342.97.7	2835 20,0	3327.42,2	3819.64,4	4311.86,6	4804.08.8	76
77	379.01,1	871.23,3	1363.45,5	1855.67,7	2347.90,0	2840.12,2	3332.34,4	3824.56 6	4316.78.8	4809.01.1	77
78	383.93,3	876.15.5	1368.37.7	1860.60,0	2352.82 2	2845.04,4	3337.26 6	3829.48,8	4321.71,1	4813.93,3	78
79	388.85.5	881.07.7	1373.30.0	1865.52,2	2357.74,4	2849.96,6	3342.18,8	3834.41,1	4326.63,3	4818.85.5	79
80	393.77,7	886,00.0	1378.22 2	1870.44,4	2362.66,6	2854.88,8	3347.11,1	3839.33 3	4331.55,5	4823.77.7	80
81	398.70,0	890.92.2	1383.14,4	1875.36,6	2367.58,8	2859.81,1	3352.03,3	3844.25,5	4336.47.7	4828.70.0	81
82	403.62,2	895.84,4	1388.06,6	1880.28,8	2372.51,1	2864.73,3	3356.95,5	3849.17.7	4341.40,0	4833.62.2	82
83	408.54,4	900.76,6	1392.98.8	1885.21,1	2377.43,3	2869.65,5	3361.87.7	3854.10,0	4346.32,2	4838.54.4	83
84	413.46,6	905.68.8	1397.91.1	1890.13,3	2382.35,5	2874.57.7	3366 80.0	3859.02,2	4351.24,4	4843.46.6	84
85	418.38 8	910 61.1	1402.83,3	1895.05,5	2387.27.7	2879.50,0	3371.72 2	3863.94 4	4356.16.6	4848.38.8	85
86	423.31,1	915.53,3	1407.75,5	1899.97 7	2392 20,0	2884.42,2	3376.64,4	3868.86,6	4361.08.8	4853.31.1	86
87	428.23,3	920.45,5	1412.67.7	1904.90,0	2397.12 2	2889,34,4	3381.56,6	3873.78.8	4366.01,1	4858.23,3	87
88	433.15,5	925.37.7	1417.60,0	1909.82,2	2402.04,4	2894.26,6	3386.48 8	3878.71,1	4370,93,3	4863.15,5	88
89	438.07.7	930.30.0	1422.52,2	1914.74,4	2406.96,6	2899.18,8	3391.41,1	3883.63,3	4375.85,5	4868.07.7	89
90	443.00,0	935.22,2	1427.44,4	1919.66,6	2411.88,8	2904.11,1	3396.33,3	3888.55,5	4380.77,7	4873.00,0	90
91	447.92,2	940.14,4	1432.36,6	1924.58 8	2416.81,1	2909.03,3	3401.25 5	3893.47,7	4385 70,0	4877.92 2	91
92	452.84,4	945 06.6	1437.28,8	1929.51 1	2421.73,3	2913.95,5	3406.17,7	3898.40,0	4390.62,2	4882.84,4	92
93	457.76,6	949.98.8	1442 21.1	1934.43,3	2426.65,5	2918.87,7	3411.10,0	3903.32 2	4395,54,4	4887.76,6	93
94	462.68,8	954.91.1	1447.13,3	1939.35,5	2431.57,7	2923.80,0	3416.02,2	3908.24,4	4400.46,6	4892.68,8	94
95	467.61.1	959.83.3	1452.05,5	1944.27.7	2436.50,0	2928.72,2	3420.94,4	3913.16,6	4405.38,8	4897.61.1	95
96	472.53 3	964.75,5	1456.97.7	1949.20,0	2441.42,2	2933.64,4	3425.86,6	3918.08,8	4410.31,1	4902.53,3	96
97	477.45.5	969.67.7	1461.90,0	1954.12,2	2446.34,4	2938.56,6	3430 78,8	3923 01.1	4415.23,3	4907.45,5	97
98	482.37.7	974.60,0	1466.82,2	1959 04,4	2451.26,6	2943.48,8	3435.71,1	3927.93,3	4420.15,5	4912.37.7	98
99	487.30,0	979.52.2	1471.74,4	1963.96 6	2456.18,8	2948.41,1	3440.63,3	3932.85,5	4425.07.7	4917.30,0	99

s.	10	11	12	13	14	15	16	17	18	19	s.
d.	$ cts.	$ cts.	$ cts.	$ cts.	$ cts.	$ cts.	$ cts.	$ cts.	$ cts.	$ cts.	d.
0	2.46.1	2.70.7	2.95.3	3.19.9	3.44.5	3.69.1	3.93.7	4.18.3	4.43.0	4 67.6	0
1	2.48,1	2.72.7	2.97.3	3.21.9	3.46.6	3.71.2	3.95.8	4.20.4	4.45.0	4.69.6	1
2	2.50,2	2.74.8	2.90.4	3.24.0	3.48,6	3.73.2	3.97.8	4.22.5	4.47.1	4.71.7	2
3	2.52,2	2.76.8	3.01.5	3.26.1	3.50.7	3.75,3	3.99,9	4.24.5	4.49.1	4.73.7	3
4	2.54,3	2.78.9	3.03.5	3.28.1	3.52.7	3.77.3	4.02,0	4.26.6	4.51,2	4.75.8	4
5	2.56,3	2.81.0	3.05,6	3.30.2	3.54.8	3.79.4	4.04.0	4.28.6	4.53,2	4.77.8	5
6	2.58,4	2.83.0	3.07.6	3.32.2	3.56.8	3.81.4	4.06.1	4.30.7	4.55.3	4.79.9	6
7	2.62,5	2.85.1	3.09.7	3.34.3	3.58.9	3.83.5	4.08.1	4.32.7	4.57.3	4.81.9	7
8	2.62,5	2.87.1	3.11.7	3.36.3	3.60.9	3.85.5	4.10.2	4.34.8	4.59.4	4.84.0	8
9	2.64,5	2.89.2	3.13.8	3.38.4	3.63.0	3.87.6	4.12.2	4.36.8	4.61,4	4.86.0	9
10	2.66,6	2.91.2	3.15.8	3.40.4	3.65.0	3.89.6	4.14.3	4.38.9	4.63.5	4.88.1	10
11	2.68,6	2.93.3	3.17.9	3.42.5	3.67.1	3.91.7	4.16.3	4.40.9	4.65.5	4.90.1	11

$4.92 7/8 TO THE POUND.

10 7/8 %

PREMIUM.

STERLING INTO DOLLARS AND CENTS, AND VICE-VERSA.

£	0	100	200	300	400	500	600	700	800	900	£
	$ cts.	$ cts.	$ cts.	$ cts.	$ cts.	$ cts.	$ cts.	$ cts.	$ cts.	$ cts.	
0	......	492.77.7	935.55.5	1478.33.3	1971.11.1	2463.88.8	2956.66.6	3449.44.4	3942.22.2	4435.00.0	0
1	4.92.7	497.70.5	990.48.3	1483.26.1	1976.03.8	2468.81.6	2961.59.4	3454.37.2	3947.15.0	4439.92.7	1
2	9.85.5	502.63.3	995.41.1	1488.18.8	1980.96.6	2473.74.4	2966.52.2	3459.30.0	3952.07.7	4444.85.5	2
3	14.78.3	507.56.1	1000.33.8	1493.11.6	1985.89.4	2478.67.2	2971.45.0	3464.22.7	3957.00.5	4449.78.3	3
4	19.71.1	512.48.8	1005.26.6	1498.04.4	1990.82.2	2483.60.0	2976.37.7	3469.15.5	3961.93.3	4454.71.1	4
5	24.63.8	517.41.6	1010.19.4	1502.97.2	1995.75.0	2488.52.7	2981.30.5	3474.08.3	3966.86.1	4459.63.8	5
6	29.56.6	522.34.4	1015.12.2	1507.90.0	2000.67.7	2493.45.5	2986.23.3	3479.01.1	3971.78.8	4464.56.6	6
7	34.49.4	527.27.2	1020.05.0	1512.82.7	2005.60.5	2498.38.3	2991.16.1	3483.93.8	3976.71.6	4469.49.4	7
8	39.42.2	532.20.0	1024.97.7	1517.75.5	2010.53.3	2503.31.1	2996.08.8	3488.86.6	3981.64.4	4474.42.2	8
9	44.35.0	537.12.7	1029.90.5	1522.68.3	2015.46.1	2508.23.8	3001.01.6	3493.79.4	3986.57.2	4479.35.0	9
10	49.27.7	542.05.5	1034.83.3	1527.61.1	2020.38.8	2513.16.6	3005.94.4	3498.72.2	3991.50.0	4484.27.7	10
11	54.20.5	546.98.3	1039.76.1	1532.53.8	2025.31.6	2518.09.4	3010.87.2	3503.65.0	3996.42.7	4489.20.5	11
12	59.13.3	551.91.1	1044.68.8	1537.46.6	2030.24.4	2523.02.2	3015.80.0	3508.57.7	4001.35.5	4494.13.3	12
13	64.06.1	556.83.8	1049.61.6	1542.39.4	2035.17.2	2527.95.0	3020.72.7	3513.50.5	4006.28.3	4499.06.1	13
14	68.98.8	561.76.6	1054.54.4	1547.32.2	2040.10.0	2532.87.7	3025.65.5	3518.43.3	4011.21.1	4503.98.8	14
15	73.91.6	566.69.4	1059.47.2	1552.25.0	2045.02.7	2537.80.5	3030.58.3	3523.36.1	4016.13.8	4508.91.6	15
16	78.84.4	571.62.2	1064.40.0	1557.17.7	2049.95.5	2542.73.3	3035.51.1	3528.28.8	4021.06.6	4513.84.4	16
17	83.77.2	576.55.0	1069.32.7	1562.10.5	2054.88.3	2547.66.1	3040.43.8	3533.21.6	4025.99.4	4518.77.2	17
18	88.70.0	581.47.7	1074.25.5	1567.03.3	2059.81.1	2552.58.8	3045.36.6	3538.14.4	4030.92.2	4523.70.0	18
19	93.62.7	586.40.5	1079.18.3	1571.96.1	2064.73.8	2557.51.6	3050.29.4	3543.07.2	4035.85.0	4528.62.7	19
20	98.55.5	591.33.3	1084.11.1	1576.88.8	2069.66.6	2562.44.4	3055.22.2	3548.00.0	4040.77.7	4533.55.5	20
21	103.48.3	596.26.1	1089.03.8	1581.81.6	2074.59.4	2567.37.2	3060.15.0	3552.92.7	4045.70.5	4538.48.3	21
22	108.41.1	601.18.8	1093.96.6	1586.74.4	2079.52.2	2572.30.0	3065.07.7	3557.85.5	4050.63.3	4543.41.1	22
23	113.33.8	606.11.6	1098.89.4	1591.67.2	2084.45.0	2577.22.7	3070.00.5	3562.78.3	4055.56.1	4548.33.8	23
24	118.26.6	611.04.4	1103.82.2	1596.60.0	2089.37.7	2582.15.5	3074.93.3	3567.71.1	4060.48.8	4553.26.6	24
25	123.19.4	615.97.2	1108.75.0	1601.52.7	2094.30.5	2587.08.3	3079.86.1	3572.63.8	4065.41.6	4558.19.4	25
26	128.12.2	620.90.0	1113.67.7	1606.45.5	2099.23.3	2592.01.1	3084.78.8	3577.56.6	4070.34.4	4563.12.2	26
27	133.05.0	625.82.7	1118.60.5	1611.38.3	2104.16.1	2596.93.8	3089.71.6	3582.49.4	4075.27.2	4568.05.0	27
28	137.97.7	630.75.5	1123.53.3	1616.31.1	2109.08.8	2601.86.6	3094.64.4	3587.42.2	4080.20.0	4572.97.7	28
29	142.90.5	635.68.3	1128.46.1	1621.23.8	2114.01.6	2606.79.4	3099.57.2	3592.35.0	4085.12.7	4577.90.5	29
30	147.83.3	640.61.1	1133.38.8	1626.16.6	2118.94.4	2611.72.2	3104.50.0	3597.27.7	4090.05.5	4582.83.3	30
31	152.76.1	645.53.8	1138.31.6	1631.09.4	2123.87.2	2616.65.0	3109.42.7	3602.20.5	4094.98.3	4587.76.1	31
32	157.68.8	650.46.6	1143.24.4	1636.02.2	2128.80.0	2621.57.7	3114.35.5	3607.13.3	4099.91.1	4592.68.8	32
33	162.61.6	655.39.4	1148.17.2	1640.95.0	2133.72.7	2626.50.5	3119.28.3	3612.06.1	4104.83.8	4597.61.6	33
34	167.54.4	660.32.2	1153.10.0	1645.87.7	2138.65.5	2631.43.3	3124.21.1	3616.98.8	4109.76.6	4602.54.4	34
35	172.47.2	665.25.0	1158.02.7	1650.80.5	2143.58.3	2636.36.1	3129.13.8	3621.91.6	4114.69.4	4607.47.2	35
36	177.40.0	670.17.7	1162.95.5	1655.73.3	2148.51.1	2641.28.8	3134.06.6	3626.84.4	4119.62.2	4612.40.0	36
37	182.32.7	675.10.5	1167.88.3	1660.66.1	2153.43.8	2646.21.6	3138.99.4	3631.77.2	4124.55.0	4617.32.7	37
38	187.25.5	680.03.3	1172.81.1	1665.58.8	2158.36.6	2651.14.4	3143.92.2	3636.70.0	4129.47.7	4622.25.5	38
39	192.18.3	684.96.1	1177.73.8	1670.51.6	2163.29.4	2656.07.2	3148.85.0	3641.62.7	4134.40.5	4627.18.3	39
40	197.11.1	689.88.8	1182.66.6	1675.44.4	2168.22.2	2661.00.0	3153.77.7	3646.55.5	4139.33.3	4632.11.1	40
41	202.03.8	694.81.6	1187.59.4	1680.37.2	2173.15.0	2665.92.7	3158.70.5	3651.48.3	4144.26.1	4637.03.8	41
42	206.96.6	699.74.4	1192.52.2	1685.30.0	2178.07.7	2670.85.5	3163.63.3	3656.41.1	4149.18.8	4641.96.6	42
43	211.89.4	704.67.2	1197.45.0	1690.22.7	2183.00.5	2675.78.3	3168.56.1	3661.33.8	4154.11.6	4646.89.4	43
44	216.82.2	709.60.0	1202.37.7	1695.15.5	2187.93.3	2680.71.1	3173.48.8	3666.26.6	4159.04.4	4651.82.2	44
45	221.75.0	714.52.7	1207.30.5	1700.08.3	2192.86.1	2685.63.8	3178.41.6	3671.19.4	4163.97.2	4656.75.0	45
46	226.67.7	719.45.5	1212.23.3	1705.01.1	2197.78.8	2690.56.6	3183.34.4	3676.12.2	4168.90.0	4661.67.7	46
47	231.60.5	724.38.3	1217.16.1	1709.93.8	2202.71.6	2695.49.4	3188.27.2	3681.05.0	4173.82.7	4666.60.5	47
48	236.53.3	729.31.1	1222.08.8	1714.86.6	2207.64.4	2700.42.2	3193.20.0	3685.97.7	4178.75.5	4671.53.3	48
49	241.46.1	734.23.8	1227.01.6	1719.79.4	2212.57.2	2705.35.0	3198.12.7	3690.90.5	4183.68.3	4676.46.1	49

s.	0	1	2	3	4	5	6	7	8	9	s.
d.	$ cts.	$ cts.	$ cts.	$ cts.	$ cts.	$ cts.	$ cts.	$ cts.	$ cts.	$ cts.	d.
0	......	.24,6	.49,2	.73,9	.98,5	1,23,1	1,47,8	1,72,4	1,97,1	2,21,7	0
1	.02.0	.26.6	.51.3	.75.9	1.00.6	1.25.2	1.49.8	1.74.5	1.99.1	2.23.8	1
2	.04.1	.28.7	.53.3	.78.0	1.02.6	1.27.3	1.51.9	1.76.5	2.01.2	2.26.8	2
3	.06.1	.30.8	.55.4	.80.0	1.04.7	1.29.3	1.54.0	1.78.6	2.03.2	2.27.9	3
4	.08.2	.32.8	.57.5	.82.1	1.06.7	1.31.4	1.56.0	1.80.7	2.05.3	2.29.9	4
5	.10.2	.34.9	.59.5	.84.1	1.08.8	1.33.4	1.58.1	1.82.7	2.07.3	2.32.0	5
6	.12.3	.36.9	.61.6	.86.2	1.10.8	1.35.5	1.60.1	1.84.8	2.09.4	2.34.0	6
7	.14.3	.39.0	.63.6	.88.3	1.12.9	1.37.5	1.62.2	1.86.8	2.11.5	2.36.1	7
8	.16.4	.41.0	.65.7	.90.3	1.15.0	1.39.6	1.64.2	1.88.9	2.13.5	2.38.2	8
9	.18.5	.43.1	.67.7	.92.4	1.17.0	1.41.7	1.66.3	1.90.9	2.15.6	2.40.2	9
10	.20.5	.45.1	.69.8	.94.4	1.19.1	1.43.7	1.68.3	1.93.0	2.17.6	2.42.3	10
11	.22.6	.47.2	.71.8	.96.5	1.21.1	1.45.8	1.70.4	1.95.0	2.19.7	2.44.3	11

$4.92 7/100 TO THE POUND.

10 7/8 %

PREMIUM.

STERLING INTO DOLLARS AND CENTS, AND VICE-VERSA.

£	0	100	200	300	400	500	600	700	800	900	£
	$ cts.	$ cts.	$ cts.	$ cts.	$ cts.	$ cts.	$ cts.	$ cts.	$ cts.	$ cts.	
50	246.38.8	739.16.6	1231.94.4	1724.72.2	2217.50.0	2710.27.7	3203.05.5	3695.83.3	4188.61.1	4681.38.8	50
51	251.31.6	744.09.4	1236.87.2	1729.65.0	2222.42.7	2715.20.5	3207.98.3	3700.76.1	4193.53.8	4686.31.6	51
52	256.24.4	749.02.2	1241.80.0	1734.57.7	2227.35.5	2720.13.3	3212.91.1	3705.68.8	4198.46.6	4691.24.4	52
53	261.17.2	753.95.0	1246.72.7	1739.50.5	2232.28.3	2725.06.1	3217.83.8	3710.61.6	4203.39.4	4696.17.2	53
54	266.10.0	758.87.7	1251.65.5	1744.43.3	2237.21.1	2729.98.8	3222.76.6	3715.54.4	4208.32.2	4701.10.0	54
55	271.02.7	763.80.5	1256.58.3	1749.36.1	2242.13.8	2734.91.6	3227.69.4	3720.47.2	4213.25.0	4706.02.7	55
56	275.95.5	768.73.3	1261.51.1	1754.28.8	2247.06.6	2739.84.4	3232.62.2	3725.40.0	4218.17.7	4710.95.5	56
57	280.88.3	773.66.1	1266.43.8	1759.21.6	2251.99.4	2744.77.2	3237.55.0	3730.32.7	4223.10.5	4715.88.3	57
58	285.81.1	778.58.8	1271.36.6	1764.14.4	2256.92.2	2749.70.0	3242.47.7	3735.25.5	4228.03.3	4720.81.1	58
59	290.73.8	783.51.6	1276.29.4	1769.07.2	2261.85.0	2754.62.7	3247.40.5	3740.18.3	4232.96.1	4725.73.8	59
60	295.66.6	788.44.4	1281.22.2	1774.00.0	2266.77.7	2759.55.5	3252.33.3	3745.11.1	4237.88.8	4730.66.6	60
61	300.59.4	793.37.2	1286.15.0	1778.92.7	2271.70.5	2764.48.3	3257.26.1	3750.03.8	4242.81.6	4735.59.4	61
62	305.52.2	798.30.0	1291.07.7	1783.85.5	2276.63.3	2769.41.1	3262.18.8	3754.96.6	4247.74.4	4740.52.2	62
63	310.45.0	803.22.7	1296.00.5	1788.78.3	2281.56.1	2774.33.8	3267.11.6	3759.89.4	4252.67.2	4745.45.0	63
64	315.37.7	808.15.5	1300.93.3	1793.71.1	2286.48.8	2779.26.6	3272.04.4	3764.82.2	4257.60.0	4750.37.7	64
65	320.30.5	813.08.3	1305.86.1	1798.63.8	2291.41.6	2784.19.4	3276.97.2	3769.75.0	4262.52.7	4755.30.5	65
66	325.23.3	818.01.1	1310.78.8	1803.56.6	2296.34.4	2789.12.2	3281.90.0	3774.67.7	4267.45.5	4760.23.3	66
67	330.16.1	822.93.8	1315.71.6	1808.49.4	2301.27.2	2794.05.0	3286.82.7	3779.60.5	4272.38.3	4765.16.1	67
68	335.08.8	827.86.6	1320.64.4	1813.42.2	2306.20.0	2798.97.7	3291.75.5	3784.53.3	4277.31.1	4770.08.8	68
69	340.01.6	832.79.4	1325.57.2	1818.35.0	2311.12.7	2803.90.5	3296.68.3	3789.46.1	4282.23.8	4775.01.6	69
70	344.94.4	837.72.2	1330.50.0	1823.27.7	2316.05.5	2808.83.3	3301.61.1	3794.38.8	4287.16.6	4779.94.4	70
71	349.87.2	842.65.0	1335.42.7	1828.20.5	2320.98.3	2813.76.1	3306.53.8	3799.31.6	4292.09.4	4784.87.2	71
72	354.80.0	847.57.7	1340.35.5	1833.13.3	2325.91.1	2818.68.8	3311.46.6	3804.24.4	4297.02.2	4789.80.0	72
73	359.72.7	852.50.5	1345.28.3	1838.06.1	2330.83.8	2823.61.6	3316.39.4	3809.17.2	4301.95.0	4794.72.7	73
74	364.65.5	857.43.3	1350.21.1	1842.98.8	2335.76.6	2828.54.4	3321.32.2	3814.10.0	4306.87.7	4799.65.5	74
75	369.58.3	862.36.1	1355.13.8	1847.91.6	2340.69.4	2833.47.2	3326.25.0	3819.02.7	4311.80.5	4804.58.3	75
76	374.51.1	867.28.8	1360.06.6	1852.84.4	2345.62.2	2838.40.0	3331.17.7	3823.95.5	4316.73.3	4809.51.1	76
77	379.43.8	872.21.6	1364.99.4	1857.77.2	2350.55.0	2843.32.7	3336.10.5	3828.88.3	4321.66.1	4814.43.8	77
78	384.36.6	877.14.4	1369.92.2	1862.70.0	2355.47.7	2848.25.5	3341.03.3	3833.81.1	4326.58.8	4819.36.6	78
79	389.29.4	882.07.2	1374.85.0	1867.62.7	2360.40.5	2853.18.3	3345.96.1	3838.73.8	4331.51.6	4824.29.4	79
80	394.22.2	887.00.0	1379.77.7	1872.55.5	2365.33.3	2858.11.1	3350.88.8	3843.66.6	4336.44.4	4829.22.2	80
81	399.15.0	891.92.7	1384.70.5	1877.48.3	2370.26.1	2863.03.8	3355.81.6	3848.59.4	4341.37.2	4834.15.0	81
82	404.07.7	896.85.5	1389.63.3	1882.41.1	2375.18.8	2867.96.6	3360.74.4	3853.52.2	4346.30.0	4839.07.7	82
83	409.00.5	901.78.3	1394.56.1	1887.33.8	2380.11.6	2872.89.4	3365.67.2	3858.45.0	4351.22.7	4844.00.5	83
84	413.93.3	906.71.1	1399.48.8	1892.26.6	2385.04.4	2877.82.2	3370.60.0	3863.37.7	4356.15.5	4848.93.3	84
85	418.86.1	911.63.8	1404.41.6	1897.19.4	2389.97.2	2882.75.0	3375.52.7	3868.30.5	4361.08.3	4853.86.1	85
86	423.78.8	916.56.6	1409.34.4	1902.12.2	2394.90.0	2887.67.7	3380.45.5	3873.23.3	4366.01.1	4858.78.8	86
87	428.71.6	921.49.4	1414.27.2	1907.05.0	2399.82.7	2892.60.5	3385.38.3	3878.16.1	4370.93.8	4863.71.6	87
88	433.64.4	926.42.2	1419.20.0	1911.97.7	2404.75.5	2897.53.3	3390.31.1	3883.08.8	4375.86.6	4868.64.4	88
89	438.57.2	931.35.0	1424.12.7	1916.90.5	2409.68.3	2902.46.1	3395.23.8	3888.01.6	4380.79.4	4873.57.2	89
90	443.50.0	936.27.7	1429.05.5	1921.83.3	2414.61.1	2907.38.8	3400.16.6	3892.94.4	4385.72.2	4878.50.0	90
91	448.42.7	941.20.5	1433.98.3	1926.76.1	2419.53.8	2912.31.6	3405.09.4	3897.87.2	4390.65.0	4883.42.7	91
92	453.35.5	946.13.3	1438.91.1	1931.68.8	2424.46.6	2917.24.4	3410.02.2	3902.80.0	4395.57.7	4888.35.5	92
93	458.28.3	951.06.1	1443.83.8	1936.61.6	2429.39.4	2922.17.2	3414.95.0	3907.72.7	4400.50.5	4893.28.3	93
94	463.21.1	955.98.8	1448.76.6	1941.54.4	2434.32.2	2927.10.0	3419.87.7	3912.65.5	4405.43.3	4898.21.1	94
95	468.13.8	960.91.6	1453.69.4	1946.47.2	2439.25.0	2932.02.7	3424.80.5	3917.58.3	4410.36.1	4903.13.8	95
96	473.06.6	965.84.4	1458.62.2	1951.40.0	2444.17.7	2936.95.5	3429.73.3	3922.51.1	4415.28.8	4908.06.6	96
97	477.99.4	970.77.2	1463.55.0	1956.32.7	2449.10.5	2941.88.3	3434.66.1	3927.43.8	4420.21.6	4912.99.4	97
98	482.92.2	975.70.0	1468.47.7	1961.25.5	2454.03.3	2946.81.1	3439.58.8	3932.36.6	4425.14.4	4917.92.2	98
99	487.85.0	980.62.7	1473.40.5	1966.18.3	2458.96.1	2951.73.8	3444.51.6	3937.29.4	4430.07.2	4922.85.0	99

S.	10	11	12	13	14	15	16	17	18	19	S.
d.	$ cts.	$ cts.	$ cts.	$ cts.	$ cts.	$ cts.	$ cts.	$ cts.	$ cts.	$ cts.	d.
0	2.46.3	2.71.0	2.95.6	3.20.3	3.44.9	3.69.5	3.94.2	4.18.8	4.43.5	4.68.1	0
1	2.48.4	2.73.0	2.97.7	3.22.3	3.47.0	3.71.6	3.96.2	4.20.9	4.45.5	4.70.1	1
2	2.50.5	2.75.1	2.99.7	3.24.4	3.49.0	3.73.7	3.98.3	4.22.9	4.47.6	4.72.2	2
3	2.52.5	2.77.1	3.01.8	3.26.4	3.51.1	3.75.7	4.00.3	4.25.0	4.49.6	4.74.3	3
4	2.54.6	2.79.2	3.03.8	3.28.5	3.53.1	3.77.8	4.02.4	4.27.0	4.51.7	4.76.3	4
5	2.56.6	2.81.3	3.05.9	3.30.5	3.55.2	3.79.8	4.04.5	4.29.1	4.53.7	4.78.4	5
6	2.58.7	2.83.3	3.08.0	3.32.6	3.57.2	3.81.9	4.06.5	4.31.2	4.55.8	4.80.4	6
7	2.60.7	2.85.4	3.10.0	3.34.6	3.59.3	3.83.9	4.08.6	4.33.2	4.57.8	4.82.5	7
8	2.62.8	2.87.4	3.12.1	3.36.7	3.61.3	3.86.0	4.10.6	4.35.3	4.59.9	4.84.6	8
9	2.64.8	2.89.5	3.14.1	3.38.8	3.63.4	3.88.0	4.12.7	4.37.3	4.62.0	4.86.6	9
10	2.66.9	2.91.5	3.16.2	3.40.8	3.65.5	3.90.1	4.14.7	4.39.4	4.64.0	4.88.6	10
11	2.69.0	2.93.6	3.18.2	3.42.9	3.67.5	3.92.2	4.16.8	4.41.4	4.66.1	4.90.7	11

$4.93 11/100 TO THE POUND.

11%

PREMIUM

STERLING INTO DOLLARS AND CENTS, AND VICE-VERSA.

£	0	100	200	300	400	500	600	700	800	900	£
	$ cts.	$ cts.	$ cts.	$ cts.	$ cts.	$ cts.	$ cts.	$ cts.	$ cts.	$ cts.	
0	......	493.33,3	986.66.6	1480.00.0	1973.33.3	2466.66.6	2960.00.0	3453.33.3	3946.66.6	4440.00.0	0
1	4.93 3	498.26.6	991.60.0	1484.93.3	1978.26.6	2471.60.0	2964.93.3	3458.26 6	3951.60.0	4444.93.3	1
2	9.86 6	503.20.0	996.53.3	1489.86.6	1983.20.0	2476.53.3	2969 86.6	3463.20 0	3956.53.3	4449.86.6	2
3	14 80.0	508.13.3	1001.46.6	1494 80.0	1988.13.3	2481 46.6	2974.80.0	3468.13.3	3961.46.6	4454.80.0	3
4	19.73.3	513.06 6	1006.40.0	1499.73 3	1993.06.6	2486.40.0	2979.73.3	3473.06.6	3966.40.0	4459.73.3	4
5	24 66.6	518.00.0	1011 33.3	1504.66.6	1998.00.0	2491.33.3	2984.66.6	3478.00.0	3971.33.3	4464.66.6	5
6	29.60.0	522.93 3	1016.26.6	1509.60.0	2002.93.3	2496.26.6	2989.60.0	3482.93.3	3976.26.6	4469.60.0	6
7	34.53.3	527.86 6	1021.20.0	1514.53 3	2007.86.6	2501.20.0	2994.53.3	3487.86.6	3981 20.0	4474.53.3	7
8	39.46.6	532.80.0	1026.13.3	1519.46.6	2012.80.0	2506.13.3	2999.46.6	3492.80.0	3986.13.3	4479.46.6	8
9	44.40.0	537.73.3	1031.06.6	1524.40.0	2017.73.3	2511.06.6	3004.40.0	3497.73.3	3991.06.6	4484.40.0	9
10	49.33.3	542.66 6	1036.00.0	1529.33.3	2022.66.6	2516.00.0	3009.33 3	3502.66.6	3996.00.0	4489.33.3	10
11	54.26.6	547.60.0	1040.93.3	1534.26.6	2027.60.0	2520.93.3	3014.26.6	3507.60.0	4000.93.3	4494.26.6	11
12	59.20 0	552.53 3	1045.86.6	1539.20.0	2032.53.3	2525.86.6	3019.20.0	3512.53.3	4005.86.6	4499.20.0	12
13	64 13.3	557.46.6	1050.80.0	1544.13.3	2037.46.6	2530.80.0	3024.13.3	3517.46.6	4010.80.0	4504 13.3	13
14	69.06.6	562.40 0	1055.73.3	1549.06.6	2042.40.0	2535.73.3	3029.06.6	3522.40.0	4015.73.3	4509.06.6	14
15	74.00.0	567.33.3	1060.66.6	1554 00.0	2047.33.3	2540.66.6	3034.00 0	3527.33.3	4020.66.6	4514.00.0	15
16	78.93.3	572 26 6	1065.60.0	1558.93.3	2052.26.6	2545.60.0	3038.93.3	3532.26.6	4025.60.0	4518.93.3	16
17	83 86.6	577.20.0	1070.53.3	1563.86.6	2057.20.0	2550.53 3	3043.86.6	3537.20.0	4030.53.3	4523.86.6	17
18	88.80 0	582.13.3	1075.46.6	1568.80 0	2062.13.3	2555.46.6	3048.80.0	3542.13.3	4035.46.6	4528.80.0	18
19	93.73.3	587.06.6	1080.40.0	1573.73.3	2067.06.6	2560.40.0	3053.73.3	3547.06.6	4040.40.0	4533.73.3	19
20	98.66.6	592.00.0	1085.33.3	1578.66 6	2072.00.0	2565.33.3	3058.66.6	3552.00.0	4045.33.3	4538 66.6	20
21	103.60.0	596 93.3	1090.26.6	1583.60.0	2076.93.3	2570.26.6	3063.60.0	3556.93.3	4050.26.6	4543 60.0	21
22	108 53.3	601.86.6	1095.20.0	1588.53.3	2081.86.6	2575.20.0	3068.53.3	3561.86.6	4055.20.0	4548.53.3	22
23	113.46.6	606.80.0	1100.13.3	1593.46.6	2086.80.0	2580.13.3	3073.46.6	3566.80.0	4060.13.3	4553.46.6	23
24	118.40.0	611.73.3	1105.06.6	1598.40.0	2091.73.3	2585.06.6	3078.40.0	3571.73.3	4065.06.6	4558.40.0	24
25	123.33.3	616.66.6	1110.00.0	1603.33.3	2096.66.6	2590.00.0	3083.33.3	3576.66.6	4070.00.0	4563.33.3	25
26	128.26.6	621.60.0	1114.93.3	1608.26.6	2101.60.0	2594.93.3	3088.26.6	3581.60.0	4074.93.3	4568.26.6	26
27	133.20.0	626 53 6	1119.86.6	1613.20.0	2106.53.3	2599.86.6	3093.20.0	3586.53.3	4079.86.6	4573.20.0	27
28	138.13.3	631.46.6	1124.80.0	1618.13.3	2111.46.6	2604.80.0	3098.13.3	3591.46.6	4084.80.0	4578.13.3	28
29	143.06.6	636.40.0	1129.73.3	1623.06.6	2116.40.0	2609.73.3	3103.06.6	3596.40.0	4089.73.3	4583.06.6	29
30	148.00.0	641.33.3	1134.66 6	1628.00.0	2121.33.3	2614.66 6	3108.00.0	3601.33.3	4094.66.6	4588.00.0	30
31	152.93.3	646.26.6	1139.60.0	1632.93.3	2126.26.6	2619.60.0	3112.93.3	3606.26.6	4099.60.0	4592.93.3	31
32	157.86.6	651.20.0	1144.53.3	1637.86.6	2131.20.0	2624 53.3	3117.86.6	3611.20.0	4104.53.3	4597.86.6	32
33	162.80.0	656.13.3	1149.46.6	1642.80.0	2136.13.3	2629.46.6	3122.80.0	3616.13.3	4109.46.6	4602.80.0	33
34	167.73.3	661.06.6	1154.40.0	1647.73.3	2141.06.6	2634.40.0	3127.73.3	3621.06.6	4114.40.0	4607.73.3	34
35	172.66.6	666.00 0	1159.33.3	1652.66.6	2146.00.0	2639.33.3	3132.66.6	3626.00.0	4119.33.3	4612.66.6	35
36	177.60.0	670.93.3	1164.26.6	1657.60.0	2150.93.3	2644.26.6	3137.60.0	3630.93.3	4124.26.6	4617.60.0	36
37	182.53.3	675.86.6	1169.20.0	1662.53.3	2155.86.6	2649.20.0	3142.53.3	3635.86.6	4129.20.0	4622.53.3	37
38	187.46.6	680.80.0	1174.13.3	1667.46.6	2160.80.0	2654.13.3	3147.46.6	3640.80.0	4134.13.3	4627.46.6	38
39	192.40.0	685.73.3	1179.06.6	1672.40.0	2165.73.3	2659.06.6	3152.40.0	3645.73.3	4139.06.6	4632.40.0	39
40	197.33.3	690.66.6	1184.00.0	1677.33.3	2170.66.6	2664.00.0	3157.33.3	3650.66.6	4144.00.0	4637.33.3	40
41	202.26.6	695.60.0	1188.93.3	1682.26.6	2175.60.0	2668.93.3	3162.26.6	3655.60.0	4148.93.3	4642.26.6	41
42	207.20.0	700.53.3	1193.86.6	1687.20.0	2180.53.3	2673 86.6	3167.20.0	3660.53.3	4153.86.6	4647.20.0	42
43	212.13.3	705.46.6	1198.80.0	1692.13.3	2185.46.6	2678.80.0	3172.13.3	3665.46.6	4158.80.0	4652.13.3	43
44	217.06.6	710.40.0	1203.73.3	1697.06.6	2190.40.0	2683.73.3	3177.06 6	3670.40.0	4163.73.3	4657.06.6	44
45	222.00.0	715.33.3	1208.66.6	1702.00.0	2195.33.3	2688.66.6	3182.00.0	3675.33.3	4168.66.6	4662.00.0	45
46	226.93.3	720.26.6	1213.60.0	1706.93.3	2200.26.6	2693 60.0	3186.93 3	3680.26.6	4173.60.0	4666.93.3	46
47	231.86.6	725.20.0	1218.53.3	1711.86.6	2205.20.0	2698.53.3	3191.86.6	3685.20.0	4178.53.3	4671.86.6	47
48	236.80.0	730.13.3	1223.46.6	1716.80.0	2210.13.3	2703 46 6	3196.80 0	3690.13.3	4183 46.6	4676.80.0	48
49	241.73.3	735.06.6	1228.40.0	1721.73.3	2215.06.6	2708.40.0	3201.73.3	3695.06.6	4188.40.0	4681.73.3	49

S.	0	1	2	3	4	5	6	7	8	9	S.
d.	$ cts.	$ cts.	$ cts.	$ cts.	$ cts.	$ cts.	$ cts.	$ cts.	$ cts.	$ cts.	d.
0	......	.24.6	.49.3	.74.0	.98.6	1.23.3	1.48.0	1.72.6	1.97.3	2.22.0	0
1	.02.0	.26.7	.51.3	.76.0	1.00.7	1.25.3	1.50.0	1.74.7	1.99.3	2.24.0	1
2	.04.1	.28.7	.53.4	.78.1	1.02.7	1.27.4	1.52.1	1.76.7	2.01.4	2.26.1	2
3	.06.1	.30.8	.55.5	.80.1	1.04.8	1.29.5	1.54.1	1.78.8	2.03.5	2.28.1	3
4	.08.2	.32.8	.57.5	.82.2	1.06.8	1.31.5	1.56.2	1.80.8	2.05.5	2.30.2	4
5	.10.2	.34.9	.59.6	.84.2	1.08.9	1.33.6	1.58.2	1.82.9	2.07.6	2.32.2	5
6	.12.3	.37.0	.61.6	.86.3	1.11.0	1.35.6	1.60.3	1.85.0	2.09.6	2.34.3	6
7	.14.3	.39.0	.63.7	.88.3	1.13.0	1.37.7	1.62.3	1.87.0	2.11.7	2.36.3	7
8	.16.4	.41.1	.65.7	.90.4	1.15.1	1.39.7	1.64.4	1.89.1	2.13.7	2.38.4	8
9	.18.5	.43.1	.67.8	.92.5	1.17.1	1.41.8	1.66.5	1.91.1	2.15.8	2.40.5	9
10	.20.5	.45.2	.69.8	.94.5	1.19.2	1.43.8	1.68.5	1.93.2	2.17.8	2.42.5	10
11	.22.6	.47.2	.71.9	.96.6	1.21.2	1.45.9	1.70.6	1.95.2	2.19.9	2.44.6	11

$4.93 44/100 TO THE POUND.

11%

PREMIUM.

STERLING INTO DOLLARS AND CENTS, AND VICE-VERSA.

£	0	100	200	300	400	500	600	700	800	900	£
	$ cts.	$ cts.	$ cts.	$ cts.	$ cts.	$ cts.	$ cts.	$ cts.	$ cts.	$ cts.	
50	246.66.6	740.00.0	1233.33.3	1726.66.6	2220.00.0	2713.33.3	3206.66.6	3700.00.0	4193.33.3	4686.66.6	50
51	251.60.0	744.93.3	1238.26.6	1731.60.0	2224.93.3	2718.26.6	3211.60.0	3704.93.3	4198.26.6	4691.60.0	51
52	256.53.3	749.86.6	1243.20.0	1736.53.3	2229.86.6	2723.20.0	3216.53.3	3709.86.6	4203.20.0	4696.53.3	52
53	261.46.6	754.80.0	1248.13.3	1741.46.6	2234.80.0	2728.13.3	3221.46.6	3714.80.0	4208.13.3	4701.46.6	53
54	266.40.0	759.73.3	1253.06.6	1746.40.0	2239.73.3	2733.06.6	3226.40.0	3719.73.3	4213.06.6	4706.40.0	54
55	271.33.3	764.66.6	1258.00.0	1751.33.3	2244.66.6	2738.00.0	3231.33.3	3724.66.6	4218.00.0	4711.33.3	55
56	276.26.6	769.60.0	1262.93.3	1756.26.6	2249.60.0	2742.93.3	3236.26.6	3729.60.0	4222.93.3	4716.26.6	56
57	281.20.0	774.53.3	1267.86.6	1761.20.0	2254.53.3	2747.86.6	3241.20.0	3731.53.3	4227.86.6	4721.20.0	57
58	286.13.3	779.46.6	1272.80.0	1766.13.3	2259.46.6	2752.80.0	3246.13.3	3739.46.6	4232.80.0	4726.13.3	58
59	291.06.6	784.40.0	1277.73.3	1771.06.6	2264.40.0	2757.73.3	3251.06.6	3744.40.0	4237.73.3	4731.06.6	59
60	296.00.0	789.33.3	1282.66.6	1776.00.0	2269.33.3	2762.66.6	3256.00.0	3749.33.3	4242.66.6	4736.00.0	60
61	300.93.3	794.26.6	1287.60.0	1780.93.3	2274.26.6	2767.60.0	3260.93.3	3754.26.6	4247.60.0	4740.93.3	61
62	305.86.6	799.20.0	1292.53.3	1785.86.6	2279.20.0	2772.53.3	3265.86.6	3759.20.0	4252.53.3	4745.86.6	62
63	310.80.0	804.13.3	1297.46.6	1790.80.0	2284.13.3	2777.46.6	3270.80.0	3764.13.3	4257.46.6	4750.80.0	63
64	315.73.3	809.06.6	1302.40.0	1795.73.3	2289.06.6	2782.40.0	3275.73.3	3769.06.6	4262.40.0	4755.73.3	64
65	320.66.6	814.00.0	1307.33.3	1800.66.6	2294.00.0	2787.33.3	3280.66.6	3774.00.0	4267.33.3	4760.66.6	65
66	325.60.0	818.93.3	1312.26.6	1805.60.0	2298.93.3	2792.26.6	3285.60.0	3778.93.3	4272.26.6	4765.60.0	66
67	330.53.3	823.86.6	1317.20.0	1810.53.3	2303.86.6	2797.20.0	3290.53.3	3783.86.6	4277.20.0	4770.53.3	67
68	335.46.6	828.80.0	1322.13.3	1815.46.6	2308.80.0	2802.13.3	3295.46.6	3788.80.0	4282.13.3	4775.46.6	68
69	340.40.0	833.73.3	1327.06.6	1820.40.0	2313.73.3	2807.06.6	3300.40.0	3793.73.3	4287.06.6	4780.40.0	69
70	345.33.3	838.66.6	1332.00.0	1825.33.3	2318.66.6	2812.00.0	3305.33.3	3798.66.6	4292.00.0	4785.33.3	70
71	350.26.6	843.60.0	1336.93.3	1830.26.6	2323.60.0	2816.93.3	3310.26.6	3803.60.0	4296.93.3	4790.26.6	71
72	355.20.0	848.53.3	1341.86.6	1835.20.0	2328.53.3	2821.86.6	3315.20.0	3808.53.3	4301.86.6	4795.20.0	72
73	360.13.3	853.46.6	1346.80.0	1840.13.3	2333.46.6	2826.80.0	3320.13.3	3813.46.6	4306.80.0	4800.13.3	73
74	365.06.6	858.40.0	1351.73.3	1845.06.6	2338.40.0	2831.73.3	3325.06.6	3818.40.0	4311.73.3	4805.06.6	74
75	370.00.0	863.33.3	1356.66.6	1850.00.0	2343.33.3	2836.66.6	3330.00.0	3823.33.3	4316.66.6	4810.00.0	75
76	374.93.3	868.26.6	1361.60.0	1854.93.3	2348.26.6	2841.60.0	3334.93.3	3828.26.6	4321.60.0	4814.93.3	76
77	379.86.6	873.20.0	1366.53.3	1859.86.6	2353.20.0	2846.53.3	3339.86.6	3833.20.0	4326.53.3	4819.86.6	77
78	384.80.0	878.13.3	1371.46.6	1864.80.0	2358.13.3	2851.46.6	3344.80.0	3838.13.3	4331.46.6	4824.80.0	78
79	389.73.3	883.06.6	1376.40.0	1869.73.3	2363.06.6	2856.40.0	3349.73.3	3843.06.6	4336.40.0	4829.73.3	79
80	394.66.6	888.00.0	1381.33.3	1874.66.6	2368.00.0	2861.33.3	3354.66.6	3848.00.0	4341.33.3	4834.66.6	80
81	399.60.0	892.93.3	1386.26.6	1879.60.0	2372.93.3	2866.26.6	3359.60.0	3852.93.3	4346.26.6	4839.60.0	81
82	404.53.3	897.86.6	1391.20.0	1884.53.3	2377.86.6	2871.20.0	3364.53.3	3857.86.6	4351.20.0	4844.53.3	82
83	409.46.6	902.80.0	1396.13.3	1889.46.6	2382.80.0	2876.13.3	3369.46.6	3862.80.0	4356.13.3	4849.46.6	83
84	414.40.0	907.73.3	1401.06.6	1894.40.0	2387.73.3	2881.06.6	3374.40.0	3867.73.3	4361.06.6	4854.40.0	84
85	419.33.3	912.66.6	1406.00.0	1899.33.3	2392.66.6	2886.00.0	3379.33.3	3872.66.6	4366.00.0	4859.33.3	85
86	424.26.6	917.60.0	1410.93.3	1904.26.6	2397.60.0	2890.93.3	3384.26.6	3877.60.0	4370.93.3	4864.26.6	86
87	429.20.0	922.53.3	1415.86.6	1909.20.0	2402.53.3	2895.86.6	3389.20.0	3882.53.3	4375.86.6	4869.20.0	87
88	434.13.3	927.46.6	1420.80.0	1914.13.3	2407.46.6	2900.80.0	3394.13.3	3887.46.6	4380.80.0	4874.13.3	88
89	439.06.6	932.40.0	1425.73.3	1919.06.6	2412.40.0	2905.73.3	3399.06.6	3892.40.0	4385.73.3	4879.06.6	89
90	444.00.0	937.33.3	1430.66.6	1924.00.0	2417.33.3	2910.66.6	3404.00.0	3897.33.3	4390.66.6	4884.00.0	90
91	448.93.3	942.26.6	1435.60.0	1928.93.3	2422.26.6	2915.60.0	3408.93.3	3902.26.6	4395.60.0	4888.93.3	91
92	453.86.6	947.20.0	1440.53.3	1933.86.6	2427.20.0	2920.53.3	3413.86.6	3907.20.0	4400.53.3	4893.86.6	92
93	458.80.0	952.13.3	1445.46.6	1938.80.0	2432.13.3	2925.46.6	3418.80.0	3912.13.3	4405.46.6	4898.80.0	93
94	463.73.3	957.06.6	1450.40.0	1943.73.3	2437.06.6	2930.40.0	3423.73.3	3917.06.6	4410.40.0	4903.73.3	94
95	468.66.6	962.00.0	1455.33.3	1948.66.6	2442.00.0	2935.33.3	3428.66.6	3922.00.0	4415.33.3	4908.66.6	95
96	473.60.0	966.93.3	1460.26.6	1953.60.0	2446.93.3	2940.26.6	3433.60.0	3926.93.3	4420.26.6	4913.60.0	96
97	478.53.3	971.86.6	1465.20.0	1958.53.3	2451.86.6	2945.20.0	3438.53.3	3931.86.6	4425.20.0	4918.53.3	97
98	483.46.6	976.80.0	1470.13.3	1963.46.6	2456.80.0	2950.13.3	3443.46.6	3936.80.0	4430.13.3	4923.46.6	98
99	488.40.0	981.73.3	1475.06.6	1968.40.0	2461.73.3	2955.06.6	3448.40.0	3941.73.3	4435.06.6	4928.40.0	99

S.	10	11	12	13	14	15	16	17	18	19	S.
d.	$ cts.	$ cts.	$ cts.	$ cts.	$ cts.	$ cts.	$ cts.	$ cts.	$ cts.	$ cts.	d.
0	2.46.6	2.71.3	2.96.0	3.20.6	3.45.3	3.70.0	3.94.6	4.19.3	4.44.0	4.68.6	0
1	2.48.7	2.73.3	2.98.0	3.22.7	3.47.3	3.72.0	3.96.7	4.21.3	4.46.0	4.70.7	1
2	2.50.7	2.75.4	3.00.1	3.24.7	3.49.4	3.74.1	3.98.7	4.23.4	4.48.1	4.72.7	2
3	2.52.8	2.77.5	3.02.1	3.26.8	3.51.5	3.76.1	4.00.8	4.25.5	4.50.1	4.74.8	3
4	2.54.8	2.79.5	3.04.2	3.28.8	3.53.5	3.78.2	4.02.8	4.27.5	4.52.2	4.76.8	4
5	2.56.9	2.81.6	3.06.2	3.30.9	3.55.6	3.80.2	4.04.9	4.29.6	4.54.2	4.78.9	5
6	2.59.0	2.83.6	3.08.3	3.33.0	3.57.6	3.82.3	4.07.0	4.31.6	4.56.3	4.81.0	6
7	2.61.0	2.85.7	3.10.3	3.35.0	3.59.7	3.84.3	4.09.0	4.33.7	4.58.3	4.83.0	7
8	2.63.1	2.87.7	3.12.4	3.37.1	3.61.7	3.86.4	4.11.1	4.35.7	4.60.4	4.85.1	8
9	2.65.1	2.89.8	3.14.5	3.39.1	3.63.8	3.88.5	4.13.1	4.37.8	4.62.5	4.87.1	9
10	2.67.2	2.91.8	3.16.5	3.41.2	3.65.8	3.90.5	4.15.2	4.39.8	4.64.5	4.89.2	10
11	2.69.2	2.93.9	3.18.6	3.43.2	3.67.9	3.92.6	4.17.2	4.41.9	4.66.6	4.91.2	11

$4.95.¹⁰⁰⁄₁₀₀ TO THE POUND.

11½ %

PREMIUM.

STERLING INTO DOLLARS AND CENTS, AND VICE-VERSA.

£	0	100	200	300	400	500	600	700	800	900	£
£	$ cts.	$ cts.	$ cts.	$ cts.	$ cts.	$ cts.	$ cts.	$ cts.	$ cts.	$ cts.	£
0	......	495.55.5	991.11.1	1486.66.6	1982.22.2	2477.77.7	2973.33.3	3468.88.8	3964.44.4	4460.00.0	0
1	4.95.5	500.51.1	996.06.6	1491.62.2	1987.17.7	2482.73.3	2978.28.8	3473.84.4	3969.40.0	4464.95.5	1
2	9.91.1	505.46.6	1001.02.2	1496.57.7	1992.13.3	2487.68.8	2983.24.4	3478.80.0	3974.35.5	4469.91.1	2
3	14.86.6	510.42.2	1005.97.7	1501.53.3	1997.08.8	2492.64.4	2988.20.0	3483.75.5	3979.31.1	4474.86.6	3
4	19.82.2	515.37.7	1010.93.3	1506.48.8	2002.04.4	2497.60.0	2993 15.5	3488.71.1	3984.26.6	4479.82.2	4
5	24.77.7	520.33.3	1015.88.8	1511.44.4	2007.00.0	2502.55.5	2998.11.1	3493.66.6	3989.22.2	4484.77.7	5
6	29.73.3	525.28.8	1020.84.4	1516.40.0	2011.95.5	2507.51.1	3003.06.6	3498.62.2	3994.17.7	4489.73.3	6
7	34.68.8	530.24.4	1025.80.0	1521.35.5	2016.91.1	2512.46.6	3008.02.2	3503.57.7	3999.13.3	4494.68.8	7
8	39.64.4	535.20.0	1030.75.5	1526.31.1	2021.86.6	2517.42.2	3012.97.7	3508.53.3	4004.08.8	4499.64.4	8
9	44 60 0	540.15.5	1035.71.1	1531.26.6	2026.82.2	2522.37.7	3017.93.3	3513.48.8	4009.04.4	4504.60.0	9
10	49.55.5	545.11.1	1040.66.6	1536.22.2	2031.77.7	2527.33.3	3022.88.8	3518.44.4	4014.00.0	4509.55.5	10
11	54.51.1	550.06.6	1045.62.2	1541.17.7	2036.73.3	2532.28.8	3027.84.4	3523.40.0	4018.95.5	4514.51.1	11
12	59.46.6	555.02.2	1050.57.7	1546.13.3	2041.68.8	2537.24.4	3032.80.0	3528.35.5	4023.91.1	4519.46.6	12
13	64.42.2	559.97.7	1055.53.3	1551.08.8	2046.64.4	2542.20.0	3037 75.5	3533.31.1	4028.86.6	4524.42.2	13
14	69.37.7	564.93.3	1060.48.8	1556.04.4	2051.60.0	2547.15.5	3042.71.1	3538.26.6	4033.82.2	4529.37.7	14
15	74.33.3	569.88.8	1065.44.4	1561.00.0	2056.55.5	2552.11.1	3047.66.6	3543 22.2	4038.77.7	4534.33.3	15
16	79.28.8	574.84.4	1070.40.0	1565.95.5	2061.51.1	2557.06.6	3052.62 2	3548.17.7	4043.73.3	4539.28.8	16
17	84 24.4	579.80.0	1075.55.5	1570.91.1	2066.46.6	2562.02.2	3057.57.7	3553.13.3	4048.68.8	4544.24.4	17
18	89.20.0	584.75.5	1080.31.1	1575.86.6	2071.42.2	2566.97.7	3062.53.3	3558.08.8	4053.64.4	4549.20.0	18
19	94.15.5	589.71.1	1085.26.6	1580.82.2	2076.37.7	2571.93.3	3067.48.8	3563.04.4	4058.60.0	4554.15.5	19
20	99.11.1	594.66 6	1090.22.2	1585.77.7	2081.33.3	2576.88.8	3072.44.4	3568.00.0	4063.55.5	4559.11.1	20
21	104.06.6	599.62.2	1095.17.7	1590.73.3	2086.28.8	2581.84 4	3077.40.0	3572.95.5	4068.51.1	4564.06.6	21
22	109.02.2	604.57.7	1100.13.3	1595.68.8	2091.24.4	2586.80.0	3082.35.5	3577.91.1	4073.46.6	4569.02.2	22
23	113.97.7	609.53.3	1105 08.8	1600.64.4	2096.20.0	2591.75.5	3087.31.1	3582.86.6	4078.42.2	4573.97.7	23
24	118.93.3	614.48.8	1110.04 4	1605.60.0	2101.15.5	2596.71.1	3092.26.6	3587.82.2	4083.37.7	4578.93.3	24
25	123.88.8	619.44.4	1115.00.0	1610.55.5	2106 11.1	2601.66.6	3097.22.2	3592.77.7	4088.33.3	4583.88.8	25
26	128.84.4	624.40.0	1119.95.5	1615.51.1	2111.06.6	2606.62.2	3102.17.7	3597.73.3	4093.28.8	4588.84.4	26
27	133.80.0	629.35.5	1124.91.1	1620.46.6	2116.02.2	2611.57.7	3107.13.3	3602.68.8	4098.24.4	4593.80.0	27
28	138.75.5	634.31.1	1129.86.6	1625.42.2	2120.97.7	2616.53.3	3112.08.8	3607.64.4	4103.20.0	4598.75.5	28
29	143.71.1	639.26.6	1134 82.2	1630.37.7	2125.93.3	2621.48.8	3117.04.4	3612.60.0	4108.15.5	4603.71.1	29
30	148.66.6	644 22.2	1139.77 7	1635.33.3	2130.88.8	2626.44.4	3122.00.0	3617.55.5	4113.11.1	4608.66.6	30
31	153.62.2	649 17.7	1144.73.3	1640.28.8	2135.84.4	2631.40.0	3126.95.5	3622.51.1	4118.06.6	4613.62.2	31
32	158.57.7	654.13.3	1149.68.8	1645.24.4	2140 80.0	2636.35.5	3131.91.1	3627.46.6	4123.02.2	4618.57.7	32
33	163.53.3	659.08.8	1154.64.4	1650.20.0	2145.75.5	2641.31.1	3136.86.6	3632.42.2	4127.97.7	4623.53.3	33
34	168.48.8	664.04.4	1159.60.0	1655.15 5	2150.71.1	2646.26.6	3141.82.2	3637.37.7	4132.93.3	4628.48.8	34
35	173.44.4	669.00.0	1164.55.5	1660.11.1	2155.66.6	2651.22.2	3146.77.7	3642.33.3	4137.88.8	4633.44.4	35
36	178.40.0	673.95.5	1169.51.1	1665.06.6	2160.62.2	2656.17.7	3151.73.3	3647.28.8	4142.84.4	4638.40.0	36
37	183.35.5	678.91.1	1174 46 6	1670.02.2	2165 57.7	2661.13.3	3156.68.8	3652.24.4	4147.80.0	4643.35.5	37
38	188.31.1	683.86.6	1179.42.2	1674.97.7	2170.53.3	2666.08.8	3161.64.4	3657.20.0	4152.75.5	4648.31.1	38
39	193.26.6	688.82.2	1184.37.7	1679.93.3	2175.48.8	2671.04.4	3166.60.0	3662.15.5	4157.71.1	4653.26.6	39
40	198.22.2	693.77.7	1189.33.3	1684.88.8	2180.44.4	2676.00.0	3171.55.5	3667.11.1	4162.66.6	4658.22.2	40
41	203.17.7	698.73.3	1194.28.8	1689.84.4	2185.40 0	2680.95.5	3176.51.1	3672.06 6	4167.62.2	4663.17.7	41
42	208.13.3	703.68.8	1199.24.4	1694 80.0	2190.35.5	2685.91.1	3181 46.6	3677.02 2	4172.57.7	4668.13.3	42
43	213.08.8	708.64.4	1204.20.0	1699.75.5	2195.31.1	2690.86 6	3186.42.2	3681.97.7	4177.53.3	4673.08.8	43
44	218.04.4	713.60.0	1209.15.5	1704.71.1	2200.26.6	2695.82.2	3191.37.7	3686 93.3	4182 48.8	4678.04.4	44
45	223.00.0	718.55.5	1214.11.1	1709.66.6	2205.22.2	2700.77.7	3196.33.3	3691.88.8	4187.44.4	4683.00.0	45
46	227.95.5	723.51.1	1219.06.6	1714.62.2	2210.17.7	2705.73.3	3201.28.8	3696.84.4	4192.40.0	4687.95.5	46
47	232.91.1	728.46.6	1224 02.2	1719.57.7	2215.13.3	2710.68.8	3206.24.4	3701.80.0	4197.35.5	4692.91.1	47
48	237.86.6	733.42.2	1228.97.7	1724.53 3	2220.08.8	2715.64.4	3211.20.0	3706.75.5	4202.31.1	4697.86.6	48
49	242.82.2	738.37.7	1233.93.3	1729.48.8	2225.04.4	2720.60.0	3216.15.5	3711.71.1	4207.26.6	4702.82 2	49

s.	0	1	2	3	4	5	6	7	8	9	s.
d.	$ cts.	$ cts.	$ cts.	$ cts.	$ cts.	$ cts.	$ cts.	$ cts.	$ cts.	$ cts.	d.
0	......	.24.7	.49.5	.74.3	.99.1	1.23.8	1.48.6	1.73.4	1.98.2	2.23.0	0
1	.02.0	.26.8	.51.6	.76.4	1.01.1	1.25.9	1.50.7	1.75.5	2.00.2	2.25.0	1
2	.04.1	.28.9	.53.6	.78.4	1.03.2	1.28.0	1.52.8	1.77.5	2.02.3	2.27.1	2
3	.06.2	.30.9	.55.7	.80.5	1.05.3	1.30.0	1.54.8	1.79.6	2.04.4	2.29.2	3
4	.08.2	.33.0	.57.8	.82.6	1.07.3	1.32.1	1.56.9	1.81.7	2.06.4	2.31.2	4
5	.10 3	.35.1	.59.8	.84.6	1.09.4	1.34.2	1.59.0	1.83.7	2.08.5	2.33.3	5
6	.12.4	.37.1	.61.9	.86.7	1.11.5	1.36.2	1.61.0	1.85.8	2.10.6	2.35.4	6
7	.14.4	.39.2	.64.0	.88.8	1.13.5	1.38.3	1.63.1	1.87.9	2.12.6	2.37.4	7
8	.16.5	.41.3	.66.0	.90.8	1.15.6	1.40.4	1.65.2	1.89.9	2.14.7	2.39.5	8
9	.18.6	.43.3	.68.1	.92.9	1.17.7	1.42.4	1.67.2	1.92.0	2.16.8	2.41.6	9
10	.20.6	.45.4	.70.2	.95.0	1.19.7	1.44.5	1.69.3	1.94.1	2.18.8	2.43.6	10
11	.22.7	.47.5	.72.2	.97.0	1.21.8	1.46.6	1.71.4	1.96.1	2.20.9	2.45.7	11

(75)

$4.95 11/100 TO THE POUND.

11½%

PREMIUM.

STERLING INTO DOLLARS AND CENTS, AND VICE-VERSA.

£	0	100	200	300	400	500	600	700	800	900	£
£	$ cts.	$ cts.	$ cts.	$ cts.	$ cts.	$ cts.	$ cts.	$ cts.	$ cts.	$ cts.	£
50	247.77.7	743.33.3	1238.88.8	1734.44.4	2230.00.0	2725.55.5	3221.11.1	3716.66.6	4212.22.2	4707.77.7	50
51	252.73.3	748.28.8	1243.84.4	1739.40.0	2234.95.5	2730.51.1	3226.06.6	3721.62.2	4217.17.7	4712.73.3	51
52	257.68.8	753.24.4	1248.80.0	1744.35.5	2239.91.1	2735.46.6	3231.02.2	3726.57.7	4222.13.3	4717.68.8	52
53	262.64.4	758.20.0	1253.75.5	1749.31.1	2244.86.6	2740.42.2	3235.97.7	3731.53.3	4227.08.8	4722.64.4	53
54	267.60.0	763.15.5	1258.71.1	1754.26.6	2249.82.2	2745.37.7	3240.93.3	3736.48.8	4232.04.4	4727.60.0	54
55	272.55.5	768.11.1	1263.66.6	1759.22.2	2254.77.7	2750.33.3	3245.88.8	3741.44.4	4237.00.0	4732.55.5	55
56	277.51.1	773.06.6	1268.62.2	1764.17.7	2259.73.3	2755.28.8	3250.84.4	3746.40.0	4241.95.5	4737.51.1	56
57	282.46.6	778.02.2	1273.57.7	1769.13.3	2264.68.8	2760.24.4	3255.80.0	3751.35.5	4246.91.1	4742.46.6	57
58	287.42.2	782.97.7	1278.53.3	1774.08.8	2269.64.4	2765.20.0	3260.75.5	3756.31.1	4251.86.6	4747.42.2	58
59	292.37.7	787.93.3	1283.48.8	1779.04.4	2274.60.0	2770.15.5	3265.71.1	3761.26.6	4256.82.2	4752.37.7	59
60	297.33.3	792.88.8	1288.44.4	1784.00.0	2279.55.5	2775.11.1	3270.66.6	3766.22.2	4261.77.7	4757.33.3	60
61	302.28.8	797.84.4	1293.40.0	1788.95.5	2284.51.1	2780.06.6	3275.62.2	3771.17.7	4266.73.3	4762.28.8	61
62	307.24.4	802.80.0	1298.35.5	1793.91.1	2289.46.6	2785.02.2	3280.57.7	3776.13.3	4271.68.8	4767.24.4	62
63	312.20.0	807.75.5	1303.31.1	1798.86.6	2294.42.2	2789.97.7	3285.53.3	3781.08.8	4276.64.4	4772.20.0	63
64	317.15.5	812.71.1	1308.26.6	1803.82.2	2299.37.7	2794.93.3	3290.48.8	3786.04.4	4281.60.0	4777.15.5	64
65	322.11.1	817.66.6	1313.22.2	1808.77.7	2304.33.3	2799.88.8	3295.44.4	3791.00.0	4286.55.5	4782.11.1	65
66	327.06.6	822.62.2	1318.17.7	1813.73.3	2309.28.8	2804.84.4	3300.40.0	3795.95.5	4291.51.1	4787.06.6	66
67	332.02.2	827.57.7	1323.13.3	1818.68.8	2314.24.4	2809.80.0	3305.35.5	3800.91.1	4296.46.6	4792.02.2	67
68	336.97.7	832.53.3	1328.08.8	1823.64.4	2319.20.0	2814.75.5	3310.31.1	3805.86.6	4301.42.2	4796.97.7	68
69	341.93.3	837.48.8	1333.04.4	1828.60.0	2324.15.5	2819.71.1	3315.26.6	3810.82.2	4306.37.7	4801.93.3	69
70	346.88.8	842.44.4	1338.00.0	1833.55.5	2329.11.1	2824.66.6	3320.22.2	3815.77.7	4311.33.3	4806.88.8	70
71	351.84.4	847.40.0	1342.95.5	1838.51.1	2334.06.6	2829.62.2	3325.17.7	3820.73.3	4316.28.8	4811.84.4	71
72	356.80.0	852.35.5	1347.91.1	1843.46.6	2339.02.2	2834.57.7	3330.13.3	3825.68.8	4321.24.4	4816.80.0	72
73	361.75.5	857.31.1	1352.86.6	1848.42.2	2343.97.7	2839.53.3	3335.08.8	3830.64.4	4326.20.0	4821.75.5	73
74	366.71.1	862.26.6	1357.82.2	1853.37.7	2348.93.3	2844.48.8	3340.04.4	3835.60.0	4331.15.5	4826.71.1	74
75	371.66.6	867.22.2	1362.77.7	1858.33.3	2353.88.8	2849.44.4	3345.00.0	3840.55.5	4336.11.1	4831.66.6	75
76	376.62.2	872.17.7	1367.73.3	1863.28.8	2358.84.4	2854.40.0	3349.95.5	3845.51.1	4341.06.6	4836.62.2	76
77	381.57.7	877.13.3	1372.68.8	1868.24.4	2363.80.0	2859.35.5	3354.91.1	3850.46.6	4346.02.2	4841.57.7	77
78	386.53.3	882.08.8	1377.64.4	1873.20.0	2368.75.5	2864.31.1	3359.86.6	3855.42.2	4350.97.7	4846.53.3	78
79	391.48.8	887.04.4	1382.60.0	1878.15.5	2373.71.1	2869.26.6	3364.82.2	3860.37.7	4355.93.3	4851.48.8	79
80	396.44.4	892.00.0	1387.55.5	1883.11.1	2378.66.6	2874.22.2	3369.77.7	3865.33.3	4360.88.8	4856.44.4	80
81	401.40.0	896.95.5	1392.51.1	1888.06.6	2383.62.2	2879.17.7	3374.73.3	3870.28.8	4365.84.4	4861.40.0	81
82	406.35.5	901.91.1	1397.46.6	1893.02.2	2388.57.7	2884.13.3	3379.68.8	3875.24.4	4370.80.0	4866.35.5	82
83	411.31.1	906.86.6	1402.42.2	1897.97.7	2393.53.3	2889.08.8	3384.64.4	3880.20.0	4375.75.5	4871.31.1	83
84	416.26.6	911.82.2	1407.37.7	1902.93.3	2398.48.8	2894.04.4	3389.60.0	3885.15.5	4380.71.1	4876.26.6	84
85	421.22.2	916.77.7	1412.33.3	1907.88.8	2403.44.4	2899.00.0	3394.55.5	3890.11.1	4385.66.6	4881.22.2	85
86	426.17.7	921.73.3	1417.28.8	1912.84.4	2408.40.0	2903.95.5	3399.51.1	3895.06.6	4390.62.2	4886.17.7	86
87	431.13.3	926.68.8	1422.24.4	1917.80.0	2413.35.5	2908.91.1	3404.46.6	3900.02.2	4395.57.7	4891.13.3	87
88	436.08.8	931.64.4	1427.20.0	1922.75.5	2418.31.1	2913.86.6	3409.42.2	3904.97.7	4400.53.3	4896.08.8	88
89	441.04.4	936.60.0	1432.15.5	1927.71.1	2423.26.6	2918.82.2	3414.37.7	3909.93.3	4405.48.8	4901.04.4	89
90	446.00.0	941.55.5	1437.11.1	1932.66.6	2428.22.2	2923.77.7	3419.33.3	3914.88.8	4410.44.4	4906.00.0	90
91	450.95.5	946.51.1	1442.06.6	1937.62.2	2433.17.7	2928.73.3	3424.28.8	3919.84.4	4415.40.0	4910.95.5	91
92	455.91.1	951.46.6	1447.02.2	1942.57.7	2438.13.3	2933.68.8	3429.24.4	3924.80.0	4420.35.5	4915.91.1	92
93	460.86.6	956.42.2	1451.97.7	1947.53.3	2443.08.8	2938.64.4	3434.20.0	3929.75.5	4425.31.1	4920.86.6	93
94	465.82.2	961.37.7	1456.93.3	1952.48.8	2448.04.4	2943.60.0	3439.15.5	3934.71.1	4430.26.6	4925.82.2	94
95	470.77.7	966.33.3	1461.88.8	1957.44.4	2453.00.0	2948.55.5	3444.11.1	3939.66.6	4435.22.2	4930.77.7	95
96	475.73.3	971.28.8	1466.84.4	1962.40.0	2457.95.5	2953.51.1	3449.06.6	3944.62.2	4440.17.7	4935.73.3	96
97	480.68.8	976.24.4	1471.80.0	1967.35.5	2462.91.1	2958.46.6	3454.02.2	3949.57.7	4445.13.3	4940.68.8	97
98	485.64.4	981.20.0	1476.75.5	1972.31.1	2467.86.6	2963.42.2	3458.97.7	3954.53.3	4450.08.8	4945.64.4	98
99	490.60.0	986.15.5	1481.71.1	1977.26.6	2472.82.2	2968.37.7	3463.93.3	3959.48.8	4455.04.4	4950.60.0	99

S.	10	11	12	13	14	15	16	17	18	19	S.
d.	$ cts.	$ cts.	$ cts.	$ cts.	$ cts.	$ cts.	$ cts.	$ cts.	$ cts.	$ cts.	d.
0	2.47.7	2.72.5	2.97.3	3.22.1	3.46.8	3.71.6	3.96.4	4.21.2	4.46.0	4.70.7	0
1	2.49.8	2.74.6	2.99.4	3.24.1	3.48.9	3.73.7	3.98.5	4.23.3	4.48.0	4.72.8	1
2	2.51.9	2.76.6	3.01.4	3.26.2	3.51.0	3.75.8	4.00.5	4.25.3	4.50.1	4.74.9	2
3	2.53.9	2.78.7	3.03.5	3.28.3	3.53.0	3.77.8	4.02.6	4.27.4	4.52.2	4.76.9	3
4	2.56.0	2.80.8	3.05.6	3.30.3	3.55.1	3.79.9	4.04.7	4.29.4	4.54.2	4.79.0	4
5	2.58.1	2.82.8	3.07.6	3.32.4	3.57.2	3.82.0	4.06.7	4.31.5	4.56.3	4.81.1	5
6	2.60.1	2.84.9	3.09.7	3.34.5	3.59.2	3.84.0	4.08.8	4.33.6	4.58.4	4.83.1	6
7	2.62.2	2.87.0	3.11.8	3.36.5	3.61.3	3.86.1	4.10.9	4.35.6	4.60.4	4.85.2	7
8	2.64.3	2.89.0	3.13.8	3.38.6	3.63.4	3.88.2	4.12.9	4.37.7	4.62.5	4.87.3	8
9	2.66.3	2.91.1	3.15.9	3.40.7	3.65.4	3.90.2	4.15.0	4.39.8	4.64.6	4.89.3	9
10	2.68.4	2.93.2	3.18.0	3.42.7	3.67.5	3.92.3	4.17.1	4.41.8	4.66.6	4.91.4	10
11	2.70.5	2.95.2	3.20.0	3.44.8	3.69.6	3.94.4	4.19.1	4.43.9	4.68.7	4.93.5	11

$4.97 11/100 TO THE POUND.
12%
PREMIUM.
STERLING INTO DOLLARS AND CENTS, AND VICE-VERSA.

£	0	100	200	300	400	500	600	700	800	900	£
	$ cts.	$ cts.	$ cts.	$ cts.	$ cts.	$ cts.	$ cts.	$ cts.	$ cts.	$ cts.	
0	....	497.77.7	995.55.5	1493.33.3	1991.11.1	2488.88.8	2986.66.6	3484.44.4	3982.22.2	4480.00.0	0
1	4.97.7	502.75.5	1000.53.3	1498.31.1	1996.08.8	2493.86.6	2991.64.4	3489.42.2	3987.20.0	4484.97.7	1
2	9.95.5	507.73.3	1005.51.1	1503.28.8	2001.06.6	2498.84.4	2996.62.2	3494.40.0	3992.17.7	4489.95.5	2
3	14.93.3	512.71.1	1010.48.8	1508.26.6	2006.04.4	2503.82.2	3001.60.0	3499.37.7	3997.15.5	4494.93.3	3
4	19.91.1	517.68.8	1015.46.6	1513.24.4	2011.02.2	2508.80.0	3006.57.7	3504.35.5	4002.13.3	4499.91.1	4
5	24.88.8	522.66.6	1020.44.4	1518.22.2	2016.00.0	2513.77.7	3011.55.5	3509.33.3	4007.11.1	4504.88.8	5
6	29.86.6	527.64.4	1025.42.2	1523.20.0	2020.97.7	2518.75.5	3016.53.3	3514.31.1	4012.08.8	4509.86.6	6
7	34.84.4	532.62.2	1030.40.0	1528.17.7	2025.95.5	2523.73.3	3021.51.1	3519.28.8	4017.06.6	4514.84.4	7
8	39.82.2	537.60.0	1035.37.7	1533.15.5	2030.93.3	2528.71.1	3026.48.8	3524.26.6	4022.04.4	4519.82.2	8
9	44.80.0	542.57.7	1040.35.5	1538.13.3	2035.91.1	2533.68.8	3031.46.6	3529.24.4	4027.02.2	4524.80.0	9
10	49.77.7	547.55.5	1045.33.3	1543.11.1	2040.88.8	2538.66.6	3036.44.4	3534.22.2	4032.00.0	4529.77.7	10
11	54.75.5	552.53.3	1050.31.1	1548.08.8	2045.86.6	2543.64.4	3041.42.2	3539.20.0	4036.97.7	4534.75.5	11
12	59.73.3	557.51.1	1055.28.8	1553.06.6	2050.84.4	2548.62.2	3046.40.0	3544.17.7	4041.95.5	4539.73.3	12
13	64.71.1	562.48.8	1060.26.6	1558.04.4	2055.82.2	2553.60.0	3051.37.7	3549.15.5	4046.93.3	4544.71.1	13
14	69.68.8	567.46.6	1065.24.4	1563.02.2	2060.80.0	2558.57.7	3056.35.5	3554.13.3	4051.91.1	4549.68.8	14
15	74.66.6	572.44.4	1070.22.2	1568.00.0	2065.77.7	2563.55.5	3061.33.3	3559.11.1	4056.88.8	4554.66.6	15
16	79.64.4	577.42.2	1075.20.0	1572.97.7	2070.75.5	2568.53.3	3066.31.1	3564.08.8	4061.86.6	4559.64.4	16
17	84.62.2	582.40.0	1080.17.7	1577.95.5	2075.73.3	2573.51.1	3071.28.8	3569.06.6	4066.84.4	4564.62.2	17
18	89.60.0	587.37.7	1085.15.5	1582.93.3	2080.71.1	2578.48.8	3076.26.6	3574.04.4	4071.82.2	4569.60.0	18
19	94.57.7	592.35.5	1090.13.3	1587.91.1	2085.68.8	2583.46.6	3081.24.4	3579.02.2	4076.80.0	4574.57.7	19
20	99.55.5	597.33.3	1095.11.1	1592.88.8	2090.66.6	2588.44.4	3086.22.2	3584.00.0	4081.77.7	4579.55.5	20
21	104.53.3	602.31.1	1100.08.8	1597.86.6	2095.64.4	2593.42.2	3091.20.0	3588.97.7	4086.75.5	4584.53.3	21
22	109.51.1	607.28.8	1105.06.6	1602.84.4	2100.62.2	2598.40.0	3096.17.7	3593.95.5	4091.73.3	4589.51.1	22
23	114.48.8	612.26.6	1110.04.4	1607.82.2	2105.60.0	2603.37.7	3101.15.5	3598.93.3	4096.71.1	4594.48.8	23
24	119.46.6	617.24.4	1115.02.2	1612.80.0	2110.57.7	2608.35.5	3106.13.3	3603.91.1	4101.68.8	4599.46.6	24
25	124.44.4	622.22.2	1120.00.0	1617.77.7	2115.55.5	2613.33.3	3111.11.1	3608.88.8	4106.66.6	4604.44.4	25
26	129.42.2	627.20.0	1124.97.7	1622.75.5	2120.53.3	2618.31.1	3116.08.8	3613.86.6	4111.64.4	4609.42.2	26
27	134.40.0	632.17.7	1129.95.5	1627.73.3	2125.51.1	2623.28.8	3121.06.6	3618.84.4	4116.62.2	4614.40.0	27
28	139.37.7	637.15.5	1134.93.3	1632.71.1	2130.48.8	2628.26.6	3126.04.4	3623.82.2	4121.60.0	4619.37.7	28
29	144.35.5	642.13.3	1139.91.1	1637.68.8	2135.46.6	2633.24.4	3131.02.2	3628.80.0	4126.57.7	4624.35.5	29
30	149.33.3	647.11.1	1144.88.8	1642.66.6	2140.44.4	2638.22.2	3136.00.0	3633.77.7	4131.55.5	4629.33.3	30
31	154.31.1	652.08.8	1149.86.6	1647.64.4	2145.42.2	2643.20.0	3140.97.7	3638.75.5	4136.53.3	4634.31.1	31
32	159.28.8	657.06.6	1154.84.4	1652.62.2	2150.40.0	2648.17.7	3145.95.5	3643.73.3	4141.51.1	4639.28.8	32
33	164.26.6	662.04.4	1159.82.2	1657.60.0	2155.37.7	2653.15.5	3150.93.3	3648.71.1	4146.48.8	4644.26.6	33
34	169.24.4	667.02.2	1164.80.0	1662.57.7	2160.35.5	2658.13.3	3155.91.1	3653.68.8	4151.46.6	4649.24.4	34
35	174.22.2	672.00.0	1169.77.7	1667.55.5	2165.33.3	2663.11.1	3160.88.8	3658.66.6	4156.44.4	4654.22.2	35
36	179.20.0	676.97.7	1174.75.5	1672.53.3	2170.31.1	2668.08.8	3165.86.6	3663.64.4	4161.42.2	4659.20.0	36
37	184.17.7	681.95.5	1179.73.3	1677.51.1	2175.28.8	2673.06.6	3170.84.4	3668.62.2	4166.40.0	4664.17.7	37
38	189.15.5	686.93.3	1184.71.1	1682.48.8	2180.26.6	2678.04.4	3175.82.2	3673.60.0	4171.37.7	4669.15.5	38
39	194.13.3	691.91.1	1189.68.8	1687.46.6	2185.24.4	2683.02.2	3180.80.0	3678.57.7	4176.35.5	4674.13.3	39
40	199.11.1	696.88.8	1194.66.6	1692.44.4	2190.22.2	2688.00.0	3185.77.7	3683.55.5	4181.33.3	4679.11.1	40
41	204.08.8	701.86.6	1199.64.4	1697.42.2	2195.20.0	2692.97.7	3190.75.5	3688.53.3	4186.31.1	4684.08.8	41
42	209.06.6	706.84.4	1204.62.2	1702.40.0	2200.17.7	2697.95.5	3195.73.3	3693.51.1	4191.28.8	4689.06.6	42
43	214.04.4	711.82.2	1209.60.0	1707.37.7	2205.15.5	2702.93.3	3200.71.1	3698.48.8	4196.26.6	4694.04.4	43
44	219.02.2	716.80.0	1214.57.7	1712.35.5	2210.13.3	2707.91.1	3205.68.8	3703.46.6	4201.24.4	4699.02.2	44
45	224.00.0	721.77.7	1219.55.5	1717.33.3	2215.11.1	2712.88.8	3210.66.6	3708.44.4	4206.22.2	4704.00.0	45
46	228.97.7	726.75.5	1224.53.3	1722.31.1	2220.08.8	2717.86.6	3215.64.4	3713.42.2	4211.20.0	4708.97.7	46
47	233.95.5	731.73.3	1229.51.1	1727.28.8	2225.06.6	2722.84.4	3220.62.2	3718.40.0	4216.17.7	4713.95.5	47
48	238.93.3	736.71.1	1234.48.8	1732.26.6	2230.04.4	2727.82.2	3225.60.0	3723.37.7	4221.15.5	4718.93.3	48
49	243.91.1	741.68.8	1239.46.6	1737.24.4	2235.02.2	2732.80.0	3230.57.7	3728.35.5	4226.13.3	4723.91.1	49

S.	0	1	2	3	4	5	6	7	8	9	S.
d.	$ cts.	$ cts.	$ cts.	$ cts.	$ cts.	$ cts.	$ cts.	$ cts.	$ cts.	$ cts.	d.
0	....	.24.8	.49.7	.74.6	.99.5	1.24.4	1.49.3	1.74.2	1.99.1	2.24.0	0
1	.02.0	.26.9	.51.8	.76.7	1.01.6	1.26.5	1.51.4	1.76.3	2.01.1	2.26.0	1
2	.04.1	.29.0	.53.9	.78.8	1.03.7	1.28.6	1.53.4	1.78.3	2.03.2	2.28.1	2
3	.06.2	.31.1	.56.0	.80.9	1.05.7	1.30.6	1.55.5	1.80.4	2.05.3	2.30.2	3
4	.08.3	.33.2	.58.0	.82.9	1.07.8	1.32.7	1.57.6	1.82.5	2.07.4	2.32.3	4
5	.10.3	.35.2	.60.1	.85.0	1.09.9	1.34.8	1.59.7	1.84.6	2.09.5	2.34.3	5
6	.12.4	.37.3	.62.2	.87.1	1.12.0	1.36.9	1.61.8	1.86.6	2.11.5	2.36.4	6
7	.14.5	.39.4	.64.3	.89.2	1.14.1	1.38.9	1.63.8	1.88.7	2.13.6	2.38.5	7
8	.16.5	.41.5	.66.4	.91.2	1.16.1	1.41.0	1.65.9	1.90.8	2.15.7	2.40.6	8
9	.18.6	.43.5	.68.4	.93.3	1.18.2	1.43.1	1.68.0	1.92.9	2.17.8	2.42.7	9
10	.20.7	.45.6	.70.5	.95.4	1.20.3	1.45.2	1.70.1	1.95.0	2.19.8	2.44.7	10
11	.22.8	.47.7	.72.6	.97.5	1.22.4	1.47.3	1.72.1	1.97.0	2.21.9	2.46.8	11

$4.97 7/100 TO THE POUND.

12%

PREMIUM.

STERLING INTO DOLLARS AND CENTS, AND VICE-VERSA.

£	0	100	200	300	400	500	600	700	800	900	£
	$ cts.	$ cts.	$ cts.	$ cts.	$ cts.	$ cts.	$ cts.	$ cts.	$ cts.	$ cts.	
50	248.88.8	746.66.6	1214.44.4	1742.22.2	2240.00.0	2737.77.7	3235.55.5	3733.33.3	4231.11.1	4728.88.8	50
51	253.86.6	751.64.4	1249.42.2	1747.20.0	2244.97.7	2742.75.5	3240.53.3	3738.31.1	4236.08.8	4733.86.6	51
52	258.84.4	756.62.2	1254.40.0	1752.17.7	2249.95.5	2747.73.3	3245.51.1	3743.28.8	4241.06.6	4738.84.4	52
53	263.82.2	761.60.0	1259.37.7	1757.15.5	2254.93.3	2752.71.1	3250.48.8	3748.26.6	4246.04.4	4743.82.2	53
54	268.80.0	766.57.7	1264.35.5	1762.13.3	2259.91.1	2757.68.8	3255.46.6	3753.24.4	4251.02.2	4748.80.0	54
55	273.77.7	771.55.5	1269.33.3	1767.11.1	2264.88.8	2762.66.6	3260.44.4	3758.22.2	4256.00.0	4753.77.7	55
56	278.75.5	776.53.3	1274.31.1	1772.08.8	2269.86.6	2767.64.4	3265.42.2	3763.20.0	4260.97.7	4758.75.5	56
57	283.73.3	781.51.1	1279.28.8	1777.06.6	2274.84.4	2772.62.2	3270.40.0	3768.17.7	4265.95.5	4763.73.3	57
58	288.71.1	786.48.8	1284.26.6	1782.04.4	2279.82.2	2777.60.0	3275.37.7	3773.15.5	4270.93.3	4768.71.1	58
59	293.68.8	791.46.6	1289.24.4	1787.02.2	2284.80.0	2782.57.7	3280.35.5	3778.13.3	4275.91.1	4773.68.8	59
60	298.66.6	796.44.4	1294.22.2	1792.00.0	2289.77.7	2787.55.5	3285.33.3	3783.11.1	4280.88.8	4778.66.6	60
61	303.64.4	801.42.2	1299.20.0	1796.97.7	2294.75.5	2792.53.3	3290.31.1	3788.08.8	4285.86.6	4783.64.4	61
62	308.62.2	806.40.0	1304.17.7	1801.95.5	2299.73.3	2797.51.1	3295.28.8	3793.06.6	4290.84.4	4788.62.2	62
63	313.60.0	811.37.7	1309.15.5	1806.93.3	2304.71.1	2802.48.8	3300.26.6	3798.04.4	4295.82.2	4793.60.0	63
64	318.57.7	816.35.5	1314.13.3	1811.91.1	2309.68.8	2807.46.6	3305.24.4	3803.02.2	4300.80.0	4798.57.7	64
65	323.55.5	821.33.3	1319.11.1	1816.88.8	2314.66.6	2812.44.4	3310.22.2	3808.00.0	4305.77.7	4803.55.5	65
66	328.53.3	826.31.1	1324.08.8	1821.86.6	2319.64.4	2817.42.2	3315.20.0	3812.97.7	4310.75.5	4808.53.3	66
67	333.51.1	831.28.8	1329.06.6	1826.84.4	2324.62.2	2822.40.0	3320.17.7	3817.95.5	4315.73.3	4813.51.1	67
68	338.48.8	836.26.6	1334.04.4	1831.82.2	2329.60.0	2827.37.7	3325.15.5	3822.93.3	4320.71.1	4818.48.8	68
69	343.46.6	841.24.4	1339.02.2	1836.80.0	2334.57.7	2832.35.5	3330.13.3	3827.91.1	4325.68.8	4823.46.6	69
70	348.44.4	846.22.2	1344.00.0	1841.77.7	2339.55.5	2837.33.3	3335.11.1	3832.88.8	4330.66.6	4828.44.4	70
71	353.42.2	851.20.0	1348.97.7	1846.75.5	2344.53.3	2842.31.1	3340.08.8	3837.86.6	4335.64.4	4833.42.2	71
72	358.40.0	856.17.7	1353.95.5	1851.73.3	2349.51.1	2847.28.8	3345.06.6	3842.84.4	4340.62.2	4838.40.0	72
73	363.37.7	861.15.5	1358.93.3	1856.71.1	2354.48.8	2852.26.6	3350.04.4	3847.82.2	4345.60.0	4843.37.7	73
74	368.35.5	866.13.3	1363.91.1	1861.68.8	2359.46.6	2857.24.4	3355.02.2	3852.80.0	4350.57.7	4848.35.5	74
75	373.33.3	871.11.1	1368.88.8	1866.66.6	2364.44.4	2862.22.2	3360.00.0	3857.77.7	4355.55.5	4853.33.3	75
76	378.31.1	876.08.8	1373.86.6	1871.64.4	2369.42.2	2867.20.0	3364.97.7	3862.75.5	4360.53.3	4858.31.1	76
77	383.28.8	881.06.6	1378.84.4	1876.62.2	2374.40.0	2872.17.7	3369.95.5	3867.73.3	4365.51.1	4863.28.8	77
78	388.26.6	886.04.4	1383.82.2	1881.60.0	2379.37.7	2877.15.5	3374.93.3	3872.71.1	4370.48.8	4868.26.6	78
79	393.24.4	891.02.2	1388.80.0	1886.57.7	2384.35.5	2882.13.3	3379.91.1	3877.68.8	4375.46.6	4873.24.4	79
80	398.22.2	896.00.0	1393.77.7	1891.55.5	2389.33.3	2887.11.1	3384.88.8	3982.66.6	4380.44.4	4878.22.2	80
81	403.20.0	900.97.7	1398.75.5	1896.53.3	2394.31.1	2892.08.8	3389.86.6	3887.64.4	4385.42.2	4883.20.0	81
82	408.17.7	905.95.5	1403.73.3	1901.51.1	2399.28.8	2897.06.6	3394.84.4	3892.62.2	4390.40.0	4888.17.7	82
83	413.15.5	910.93.3	1408.71.1	1906.48.8	2404.26.6	2902.04.4	3399.82.2	3897.60.0	4395.37.7	4893.15.5	83
84	418.13.3	915.91.1	1413.68.8	1911.46.6	2409.24.4	2907.02.2	3404.80.0	3902.57.7	4400.35.5	4898.13.3	84
85	423.11.1	920.88.8	1418.66.6	1916.44.4	2414.22.2	2912.00.0	3409.77.7	3907.55.5	4405.33.3	4903.11.1	85
86	428.08.8	925.86.6	1423.64.4	1921.42.2	2419.20.0	2916.97.7	3414.75.5	3912.53.3	4410.31.1	4908.08.8	86
87	433.06.6	930.84.4	1428.62.2	1926.40.0	2424.17.7	2921.95.5	3419.73.3	3917.51.1	4415.28.8	4913.06.6	87
88	438.04.4	935.82.2	1433.60.0	1931.37.7	2429.15.5	2926.93.3	3424.71.1	3922.48.8	4420.26.6	4918.04.4	88
89	443.02.2	940.80.0	1438.57.7	1936.35.5	2434.13.3	2931.91.1	3429.68.8	3927.46.6	4425.24.4	4923.02.2	89
90	448.00.0	945.77.7	1443.55.5	1941.33.3	2439.11.1	2936.88.8	3434.66.6	3932.44.4	4430.22.2	4928.00.0	90
91	452.97.7	950.75.5	1448.53.3	1946.31.1	2444.08.8	2941.86.6	3439.64.4	3937.42.2	4435.20.0	4932.97.7	91
92	457.95.5	955.73.3	1453.51.1	1951.28.8	2449.06.6	2946.84.4	3444.62.2	3942.40.0	4440.17.7	4937.95.5	92
93	462.93.3	960.71.1	1458.48.8	1956.26.6	2454.04.4	2951.82.2	3449.60.0	3947.37.7	4445.15.5	4942.93.3	93
94	467.91.1	965.68.8	1463.46.6	1961.24.4	2459.02.2	2956.80.0	3454.57.7	3952.35.5	4450.13.3	4947.91.1	94
95	472.88.8	970.66.6	1468.44.4	1966.22.2	2464.00.0	2961.77.7	3459.55.5	3957.33.3	4455.11.1	4952.88.8	95
96	477.86.6	975.64.4	1473.42.2	1971.20.0	2468.97.7	2966.75.5	3464.53.3	3962.31.1	4460.08.8	4957.86.6	96
97	482.84.4	980.62.2	1478.40.0	1976.17.7	2473.95.5	2971.73.3	3469.51.1	3967.28.8	4465.06.6	4962.84.4	97
98	487.82.2	985.60.0	1483.37.7	1981.15.5	2478.93.3	2976.71.1	3474.48.8	3972.26.6	4470.04.4	4967.82.2	98
99	492.80.0	990.57.7	1488.35.5	1986.13.3	2483.91.1	2981.68.8	3479.46.6	3977.24.4	4475.02.2	4972.80.0	99

s.	10	11	12	13	14	15	16	17	18	19	s.
d.	$ cts.	$ cts.	$ cts.	$ cts.	$ cts.	$ cts.	$ cts.	$ cts.	$ cts.	$ cts.	d.
0	2.48.8	2.73.7	2.98.6	3.23.5	3.48.4	3.73.3	3.98.2	4.23.1	4.48.0	4.72.8	0
1	2.50.9	2.75.8	3.00.7	3.25.6	3.50.5	3.75.4	4.00.3	4.25.1	4.50.0	4.74.9	1
2	2.53.0	2.77.9	3.02.8	3.27.7	3.52.6	3.77.4	4.02.3	4.27.2	4.52.1	4.77.0	2
3	2.55.1	2.80.0	3.04.9	3.29.7	3.54.6	3.79.5	4.04.4	4.29.3	4.54.2	4.79.1	3
4	2.57.2	2.82.0	3.06.9	3.31.8	3.56.7	3.81.6	4.06.5	4.31.4	4.56.3	4.81.2	4
5	2.59.2	2.84.1	3.09.0	3.33.9	3.58.8	3.83.7	4.08.6	4.33.5	4.58.3	4.83.2	5
6	2.61.3	2.86.2	3.11.1	3.36.0	3.60.9	3.85.8	4.10.6	4.35.5	4.60.4	4.85.3	6
7	2.63.4	2.88.3	3.13.2	3.38.1	3.62.9	3.87.8	4.12.7	4.37.6	4.62.5	4.87.4	7
8	2.65.5	2.90.4	3.15.2	3.40.1	3.65.0	3.89.9	4.14.8	4.39.7	4.64.6	4.89.5	8
9	2.67.5	2.92.4	3.17.3	3.42.2	3.67.1	3.92.0	4.16.9	4.41.8	4.66.7	4.91.5	9
10	2.69.6	2.94.5	3.19.4	3.44.3	3.69.2	3.94.1	4.19.0	4.43.8	4.68.7	4.93.6	10
11	2.71.7	2.96.6	3.21.5	3.46.4	3.71.3	3.96.1	4.21.0	4.45.9	4.70.8	4.95.7	11

$5.00₁₀₀ TO THE POUND.

12½%

PREMIUM

STERLING INTO DOLLARS AND CENTS, AND VICE-VERSA.

£	0	100	200	300	400	500	600	700	800	900	£
£	$	$	$	$	$	$	$	$	$	$	£
0	......	500	1000	1500	2000	2500	3000	3500	4000	4500	0
1	5	505	1005	1505	2005	2505	3005	3505	4005	4505	1
2	10	510	1010	1510	2010	2510	3010	3510	4010	4510	2
3	15	515	1015	1515	2015	2515	3015	3515	4015	4515	3
4	20	520	1020	1520	2020	2520	3020	3520	4020	4520	4
5	25	525	1025	1525	2025	2525	3025	3525	4025	4525	5
6	30	530	1030	1530	2030	2530	3030	3530	4030	4530	6
7	35	535	1035	1535	2035	2535	3035	3535	4035	4535	7
8	40	540	1040	1540	2040	2540	3040	3540	4040	4540	8
9	45	545	1045	1545	2045	2545	3045	3545	4045	4545	9
10	50	550	1050	1550	2050	2550	3050	3550	4050	4550	10
11	55	555	1055	1555	2055	2555	3055	3555	4055	4555	11
12	60	560	1060	1560	2060	2560	3060	3560	4060	4560	12
13	65	565	1065	1565	2065	2565	3065	3565	4065	4565	13
14	70	570	1070	1570	2070	2570	3070	3570	4070	4570	14
15	75	575	1075	1575	2075	2575	3075	3575	4075	4575	15
16	80	580	1080	1580	2080	2580	3080	3580	4080	4580	16
17	85	585	1085	1585	2085	2585	3085	3585	4085	4585	17
18	90	590	1090	1590	2090	2590	3090	3590	4090	4590	18
19	95	595	1095	1595	2095	2595	3095	3595	4095	4595	19
20	100	600	1100	1600	2100	2600	3100	3600	4100	4600	20
21	105	605	1105	1605	2105	2605	3105	3605	4105	4605	21
22	110	610	1110	1610	2110	2610	3110	3610	4110	4610	22
23	115	615	1115	1615	2115	2615	3115	3615	4115	4615	23
24	120	620	1120	1620	2120	2620	3120	3620	4120	4620	24
25	125	625	1125	1625	2125	2625	3125	3625	4125	4625	25
26	130	630	1130	1630	2130	2630	3130	3630	4130	4630	26
27	135	635	1135	1635	2135	2635	3135	3635	4135	4635	27
28	140	640	1140	1640	2140	2640	3140	3640	4140	4640	28
29	145	645	1145	1645	2145	2645	3145	3645	4145	4645	29
30	150	650	1150	1650	2150	2650	3150	3650	4150	4650	30
31	155	655	1155	1655	2155	2655	3155	3655	4155	4655	31
32	160	660	1160	1660	2160	2660	3160	3660	4160	4660	32
33	165	665	1165	1665	2165	2665	3165	3665	4165	4665	33
34	170	670	1170	1670	2170	2670	3170	3670	4170	4670	34
35	175	675	1175	1675	2175	2675	3175	3675	4175	4675	35
36	180	680	1180	1680	2180	2680	3180	3680	4180	4680	36
37	185	685	1185	1685	2185	2685	3185	3685	4185	4685	37
38	190	690	1190	1690	2190	2690	3190	3690	4190	4690	38
39	195	695	1195	1695	2195	2695	3195	3695	4195	4695	39
40	200	700	1200	1700	2200	2700	3200	3700	4200	4700	40
41	205	705	1205	1705	2205	2705	3205	3705	4205	4705	41
42	210	710	1210	1710	2210	2710	3210	3710	4210	4710	42
43	215	715	1215	1715	2215	2715	3215	3715	4215	4715	43
44	220	720	1220	1720	2220	2720	3220	3720	4220	4720	44
45	225	725	1225	1725	2225	2725	3225	3725	4225	4725	45
46	230	730	1230	1730	2230	2730	3230	3730	4230	4730	46
47	235	735	1235	1735	2235	2735	3235	3735	4235	4735	47
48	240	740	1240	1740	2240	2740	3240	3740	4240	4740	48
49	245	745	1245	1745	2245	2745	3245	3745	4245	4745	49

S.	0	1	2	3	4	5	6	7	8	9	S.
d.	$ cts.	$ cts.	$ cts.	$ cts.	$ cts.	$ cts.	$ cts.	$ cts.	$ ots.	$ cts.	d.
0	......	.25.0	.50.0	.75.0	1.00.0	1.25.0	1.50.0	1.75.0	2.00.0	2.25.0	0
1	.02 0	.27.0	.52.0	.77.0	1.02.0	1.27.0	1.52.0	1.77.0	2.02.0	2.27.0	1
2	.04.1	.29.1	.54.1	.79.1	1.04.1	1.29.1	1.54.1	1.79.1	2.04.1	2.29.1	2
3	.06.2	.31.2	.56.2	.81.2	1.06.2	1.31.2	1.56.2	1.81.2	2.06.2	2.31.2	3
4	.08 3	.33.3	.58.3	.83.3	1.08.3	1.33.3	1.58.3	1.83.3	2.08.3	2.33.3	4
5	.10.4	.35.4	.60.4	.85.4	1.10.4	1.35.4	1.60.4	1.85.4	2.10.4	2.35.4	5
6	.12.5	.37.5	.62.5	.87.5	1.12.5	1.37.5	1.62.5	1.87.5	2.12.5	2.37.5	6
7	.14.6	.39.6	.64.6	.89.6	1.14.6	1.39.6	1.64.6	1.89.6	2.14.6	2.39.6	7
8	.16.7	.41.7	.66.7	.91.7	1.16.7	1.41.7	1.66.7	1.91.7	2.16.7	2.41.7	8
9	.18.8	.43.8	.68.8	.93.8	1.18.8	1.43.8	1.68.8	1.93.8	2.18.8	2.43.8	9
10	.20.8	.45.8	.70.8	.95.8	1.20.8	1.45.8	1.70.8	1.95.8	2.20.8	2.45.8	10
11	.22.9	.47.9	.72.9	.97.9	1.22.9	1.47.9	1.72.9	1.97.9	2.22.9	2.47.9	11

$5.00 1⁄100 TO THE POUND.

12½%

PREMIUM.

STERLING INTO DOLLARS AND CENTS, AND VICE-VERSA.

£	0	100	200	300	400	500	600	700	800	900	£
£	$	$	$	$	$	$	$	$	$	$	£
50	250	750	1250	1750	2250	2750	3250	3750	4250	4750	50
51	255	755	1255	1755	2255	2755	3255	3755	4255	4755	51
52	260	760	1260	1760	2260	2760	3260	3760	4260	4760	52
53	265	765	1265	1765	2265	2765	3265	3765	4265	4765	53
54	270	770	1270	1770	2270	2770	3270	3770	4270	4770	54
55	275	775	1275	1775	2275	2775	3275	3775	4275	4775	55
56	280	780	1280	1780	2280	2780	3280	3780	4280	4780	56
57	285	785	1285	1785	2285	2785	3285	3785	4285	4785	57
58	290	790	1290	1790	2290	2790	3290	3790	4290	4790	58
59	295	795	1295	1795	2295	2795	3295	3795	4295	4795	59
60	300	800	1300	1800	2300	2800	3300	3800	4300	4800	60
61	305	805	1305	1805	2305	2805	3305	3805	4305	4805	61
62	310	810	1310	1810	2310	2810	3310	3810	4310	4810	62
63	315	815	1315	1815	2315	2815	3315	3815	4315	4815	63
64	320	820	1320	1820	2320	2820	3320	3820	4320	4820	64
65	325	825	1325	1825	2325	2825	3325	3825	4325	4825	65
66	330	830	1330	1830	2330	2830	3330	3830	4330	4830	66
67	335	835	1335	1835	2335	2835	3335	3835	4335	4835	67
68	340	840	1340	1840	2340	2840	3340	3840	4340	4840	68
69	345	845	1345	1845	2345	2845	3345	3845	4345	4845	69
70	350	850	1350	1850	2350	2850	3350	3850	4350	4850	70
71	355	855	1355	1855	2355	2855	3355	3855	4355	4855	71
72	360	860	1360	1860	2360	2860	3360	3860	4360	4860	72
73	365	865	1365	1865	2365	2865	3365	3865	4365	4865	73
74	370	870	1370	1870	2370	2870	3370	3870	4370	4870	74
75	375	875	1375	1875	2375	2875	3375	3875	4375	4875	75
76	380	880	1380	1880	2380	2880	3380	3880	4380	4880	76
77	385	885	1385	1885	2385	2885	3385	3885	4385	4885	77
78	390	890	1390	1890	2390	2890	3390	3890	4390	4890	78
79	395	895	1395	1895	2395	2895	3395	3895	4395	4895	79
80	400	900	1400	1900	2400	2900	3400	3900	4400	4900	80
81	405	905	1405	1905	2405	2905	3405	3905	4405	4905	81
82	410	910	1410	1910	2410	2910	3410	3910	4410	4910	82
83	415	915	1415	1915	2415	2915	3415	3915	4415	4915	83
84	420	920	1420	1920	2420	2920	3420	3920	4420	4920	84
85	425	925	1425	1925	2425	2925	3425	3925	4425	4925	85
86	430	930	1430	1930	2430	2930	3430	3930	4430	4930	86
87	435	935	1435	1935	2435	2935	3435	3935	4435	4935	87
88	440	940	1440	1940	2440	2940	3440	3940	4440	4940	88
89	445	945	1445	1945	2445	2945	3445	3945	4445	4945	89
90	450	950	1450	1950	2450	2950	3450	3950	4450	4950	90
91	455	955	1455	1955	2455	2955	3455	3955	4455	4955	91
92	460	960	1460	1960	2460	2960	3460	3960	4460	4960	92
93	465	965	1465	1965	2465	2965	3465	3965	4465	4965	93
94	470	970	1470	1970	2470	2970	3470	3970	4470	4970	94
95	475	975	1475	1975	2475	2975	3475	3975	4475	4975	95
96	480	980	1480	1980	2480	2980	3480	3980	4480	4980	96
97	485	985	1485	1985	2485	2985	3485	3985	4485	4985	97
98	490	990	1490	1990	2490	2990	3490	3990	4490	4990	98
99	495	995	1495	1995	2495	2995	3495	3995	4495	4995	99

S.	10	11	12	13	14	15	16	17	18	19	S.
d.	$ cts.	$ cts.	$ cts.	$ cts.	$ cts.	$ cts.	$ cts.	$ cts.	$ cts.	$ cts.	d.
0	2.50.0	2.75.0	3.00.0	3.25.0	3.50.0	3.75.0	4.00.0	4.25.0	4.50.0	4.75.0	0
1	2.52.0	2.77.0	3.02.0	3.27.0	3.52.0	3.77.0	4.02.0	4.27.0	4.52.0	4.77.0	1
2	2.54.1	2.79.1	3.04.1	3.29.1	3.54.1	3.79.1	4.04.1	4.29.1	4.54.1	4.79.1	2
3	2.56.2	2.81.2	3.06.2	3.31.2	3.56.2	3.81.2	4.06.2	4.31.2	4.56.2	4.81.2	3
4	2.58.3	2.83.3	3.08.3	3.33.3	3.58.3	3.83.3	4.08.3	4.33.3	4.58.3	4.83.3	4
5	2.60.4	2.85.4	3.10.4	3.35.4	3.60.4	3.85.4	4.10.4	4.35.4	4.60.4	4.85.4	5
6	2.62.5	2.87.5	3.12.5	3.37.5	3.62.5	3.87.5	4.12.5	4.37.5	4.62.5	4.87.5	6
7	2.64.6	2.89.6	3.14.6	3.39.6	3.64.6	3.89.6	4.14.6	4.39.6	4.64.6	4.89.6	7
8	2.66.7	2.91.7	3.16.7	3.41.7	3.66.7	3.91.7	4.16.7	4.41.7	4.66.7	4.91.7	8
9	2.68.8	2.93.8	3.18.8	3.43.8	3.68.8	3.93.8	4.18.8	4.43.8	4.68.8	4.93.8	9
10	2.70.8	2.95.8	3.20.8	3.45.8	3.70.8	3.95.8	4.20.8	4.45.8	4.70.8	4.95.8	10
11	2.72.9	2.97.9	3.22.9	3.47.9	3.72.9	3.97.9	4.22.9	4.47.9	4.72.9	4.97.9	11

EXPLANATIONS TO PART II.

STERLING EXCHANGE TABLES,

ADVANCING BY SIXTEENTHS.

The following Sterling Exchange Tables have been compiled in the belief that they will fill a long existing want felt by bankers and business men who are dealing in large amounts. They are put in what is believed to be the most concise and convenient form, each sheet embracing eight rates, which is at once a great saving of space; and the range of quotations, viz., $5\frac{9}{16}$ to $12\frac{1}{2}$, will be found amply sufficient for all practical purposes.

For conversion of Sterling into Dollars and Cents, the arrangement of the form is from £100 to £10000 (£100 to £5000 on left-hand, and £5100 to £10000 on right-hand side of each sheet); but by a removal of the decimal points the amounts can be readily adjusted to units and tens, or hundreds if desired; thus, £5100 at $8\frac{9}{16} = \$24607.50$, £510 = $\$2460.75$, £51 = $\$246.07$, and £51000 = $\$246075.00$. The arrangement of shillings and pence is from 1 to 15s. on left-hand, and the balance of shillings and pence on right-hand side at foot of each column.

Example: To find value of £7561 17s. 2d. at $8\frac{15}{16}$ (see pp. 94 and 95):—

£7500	=	$36312 50
61	=	295.34
17s.	=	4.11
2d.	=	.01
Result: £7561 17s. 2d.	=	$36611.99

A short table is added for converting Dollars and Cents into Sterling, in which the range of quotations is shortened as much as possible, but is believed to be adequate to the requirements of these tables, which are not so much used. The form is necessarily different from that in the preceding tables, but will cover any amount from $1 to $10000. The amounts advance by units from $1 to $10, and by tens from $10 to $10000, giving up to $500 on left-hand side and from $510 to $10000 on right-hand side of each sheet. It will be observed that while there are eight rates to a page for Dollars, only one rate is given for Cents (viz., 1c. to 49c. on left-hand and 50c. to 99c. on right-hand side of each sheet), as a variation of $\frac{1}{2}\%$ in the rate will not make a difference of over $\frac{1}{4}$ of a penny on $1.

Example: To find value of $897.33 at $9\frac{9}{16}$ (see pp. 112 and 113):—

$890	=	£183	3s.	$9\frac{3}{4}$d.
7	=	1	8s.	$9\frac{3}{4}$d.
.33	=		1s.	$4\frac{1}{4}$d.
Result: $897.33	=	£184	13s.	$11\frac{3}{4}$d.

STERLING EXCHANGE TABLES.

5 9/16 to 6%

STERLING INTO DOLLARS AND CENTS.

Stg. £	5 1/16 $ cts.	5 5/8 $ cts.	5 11/16 $ cts.	5 3/4 $ cts.	5 13/16 $ cts.	5 7/8 $ cts.	5 15/16 $ cts.	6 $ cts.	Stg. £
100	469.16.6	469.44.4	469.72.2	470.00.0	470.27.7	470.55.5	470.83.3	471.11.1	100
200	938.33.3	938.88.8	939.44.4	940.00.0	940.55.5	941.11.1	941.66.6	942.22.2	200
300	1407.50.0	1408.33.3	1409.16.6	1410.00.0	1410.83.3	1411.66.6	1412.50.0	1413.33.3	300
400	1876.66.6	1877.77.7	1878.88.8	1880.00.0	1881.11.1	1882.22.2	1883.33.3	1884.44.4	400
500	2345.83.3	2347.22.2	2348.61.1	2350.00.0	2351.38.8	2352.77.7	2354.16.6	2355.55.5	500
600	2815.00.0	2816.66.6	2818.33.3	2820.00.0	2821.66.6	2823.33.3	2825.00.0	2826.66.6	600
700	3284.16.6	3286.11.1	3288.05.5	3290.00.0	3291.94.4	3293.88.8	3295.83.3	3297.77.7	700
800	3753.33.3	3755.55.5	3757.77.7	3760.00.0	3762.22.2	3764.44.4	3766.66.6	3768.88.8	800
900	4222.50.0	4225.00.0	4227.50.0	4230.00.0	4232.50.0	4235.00.0	4237.50.0	4240.00.0	900
1000	4691.66.6	4694.44.4	4697.22.2	4700.00.0	4702.77.7	4705.55.5	4708.33.3	4711.11.1	1000
1100	5160.83.3	5163.88.8	5166.94.4	5170.00.0	5173.05.5	5176.11.1	5179.16.6	5182.22.2	1100
1200	5630.00.0	5633.33.3	5636.66.6	5640.00.0	5643.33.3	5646.66.6	5650.00.0	5653.33.3	1200
1300	6099.16.6	6102.77.7	6106.38.8	6110.00.0	6113.61.1	6117.22.2	6120.83.3	6124.44.4	1300
1400	6568.33.3	6572.22.2	6576.11.1	6580.00.0	6583.88.8	6587.77.7	6591.66.6	6595.55.5	1400
1500	7037.50.0	7041.66.6	7045.83.3	7050.00.0	7054.16.6	7058.33.3	7062.50.0	7066.66.6	1500
1600	7506.66.6	7511.11.1	7515.55.5	7520.00.0	7524.44.4	7528.88.8	7533.33.3	7537.77.7	1600
1700	7975.83.3	7980.55.5	7985.27.7	7990.00.0	7994.72.2	7999.44.4	8004.16.6	8008.88.8	1700
1800	8445.00.0	8450.00.0	8455.00.0	8460.00.0	8465.00.0	8470.00.0	8475.00.0	8480.00.0	1800
1900	8914.16.6	8919.44.4	8924.72.2	8930.00.0	8935.27.7	8940.55.5	8945.83.3	8951.11.1	1900
2000	9383.33.3	9388.88.8	9394.44.4	9400.00.0	9405.55.5	9411.11.1	9416.66.6	9422.22.2	2000
2100	9852.50.0	9858.33.3	9864.16.6	9870.00.0	9875.83.3	9881.66.6	9887.50.0	9893.33.3	2100
2200	10321.66.6	10327.77.7	10333.88.8	10340.00.0	10346.11.1	10352.22.2	10358.33.3	10364.44.4	2200
2300	10790.83.3	10797.22.2	10803.61.1	10810.00.0	10816.38.8	10822.77.7	10829.16.6	10835.55.5	2300
2400	11260.00.0	11266.66.6	11273.33.3	11280.00.0	11286.66.6	11293.33.3	11300.00.0	11306.66.6	2400
2500	11729.16.6	11736.11.1	11743.05.5	11750.00.0	11756.94.4	11763.88.8	11770.83.3	11777.77.7	2500
2600	12198.33.3	12205.55.5	12212.77.7	12220.00.0	12227.22.2	12234.44.4	12241.66.6	12248.88.8	2600
2700	12667.50.0	12675.00.0	12682.50.0	12690.00.0	12697.50.0	12705.00.0	12712.50.0	12720.00.0	2700
2800	13136.66.6	13144.44.4	13152.22.2	13160.00.0	13167.77.7	13175.55.5	13183.33.3	13191.11.1	2800
2900	13605.83.3	13613.88.8	13621.94.4	13630.00.0	13638.05.5	13646.11.1	13654.16.6	13662.22.2	2900
3000	14075.00.0	14083.33.3	14091.66.6	14100.00.0	14108.33.3	14116.66.6	14125.00.0	14133.33.3	3000
3100	14544.16.6	14552.77.7	14561.38.8	14570.00.0	14578.61.1	14587.22.2	14595.83.3	14604.44.4	3100
3200	15013.33.3	15022.22.2	15031.11.1	15040.00.0	15048.88.8	15057.77.7	15066.66.6	15075.55.5	3200
3300	15482.50.0	15491.66.6	15500.88.3	15510.00.0	15519.16.6	15528.33.3	15537.50.0	15546.66.6	3300
3400	15951.66.6	15961.11.1	15970.55.5	15980.00.0	15989.44.4	15998.88.8	16008.33.3	16017.77.7	3400
3500	16420.83.3	16430.55.5	16440.27.7	16450.00.0	16459.72.2	16469.44.4	16479.16.6	16488.88.8	3500
3600	16890.00.0	16900.00.0	16910.00.0	16920.00.0	16930.00.0	16940.00.0	16950.00.0	16960.00.0	3600
3700	17359.16.6	17369.44.4	17379.72.2	17390.00.0	17400.27.7	17410.55.5	17420.83.3	17431.11.1	3700
3800	17828.33.3	17838.88.8	17849.44.4	17860.00.0	17870.55.5	17881.11.1	17891.66.6	17902.22.2	3800
3900	18297.50.0	18308.33.3	18319.16.6	18330.00.0	18340.83.3	18351.66.6	18362.50.0	18373.33.3	3900
4000	18766.66.6	18777.77.7	18788.88.8	18800.00.0	18811.11.1	18822.22.2	18833.33.3	18844.44.4	4000
4100	19235.83.3	19247.22.2	19258.61.1	19270.00.0	19281.38.8	19292.77.7	19304.16.6	19315.55.5	4100
4200	19705.00.0	19716.66.6	19728.33.3	19740.00.0	19751.66.6	19763.33.3	19775.00.0	19786.66.6	4200
4300	20174.16.6	20186.11.1	20196.05.5	20210.00.0	20221.94.4	20233.88.8	20245.83.3	20257.77.7	4300
4400	20643.33.3	20655.55.5	20667.77.7	20680.00.0	20692.22.2	20704.44.4	20716.66.6	20728.88.8	4400
4500	21112.50.0	21125.00.0	21137.50.0	21150.00.0	21162.50.0	21175.00.0	21187.50.0	21200.00.0	4500
4600	21581.66.6	21594.44.4	21607.22.2	21620.00.0	21632.77.7	21645.55.5	21658.33.3	21671.11.1	4600
4700	22050.83.3	22063.88.8	22076.94.4	22090.00.0	22103.05.5	22116.11.1	22129.16.6	22142.22.2	4700
4800	22520.00.0	22533.33.3	22546.66.6	22560.00.0	22573.33.3	22586.66.6	22600.00.0	22613.33.3	4800
4900	22989.16.6	23002.77.7	23016.38.8	23030.00.0	23043.61.1	23057.22.2	23070.83.3	23084.44.4	4900
5000	23458.33.3	23472.22.2	23486.11.1	23500.00.0	23513.88.8	23527.77.7	23541.66.6	23555.55.5	5000

s.	$ cts.	$ cts.	$ cts.	$ cts.	$ cts.	$ cts.	$ cts.	$ cts.	s.
1	.23.4	.23.5	.23.5	.23.5	.23.5	.23.5	.23.5	.23.5	1
2	.46.9	.46.9	.47.0	.47.0	.47.0	.47.0	.47.1	.47.1	2
3	.70.4	.70.4	.70.4	.70.5	.70.5	.70.6	.70.6	.70.7	3
4	.93.8	.93.9	.93.9	.94.0	.94.0	.94.1	.94.2	.94.2	4
5	1.17.3	1.17.4	1.17.4	1.17.5	1.17.6	1.17.6	1.17.7	1.17.8	5
6	1.40.7	1.40.8	1.40.9	1.41.0	1.41.1	1.41.2	1.41.2	1.41.3	6
7	1.64.2	1.64.3	1.64.4	1.64.5	1.64.6	1.64.7	1.64.8	1.64.9	7
8	1.87.7	1.87.8	1.87.9	1.88.0	1.88.1	1.88.2	1.88.3	1.88.4	8
9	2.11.1	2.11.2	2.11.4	2.11.5	2.11.6	2.11.7	2.11.9	2.12.0	9
10	2.34.6	2.34.7	2.34.9	2.35.0	2.35.1	2.35.3	2.35.4	2.35.5	10
11	2.58.0	2.58.2	2.58.3	2.58.5	2.58.6	2.58.8	2.58.9	2.59.1	11
12	2.81.5	2.81.7	2.81.8	2.82.0	2.82.2	2.82.3	2.82.5	2.82.7	12
13	3.04.9	3.05.1	3.05.3	3.05.5	3.05.7	3.05.9	3.06.0	3.06.2	13
14	3.28.4	3.28.6	3.28.8	3.29.0	3.29.2	3.29.4	3.29.6	3.29.8	14
15	3.51.9	3.52.1	3.52.3	3.52.5	3.52.7	3.52.9	3.53.1	3.53.3	15

STERLING EXCHANGE TABLES.

5 9/16 TO 6 %

STERLING INTO DOLLARS AND CENTS.

Stg. £	5 9/16	5 5/8	5 11/16	5 3/4	5 13/16	5 7/8	5 15/16	6	Stg. £
	$ cts.	$ cts.	$ cts.	$ cts.	$ cts.	$ cts.	$ cts.	$ cts.	
5100	23927.50.0	23941.66.6	23955.83.3	23970.00.0	23984.16.6	23998.33.3	21012.50.0	24026.66.6	5100
5200	24396.66.6	24411.11.1	24425.55.5	24440.00.0	24454.44.4	24468.88.8	24483.33.3	24497.77.7	5200
5300	24865.83.3	24880.55.5	24895.27.7	24910.00.0	24924.72.2	24939.44.4	24954.16.6	24968.88.8	5300
5400	25335.00.0	25350.00.0	25365.00.0	25380.00.0	25395.00.0	25410.00.0	25425.00.0	25440.00.0	5400
5500	25804.16.6	25819.44.4	25834.72.2	25850.00.0	25865.27.7	25880.55.5	25895.83.3	25911.11.1	5500
5600	26273.33.3	26288.88.8	26304.44.4	26320.00.0	26335.55.5	26351.11.1	26366.66.6	26382.22.2	5600
5700	26742.50.0	26758.33.3	26774.16.6	26790.00.0	26805.83.3	26821.66.6	26837.50.0	26853.33.3	5700
5800	27211.66.6	27227.77.7	27243.88.8	27260.00.0	27276.11.1	27292.22.2	27308.33.3	27324.44.4	5800
5900	27680.83.3	27697.22.2	27713.61.1	27730.00.0	27746.38.8	27762.77.7	27779.16.6	27795.55.5	5900
6000	28150.00.0	28166.66.6	28183.33.3	28200.00.0	28216.66.6	28233.33.3	28250.00.0	28266.66.6	6000
6100	28619.16.6	28636.11.1	28653.05.5	28670.00.0	28686.94.4	28703.88.8	28720.83.3	28737.77.7	6100
6200	29088.33.3	29105.55.5	29122.77.7	29140.00.0	29157.22.2	29174.44.4	29191.66.6	29208.88.8	6200
6300	29557.50.0	29575.00.0	29592.50.0	29610.00.0	29627.50.0	29645.00.0	29662.50.0	29680.00.0	6300
6400	30026.66.6	30044.44.4	30062.22.2	30080.00.0	30097.77.7	30115.55.5	30133.33.3	30151.11.1	6400
6500	30495.83.3	30513.88.8	30531.94.4	30550.00.0	30568.05.5	30586.11.1	30604.16.6	30622.22.2	6500
6600	30965.00.0	30983.33.3	31001.66.6	31020.00.0	31038.33.3	31056.66.6	31075.00.0	31093.33.3	6600
6700	31434.16.6	31452.77.7	31471.38.8	31490.00.0	31508.61.1	31527.22.2	31545.83.3	31564.44.4	6700
6800	31903.33.3	31922.22.2	31941.11.1	31960.00.0	31978.88.8	31997.77.7	32016.66.6	32035.55.5	6800
6900	32372.50.0	32391.66.6	32410.83.3	32430.00.0	32449.16.6	32468.33.3	32487.50.0	32506.66.6	6900
7000	32841.66.6	32861.11.1	32880.55.5	32900.00.0	32919.44.4	32938.88.8	32958.33.3	32977.77.7	7000
7100	33310.83.3	33330.55.5	33350.27.7	33370.00.0	33389.72.2	33409.44.4	33429.16.6	33448.88.8	7100
7200	33780.00.0	33800.00.0	33820.00.0	33840.00.0	33860.00.0	33880.00.0	33900.00.0	33920.00.0	7200
7300	34249.16.6	34269.44.4	34289.72.2	34310.00.0	34330.27.7	34350.55.5	34370.83.3	34391.11.1	7300
7400	34718.33.3	34738.88.8	34759.44.4	34780.00.0	34800.55.5	34821.11.1	34841.66.6	34862.22.2	7400
7500	35187.50.0	35208.33.3	35229.16.6	35250.00.0	35270.83.3	35291.66.6	35312.50.0	35333.33.3	7500
7600	35656.66.6	35677.77.7	35698.88.8	35720.00.0	35741.11.1	35762.22.2	35783.33.3	35804.44.4	7600
7700	36125.83.3	36147.22.2	36168.61.1	36190.00.0	36211.38.8	36232.77.7	36254.16.6	36275.55.5	7700
7800	36595.00.0	36616.66.6	36638.33.3	36660.00.0	36681.66.6	36703.33.3	36725.00.0	36746.66.6	7800
7900	37064.16.6	37086.11.1	37108.05.5	37130.00.0	37151.94.4	37173.88.8	37195.83.3	37217.77.7	7900
8000	37533.33.3	37555.55.5	37577.77.7	37690.00.0	37622.22.2	37614.44.4	37666.66.6	37688.88.8	8000
8100	38002.50.0	38025.00.0	38047.50.0	38070.00.0	38092.50.0	38115.00.0	38137.50.0	38160.00.0	8100
8200	38471.66.6	38494.44.4	38517.22.2	38540.00.0	38562.77.7	38585.55.5	38608.33.3	38631.11.1	8200
8300	38940.83.3	38963.88.8	38986.94.4	39010.00.0	39033.05.5	39056.11.1	39079.16.6	39102.22.2	8300
8400	39410.00.0	39433.33.3	39456.66.6	39480.00.0	39503.33.3	39526.66.6	39550.00.0	39573.33.3	8400
8500	39879.16.6	39902.77.7	39926.38.8	39950.00.0	39973.61.1	39997.22.2	40020.83.3	40044.44.4	8500
8600	40348.33.3	40372.22.2	40396.11.1	40420.00.0	40443.88.8	40467.77.7	40491.66.6	40515.55.5	8600
8700	40817.50.0	40841.66.6	40865.83.3	40890.00.0	40914.16.6	40938.33.3	40962.50.0	40986.66.6	8700
8800	41286.66.6	41311.11.1	41335.55.5	41360.00.0	41384.44.4	41408.88.8	41433.33.3	41457.77.7	8800
8900	41755.83.3	41780.55.5	41805.27.7	41830.00.0	41854.72.2	41879.44.4	41904.16.6	41928.88.8	8900
9000	42225.00.0	42250.00.0	42275.00.0	42300.00.0	42325.00.0	42350.00.0	42375.00.0	42400.00.0	9000
9100	42694.16.6	42719.44.4	42744.72.2	42770.00.0	42795.27.7	42820.55.5	42845.83.3	42871.11.1	9100
9200	43163.33.3	43188.88.8	43214.44.4	43240.00.0	43265.55.5	43291.11.1	43316.66.6	43342.22.2	9200
9300	43632.50.0	43658.33.3	43684.16.6	43710.00.0	43735.83.3	43761.66.6	43787.50.0	43813.33.3	9300
9400	44101.66.6	44127.77.7	44153.88.8	44180.00.0	44206.11.1	44232.22.2	44258.33.3	44284.44.4	9400
9500	44570.83.3	44597.22.2	44623.61.1	44650.00.0	44676.38.8	44702.77.7	44729.16.6	44755.55.5	9500
9600	45040.00.0	45066.66.6	45093.33.3	45120.00.0	45146.66.6	45173.33.3	45200.00.0	45226.66.6	9600
9700	45509.16.6	45536.11.1	45563.05.5	45590.00.0	45616.94.4	45643.88.8	45670.83.3	45697.77.7	9700
9800	45978.33.3	46005.55.5	46032.77.7	46060.00.0	46087.22.2	46114.44.4	46141.66.6	46168.88.8	9800
9900	46447.50.0	46475.00.0	46502.50.0	46530.00.0	46557.50.0	46585.00.0	46612.50.0	46640.00.0	9900
10000	46916.66.6	46944.44.4	46972.22.2	47000.00.0	47027.77.7	47055.55.5	47083.33.3	47111.11.1	10000

s.	$ cts.	$ cts.	$ cts.	$ cts.	$ cts.	$ cts.	$ cts.	$ cts.	s.
16	3.75.3	3.75.5	3.75.8	3.76.0	3.76.2	3.76.4	3.76.7	3.76.9	16
17	3.98.8	3.99.0	3.99.3	3.99.5	3.99.7	4.00.0	4.00.2	4.00.4	17
18	4.22.2	4.22.5	4.22.7	4.23.0	4.23.2	4.23.5	4.23.7	4.24.0	18
19	4.45.7	4.46.0	4.46.2	4.46.5	4.46.8	4.47.0	4.47.3	4.47.5	19
1d	.01.9	.01.9	.01.9	.01.9	.01.9	.02.0	.02.0	.02.0	1d
2	.03.9	.03.9	.05.9	.03.9	.03.9	.03.9	.03.9	.03.9	2
3	.05.9	.05.9	.05.9	.05.9	.05.9	.05.9	.05.9	.05.9	3
4	.07.8	.07.8	.07.8	.07.8	.07.8	.07.8	.07.8	.07.8	4
5	.09.8	.09.8	.09.8	.09.8	.09.8	.09.8	.09.8	.09.8	5
6	.11.7	.11.7	.11.7	.11.7	.11.7	.11.8	.11 8	.11.8	6
7	.13.7	.13.7	.13.7	.13.7	.13.7	.13.7	.13.7	.13.7	7
8	.15.6	.15.6	.15.6	.15.7	.15.7	.15.7	.15.7	.15.7	8
9	.17.6	.17.6	.17.6	.17.6	.17.6	.17.6	.17.6	.17.7	9
10	.19.5	.19.6	.19.6	.19.6	.19.6	.19.6	.19.6	.19.6	10
11	.21.5	.21.5	.21.5	.21.5	.21.5	.21.6	.21.6	.21.6	11

STERLING EXCHANGE TABLES.

$6\frac{1}{16}$ TO $6\frac{1}{2}$ %

STERLING INTO DOLLARS AND CENTS.

Stg. £	$6\frac{1}{16}$	$6\frac{1}{8}$	$6\frac{3}{16}$	$6\frac{1}{4}$	$6\frac{5}{16}$	$6\frac{3}{8}$	$6\frac{7}{16}$	$6\frac{1}{2}$	Stg. £
	$ cts.	$ cts.	$ cts.	$ cts.	$ cts.	$ cts.	$ cts.	$ cts.	
100	471.38.8	471.66.6	471.94.4	472.22.2	472.50.0	472.77.7	473.05.5	473.33.3	100
200	942.77.7	943.33.3	943.88.8	944.44.4	945.00.0	945.55.5	946.11.1	946.66.6	200
300	1414.16.6	1415.00.0	1415.83.3	1416.66.6	1417.50.0	1418.33.3	1419.16.6	1420.00.0	300
400	1885.55.5	1886.66.6	1887.77.7	1888.88.8	1890.00.0	1891.11.1	1892.22.2	1893.33.3	400
500	2356.94.4	2358.33.3	2359.72.2	2361.11.1	2362.50.0	2363.88.8	2365.27.7	2366.66.6	500
600	2828.33.3	2830.00.0	2831.66.6	2833.33.3	2835.00.0	2836.66.6	2838.33.3	2840.00.0	600
700	3299.72.2	3301.66.6	3303.61.1	3305.55.5	3307.50.0	3309.44.4	3311.38.8	3313.33.3	700
800	3771.11.1	3773.33.3	3775.55.5	3777.77.7	3780.00.0	3782.22.2	3784.44.4	3786.66.6	800
900	4242.50.0	4245.00.0	4247.50.0	4250.00.0	4252.50.0	4255.00.0	4257.50.0	4260.00.0	900
1000	4713.88.8	4716.66.6	4719.44.4	4722.22.2	4725.00.0	4727.77.7	4730.55.5	4733.33.3	1000
1100	5185.27.7	5188.33.3	5191.38.8	5194.44.4	5197.50.0	5200.55.5	5203.61.1	5206.66.6	1100
1200	5656.66.6	5660.00.0	5663.33.3	5666.66.6	5670.00.0	5673.33.3	5676.66.6	5680.00.0	1200
1300	6128.05.5	6131.66.6	6135.27.7	6138.88.8	6142.50.0	6146.11.1	6149.72.2	6153.33.3	1300
1400	6599.44.4	6603.33.3	6607.22.2	6611.11.1	6615.00.0	6618.88.8	6622.77.7	6626.66.6	1400
1500	7070.83.3	7075.00.0	7079.16.6	7083.33.3	7087.50.0	7091.66.6	7095.83.3	7100.00.0	1500
1600	7542.22.2	7546.66.6	7551.11.1	7555.55.5	7560.00.0	7564.44.4	7568.88.8	7573.33.3	1600
1700	8013.61.1	8018.33.3	8023.05.5	8027.77.7	8032.50.0	8037.22.2	8041.94.4	8046.66.6	1700
1800	8485.00.0	8490.00.0	8495.00.0	8500.00.0	8505.00.0	8510.00.0	8515.00.0	8520.00.0	1800
1900	8956.38.8	8961.66.6	8966.94.4	8972.22.2	8977.50.0	8982.77.7	8988.05.5	8993.33.3	1900
2000	9427.77.7	9433.33.3	9438.88.8	9444.44.4	9450.00.0	9455.55.5	9461.11.1	9466.66.6	2000
2100	9899.16.6	9905.00.0	9910.83.3	9916.66.6	9922.50.0	9928.33.3	9934.16.6	9940.00.0	2100
2200	10370.55.5	10376.66.6	10382.77.7	10388.88.8	10395.00.0	10401.11.1	10407.22.2	10413.33.3	2200
2300	10841.94.4	10848.33.3	10854.72.2	10861.11.1	10867.50.0	10873.88.8	10880.27.7	10886.66.6	2300
2400	11313.33.3	11320.00.0	11326.66.6	11333.33.3	11340.00.0	11346.66.6	11353.33.3	11360.00.0	2400
2500	11784.72.2	11791.66.6	11798.61.1	11805.55.5	11812.50.0	11819.44.4	11826.38.8	11833.33.3	2500
2600	12256.11.1	12263.33.3	12270.55.5	12277.77.7	12285.00.0	12292.22.2	12299.44.4	12306.66.6	2600
2700	12727.50.0	12735.00.0	12742.50.0	12750.00.0	12757.50.0	12765.00.0	12772.50.0	12780.00.0	2700
2800	13198.88.8	13206.66.6	13214.44.4	13222.22.2	13230.00.0	13237.77.7	13245.55.5	13253.33.3	2800
2900	13670.27.7	13678.33.3	13686.38.8	13694.44.4	13702.50.0	13710.55.5	13718.61.1	13726.66.6	2900
3000	14141.66.6	14150.00.0	14158.33.3	14166.66.6	14175.00.0	14183.33.3	14191.66.6	14200.00.0	3000
3100	14613.05.5	14621.66.6	14630.27.7	14638.88.8	14647.50.0	14656.11.1	14664.72.2	14673.33.3	3100
3200	15084.44.4	15093.33.3	15102.22.2	15111.11.1	15120.00.0	15128.88.8	15137.77.7	15146.66.6	3200
3300	15555.83.3	15565.00.0	15574.16.6	15583.33.3	15592.50.0	15601.66.6	15610.83.3	15620.00.0	3300
3400	16027.22.2	16036.66.6	16046.11.1	16055.55.5	16065.00.0	16074.44.4	16083.88.8	16093.33.3	3400
3500	16498.61.1	16508.33.3	16518.05.5	16527.77.7	16537.50.0	16547.22.2	16556.94.4	16566.66.6	3500
3600	16970.00.0	16980.00.0	16990.00.0	17000.00.0	17010.00.0	17020.00.0	17030.00.0	17040.00.0	3600
3700	17441.38.8	17451.66.6	17461.94.4	17472.22.2	17482.50.0	17492.77.7	17503.05.5	17513.33.3	3700
3800	17912.77.7	17923.33.3	17933.88.8	17944.44.4	17955.00.0	17965.55.5	17976.11.1	17986.66.6	3800
3900	18384.16.6	18395.00.0	18405.83.3	18416.66.6	18427.50.0	18438.33.3	18449.16.6	18460.00.0	3900
4000	18855.55.5	18866.66.6	18877.77.7	18888.88.8	18900.00.0	18911.11.1	18922.22.2	18933.33.3	4000
4100	19326.94.4	19338.33.3	19349.72.2	19361.11.1	19372.50.0	19383.88.8	19395.27.7	19406.66.6	4100
4200	19798.33.3	19810.00.0	19821.66.6	19833.33.3	19845.00.0	19856.66.6	19868.33.3	19880.00.0	4200
4300	20269.72.2	20281.66.6	20293.61.1	20305.55.5	20317.50.0	20329.44.4	20341.38.8	20353.33.3	4300
4400	20741.11.1	20753.33.3	20765.55.5	20777.77.7	20790.00.0	20802.22.2	20814.44.4	20826.66.6	4400
4500	21212.50.0	21225.00.0	21237.50.0	21250.00.0	21262.50.0	21275.00.0	21287.50.0	21300.00.0	4500
4600	21683.88.8	21696.66.6	21709.44.4	21722.22.2	21735.00.0	21747.77.7	21760.55.5	21773.33.3	4600
4700	22155.27.7	22168.33.3	22181.38.8	22194.44.4	22207.50.0	22220.55.5	22233.61.1	22246.66.6	4700
4800	22626.66.6	22640.00.0	22653.33.3	22666.66.6	22680.00.0	22693.33.3	22706.66.6	22720.00.0	4800
4900	23098.05.5	23111.66.6	23125.27.7	23138.88.8	23152.50.0	23166.11.1	23179.72.2	23193.33.3	4900
5000	23569.44.4	23583.33.3	23597.22.2	23611.11.1	23625.00.0	23638.88.8	23652.77.7	23666.66.6	5000

s.	$ cts.	$ cts.	$ cts.	$ cts.	$ cts.	$ cts.	$ cts.	$ cts.	s.
1	.23.6	.23.6	.23.6	.23.6	.23.6	.23.6	.23.6	.23.7	1
2	.47.1	.47.2	.47.2	.47.2	.47.2	.47.3	.47.3	.47.3	2
3	.70.7	.70.7	.70.8	.70.8	.70.9	.70.9	.70.9	.71.0	3
4	.94.3	.94.3	.94.4	.94.4	.94.5	.94.5	.94.6	.94.7	4
5	1.17.8	1.17.9	1.18.0	1.18.0	1.18.1	1.18.2	1.18.3	1.18.3	5
6	1.41.4	1.41.5	1.41.6	1.41.7	1.41.7	1.41.8	1.41.9	1.42.0	6
7	1.65.0	1.65.1	1.65.2	1.65.3	1.65.4	1.65.5	1.65.6	1.65.7	7
8	1.88.5	1.88.7	1.88.8	1.88.9	1.89.0	1.89.1	1.89.2	1.89.3	8
9	2.12.1	2.12.2	2.12.4	2.12.5	2.12.6	2.12.7	2.12.9	2.13.0	9
10	2.35.7	2.35.8	2.36.0	2.36.1	2.36.2	2.36.4	2.36.5	2.36.7	10
11	2.59.2	2.59.4	2.59.6	2.59.7	2.59.9	2.60.0	2.60.2	2.60.3	11
12	2.82.8	2.83.0	2.83.2	2.83.3	2.83.5	2.83.7	2.83.8	2.84.0	12
13	3.06.4	3.06.6	3.06.8	3.06.9	3.07.1	3.07.3	3.07.5	3.07.7	13
14	3.30.0	3.30.2	3.30.4	3.30.5	3.30.7	3.30.9	3.31.1	3.31.3	14
15	3.53.5	3.53.7	3.53.9	3.54.1	3.54.4	3.54.6	3.54.8	3.55.0	15

STERLING EXCHANGE TABLES.

$6\frac{1}{16}$ TO $6\frac{1}{2}$ %

STERLING INTO DOLLARS AND CENTS.

Stg.	$6\frac{1}{16}$	$6\frac{1}{8}$	$6\frac{3}{16}$	$6\frac{1}{4}$	$6\frac{5}{16}$	$6\frac{3}{8}$	$6\frac{7}{16}$	$6\frac{1}{2}$	Stg.
£	$ cts.	$ cts.	$ cts.	$ cts.	$ cts.	$ cts.	$ cts.	$ cts.	£
6100..	24040.83.3	24055.00.0	24069.16.6	24083.33.3	24097.50.0	24111.66.6	24125.83.3	24140.00.0	..6100
5200..	24512.22.2	24526.66.6	24541.11.1	24555.55.5	24570.00.0	24584.44.4	24598.88.8	24613.33.3	..5200
5300..	24983.61.1	24998.33.3	25013.05.5	25027.77.7	25042.50.0	25057.22.2	25071.94.4	25086.66.6	..5300
5400..	25455.00.0	25470.00.0	25485.00.0	25500.00.0	25515.00.0	25530.00.0	25545.00.0	25560.00.0	..5400
5500..	25926.38.8	25941.66.6	25956.94.4	25972.22.2	25987.50.0	26002.77.7	26018.05.5	26033.33.3	..5500
5600..	26397.77.7	26413.33.3	26428.88.8	26444.44.4	26460.00.0	26475.55.5	26491.11.1	26506.66.6	..5600
5700..	26869.16.6	26885.00.0	26900.83.3	26916.66.6	26932.50.0	26948.33.3	26964.16.6	26980.00.0	..5700
5800..	27340.55.5	27356.66.6	27372.77.7	27388.88.8	27405.00.0	27421.11.1	27437.22.2	27453.33.3	..5800
5900..	27811.94.4	27828.33.3	27844.72.2	27861.11.1	27877.50.0	27893.88.8	27910.27.7	27926.66.6	..5900
6000..	28283.33.3	28300.00.0	28316.66.6	28333.33.3	28350.00.0	28366.66.6	28383.33.3	28400.00.0	..6000
6100..	28754.72.2	28771.66.6	28788.61.1	28805.55.5	28822.50.0	28839.44.4	28856.38.8	28873.33.3	..6100
6200..	29226.11.1	29243.33.3	29260.55.5	29277.77.7	29295.00.0	29312.22.2	29329.44.4	29346.66.6	..6200
6300..	29697.50.0	29715.00.0	29732.50.0	29750.00.0	29767.50.0	29785.00.0	29802.50.0	29820.00.0	..6300
6400..	30168.88.8	30186.66.6	30204.44.4	30222.22.2	30240.00.0	30257.77.7	30275.55.5	30293.33.3	..6400
6500..	30640.27.7	30658.33.3	30676.38.8	30694.44.4	30712.50.0	30730.55.5	30748.61.1	30766.66.6	..6500
6600..	31111.66.6	31130.00.0	31148.33.3	31166.66.6	31185.00.0	31203.33.3	31221.66.6	31240.00.0	..6600
6700..	31583.05.5	31601.66.6	31620.27.7	31638.88.8	31657.50.0	31676.11.1	31694.72.2	31713.33.3	..6700
6800..	32054.44.4	32073.33.3	32092.22.2	32111.11.1	32130.00.0	32148.88.8	32167.77.7	32186.66.6	..6800
6900..	32525.83.3	32545.00.0	32564.16.6	32583.33.3	32602.50.0	32621.66.6	32640.83.3	32660.00.0	..6900
7000..	32997.22.2	33016.66.6	33036.11.1	33055.55.5	33075.00.0	33094.44.4	33113.88.8	33133.33.3	..7000
7100..	33468.61.1	33488.33.3	33508.05.5	33527.77.7	33547.50.0	33567.22.2	33586.94.4	33606.66.6	..7100
7200..	33940.00.0	33960.00.0	33980.00.0	34000.00.0	34020.00.0	34040.00.0	34060.00.0	34080.00.0	..7200
7300..	34411.38.8	34431.66.6	34451.94.4	34472.22.2	34492.50.0	34512.77.7	34533.05.5	34553.33.3	..7300
7400..	34882.77.7	34903.33.3	34923.88.8	34944.44.4	34965.00.0	34985.55.5	35006.11.1	35026.66.6	..7400
7500..	35354.16.6	35375.00.0	35395.83.3	35416.66.6	35437.50.0	35458.33.3	35479.16.6	35500.00.0	..7500
7600..	35825.55.5	35846.66.6	35867.77.7	35888.88.8	35910.00.0	35931.11.1	35952.22.2	35973.33.3	..7600
7700..	36296.94.4	36318.33.3	36339.72.2	36361.11.1	36382.50.0	36403.88.8	36425.27.7	36446.66.6	..7700
7800..	36768.33.3	36790.00.0	36811.66.6	36833.33.3	36855.00.0	36876.66.6	36898.33.3	36920.00.0	..7800
7900..	37239.72.2	37261.66.6	37283.61.1	37305.55.5	37327.50.0	37349.44.4	37371.38.8	37393.33.3	..7900
8000..	37711.11.1	37733.33.3	37755.55.5	37777.77.7	37800.00.0	37822.22.2	37844.44.4	37866.66.6	..8000
8100..	38182.50.0	38205.00.0	38227.50.0	38250.00.0	38272.50.0	38295.00.0	38317.50.0	38340.00.0	..8100
8200..	38653.88.8	38676.66.6	38699.44.4	38722.22.2	38745.00.0	38767.77.7	38790.55.5	38813.33.3	..8200
8300..	39125.27.7	39148.33.3	39171.38.8	39194.44.4	39217.50.0	39240.55.5	39263.61.1	39286.66.6	..8300
8400..	39596.66.6	39620.00.0	39643.33.3	39666.66.6	39690.00.0	39713.33.3	39736.66.6	39760.00.0	..8400
8500..	40068.05.5	40091.66.6	40115.27.7	40138.88.8	40162.50.0	40186.11.1	40209.72.2	40233.33.3	..8500
8600..	40539.44.4	40563.33.3	40587.22.2	40611.11.1	40635.00.0	40658.88.8	40682.77.7	40706.66.6	..8600
8700..	41010.83.3	41035.00.0	41059.16.6	41083.33.3	41107.50.0	41131.66.6	41155.83.3	41180.00.0	..8700
8800..	41482.22.2	41506.66.6	41531.11.1	41555.55.5	41580.00.0	41605.44.4	41628.88.8	41653.33.3	..8800
8900..	41953.61.1	41978.33.3	42003.05.5	42027.77.7	42052.50.0	42077.22.2	42101.94.4	42126.66.6	..8900
9000..	42425.00.0	42450.00.0	42475.00.0	42500.00.0	42525.00.0	42550.00.0	42575.00.0	42600.00.0	..9000
9100..	42896.38.8	42921.66.6	42946.94.4	42972.22.2	42997.50.0	43022.77.7	43048.05.5	43073.33.3	..9100
9200..	43367.77.7	43393.33.3	43418.88.8	43444.44.4	43470.00.0	43495.55.5	43521.11.1	43546.66.6	..9200
9300..	43839.16.6	43865.00.0	43890.83.3	43916.66.6	43942.50.0	43968.33.3	43994.16.6	44020.00.0	..9300
9400..	44310.55.5	44336.66.6	44362.77.7	44388.88.8	44415.00.0	44441.11.1	44467.22.2	44493.33.3	..9400
9500..	44781.94.4	44808.33.3	44834.72.2	44861.11.1	44887.50.0	44913.88.8	44940.27.7	44966.66.6	..9500
9600..	45253.33.3	45280.00.0	45306.66.6	45333.33.3	45360.00.0	45386.66.6	45413.33.3	45440.00.0	..9600
9700..	45724.72.2	45751.66.6	45778.61.1	45805.55.5	45832.50.0	45859.44.4	45886.38.8	45913.33.3	..9700
9800..	46196.11.1	46223.33.3	46250.55.5	46277.77.7	46305.00.0	46332.22.2	46359.44.4	46386.66.6	..9800
9900..	46667.50.0	46695.00.0	46722.50.0	46750.00.0	46777.50.0	46805.00.0	46832.50.0	46860.00.0	..9900
10000..	47138.88.8	47166.66.6	47194.44.4	47222.22.2	47250.00.0	47277.77.7	47305.55.5	47333.33.3	..10000

s.	$ cts.	$ cts.	$ cts.	$ cts.	$ cts.	$ cts.	$ cts.	$ cts.	s.
16..	3.77.1	3.77.3	3.77.5	3.77.8	3.78.0	3.78.2	3.78.4	3.78.7	..16
17..	4.00.7	4.00.9	4.01.1	4.01.4	4.01.6	4.01.9	4.02.1	4.02.3	..17
18..	4.24.2	4.24.5	4.24.7	4.25.0	4.25.2	4.25.5	4.25.7	4.26.0	..18
19..	4.47.8	4.48.1	4.48.3	4.48.6	4.48.9	4.49.1	4.49.4	4.49.7	..19
1d..	.02.0	.02.0	.02.0	.02.0	.02.0	.02.0	.02.0	.02.0	..1d.
2..	.03.9	.03.9	.03.9	.03.9	.03.9	.03.9	.03.9	.03.9	..2
3..	.05.9	.05.9	.05.9	.05.9	.05.9	.05.9	.05.9	.05.9	..3
4..	.07.8	.07.9	.07.9	.07.9	.07.9	.07.9	.07.9	.07.9	..4
5..	.09.8	.09.8	.09.8	.09.8	.09.8	.09.8	.09.8	.09.9	..5
6..	.11.8	.11.8	.11.8	.11.8	.11.8	.11.8	.11.8	.11.8	..6
7..	.13.7	.13.7	.13.8	.13.8	.13.8	.13.8	.13.8	.13.8	..7
8..	.15.7	.15.7	.15.7	.15.7	.15.7	.15.7	.15.8	.15.8	..8
9..	.17.7	.17.7	.17.7	.17.7	.17.7	.17.7	.17.7	.17.7	..9
10..	.19.6	.19.6	.19.7	.19.7	.19.7	.19.7	.19.7	.19.7	..10
11..	.21.6	.21.6	.21.6	.21.6	.21.6	.21.7	.21.7	.21.7	..11

STERLING EXCHANGE TABLES.

6 9/16 TO 7 %

STERLING INTO DOLLARS AND CENTS.

Stg.	6 9/16	6 5/8	6 11/16	6 3/4	6 13/16	6 7/8	6 15/16	7	Stg.
£	$ cts.	$ cts.	$ cts.	$ cts.	$ cts.	$ cts.	$ cts.	$ cts.	£
100..	473.61.1	473.88.8	474.16.6	474.44.4	474.72.2	475.00.0	475.27.7	475.55.5	100
200..	947.22.2	947.77.7	948.33.3	948.88.8	949.44.4	950.00.0	950.55.5	951.11.1	200
300..	1420.83.3	1421.66.6	1422.50.0	1423.33.3	1424.16.6	1425.00.0	1425.83.3	1426.66.6	300
400..	1894.44.4	1895.55.5	1896.66.6	1897.77.7	1898.88.8	1900.00.0	1901.11.1	1902.22.2	400
500..	2368.05.5	2369.44.4	2370.83.3	2372.22.2	2373.61.1	2375.00.0	2376.38.8	2377.77.7	500
600..	2841.66.6	2843.33.3	2845.00.0	2846.66.6	2848.33.3	2850.00.0	2851.66.6	2853.33.3	600
700..	3315.27.7	3317.22.2	3319.16.6	3321.11.1	3323.05.5	3325.00.0	3326.94.4	3328.88.8	700
800..	3788.88.8	3791.11.1	3793.33.3	3795.55.5	3797.77.7	3800.00.0	3802.22.2	3804.44.4	800
900..	4262.50.0	4265.00.0	4267.50.0	4270.00.0	4272.50.0	4275.00.0	4277.50.0	4280.00.0	900
1000..	4736.11.1	4738.88.8	4741.66.6	4744.44.4	4747.22.2	4750.00.0	4752.77.7	4755.55.5	1000
1100..	5209.72.2	5212.77.7	5215.83.3	5218.88.8	5221.94.4	5225.00.0	5228.05.5	5231.11.1	1100
1200..	5683.33.3	5686.66.6	5690.00.0	5693.33.3	5696.66.6	5700.00.0	5703.33.3	5706.66.6	1200
1300..	6156.94.4	6160.55.5	6164.16.6	6167.77.7	6171.38.8	6175.00.0	6178.61.1	6182.22.2	1300
1400..	6630.55.5	6634.44.4	6638.33.3	6642.22.2	6646.11.1	6650.00.0	6653.88.8	6657.77.7	1400
1500..	7104.16.6	7108.33.3	7112.50.0	7116.66.6	7120.83.3	7125.00.0	7129.16.6	7133.33.3	1500
1600..	7577.77.7	7582.22.2	7586.66.6	7591.11.1	7595.55.5	7600.00.0	7604.44.4	7608.88.8	1600
1700..	8051.38.8	8056.11.1	8060.83.3	8065.55.5	8070.27.7	8075.00.0	8079.72.2	8084.44.4	1700
1800..	8525.00.0	8530.00.0	8535.00.0	8540.00.0	8545.00.0	8550.00.0	8555.00.0	8560.00.0	1800
1900..	8998.61.1	9003.88.8	9009.16.6	9014.44.4	9019.72.2	9025.00.0	9030.27.7	9035.55.5	1900
2000..	9472.22.2	9477.77.7	9483.33.3	9488.88.8	9494.44.4	9500.00.0	9505.55.5	9511.11.1	2000
2100..	9945.83.3	9951.66.6	9957.50.0	9963.33.3	9969.16.6	9975.00.0	9980.83.3	9986.66.6	2100
2200..	10419.44.4	10425.55.5	10431.66.6	10437.77.7	10443.88.8	10450.00.0	10456.11.1	10462.22.2	2200
2300..	10893.05.5	10899.44.4	10905.83.3	10912.22.2	10918.61.1	10925.00.0	10931.38.8	10937.77.7	2300
2400..	11366.66.6	11373.33.3	11380.00.0	11386.66.6	11393.33.3	11400.00.0	11406.66.6	11413.33.3	2400
2500..	11840.27.7	11847.22.2	11854.16.6	11861.11.1	11868.05.5	11875.00.0	11881.94.4	11888.88.8	2500
2600..	12313.88.8	12321.11.1	12328.33.3	12335.55.5	12342.77.7	12350.00.0	12357.22.2	12364.44.4	2600
2700..	12787.50.0	12795.00.0	12802.50.0	12810.00.0	12817.50.0	12825.00.0	12832.50.0	12840.00.0	2700
2800..	13261.11.1	13268.88.8	13276.66.6	13284.44.4	13292.22.2	13300.00.0	13307.77.7	13315.55.5	2800
2900..	13734.72.2	13742.77.7	13750.83.3	13758.88.8	13766.94.4	13775.00.0	13783.05.5	13791.11.1	2900
3000..	14208.33.3	14216.66.6	14225.00.0	14233.33.3	14241.66.6	14250.00.0	14258.33.3	14266.66.6	3000
3100.	14681.94.4	14690.55.5	14699.16.6	14707.77.7	14716.38.8	14725.00.0	14733.61.1	14742.22.2	3100
3200..	15155.55.5	15164.44.4	15173.33.3	15182.22.2	15191.11.1	15200.00.0	15208.88.8	15217.77.7	3200
3300..	15629.16.6	15638.33.3	15647.50.0	15656.66.6	15665.83.3	15675.00.0	15684.16.6	15693.33.3	3300
3400..	16102.77.7	16112.22.2	16121.66.6	16131.11.1	16140.55.5	16150.00.0	16159.44.4	16168.88.8	3400
3500..	16576.38.8	16586.11.1	16595.83.3	16605.55.5	16615.27.7	16625.00.0	16634.72.2	16644.44.4	3500
3600..	17050.00.0	17060.00.0	17070.00.0	17080.00.0	17090.00.0	17100.00.0	17110.00.0	17120.00.0	3600
3700..	17523.61.1	17533.88.8	17544.16.6	17554.44.4	17564.72.2	17575.00.0	17585.27.7	17595.55.5	3700
3800..	17997.22.2	18007.77.7	18018.33.3	18028.88.8	18039.44.4	18050.00.0	18060.55.5	18071.11.1	3800
3900..	18470.83.3	18481.66.6	18492.50.0	18503.33.3	18514.16.6	18525.00.0	18535.83.3	18546.66.6	3900
4000..	18944.44.4	18955.55.5	18966.66.6	18977.77.7	18988.88.8	19000.00.0	19011.11.1	19022.22.2	4000
4100..	19418.05.5	19429.44.4	19440.83.3	19452.22.2	19463.61.1	19475.00.0	19486.38.8	19497.77.7	4100
4200..	19891.66.6	19903.33.3	19915.00.0	19926.66.6	19938.33.3	19950.00.0	19961.66.6	19973.33.3	4200
4300..	20365.27.7	20377.22.2	20389.16.6	20401.11.1	20413.05.5	20425.00.0	20436.94.4	20448.88.8	4300
4400..	20838.88.8	20851.11.1	20863.33.3	20875.55.5	20887.77.7	20900.00.0	20912.22.2	20924.44.4	4400
4500..	21312.50.0	21325.00.0	21337.50.0	21350.00.0	21362.50.0	21375.00.0	21387.50.0	21400.00.0	4500
4600..	21786.11.1	21798.88.8	21811.66.6	21824.44.4	21837.22.2	21850.00.0	21862.77.7	21875.55.5	4600
4700..	22259.72.2	22272.77.7	22285.83.3	22298.88.8	22311.94.4	22325.00.0	22338.05.5	22351.11.1	4700
4800..	22733.33.3	22746.66.6	22760.00.0	22773.33.3	22786.66.6	22800.00.0	22813.33.3	22826.66.6	4800
4900..	23206.94.4	23220.55.5	23234.16.6	23247.77.7	23261.38.8	23275.00.0	23288.61.1	23302.22.2	4900
5000..	23680.55.5	23694.44.4	23708.33.3	23722.22.2	23736.11.1	23750.00.0	23763.88.8	23777.77.7	5000

s.	$ cts.	$ cts.	$ cts.	$ cts.	$ cts.	$ cts.	$ cts.	$ cts.	s.
1..	.23.7	.23.7	.23.7	.23.7	.23.7	.23.7	.23.8	.23.8	1
2..	.47.4	.47.4	.47.4	.47.4	.47.5	.47.5	.47.5	.47.5	2
3..	.71.0	.71.1	.71.1	.71.2	.71.2	.71.2	.71.3	.71.3	3
4..	.94.7	.94.8	.94.8	.94.9	.94.9	.95.0	.96.0	.96.1	4
5..	1.18.4	1.18.5	1.18.5	1.18.6	1.18.7	1.18.7	1.18.8	1.18.9	5
6..	1.42.1	1.42.2	1.42.2	1.42.3	1.42.4	1.42.5	1.42.6	1.42.7	6
7..	1.65.8	1.65.9	1.65.9	1.66.0	1.66.1	1.66.2	1.66.3	1.66.4	7
8..	1.89.4	1.89.5	1.89.7	1.89.8	1.89.9	1.90.0	1.90.1	1.90.2	8
9..	2.13.1	2.13.2	2.13.4	2.13.5	2.13.6	2.13.7	2.13.9	2.14.0	9
10..	2.36.8	2.36.9	2.37.1	2.37.2	2.37.4	2.37.5	2.37.6	2.37.8	10
11..	2.60.5	2.60.6	2.60.8	2.60.9	2.61.1	2.61.2	2.61.4	2.61.5	11
12..	2.84.2	2.84.3	2.84.5	2.84.7	2.84.8	2.85.0	2.85.2	2.85.3	12
13..	3.07.8	3.08.0	3.08.2	3.08.4	3.08.6	3.08.7	3.08.9	3.09.1	13
14..	3.31.5	3.31.7	3.31.9	3.32.1	3.32.3	3.32.6	3.32.7	3.32.9	14
15..	3.55.2	3.55.4	3.55.6	3.55.8	3.56.0	3.56.2	3.56.4	3.56.7	15

STERLING EXCHANGE TABLES.

6 9/16 TO 7 %

STERLING INTO DOLLARS AND CENTS.

Stg. £	6 9/16 $ cts.	6 5/8 $ cts.	6 11/16 $ cts.	6 3/4 $ cts.	6 13/16 $ cts.	6 7/8 $ cts.	6 15/16 $ cts.	7 $ cts.	Stg. £
5100..	24154.16.6	24168.33.3	24182.50.0	24196.66.6	24210.83.3	24225.00.0	24239.16.6	24253.33.3	..5100
5200..	24627.77.7	24642.22.2	24656.66.6	24671.11.1	24685.55.5	24700.00.0	24714.44.4	24728.88.8	..5200
5300..	25101.38.8	25116.11.1	25130.83.3	25145.55.5	25160.27.7	25175.00.0	25189.72.2	25204.44.4	..5300
5400..	25575.00.0	25590.00.0	25605.00.0	25620.00.0	25635.00.0	25650.00.0	25665.00.0	25680.00.0	..5400
5500..	26048.61.1	26063.88.8	26079.16.6	26094.44.4	26109.72.2	26125.00.0	26140.27.7	26155.55.5	..5500
5600..	26522.22.2	26537.77.7	26553.33.3	26568.88.8	26584.44.4	26600.00.0	26615.55.5	26631.11.1	..5600
5700..	26995.83.3	27011.66.6	27027.50.0	27043.33.3	27059.16.6	27075.00.0	27090.83.3	27106.66.6	..5700
5800..	27469.44.4	27485.55.5	27501.66.6	27517.77.7	27533.88.8	27550.00.0	27566.11.1	27582.22.2	..5800
5900..	27943.05.5	27959.44.4	27975.83.3	27992.22.2	28008.61.1	28025.00.0	28041.38.8	28057.77.7	..5900
6000..	28416.66.6	28433.33.3	28450.00.0	28466.66.6	28483.33.3	28500.00.0	28516.66.6	28533.33.3	..6000
6100..	28890.27.7	28907.22.2	28924.16.6	28941.11.1	28958.05.5	28975.00.0	28991.94.4	29008.88.8	..6100
6200..	29363.88.8	29381.11.1	29398.33.3	29415.55.5	29432.77.7	29450.00.0	29467.22.2	29484.44.4	..6200
6300..	29837.50.0	29855.00.0	29872.50.0	29890.00.0	29907.50.0	29925.00.0	29942.50.0	29960.00.0	..6300
6400..	30311.11.1	30328.88.8	30346.66.6	30364.44.4	30382.22.2	30400.00.0	30417.77.7	30435.55.5	..6400
6500..	30784.72.2	30802.77.7	30820.83.3	30838.88.8	30856.94.4	30875.00.0	30893.05.5	30911.11.1	..6500
6600..	31258.33.3	31276.66.6	31295.00.0	31313.33.3	31331.66.6	31350.00.0	31368.33.3	31386.66.6	..6600
6700..	31731.94.4	31750.55.5	31769.16.6	31787.77.7	31806.38.8	31825.00.0	31843.61.1	31862.22.2	..6700
6800..	32205.55.5	32224.44.4	32243.33.3	32262.22.2	32281.11.1	32300.00.0	32318.88.8	32337.77.7	..6800
6900..	32679.16.6	32698.33.3	32717.50.0	32736.66.6	32755.83.3	32775.00.0	32794.16.6	32813.33.3	..6900
7000..	33152.77.7	33172.22.2	33191.66.6	33211.11.1	33230.55.5	33250.00.0	33269.44.4	33288.88.8	..7000
7100..	33626.38.8	33646.11.1	33665.83.3	33685.55.5	33705.27.7	33725.00.0	33744.72.2	33764.44.4	..7100
7200..	34100.00.0	34120.00.0	34140.00.0	34160.00.0	34180.00.0	34200.00.0	34220.00.0	34240.00.0	..7200
7300..	34573.61.1	34593.88.8	34614.16.6	34634.44.4	34654.72.2	34675.00.0	34695.27.7	34715.55.5	..7300
7400..	35047.22.2	35067.77.7	35088.33.3	35108.88.8	35129.44.4	35150.00.0	35170.55.5	35191.11.1	..7400
7500..	35520.83.3	35541.66.6	35562.50.0	35583.33.3	35604.16.6	35625.00.0	35645.83.3	35666.66.6	..7500
7600..	35994.44.4	36015.55.5	36036.66.6	36057.77.7	36078.88.8	36100.00.0	36121.11.1	36142.22.2	..7600
7700..	36468.05.5	36489.44.4	36510.83.3	36532.22.2	36553.61.1	36575.00.0	36596.38.8	36617.77.7	..7700
7800..	36941.66.6	36963.33.3	36985.00.0	37006.66.6	37028.33.3	37050.00.0	37071.66.6	37093.33.3	..7800
7900..	37415.27.7	37437.22.2	37459.16.6	37481.11.1	37503.05.5	37525.00.0	37546.94.4	37568.88.8	..7900
8000..	37888.88.8	37911.11.1	37933.33.3	37955.55.5	37977.77.7	38000.00.0	38022.22.2	38044.44.4	..8000
8100..	38362.50.0	38385.00.0	38407.50.0	38430.00.0	38452.50.0	38475.00.0	38497.50.0	38520.00.0	..8100
8200..	38836.11.1	38858.88.8	38881.66.6	38904.44.4	38927.22.2	38950.00.0	38972.77.7	38995.55.5	..8200
8300..	39309.72.2	39332.77.7	39355.83.3	39378.88.8	39401.94.4	39425.00.0	39448.05.5	39471.11.1	..8300
8400..	39783.33.3	39806.66.6	39830.00.0	39853.33.3	39876.66.6	39900.00.0	39923.33.3	39946.66.6	..8400
8500..	40256.94.4	40280.55.5	40304.16.6	40327.77.7	40351.38.8	40375.00.0	40398.61.1	40422.22.2	..8500
8600..	40730.55.5	40754.44.4	40778.33.3	40802.22.2	40826.11.1	40850.00.0	40873.88.8	40897.77.7	..8600
8700..	41204.16.6	41228.33.3	41252.50.0	41276.66.6	41300.83.3	41325.00.0	41349.16.6	41373.33.3	..8700
8800..	41677.77.7	41702.22.2	41726.66.6	41751.11.1	41775.55.5	41800.00.0	41824.44.4	41848.88.8	..8800
8900..	42151.38.8	42176.11.1	42200.83.3	42225.55.5	42250.27.7	42275.00.0	42299.72.2	42324.44.4	..8900
9000..	42625.00.0	42650.00.0	42675.00.0	42700.00.0	42725.00.0	42750.00.0	42775.00.0	42800.00.0	..9000
9100..	43098.61.1	43123.88.8	43149.16.6	43174.44.4	43199.72.2	43225.00.0	43250.27.7	43275.55.5	..9100
9200..	43572.22.2	43597.77.7	43623.33.3	43648.88.8	43674.44.4	43700.00.0	43725.55.5	43751.11.1	..9200
9300..	44045.83.3	44071.66.6	44097.50.0	44123.33.3	44149.16.6	44175.00.0	44200.83.3	44226.66.6	..9300
9400..	44519.44.4	44545.55.5	44571.66.6	44597.77.7	44623.88.8	44650.00.0	44676.11.1	44702.22.2	..9400
9500..	44993.05.5	45019.44.4	45045.83.3	45072.22.2	45098.61.1	45125.00.0	45151.38.8	45177.77.7	..9500
9600..	45466.66.6	45493.33.3	45520.00.0	45546.66.6	45573.33.3	45600.00.0	45626.66.6	45653.33.3	..9600
9700..	45940.27.7	45967.22.2	45994.16.6	46021.11.1	46048.05.5	46075.00.0	46101.94.4	46128.88.8	..9700
9800..	46413.88.8	46441.11.1	46468.33.3	46495.55.5	46522.77.7	46550.00.0	46577.22.2	46604.44.4	..9800
9900..	46887.50.0	46915.00.0	46942.50.0	46970.00.0	46997.50.0	47025.00.0	47052.50.0	47080.00.0	..9900
10000..	47361.11.1	47388.88.8	47416.66.6	47444.44.4	47472.22.2	47500.00.0	47527.77.7	47555.55.5	..10000

s.	$ cts.	$ cts.	$ cts.	$ cts.	$ cts.	$ cts.	$ cts.	$ cts.	s.
16..	3.78.9	3.79.1	3.79.3	3.79.6	3.79.8	3.80.0	3.80.2	3.80.4	..16
17..	4.02.6	4.02.8	4.03.0	4.03.2	4.03.5	4.03.7	4.04.0	4.04.2	..17
18..	4.26.2	4.26.5	4.26.7	4.27.0	4.27.2	4.27.5	4.27.7	4.28.0	..18
19..	4.50.0	4.50.2	4.50.4	4.50.7	4.51.0	4.51.2	4.51.5	4.51.8	..19
1d.	.02.0	.02.0	.02.0	.02.0	.02.0	.02.0	.02.0	.02.0	..1d.
2..	.03.9	.03.9	.03.9	.03.9	.03.9	.03.9	.04.0	04.0	..2
3..	.05.9	.05.9	.05.9	.05.9	.05.9	.05.9	.05.9	.05.9	..3
4..	.07.9	.07.9	.07.9	.07.9	.07.9	.07.9	.07.9	.07.9	..4
5..	.09.9	.09.9	.09.9	.09.9	.09.9	.09.9	.09.9	.09.9	..5
6..	.11.8	.11.8	.11.8	.11.9	.11.9	.11.9	.11.9	.11.9	..6
7..	.13.8	.13.8	.13.8	.13.8	.13.8	.13.8	.13.9	.13.9	..7
8..	.15.8	.15.8	.15.8	.15.8	.15.8	.15.8	.15.8	.15.8	..8
9..	.17.7	.17.8	.17.8	.17.8	.17.8	.17.8	.17.8	.17.8	..9
10..	.19.7	.19.7	.19.7	.19.8	.19.8	.19.8	.19.8	.19.8	..10
11..	.21.7	.21.7	.21.7	.21.7	.21.7	.21.8	.21.8	.21.8	..11

STERLING EXCHANGE TABLES.

7 1/16 TO 7 1/2 %

STERLING INTO DOLLARS AND CENTS.

Stg. £	7 1/16 $ cts.	7 1/8 $ cts.	7 3/16 $ cts.	7 1/4 $ cts.	7 5/16 $ cts.	7 3/8 $ cts.	7 7/16 $ cts.	7 1/2 $ cts.	Stg. £
100	475.83.3	476.11.1	476.38.8	476.66.6	476.94.4	477.22.2	477.50.0	477.77.7	100
200	951.66.6	952.22.2	952.77.7	953.33.3	953.88.8	954.44.4	955.00.0	955.55.5	200
300	1427.50.0	1428.33.3	1429.16.6	1430.00.0	1430.83.3	1431.66.6	1432.50.0	1433.33.3	300
400	1903.33.3	1904.44.4	1905.55.5	1906.66.6	1907.77.7	1908.88.8	1910.00.0	1911.11.1	400
500	2379.16.6	2380.55.5	2381 94.4	2383.33.3	2384.72.2	2386.11.1	2387.50.0	2388.88.8	500
600	2855.00.0	2856.66.6	2858.33.3	2860.00.0	2861.66.6	2863.33.3	2865.00.0	2866.66.6	600
700	3330.83.3	3332.77.7	3334.72.2	3336.66.6	3338.61.1	3340.55.5	3342.50.0	3344.44.4	700
800	3806.66.6	3808.88.8	3811.11.1	3813.33.3	3815.55.5	3817.77.7	3820.00.0	3822.22.2	800
900	4282.50.0	4285.00.0	4287.50.0	4290.00.0	4292.50.0	4295.00.0	4297.50.0	4300.00.0	900
1000	4758.33.3	4761.11.1	4763.88.8	4766.66.6	4769.44.4	4772.22.2	4775.00.0	4777.77.7	1000
1100	5234.16.6	5237.22.2	5240.27.7	5243.33.3	5246.38.8	5249.44.4	5252.50.0	5255.55.5	1100
1200	5710.00.0	5713.33.3	5716.66.6	5720.00.0	5723.33.3	5726.66.6	5730.00.0	5733.33.3	1200
1300	6185.83.3	6189.44.4	6193.05 5	6196.66.6	6200.27.7	6203.88.8	6207.50.0	6211.11.1	1300
1400	6661.66.6	6665.55.5	6669.44.4	6673.33.3	6677.22.2	6681.11.1	6685.00.0	6688.88.8	1400
1500	7137.50.0	7141.66.6	7145.83.3	7150.00.0	7154.16.6	7158.33.3	7162.50.0	7166.66.6	1500
1600	7613.33.3	7617.77.7	7622.22.2	7626.66.6	7631.11.1	7635.55.5	7640.00.0	7644.44.4	1600
1700	8089.16.6	8093.88.8	8098.61.1	8103.33.3	8108.05.5	8112.77.7	8117.50.0	8122.22.2	1700
1800	8565.00.0	8570.00.0	8575.00.0	8580.00.0	8585.00.0	8590.00.0	8595.00.0	8600.00.0	1800
1900	9040.83.3	9046.11.1	9051.38.8	9056.66.6	9061.94.4	9067.22.2	9072.50.0	9077.77.7	1900
2000	9516.66.6	9522.22.2	9527.77.7	9533.33.3	9538.88.8	9544.44 4	9550.00.0	9555.55.5	2000
2100	9992.50.0	9998.33.3	10004.16.6	10010.00.0	10015.83.3	10021.66.6	10027.50.0	10033.33.3	2100
2200	10468.33.3	10474.44.4	10480.55.5	10486.66.6	10492.77.7	10498.88.8	10505.00.0	10511.11.1	2200
2300	10944.16.6	10950.55.5	10956.94.4	10963.33.3	10969.72.2	10976.11.1	10982.50.0	10988.88.8	2300
2400	11420.00.0	11426.66.6	11433.33.3	11440.00.0	11446.66.6	11453.33.3	11460.00.0	11466.66.6	2400
2500	11895.83.3	11902.77.7	11909.72.2	11916.66.6	11923.61.1	11930.55.5	11937.50.0	11944.44.4	2500
2600	12371.66.6	12378.88.8	12386.11.1	12393.33.3	12400.55.5	12407.77.7	12415.00.0	12422.22.2	2600
2700	12847.50 0	12855.00.0	12862.50.0	12870.00.0	12877.50.0	12885.00.0	12892.50.0	12900.00.0	2700
2800	13323.33.3	13331.11.1	13338.88.8	13346.66.6	13354.44.4	13362.22.2	13370.00.0	13377.77.7	2800
2900	13799.16.6	13807.22.2	13815.27.7	13823.33.3	13831.38.8	13839.44.4	13847.50.0	13855.55.5	2900
3000	14275.00 0	14283.33.3	14291.66.6	14300.00.0	14308.33.3	14316.66.6	14325.00.0	14333.33.3	3000
3100	14750.83.3	14759.44.4	14768.05.5	14776.66.6	14785.27.7	14793.88.8	14802.50.0	14811.11.1	3100
3200	15226.66.6	15235.55.5	15244.44.4	15253.33.3	15262.22.2	15271.11.1	15280.00.0	15288.88.8	3200
3300	15702.50.0	15711.66.6	15720.83.3	15730.00.0	15739.16.6	15748.33.3	15757.50.0	15766.66.6	3300
3400	16178.33.3	16187.77.7	16197.22.2	16206.66.6	16216.11.1	16225.55.5	16235.00.0	16244.44.4	3400
3500	16654.16.6	16663.88.8	16673.61.1	16683.33.3	16693.05.5	16702.77.7	16712.50.0	16722.22.2	3500
3600	17130.00.0	17140.00.0	17150.00.0	17160.00.0	17170.00.0	17180.00.0	17190.00.0	17200.00.0	3600
3700	17605.83.3	17616.11.1	17626.38.8	17636.66.6	17646.94.4	17657.22.2	17667.50.0	17677.77.7	3700
3800	18081.66.6	18092.22.2	18102.77.7	18113.33.3	18123.88.8	18134.44.4	18145.00.0	18155.55.5	3800
3900	18557.50.0	18568.33.3	18579.16.6	18590 00.0	18600.83.3	18611.66.6	18622.50.0	18633.33.3	3900
4000	19033.33.3	19044.44.4	19055.55.5	19066.66.6	19077.77.7	19088.88.8	19100.00.0	19111.11.1	4000
4100	19509.16.6	19520.55.5	19531.94.4	19543.33.3	19554.72.2	19566.11.1	19577.50.0	19588.88.8	4100
4200	19985.00.0	19996.66.6	20008.33.3	20020.00.0	20031 66.6	20043.33.3	20055.00.0	20066.66.6	4200
4300	20460.83.3	20472.77.7	20484.72.2	20496.66.6	20508.61.1	20520.55.5	20532.50.0	20544.44.4	4300
4400	20936.66.6	20948.88.8	20961.11.1	20973.33.3	20985.55.5	20997.77.7	21010.00.0	21022.22.2	4400
4500	21412.50.0	21425.00.0	21437.50.0	21450.00.0	21462.50.0	21475.00.0	21487.50.0	21500.00.0	4500
4600	21888.33.3	21901.11.1	21913.88.8	21926.66.6	21939.44.4	21952.22.2	21965.00.0	21977.77.7	4600
4700	22364 16.6	22377.22.2	22390.27.7	22403.33.3	22416.38.8	22429.44.4	22442.50.0	22455.55.5	4700
4800	22840.00.0	22853.33.3	22866.66.6	22880.00.0	22893.33.3	22906.66.6	22920.00.0	22933.33.3	4800
4900	23315.83.3	23329.44.4	23343.05.5	23356.66 6	23370.27.7	23383.88 8	23397.50.0	23411.11.1	4900
5000	23791.66.6	23805.55.5	23819.44.4	23833.33.3	23847.22.2	23861.11.1	23875.00.0	23888.88.8	5000

s.	$ cts.	$ cts.	$ cts.	$ cts.	$ cts.	$ cts.	$ cts.	$ cts.	s.
1	.23,8	.23,8	.23,8	.23,8	.23,8	.23,9	.23,9	.23,9	1
2	.47,6	.47,6	.47,6	.47,7	.47,7	.47,7	.47,7	.47,8	2
3	.71,4	.71,4	.71,4	.71,5	.71,5	.71,6	.71,6	.71,7	3
4	.95,2	.95,2	.95,3	.95,3	.95,4	.95,4	.95,5	.95,5	4
5	1.18,9	1.19,0	1.19,1	1.19,2	1.19,2	1.19,3	1.19,4	1.19,4	5
6	1.42,7	1.42,8	1.42,9	1.43,0	1.43,1	1.43,2	1.43,2	1.43,3	6
7	1.66,5	1.66,6	1.66,7	1.66,8	1.66,9	1.67,0	1.67,1	1.67,2	7
8	1.90,3	1.90,4	1.90,5	1.90,7	1.90,8	1.90,9	1.91,0	1.91,1	8
9	2.14,1	2.14,2	2.14,4	2.14,5	2.14,6	2.14,7	2.14,9	2.15,0	9
10	2.37,9	2.38,0	2.38,2	2.38,3	2.38,5	2.38,6	2.38,7	2.38,9	10
11	2.61,7	2.61,9	2.62,0	2.62,2	2.62,3	2.62,5	2.62,6	2.62,8	11
12	2.85,5	2.85,7	2.85,8	2.86,0	2.86,2	2.86,3	2.86,5	2.86,7	12
13	3.09,3	3.09,5	3.09,6	3.09,8	3.10,0	3.10,2	3.10,4	3.10,5	13
14	3.33,1	3.33,3	3.33,5	3.33,7	3.33,9	3.34,0	3.34,2	3.34,4	14
15	3.56,9	3.57,1	3.57,3	3.57,5	3.57,7	3.57,9	3.58,1	3.58,3	15

STERLING EXCHANGE TABLES.

$7\tfrac{1}{16}$ TO $7\tfrac{1}{2}$ %

STERLING INTO DOLLARS AND CENTS.

Stg.	$7\tfrac{1}{16}$	$7\tfrac{1}{8}$	$7\tfrac{3}{16}$	$7\tfrac{1}{4}$	$7\tfrac{5}{16}$	$7\tfrac{3}{8}$	$7\tfrac{7}{16}$	$7\tfrac{1}{2}$	Stg.
£	$ cts.	$ cts.	$ cts.	$ cts.	$ cts.	$ cts.	$ cts.	$ cts.	£
5100	24267.50.0	24281.66.6	24295.83.3	24310.00.0	24324.16.6	24338.33.3	24352.50.0	24366.66.6	..5100
5200	24743.33.3	24757.77.7	24772.22.2	24786.66.6	24801.11.1	24815.55.5	24830.00.0	24844.44.4	..5200
5300	25219.16.6	25233.88.8	25248.61.1	25263.33.3	25278.05.5	25292.77.7	25307.50.0	25322.22.2	..5300
5400	25695.00.0	25710.00.0	25725.00.0	25740.00.0	25755.00.0	25770.00.0	25785.00.0	25800.00.0	..5400
5500	26170.83.3	26186.11.1	26201.38.8	26216.66.6	26231.94.4	26247.22.2	26262.50.0	26277.77.7	..5500
5600	26646.66.6	26662.22.2	26677.77.7	26693.33.3	26708.88.8	26724.44.4	26740.00.0	26755.55.5	..5600
5700	27122.50.0	27138.33.3	27154.16.6	27170.00.0	27185.83.3	27201.66.6	27217.50.0	27233.33.3	..5700
5800	27598.33.3	27614.44.4	27630.55.5	27646.66.6	27662.77.7	27678.88.8	27695.00.0	27711.11.1	..5800
5900	28074.16.6	28090.55.5	28106.94.4	28123.33.3	28139.72.2	28156.11.1	28172.50.0	28188.88.8	..5900
6000	28550.00.0	28566.66.6	28583.33.3	28600.00.0	28616.66.6	28633.33.3	28650.00.0	28666.66.6	..6000
6100	29025.83.3	29042.77.7	29059.72.2	29076.66.6	29093.61.1	29110.55.5	29127.50.0	29144.44.4	..6100
6200	29501.66.6	29518.88.8	29536.11.1	29553.33.3	29570.55.5	29587.77.7	29605.00.0	29622.22.2	..6200
6300	29977.50.0	29995.00.0	30012.50.0	30030.00.0	30047.50.0	30065.00.0	30082.50.0	30100.00.0	..6300
6400	30453.33.3	30471.11.1	30488.88.8	30506.66.6	30524.44.4	30542.22.2	30560.00.0	30577.77.7	..6400
6500	30929.16.6	30947.22.2	30965.27.7	30983.33.3	31001.38.8	31019.44.4	31037.50.0	31055.55.5	..6500
6600	31405.00.0	31423.33.3	31441.66.6	31460.00.0	31478.33.3	31496.66.6	31515.00.0	31533.33.3	..6600
6700	31880.83.3	31899.44.4	31918.05.5	31936.66.6	31955.27.7	31973.88.8	31992.50.0	32011.11.1	..6700
6800	32356.66.6	32375.55.5	32394.44.4	32413.33.3	32432.22.2	32451.11.1	32470.00.0	32488.88.8	..6800
6900	32832.50.0	32851.66.6	32870.83.3	32890.00.0	32909.16.6	32928.33.3	32947.50.0	32966.66.6	..6900
7000	33308.33.3	33327.77.7	33347.22.2	33366.66.6	33386.11.1	33405.55.5	33425.00.0	33444.44.4	..7000
7100	33784.16.6	33803.88.8	33823.61.1	33843.33.3	33863.05.5	33882.77.7	33902.50.0	33922.22.2	..7100
7200	34260.00.0	34280.00.0	34300.00.0	34320.00.0	34340.00.0	34360.00.0	34380.00.0	34400.00.0	..7200
7300	34735.83.3	34756.11.1	34776.38.8	34796.66.6	34816.94.4	34837.22.2	34857.50.0	34877.77.7	..7300
7400	35211.66.6	35232.22.2	35252.77.7	35273.33.3	35293.88.8	35314.44.4	35335.00.0	35355.55.5	..7400
7500	35687.50.0	35708.33.3	35729.16.6	35750.00.0	35770.83.3	35791.66.6	35812.50.0	35833.33.3	..7500
7600	36163.33.3	36184.44.4	36205.55.5	36226.66.6	36247.77.7	36268.88.8	36290.00.0	36311.11.1	..7600
7700	36639.16.6	36660.55.5	36681.94.4	36703.33.3	36724.72.2	36746.11.1	36767.50.0	36788.88.8	..7700
7800	37115.00.0	37136.66.6	37158.33.3	37180.00.0	37201.66.6	37223.33.3	37245.00.0	37266.66.6	..7800
7900	37590.83.3	37612.77.7	37634.72.2	37656.66.6	37678.61.1	37700.55.5	37722.50.0	37744.44.4	..7900
8000	38066.66.6	38088.88.8	38111.11.1	38133.33.3	38155.55.5	38177.77.7	38200.00.0	38222.22.2	..8000
8100	38542.50.0	38565.00.0	38587.50.0	38610.00.0	38632.50.0	38655.00.0	38677.50.0	38700.00.0	..8100
8200	39018.33.3	39041.11.1	39063.88.8	39086.66.6	39109.44.4	39132.22.2	39155.00.0	39177.77.7	..8200
8300	39494.16.6	39517.22.2	39540.27.7	39563.33.3	39586.38.8	39609.44.4	39632.50.0	39655.55.5	..8300
8400	39970.00.0	39993.33.3	40016.66.6	40040.00.0	40063.33.3	40086.66.6	40110.00.0	40133.33.3	..8400
8500	40445.83.3	40469.44.4	40493.05.5	40516.66.6	40540.27.7	40563.88.8	40587.50.0	40611.11.1	..8500
8600	40921.66.6	40945.55.5	40969.44.4	40993.33.3	41017.22.2	41041.11.1	41065.00.0	41088.88.8	..8600
8700	41397.50.0	41421.66.6	41445.83.3	41470.00.0	41494.16.6	41518.33.3	41542.50.0	41566.66.6	..8700
8800	41873.33.3	41897.77.7	41922.22.2	41946.66.6	41971.11.1	41995.55.5	42020.00.0	42044.44.4	..8800
8900	42349.16.6	42373.88.8	42398.61.1	42423.33.3	42448.05.5	42472.77.7	42497.50.0	42522.22.2	..8900
9000	42825.00.0	42850.00.0	42875.00.0	42900.00.0	42925.00.0	42950.00.0	42975.00.0	43000.00.0	..9000
9100	43300.83.3	43326.11.1	43351.38.8	43376.66.6	43401.94.4	43427.22.2	43452.50.0	43477.77.7	..9100
9200	43776.66.6	43802.22.2	43827.77.7	43853.33.3	43878.88.8	43904.44.4	43930.00.0	43955.55.5	..9200
9300	44252.50.0	44278.33.3	44304.16.6	44330.00.0	44355.83.3	44381.66.6	44407.50.0	44433.33.3	..9300
9400	44728.33.3	44754.44.4	44780.55.5	44806.66.6	44832.77.7	44858.88.8	44885.00.0	44911.11.1	..9400
9500	45204.16.6	45230.55.5	45256.94.4	45283.33.3	45309.72.2	45336.11.1	45362.50.0	45388.88.8	..9500
9600	45680.00.0	45706.66.6	45733.33.3	45760.00.0	45786.66.6	45813.33.3	45840.00.0	45866.66.6	..9600
9700	46155.83.3	46182.77.7	46209.72.2	46236.66.6	46263.61.1	46290.55.5	46317.50.0	46344.44.4	..9700
9800	46631.66.6	46658.88.8	46686.11.1	46713.33.3	46740.55.5	46767.77.7	46795.00.0	46822.22.2	..9800
9900	47107.50.0	47135.00.0	47162.50.0	47190.00.0	47217.50.0	47245.00.0	47272.50.0	47300.00.0	..9900
10000	47583.33.3	47611.11.1	47638.88.8	47666.66.6	47694.44.4	47722.22.2	47750.00.0	47777.77.7	..10000

s.	$ cts.	$ cts.	$ cts.	$ cts.	$ cts.	$ cts.	$ cts.	$ cts.	s.
16..	3.80.7	3.80.9	3.81.1	3.81.3	3.81.5	3.81.8	3.82.0	3.82.2	..16
17..	4.04.4	4.04.7	4.04.9	4.05.2	4.05.4	4.05.6	4.05.9	4.06.1	..17
18..	4.28.2	4.28.5	4.28.7	4.29.0	4.29.2	4.29.5	4.29.7	4.30.0	..18
19..	4.52.0	4.52.3	4.52.6	4.52.8	4.53.1	4.53.4	4.53.6	4.53.9	..19
1d.	.02.0	.02.0	.02.0	.02.0	.02.0	.02.0	.02.0	.02.0	..1d.
2..	.04.0	.04.0	.04.0	.04.0	.04.0	.04.0	.04.0	.04.0	..2
3..	.05.9	.05.9	.05.9	.05.9	.06.0	.06.0	.06.0	.06.0	..3
4..	.07.9	.07.9	.07.9	.07.9	.07.9	.07.9	.07.9	.08.0	..4
5..	.09.9	.09.9	.09.9	.09.9	.09.9	.09.9	.09.9	.09.9	..5
6..	.11.9	.11.9	.11.9	.11.9	.11.9	.11.9	.11.9	.11.9	..6
7..	.13.9	.13.9	.13.9	.13.9	.13.9	.13.9	.13.9	.13.9	..7
8..	.15.9	.15.9	.15.9	.15.9	.15.9	.15.9	.15.9	.15.9	..8
9..	.17.9	.17.9	.17.9	.17.9	.17.9	.17.9	.17.0	.17.9	..9
10..	.19.9	.19.9	.19.9	.19.9	.19.9	.19.9	.19.9	.19.9	..10
11..	.21.8	.21.8	.21.8	.21.8	.21.8	.21.9	.21.9	.21.9	..11

STERLING EXCHANGE TABLES.

$7\frac{9}{16}$ TO 8 %

STERLING INTO DOLLARS AND CENTS.

Stg. £	$7\frac{9}{16}$	$7\frac{5}{8}$	$7\frac{11}{16}$	$7\frac{3}{4}$	$7\frac{13}{16}$	$7\frac{7}{8}$	$7\frac{15}{16}$	8	Stg. £
	$ cts.	$ cts.	$ cts.	$ cts.	$ cts.	$ cts.	$ cts.	$ cts.	
100	478.05.5	478.33.3	478.61.1	478.88.8	479.16.6	479.44.4	479.72.2	480.00.0	100
200	956.11.1	956.66.6	957.22.2	957.77.7	958.33.3	958.88.8	959.44.4	960.00.0	200
300	1434.16.6	1435.00.0	1435.83.3	1436.66.6	1437.50.0	1438.33.3	1439.16.6	1440.00.0	300
400	1912.22.2	1913.33.3	1914.44.4	1915.55.5	1916.66.6	1917.77.7	1918.88.8	1920.00.0	400
500	2390.27.7	2391.66.6	2393.05.5	2394.44.4	2395.83.3	2397.22.2	2398.61.1	2400.00.0	500
600	2868.33.3	2870.00.0	2871.66.6	2873.33.3	2875.00.0	2876.66.6	2878.33.3	2880.00.0	600
700	3346.38.8	3348.33.3	3350.27.7	3352.22.2	3354.16.6	3356.11.1	3358.05.5	3360.00.0	700
800	3824.44.4	3826.66.6	3828.88.8	3831.11.1	3833.33.3	3835.55.5	3837.77.7	3840.00.0	800
900	4302.50.0	4305.00.0	4307.50.0	4310.00.0	4312.50.0	4315.00.0	4317.50.0	4320.00.0	900
1000	4780.55.5	4783.33.3	4786.11.1	4788.88.8	4791.66.6	4794.44.4	4797.22.2	4800.00.0	1000
1100	5258.61.1	5261.66.6	5264.72.2	5267.77.7	5270.83.3	5273.88.8	5276.94.4	5280.00.0	1100
1200	5736.66.6	5740.00.0	5743.33.3	5746.66.6	5750.00.0	5753.33.3	5756.66.6	5760.00.0	1200
1300	6214.72.2	6218.33.3	6221.94.4	6225.55.5	6229.16.6	6232.77.7	6236.38.8	6240.00.0	1300
1400	6692.77.7	6696.66.6	6700.55.5	6704.44.4	6708.33.3	6712.22.2	6716.11.1	6720.00.0	1400
1500	7170.83.3	7175.00.0	7179.16.6	7183.33.3	7187.50.0	7191.66.6	7195.83.3	7200.00.0	1500
1600	7648.88.8	7653.33.3	7657.77.7	7662.22.2	7666.66.6	7671.11.1	7675.55.5	7680.00.0	1600
1700	8126.94.4	8131.66.6	8136.38.8	8141.11.1	8145.83.3	8150.55.5	8155.27.7	8160.00.0	1700
1800	8605.00.0	8610.00.0	8615.00.0	8620.00.0	8625.00.0	8630.00.0	8635.00.0	8640.00.0	1800
1900	9083.05.5	9088.33.3	9093.61.1	9098.88.8	9104.16.6	9109.44.4	9114.72.2	9120.00.0	1900
2000	9561.11.1	9566.66.6	9572.22.2	9577.77.7	9583.33.3	9588.88.8	9594.44.4	9600.00.0	2000
2100	10039.16.6	10045.00.0	10050.83.3	10056.66.6	10062.50.0	10068.33.3	10074.16.6	10080.00.0	2100
2200	10517.22.2	10523.33.3	10529.44.4	10535.55.5	10541.66.6	10547.77.7	10553.88.8	10560.00.0	2200
2300	10995.27.7	11001.66.6	11008.05.5	11014.44.4	11020.83.3	11027.22.2	11033.61.1	11040.00.0	2300
2400	11473.33.3	11480.00.0	11486.66.6	11493.33.3	11500.00.0	11506.66.6	11513.33.3	11520.00.0	2400
2500	11951.38.8	11958.33.3	11965.27.7	11972.22.2	11979.16.6	11986.11.1	11993.05.5	12000.00.0	2500
2600	12429.44.4	12436.66.6	12443.88.8	12451.11.1	12458.33.3	12465.55.5	12472.77.7	12480.00.0	2600
2700	12907.50.0	12915.00.0	12922.50.0	12930.00.0	12937.50.0	12945.00.0	12952.50.0	12960.00.0	2700
2800	13385.55.5	13393.33.3	13401.11.1	13408.88.8	13416.66.6	13424.44.4	13432.22.2	13440.00.0	2800
2900	13863.61.1	13871.66.6	13879.72.2	13887.77.7	13895.83.3	13903.88.8	13911.94.4	13920.00.0	2900
3000	14341.66.6	14350.00.0	14358.33.3	14366.66.6	14375.00.0	14383.33.3	14391.66.6	14400.00.0	3000
3100	14819.72.2	14828.33.3	14836.94.4	14845.55.5	14854.16.6	14862.77.7	14871.38.8	14880.00.0	3100
3200	15297.77.7	15306.66.6	15315.55.5	15324.44.4	15333.33.3	15342.22.2	15351.11.1	15360.00.0	3200
3300	15775.83.3	15785.00.0	15794.16.6	15803.33.3	15812.50.0	15821.66.6	15830.83.3	15840.00.0	3300
3400	16253.88.8	16263.33.3	16272.77.7	16282.22.2	16291.66.6	16301.11.1	16310.55.5	16320.00.0	3400
3500	16731.94.4	16741.66.6	16751.38.8	16761.11.1	16770.83.3	16780.55.5	16790.27.7	16800.00.0	3500
3600	17210.00.0	17220.00.0	17230.00.0	17240.00.0	17250.00.0	17260.00.0	17270.00.0	17280.00.0	3600
3700	17688.05.5	17698.33.3	17708.61.1	17718.88.8	17729.16.6	17739.44.4	17749.72.2	17760.00.0	3700
3800	18166.11.1	18176.66.6	18187.22.2	18197.77.7	18208.33.3	18218.88.8	18229.44.4	18240.00.0	3800
3900	18644.16.6	18655.00.0	18665.83.3	18676.66.6	18687.50.0	18698.33.3	18709.16.6	18720.00.0	3900
4000	19122.22.2	19133.33.3	19144.44.4	19155.55.5	19166.66.6	19177.77.7	19188.88.8	19200.00.0	4000
4100	19600.27.7	19611.66.6	19623.05.5	19634.44.4	19645.83.3	19657.22.2	19668.61.1	19680.00.0	4100
4200	20078.33.3	20090.00.0	20101.66.6	20113.33.3	20125.00.0	20136.66.6	20148.33.3	20160.00.0	4200
4300	20556.38.8	20568.33.3	20580.27.7	20592.22.2	20604.16.6	20616.11.1	20628.05.5	20640.00.0	4300
4400	21034.44.4	21046.66.6	21058.88.8	21071.11.1	21083.33.3	21095.55.5	21107.77.7	21120.00.0	4400
4500	21512.50.0	21525.00.0	21537.50.0	21550.00.0	21562.50.0	21575.00.0	21587.50.0	21600.00.0	4500
4600	21990.55.5	22003.33.3	22016.11.1	22028.88.8	22041.66.6	22054.44.4	22067.22.2	22080.00.0	4600
4700	22468.61.1	22481.66.6	22494.72.2	22507.77.7	22520.83.3	22533.88.8	22546.94.4	22560.00.0	4700
4800	22946.66.6	22960.00.0	22973.33.3	22986.66.6	23000.00.0	23013.33.3	23026.66.6	23040.00.0	4800
4900	23424.72.2	23438.33.3	23451.94.4	23465.55.5	23479.16.6	23492.77.7	23506.38.8	23520.00.0	4900
5000	23902.77.7	23916.66.6	23930.55.5	23944.44.4	23958.33.3	23972.22.2	23986.11.1	24000.00.0	5000

s.	$7\frac{9}{16}$	$7\frac{5}{8}$	$7\frac{11}{16}$	$7\frac{3}{4}$	$7\frac{13}{16}$	$7\frac{7}{8}$	$7\frac{15}{16}$	8	s.
	$ cts.	$ cts.	$ cts.	$ cts.	$ cts.	$ cts.	$ cts.	$ cts.	
1	.23.9	.23.9	.23.9	.23.9	.23.9	.24.0	.24.0	.24.0	1
2	.47.8	.47.8	.47.9	.47.9	.47.9	.47.9	.48.0	.48.0	2
3	.71.7	.71.7	.71.8	.71.8	.71.9	.71.9	.71.9	.72.0	3
4	.95.6	.95.7	.95.7	.95.8	.95.8	.95.9	.95.9	.96.0	4
5	1.19.5	1.19.6	1.19.6	1.19.7	1.19.8	1.19.9	1.19.9	1.20.0	5
6	1.43.4	1.43.5	1.43.6	1.43.7	1.43.7	1.43.8	1.43.9	1.44.0	6
7	1.67.3	1.67.4	1.67.5	1.67.6	1.67.7	1.67.8	1.67.9	1.68.0	7
8	1.91.2	1.91.3	1.91.4	1.91.5	1.91.7	1.91.8	1.91.9	1.92.0	8
9	2.15.1	2.15.2	2.15.4	2.15.5	2.15.6	2.15.7	2.15.9	2.16.0	9
10	2.39.0	2.39.2	2.39.3	2.39.4	2.39.6	2.39.7	2.39.9	2.40.0	10
11	2.62.9	2.63.1	2.63.2	2.63.4	2.63.5	2.63.7	2.63.8	2.64.0	11
12	2.86.8	2.87.0	2.87.2	2.87.3	2.87.5	2.87.7	2.87.8	2.88.0	12
13	3.10.7	3.10.9	3.11.1	3.11.3	3.11.4	3.11.6	3.11.8	3.12.0	13
14	3.34.6	3.34.8	3.35.0	3.35.2	3.35.4	3.35.6	3.35.8	3.36.0	14
15	3.58.5	3.58.7	3.58.9	3.59.1	3.59.4	3.59.6	3.59.8	3.60.0	15

STERLING EXCHANGE TABLES.

$7\frac{9}{16}$ TO 8 %

STERLING INTO DOLLARS AND CENTS.

Stg.	$7\frac{9}{16}$	$7\frac{5}{8}$	$7\frac{11}{16}$	$7\frac{3}{4}$	$7\frac{13}{16}$	$7\frac{7}{8}$	$7\frac{15}{16}$	8	Stg.
£	$ cts.	$ cts.	$ cts.	$ cts.	$ cts.	$ cts.	$ cts.	$ cts.	£
5100..	24380.83,3	24395.00,0	24409.16,6	24423.33,3	24437.50,0	24451.66,6	24465.83,3	24480.00,0	..5100
5200..	24858.88,8	24873.33,3	24887.77,7	24902.22,2	24916.66,6	24931.11,1	24945.55,5	24960.00,0	..5200
5300..	25336.94,4	25351.66,6	25366.38,8	25381.11,1	25395.83,3	25410.55,5	25425.27,7	25440.00,0	..5300
5400..	25815.00,0	25830.00,0	25845.00,0	25860.00,0	25875.00,0	25890.00,0	25905.00,0	25920.00,0	..5400
5500..	26293.05,5	26308.33,3	26323.61,1	26338.88,8	26354.16,6	26369.44,4	26384.72,2	26400.00,0	..5500
5600..	26771.11,1	26786.66,6	26802.22,2	26817.77,7	26833.33,3	26848.88,8	26864.44,4	26880.00,0	..5600
5700..	27249.16,6	27265.00,0	27280.83,3	27296.66,6	27312.50,0	27328.33,3	27344.16,6	27360.00,0	..5700
5800..	27727.22,2	27743.33,3	27759.44,4	27775.55,5	27791.66,6	27807.77,7	27823.88,8	27840.00,0	..5800
5900..	28205.27,7	28221.66,6	28238.05,5	28254.44,4	28270.83,3	28287.22,2	28303.61,1	28320.00,0	..5900
6000..	28683.33,3	28700.00,0	28716.66,6	28733.33,3	28750.00,0	28766.66,6	28783.33,3	28800.00,0	..6000
6100..	29161.38,8	29178.33,3	29195.27,7	29212.22,2	29229.16,6	29246.11,1	29263.05,5	29280.00,0	..6100
6200..	29639.44,4	29656.66,6	29673.88,8	29691.11,1	29708.33,3	29725.55,5	29742.77,7	29760.00,0	..6200
6300..	30117.50,0	30135.00,0	30152.50,0	30170.00,0	30187.50,0	30205.00,0	30222.50,0	30240.00,0	..6300
6400..	30595.55,5	30613.33,3	30631.11,1	30648.88,8	30666.66,6	30684.44,4	30702.22,2	30720.00,0	..6400
6500..	31073.61,1	31091.66,6	31109.72,2	31127.77,7	31145.83,3	31163.88,8	31181.94,4	31200.00,0	..6500
6600..	31551.66,6	31570.00,0	31588.33,3	31606.66,6	31625.00,0	31643.33,3	31661.66,6	31680.00,0	..6600
6700..	32029.72,2	32048.33,3	32066.94,4	32085.55,5	32104.16,6	32122.77,7	32141.38,8	32160.00,0	..6700
6800..	32507.77,7	32526.66,6	32545.55,5	32564.44,4	32583.33,3	32602.22,2	32621.11,1	32640.00,0	..6800
6900..	32985.83,3	33005.00,0	33024.16,6	33043.33,3	33062.50,0	33081.66,6	33100.83,3	33120.00,0	..6900
7000..	33463.88,8	33483.33,3	33502.77,7	33522.22,2	33541.66,6	33561.11,1	33580.55,5	33600.00,0	..7000
7100..	33941.94,4	33961.66,6	33981.38,8	34001.11,1	34020.83,3	34040.55,5	34060.27,7	34080.00,0	..7100
7200..	34420.00,0	34440.00,0	34460.00,0	34480.00,0	34500.00,0	34520.00,0	34540.00,0	34560.00,0	..7200
7300..	34898.05,5	34918.33,3	34938.61,1	34958.88,8	34979.16,6	34999.44,4	35019.72,2	35040.00,0	..7300
7400..	35376.11,1	35396.66,6	35417.22,2	35437.77,7	35458.33,3	35478.88,8	35499.44,4	35520.00,0	..7400
7500..	35854.16,6	35875.00,0	35895.83,3	35916.66,6	35937.50,0	35958.33,3	35979.16,6	36000.00,0	..7500
7600..	36332.22,2	36353.33,3	36374.44,4	36395.55,5	36416.66,6	36437.77,7	36458.88,8	36480.00,0	..7600
7700..	36810.27,7	36831.66,6	36853.05,5	36874.44,4	36895.83,3	36917.22,2	36938.61,1	36960.00,0	..7700
7800..	37288.33,3	37310.00,0	37331.66,6	37353.33,3	37375.00,0	37396.66,6	37418.33,3	37440.00,0	..7800
7900..	37766.38,8	37788.33,3	37810.27,7	37832.22,2	37854.16,6	37876.11,1	37898.05,5	37920.00,0	..7900
8000..	38244.44,4	38266.66,6	38288.88,8	38311.11,1	38333.33,3	38355.55,5	38377.77,7	38400.00,0	..8000
8100..	38722.50,0	38745.00,0	38767.50,0	38790.00,0	38812.50,0	38835.00,0	38857.50,0	38880.00,0	..8100
8200..	39200.55,5	39223.33,3	39246.11,1	39268.88,8	39291.66,6	39314.44,4	39337.22,2	39360.00,0	..8200
8300..	39678.61,1	39701.66,6	39724.72,2	39747.77,7	39770.83,3	39793.88,8	39816.94,4	39840.00,0	..8300
8400..	40156.66,6	40180.00,0	40203.33,3	40226.66,6	40250.00,0	40273.33,3	40296.66,6	40320.00,0	..8400
8500..	40634.72,2	40658.33,3	40681.94,4	40705.55,5	40729.16,6	40752.77,7	40776.38,8	40800.00,0	..8500
8600..	41112.77,7	41136.66,6	41160.55,5	41184.44,4	41208.33,3	41232.22,2	41256.11,1	41280.00,0	..8600
8700..	41590.83,3	41615.00,0	41639.16,6	41663.33,3	41687.50,0	41711.66,6	41735.83,3	41760.00,0	..8700
8800..	42068.88,8	42093.33,3	42117.77,7	42142.22,2	42166.66,6	42191.11,1	42215.55,5	42240.00,0	..8800
8900..	42546.94,4	42571.66,6	42596.38,8	42621.11,1	42645.83,3	42670.55,5	42695.27,7	42720.00,0	..8900
9000..	43025.00,0	43050.00,0	43075.00,0	43100.00,0	43125.00,0	43150.00,0	43175.00,0	43200.00,0	..9000
9100..	43503.05,5	43528.33,3	43553.61,1	43578.88,8	43604.16,6	43629.44,4	43654.72,2	43680.00,0	..9100
9200..	43981.11,1	44006.66,6	44032.22,2	44057.77,7	44083.33,3	44108.88,8	44134.44,4	44160.00,0	..9200
9300..	44459.16,6	44485.00,0	44510.83,3	44536.66,6	44562.50,0	44588.33,3	44614.16,6	44640.00,0	..9300
9400..	44937.22,2	44963.33,3	44989.44,4	45015.55,5	45041.66,6	45067.77,7	45093.88,8	45120.00,0	..9400
9500..	45415.27,7	45441.66,6	45468.05,5	45494.44,4	45520.83,3	45547.22,2	45573.61,1	45600.00,0	..9500
9600..	45893.33,3	45920.00,0	45946.66,6	45973.33,3	46000.00,0	46026.66,6	46053.33,3	46080.00,0	..9600
9700..	46371.38,8	46398.33,3	46425.27,7	46452.22,2	46479.16,6	46506.11,1	46533.05,5	46560.00,0	..9700
9800..	46849.44,4	46876.66,6	46903.88,8	46931.11,1	46958.33,3	46985.55,5	47012.77,7	47040.00,0	..9800
9900..	47327.50,0	47355.00,0	47382.50,0	47410.00,0	47437.50,0	47465.00,0	47492.50,0	47520.00,0	..9900
10000..	47805.55,5	47833.33,3	47861.11,1	47888.88,8	47916.66,6	47944.44,4	47972.22,2	48000.00,0	..10000

s.	$ cts.	$ cts.	$ cts.	$ cts.	$ cts.	$ cts.	$ cts.	$ cts.	s.
16..	3.82,4	3.82,7	3.82,9	3.83,1	3.83,3	3.83,5	3.83,8	3.84,0	..16
17..	4.06,3	4.06,6	4.06,8	4.07,0	4.07,3	4.07,5	4.07,8	4.08,0	..17
18..	4.30,2	4.30,5	4.30,7	4.31,0	4.31,2	4.31,5	4.31,7	4.32,0	..18
19..	4.54,1	4.54,4	4.54,7	4.51,9	4.55,2	4.55,5	4.55,7	4.56,0	..19
1d.	.02,0	.02,0	.02,0	.02,0	.02,0	.02,0	.02,0	.02,0	..1d.
2..	.04,0	.04,0	.04,0	.04,0	.04,0	.04,0	.04,0	.04,0	..2
3..	.06,0	.06,0	.06,0	.06,0	.06,0	.06,0	.06,0	.06,0	..3
4..	.08,0	.08,0	.08,0	.08,0	.08,0	.03,0	.08,0	.08,0	..4
5..	.09,9	.10,0	.10,0	.10,0	.10,0	.10,0	.10,0	.10,0	..5
6..	.11,9	.11,9	.12,0	.12,0	.12,0	.12,0	.12,0	.12,0	..6
7..	.13,9	.13,9	.13,9	.14,0	.14,0	.14,0	.14,0	.14,0	..7
8..	.15,9	.15,9	.16,0	.16,0	.16,0	.16,0	.16,0	.16,0	..8
9..	.17,9	.17,9	.17,9	.17,9	.18,0	.18,0	.18,0	.18,0	..9
10..	.19,9	.19,9	.19,9	.19,9	.20,0	.20,0	.20,0	.20,0	..10
11..	.21,9	.21,9	.21,9	.21,9	.22,0	.22,0	.22,0	.22,0	..11

STERLING EXCHANGE TABLES.

$8\frac{1}{16}$ TO $8\frac{1}{2}\%$

STERLING INTO DOLLARS AND CENTS.

Stg.	$8\frac{1}{16}$	$8\frac{1}{8}$	$8\frac{3}{16}$	$8\frac{1}{4}$	$8\frac{5}{16}$	$8\frac{3}{8}$	$8\frac{7}{16}$	$8\frac{1}{2}$	Stg.
£	$ cts.	$ cts.	$ cts.	$ cts.	$ cts.	$ cts.	$ cts.	$ cts.	£
100..	480.27.7	480.55.5	480.83,3	481.11.1	481.38.8	481.66.6	481.94.4	482.22.2	..100
200..	960.55.5	961.11.1	961.66.6	962.22.2	062.77.7	963.33.3	963.88.8	964.44.4	..200
300..	1440.83,3	1441.66.6	1442.50.0	1443.33,3	1444.16.6	1445.00.0	1445.83.3	1446.66.6	..300
400..	1921.11.1	1922.22.2	1923.33,3	1924.44.4	1925.55.5	1926.66.6	1927.77.7	1928.88.8	..400
500..	2401.38.8	2402.77.7	2404.16.6	2405.55.5	2406.94.4	2408.33.3	2409.72.2	2411.11.1	..500
600..	2881.66.6	2883.33.3	2885.00.0	2886.66.6	2888.33.3	2890.00.0	2891.66.6	2893.33,3	..600
700..	3361.94.4	3363.88.8	3365.83,3	3367.77.7	3369.72.2	3371.66.6	3373.61.1	3375.55.5	..700
800..	3842.22.2	3844.44.4	3846.66.6	3848.68.8	3851.11.1	3853.33.3	3855.55.5	3857.77.7	..800
900..	4322.50.0	4325.00.0	4327.50.0	4330.00.0	4332.50.0	4335.00.0	4337.50.0	4340.00.0	..900
1000..	4802.77.7	4805.55.5	4808.33.3	4811.11.1	4813.88.8	4816.66.6	4819.44.4	4822.22.2	..1000
1100..	5283.05.5	5286.11.1	5289.16.6	5292.22.2	5295.27.7	5298.33.3	5301.38.8	5304.44 4	..1100
1200..	5763.33.3	5766.66.6	5770.00.0	5773.33,3	5776.66.6	5780.00.0	5783.33.3	5786.66.6	..1200
1300..	6243.61.1	6247.22.2	6250.83,3	6254.44.4	6258.05.5	6261.66.6	6265.27.7	6268.88.8	..1300
1400..	6723.88,8	6727.77.7	6731.66.6	6735.55.5	6739.44.4	6743.33.3	6747.22.2	6751.11.1	..1400
1500..	7204.16.6	7208.33.3	7212.50 0	7216.66.6	7220.83,3	7225.00.0	7229.16.6	7233.33,3	..1500
1600..	7684.44.4	7688.88.8	7693.33,3	7697.77.7	7702.22.2	7706.66.6	7711.11.1	7715.55.5	..1600
1700..	8164.72.2	8169.44.4	8174.16.6	8178.88.8	8183.61.1	8188.33.3	8193.05.5	8197.77.7	..1700
1800..	8645.00.0	8650.00.0	8655.00.0	8660.00.0	8665.00.0	8670 00.0	8675.00.0	8680.00.0	..1800
1900..	9125.27.7	9130.55.5	9135.83,3	9141.11.1	9146.38.8	9151.66.6	9156.94.4	9162.22.2	..1900
2000 .	9605.55.5	9611.11.1	9616.66.6	9622.22.2	9627.77.7	9633.33.3	9638.88.8	9644.44.4	..2000
2100..	10085.83,3	10091.66.6	10097.50.0	10103.33.3	10109.16.6	10115.00.0	10120.83.3	10126.66.6	..2100
2200..	10566.11.1	10572.22.2	10578.33.3	10584.44.4	10590.55.5	10596.66.6	10602.77.7	10608.88.8	..2200
2300..	11046.38.8	11052.77.7	11059.16.6	11065.55.5	11071.94.4	11078.33.3	11084.72.2	11091.11.1	..2300
2400..	11526.66.6	11533.33.3	11540.00.0	11546.66.6	11553.33.3	11560.00.0	11566.66.6	11573.33,3	..2400
2500..	12006.94.4	12013.88.8	12020.83.3	12027.77.7	12034.72.2	12041.66.6	12048.61.1	12055 55.5	..2500
2600..	12487.22.2	12494.44.4	12501.66.6	12508.88.8	12516.11.1	12523.33.3	12530.55.5	12537.77.7	..2600
2700..	12967.50.0	12975.00.0	12982.50.0	12990.00.0	12997.50.0	13005.00.0	13012.50.0	13020.00.0	..2700
2800..	13447.77.7	13455.55.5	13463.33.3	13471.11.1	13478.88.8	13486.66.6	13494.44.4	13502.22.2	..2800
2900..	13928.05.5	13936.11.1	13944.16.6	13952.22.2	13960.27.7	13968.33.3	13976.38.8	13984.44.4	..2900
3000..	14408.33.3	14416.66.6	14425.00.0	14433.33.3	14441.66.6	14450.00.0	14458.33.3	14466.66.6	3000
3100.	14888.61.1	14897.22.2	14905.83.3	14914.44.4	14923.05.5	14931.66.6	14940.27.7	14948.88.8	..3100
3200..	15368.88.8	15377.77.7	15386.66.6	15395.55.5	15404.44.4	15413.33.3	15422.22.2	15431.11.1	..3200
3300..	15849.16.6	15858.33.3	15867.50.0	15876.66.6	15885.83.3	15895.00.0	15904.16.6	15913.33.3	..3300
3400..	16329.44.4	16338.88.8	16348.33.3	16357.77.7	16367.22.2	16376.66.6	16386.11.1	16395.55.5	..3400
3500..	16809.72.2	16819.44.4	16829.16.6	16838.88.8	16848.61.1	16858.33.3	16868.05.5	16877.77.7	..3500
3600..	17290.00.0	17300.00.0	17310.00.0	17320.00.0	17330.00.0	17340.00.0	17350.00.0	17360.00.0	..3600
3700..	17770.27.7	17780.55.5	17790.83.3	17801.11.1	17811.38.8	17821.66.6	17831.94.4	17842.22.2	..3700
3800..	18250.55.5	18261.11.1	18271.66.6	18282.22.2	18292.77.7	18303.33.3	18313.88.8	18324.44.4	..3800
3900..	18730.83.3	18741.66.6	18752.50.0	18763.33.3	18774.16.6	18785.00.0	18795.83.3	18806.66.6	..3900
4000..	19211.11.1	19222.22.2	19233.33,3	19244.44.4	19255.55.5	19266.66.6	19277.77.7	19288.88.8	..4000
4100..	19691.38.8	19702.77.7	19714.16.6	19725.55.5	19736.94.4	19748.33.3	19759.72.2	19771.11.1	..4100
4200..	20171.66.6	20183.33.3	20195.00.0	20206.66.6	20218.33.3	20230 00.0	20241.66.6	20253.33.3	..4200
4300..	20651.94.4	20663.88.8	20675.83.3	20687.77.7	20699.72.2	20711.66.6	20723.61.1	20735.55.5	..4300
4400..	21132.22.2	21144.44.4	21156.66.6	21168.88.8	21181.11.1	21193.33.3	21205.55.5	21217.77.7	..4400
4500..	21612.50.0	21625.00.0	21637.50.0	21650.00.0	21662.50.0	21675.00.0	21687 50.0	21700.00 0	..4500
4600..	22092.77.7	22105.55.5	22118.33.3	22131.11.1	22143.88.8	22156.66.6	22169.44.4	22182.22.2	..4600
4700..	22573.05.5	22586.11.1	22599.16.6	22612.22.2	22625.27.7	22638.33.3	22651.38.8	22664.44.4	..4700
4800..	23053.33.3	23066.66.6	23080.00.0	23093.33.3	23106.66.6	23120.00.0	23133.33.3	23146.66.6	..4800
4900..	23533.61.1	23547.22.2	23560.83.3	23574.44.4	23588.05.5	23601.66.6	23615.27.7	23628.88.8	..4900
5000..	24013.88.8	24027.77.7	24041.66.6	24055.55.5	24069.44.4	24083.33.3	24097.22.2	24111.11.1	..5000
s.	$ cts.	$ cts.	$ cts.	$ cts.	$ cts.	$ cts.	$ cts.	$ cts.	s.
1..	.24.0	.24.0	.24.0	.24.0	.24.1	.24.1	.24.1	.24.1	..1
2..	.48.0	.48.0	.48.1	.48.1	.48.1	.48.2	.48.2	.48.2	..2
3..	.72.0	.72.1	.72.1	.72.2	.72.2	.72.2	.72.3	.72.3	..3
4..	.96.0	.96.1	.96.2	.96 2	.96.3	.96.3	.96.4	.96.4	..4
5..	1.20.1	1.20.1	1.20.2	1.20.3	1.20.3	1.20.4	1.20.5	1.20.6	..5
6..	1.44.1	1.44.2	1.44.2	1.44.3	1.44.4	1.44.5	1.44.6	1.44.7	..6
7..	1.68.1	1.68.2	1.68.3	1.68.4	1.68.5	1.68.6	1.68.7	1.68.8	..7
8..	1.92.1	1.92.2	1.92.3	1.92.4	1.92.5	1.92.7	1.92.8	1.92.9	..8
9..	2.16.1	2.16.2	2.16.4	2.16.5	2.16.6	2.16.7	2.16.9	2 17.0	..9
10..	2.40 1	2.40.3	2.40.4	2.40.5	2.40.7	2.40.8	2.41.0	2.41.1	..10
11..	2.64.1	2.64.3	2.64.4	2.64.6	2.64.8	2.64.9	2 65.1	2.65.2	..11
12..	2.88.2	2.88.3	2.88.5	2.88.7	2.88.8	2.89.0	2.89.2	2.89.3	..12
13..	3.12.2	3.12.4	3.12.5	3.12.7	3.12.9	3.13.1	3.13.3	3.13.4	..13
14..	3.36.2	3.36.4	3.36.6	3.36.8	3.37.0	3.37.2	3.37.4	3.37.5	..14
15..	3.60.2	3.60.4	3.60.6	3.60.8	3.61.0	3.61.2	3.61.4	3.61.7	..15

STERLING EXCHANGE TABLES.

$8\frac{1}{16}$ TO $8\frac{1}{2}\%$

STERLING INTO DOLLARS AND CENTS.

Stg.	$8\frac{1}{16}$	$8\frac{1}{8}$	$8\frac{3}{16}$	$8\frac{1}{4}$	$8\frac{5}{16}$	$8\frac{3}{8}$	$8\frac{7}{16}$	$8\frac{1}{2}$	Stg.
£	$ cts.	$ cts.	$ cts.	$ cts.	$ cts.	$ cts.	$ cts.	$ cts.	£
5100..	24494.16.6	24508.33.3	24522.50.0	24536.66.6	24550.83.3	24565.00.0	24579.16.6	24593.33.3	..5100
5200..	24974.44.4	24988.88.8	25003.33.3	25017.77.7	25032.22.2	25046.66.6	25061.11.1	25075.55.5	..5200
5300..	25454.72.2	25469.44.4	25484.16.6	25498.88.8	25513.61.1	25528.33.3	25543.05.5	25557.77.7	..5300
5400..	25935.00.0	25950.00.0	25965.00.0	25980.00.0	25995.00.0	26010.00.0	26025.00.0	26040.00.0	..5400
5500..	26415.27.7	26430.55.5	26445.83.3	26461.11.1	26476.38.8	26491.66.6	26506.94.4	26522.22.2	..5500
5600..	26895.55.5	26911.11.1	26926.66.6	26942.22.2	26957.77.7	26973.33.3	26988.88.8	27004.44.4	..5600
5700..	27375.83.3	27391.66.6	27407.50.0	27423.33.3	27439.16.6	27455.00.0	27470.83.3	27486.66.6	..5700
5800..	27856.11.1	27872.22.2	27888.33.3	27904.44.4	27920.55.5	27936.66.6	27952.77.7	27968.88.8	..5800
5900..	28336.38.8	28352.77.7	28369.16.6	28385.55.5	28401.94.4	28418.33.3	28434.72.2	28451.11.1	..5900
6000..	28816.66.6	28833.33.3	28850.00.0	28866.66.6	28883.33.3	28900.00.0	28916.66.6	28933.33.3	..6000
6100..	29296.94.4	29313.88.8	29330.83.3	29347.77.7	29364.72.2	29381.66.6	29398.61.1	29415.55.5	..6100
6200..	29777.22.2	29794.44.4	29811.66.6	29829.88.8	29846.11.1	29863.33.3	29880.55.5	29897.77.7	..6200
6300..	30257.50.0	30275.00.0	30292.50.0	30310.00.0	30327.50.0	30345.00.0	30362.50.0	30380.00.0	..6300
6400..	30737.77.7	30755.55.5	30773.33.3	30791.11.1	30808.88.8	30826.66.6	30844.44.4	30862.22.2	..6400
6500..	31218.05.5	31236.11.1	31254.16.6	31272.22.2	31290.27.7	31308.33.3	31326.38.8	31344.44.4	..6500
6600..	31698.33.3	31716.66.6	31735.00.0	31753.33.3	31771.66.6	31790.00.0	31808.33.3	31826.66.6	..6600
6700..	32178.61.1	32197.22.2	32215.83.3	32234.44.4	32253.05.5	32271.66.6	32290.27.7	32308.88.8	..6700
6800..	32658.88.8	32677.77.7	32696.66.6	32715.55.5	32734.44.4	32753.33.3	32772.22.2	32791.11.1	..6800
6900..	33139.16.6	33158.33.3	33177.50.0	33196.66.6	33215.83.3	33235.00.0	33254.16.6	33273.33.3	..6900
7000..	33619.44.4	33638.88.8	33658.33.3	33677.77.7	33697.22.2	33716.66.6	33736.11.1	33755.55.5	..7000
7100..	34099.72.2	34119.44.4	34139.16.6	34158.88.8	34178.61.1	34198.33.3	34218.05.5	34237.77.7	..7100
7200..	34580.00.0	34600.00.0	34620.00.0	34640.00.0	34660.00.0	34680.00.0	34700.00.0	34720.00.0	..7200
7300..	35060.27.7	35080.55.5	35100.83.3	35121.11.1	35141.38.8	35161.66.6	35181.94.4	35202.22.2	..7300
7400..	35540.55.5	35561.11.1	35581.66.6	35602.22.2	35622.77.7	35643.33.3	35663.88.8	35684.44.4	..7400
7500..	36020.83.3	36041.66.6	36062.50.0	36083.33.3	36104.16.6	36125.00.0	36145.83.3	36166.66.6	..7500
7600..	36501.11.1	36522.22.2	36543.33.3	36564.44.4	36585.55.5	36606.66.6	36627.77.7	36648.88.8	..7600
7700..	36981.38.8	37002.77.7	37024.16.6	37045.55.5	37066.94.4	37088.33.3	37109.72.2	37131.11.1	..7700
7800..	37461.66.6	37483.33.3	37505.00.0	37526.66.6	37548.33.3	37570.00.0	37591.66.6	37613.33.3	..7800
7900..	37941.94.4	37963.88.8	37985.83.3	38007.77.7	38029.72.2	38051.66.6	38073.61.1	38095.55.5	..7900
8000..	38422.22.2	38444.44.4	38466.66.6	38488.88.8	38511.11.1	38533.33.3	38555.55.5	38577.77.7	..8000
8100..	38902.50.0	38925.00.0	38947.50.0	38970.00.0	38992.50.0	39015.00.0	39037.50.0	39060.00.0	..8100
8200..	39382.77.7	39405.55.5	39428.33.3	39451.11.1	39473.88.8	39496.66.6	39519.44.4	39542.22.2	..8200
8300..	39863.05.5	39886.11.1	39909.16.6	39932.22.2	39955.27.7	39978.33.3	40001.38.8	40024.44.4	..8300
8400..	40343.33.3	40366.66.6	40390.00.0	40413.33.3	40436.66.6	40460.00.0	40483.33.3	40506.66.6	..8400
8500..	40823.61.1	40847.22.2	40870.83.3	40894.44.4	40918.05.5	40941.66.6	40965.27.7	40988.88.8	..8500
8600..	41303.88.8	41327.77.7	41351.66.6	41375.55.5	41399.44.4	41423.33.3	41447.22.2	41471.11.1	..8600
8700..	41784.16.6	41808.33.3	41832.50.0	41856.66.6	41880.83.3	41905.00.0	41929.16.6	41953.33.3	..8700
8800..	42264.44.4	42288.88.8	42313.33.3	42337.77.7	42362.22.2	42386.66.6	42411.11.1	42435.55.5	..8800
8900..	42744.72.2	42769.44.4	42794.16.6	42818.88.8	42843.61.1	42868.33.3	42893.05.5	42917.77.7	..8900
9000..	43225.00.0	43250.00.0	43275.00.0	43300.00.0	43325.00.0	43350.00.0	43375.00.0	43400.00.0	..9000
9100..	43705.27.7	43730.55.5	43755.83.3	43781.11.1	43806.38.8	43831.66.6	43856.94.4	43882.22.2	..9100
9200..	44185.55.5	44211.11.1	44236.66.6	44262.22.2	44287.77.7	44313.33.3	44338.88.8	44364.44.4	..9200
9300..	44665.83.3	44691.66.6	44717.50.0	44743.33.3	44769.16.6	44795.00.0	44820.83.3	44846.66.6	..9300
9400..	45146.11.1	45172.22.2	45198.33.3	45224.44.4	45250.55.5	45276.66.6	45302.77.7	45328.88.8	..9400
9500..	45626.38.8	45652.77.7	45679.16.6	45705.55.5	45731.94.4	45758.33.3	45784.72.2	45811.11.1	..9500
9600..	46106.66.6	46133.33.3	46160.00.0	46186.66.6	46213.33.3	46240.00.0	46266.66.6	46293.33.3	..9600
9700..	46586.94.4	46613.88.8	46640.83.3	46667.77.7	46694.72.2	46721.66.6	46748.61.1	46775.55.5	..9700
9800..	47067.22.2	47094.44.4	47121.66.6	47148.88.8	47176.11.1	47203.33.3	47230.55.5	47257.77.7	..9800
9900..	47547.50.0	47575.00.0	47602.50.0	47630.00.0	47657.50.0	47685.00.0	47712.50.0	47740.00.0	..9900
10000..	48027.77.7	48055.55.5	48083.33.3	48111.11.1	48138.88.8	48166.66.6	48194.44.4	48222.22.2	..10000

s.	$ cts.	$ cts.	$ cts.	$ cts.	$ cts.	$ cts.	$ cts.	$ cts.	s.
16..	3.84.2	3.84.4	3.84.7	3.84.9	3.85.1	3.85.3	3.85.5	3.85.8	..16
17..	4.08.2	4.08.5	4.08.7	4.08.9	4.09.2	4.09.4	4.09.6	4.09.9	..17
18..	4.32.2	4.32.5	4.32.7	4.33.0	4.33.2	4.33.5	4.33.7	4.34.0	..18
19..	4.56.3	4.56.5	4.56.8	4.57.0	4.57.3	4.57.6	4.57.8	4.58.1	..19
1d.	.02.0	.02.0	.02.0	.02.0	.02.0	.02.0	.02.0	.02.0	..1d.
2..	.04.0	.04.0	.04.0	.04.0	.04.0	.04.0	.04.0	.01.0	..2
3..	.06.0	.06.0	.06.0	.06.0	.06.0	.06.0	.06.0	.06.0	..3
4..	.08.0	.08.0	.08.0	.08.0	.08.0	.08.0	.08.0	.08.0	..4
5..	.10.0	.10.0	.10.0	.10.0	.10.0	.10.0	.10.0	.10.0	..5
6..	.12.0	.12.0	.12.0	.12.0	.12.0	.12.0	.12.0	.12.0	..6
7..	.14.0	.14.0	.14.0	.14.0	.14.0	.14.0	.14.0	.14.1	..7
8..	.16.0	.16.0	.16.0	.16.0	.16.0	.16.0	.16.1	.16.1	..8
9..	.18.0	.18.0	.18.0	.18.0	.18.0	.18.1	.18.1	.18.1	..9
10..	.20.0	.20.0	.20.0	.20.0	.20.0	.20.1	.20.1	.20.1	..10
11..	.22.0	.22.0	.22.0	.22.0	.22.1	.22.1	.22.1	.22.1	..11

STERLING EXCHANGE TABLES.

$8\frac{9}{16}$ TO 9 %

STERLING INTO DOLLARS AND CENTS.

Stg. £	$8\frac{9}{16}$ $ cts.	$8\frac{5}{8}$ $ cts.	$8\frac{11}{16}$ $ cts.	$8\frac{3}{4}$ $ cts.	$8\frac{13}{16}$ $ cts.	$8\frac{7}{8}$ $ cts.	$8\frac{15}{16}$ $ cts.	9 $ cts.	Stg. £
100	482.50.0	482.77.7	483.05.5	483.33.3	483.61.1	483.88.8	484.16.6	484.44.4	100
200	965.00.0	965.55.5	966.11.1	966.66.6	967.22.2	967 77.7	968.33.3	968.88.8	200
300	1447.50.0	1448.33.3	1449.16.6	1450.00.0	1450.83.3	1451.66.6	1452.50.0	1453.33.3	300
400	1930.00.0	1931.11.1	1932.22.2	1933.33.3	1934.44.4	1935.55.5	1936.66.6	1937.77.7	400
500	2412.50.0	2413.88.8	2415.27.7	2416.66.6	2418.05 5	2419.44.4	2420.83 3	2422.22.2	500
600	2895.00.0	2896.66.6	2898.33.3	2900.00.0	2901.66.6	2903.33.3	2905.00.0	2906.66.6	600
700	3377.50.0	3379.44.4	3381.38 8	3383.33.3	3385.27.7	3387.22.2	3389.16.6	3391.11.1	700
800	3860.00.0	3862.22.2	3864.44.4	3866.66.6	3868.88.8	3871.11.1	3873.33.3	3875.55.5	800
900	4342.50.0	4345.00.0	4347.50.0	4350.00.0	4352.50.0	4355.00.0	4357.50.0	4360.00.0	900
1000	4825.00.0	4827.77.7	4830.55.5	4833.33.3	4836.11.1	4838.88.8	4841.66 6	4844.44.4	1000
1100	5307.50.0	5310.55.5	5313.61.1	5316.66 6	5319.72.2	5322.77.7	5325.83.	5328.88.8	1100
1200	5790.00.0	5793.33.3	5796.66.6	5800.00.0	5803.33.3	5806.66.6	5810.00.0	5813.33.3	1200
1300	6272.50.0	6276.11.1	6279.72.2	6283.33.3	6286.94.4	6290.55.5	6294.16.6	6297.77.7	1300
1400	6755.00.0	6758.88.8	6762.77.7	6766.66.6	6770.55.5	6774.44.4	6778.33.3	6782.22.2	1400
1500	7237.50.0	7241.66.6	7245.83.3	7250.00.0	7254.16.6	7258.33 3	7262.50.0	7266.66.6	1500
1600	7720.00.0	7724.44.4	7728.88.8	7733.33.3	7737.77.7	7742.22.2	7746.66.6	7751.11.1	1600
1700	8202.50.0	8207.22.2	8211.94.4	8216.66.6	8221.38.8	8226.11.1	8230.83.3	8235.55.5	1700
1800	8685.00.0	8690.00.0	8695.00.0	8700.00.0	8705.00.0	8710.00.0	8715.00.0	8720.00 0	1800
1900	9167.50.0	9172.77.7	9178.05.5	9183.33.3	9188.61.1	9193.88.8	9199.16.6	9204 44.4	1900
2000	9650.00.0	9655.55.5	9661.11.1	9666.66.6	9672.22.2	9677.77.7	9683.33.3	9688.88.8	2000
2100	10132.50.0	10138.33.3	10144.16.6	10150.00.0	10155.83.3	10161.66.6	10167.50.0	10173.33.3	2100
2200	10615.00.0	10621.11.1	10627.22.2	10633.33.3	10639.44.4	10645.55.5	10651.66.6	10657.77.7	2200
2300	11097.50.0	11103.88.8	11110.27.7	11116.66.6	11123.05.5	11129.44.4	11135.83.3	11142.22.2	2300
2400	11580.00.0	11586.66.6	11593.33.3	11600.00.0	11606.66.6	11613.33.3	11620.00.0	11626 66.6	2400
2500	12062.50.0	12069.44.4	12076.38.8	12083.33.3	12090.27.7	12097.22.2	12104.16.6	12111.11.1	2500
2600	12545.00.0	12552.22.2	12559.44.4	12566.66.6	12573.88.8	12581.11.1	12588.33.3	12595.55.5	2600
2700	13027.50.0	13035.00.0	13042.50.0	13050.00.0	13057.50.0	13065.00.0	13072.50.0	13080.00.0	2700
2800	13510.00.0	13517.77.7	13525.55.5	13533.33.3	13541.11.1	13548.88.8	13556.66.6	13564.44.4	2800
2900	13992.50.0	14000.55.5	14008.61.1	14016.66.6	14024.72.2	14032.77.7	14040.83.3	14048.88.8	2900
3000	14475.00.0	14483.33.3	14491.66.6	14500.00.0	14508.33.3	14516.66.6	14525.00.0	14533.33.3	3000
3100	14957.50.0	14966.11.1	14974.72.2	14983.33.3	14991.94.4	15000.55.5	15009 16 6	15017.77.7	3100
3200	15440.00.0	15448.88.8	15457.77.7	15466.66.6	15475.55.5	15484.44.4	15493.33.3	15502.22.2	3200
3300	15922.50.0	15931.66.6	15940.83.3	15950.00.0	15959.16.6	15968.33.3	15977.50.0	15986.66 6	3300
3400	16405.00.0	16414.44.4	16423.88.8	16433.33.3	16442.77.7	16452.22.2	16461.66.6	16471.11.1	3400
3500	16887.50.0	16897.22.2	16906.94.4	16916.66.6	16926.38.8	16936.11.1	16945.83.3	16955.55.5	3500
3600	17370.00.0	17380.00.0	17390.00.0	17400.00.0	17410.00.0	17420.00.0	17430.00.0	17440.00.0	3600
3700	17852.50.0	17862.77.7	17873.05.5	17883.33.3	17893.61.1	17903.88.8	17914.16.6	17924.44.4	3700
3800	18335.00.0	18345.55.5	18356.11.1	18366.66.6	18377.22.2	18387.77.7	18398.33.3	18408.88.8	3800
3900	18817.50.0	18828.33.3	18839.16.6	18850.00.0	18860.83.3	18871.66.6	18882.50.0	18893.33.3	3900
4000	19300.00.0	19311.11.1	19322.22.2	19333.33.3	19344.44.4	19355.55.5	19366.66.6	19377.77.7	4000
4100	19782.50.0	19793.88.8	19805.27.7	19816.66.6	19828.05.5	19839.44.4	19850.83.3	19862.22.2	4100
4200	20265.00.0	20276.66.6	20288.33.3	20300.00.0	20311.66.6	20323.33.3	20335.00.0	20346.66.6	4200
4300	20747.50.0	20759.44.4	20771.38.8	20783.33.3	20795.27.7	20807.22.2	20819.16.6	20831.11.1	4300
4400	21230.00.0	21242.22.2	21254.44.4	21266.66.6	21278.88.8	21291.11.1	21303.33.3	21315.55.5	4400
4500	21712.50.0	21725.00.0	21737.50.0	21750.00.0	21762.50.0	21775.00.0	21787.50.0	21800.00.0	4500
4600	22195.00.0	22207.77.7	22220.55.5	22233.33.3	22246.11.1	22258.88.8	22271.66.6	22284.44.4	4600
4700	22677.50.0	22690.55.5	22703.61.1	22716.66.6	22729.72.2	22742.77.7	22755.83.3	22768.88.8	4700
4800	23160.00.0	23173 33.3	23186.66.6	23200.00.0	23213.33.3	23226.66.6	23240.00.0	23253.33.3	4800
4900	23642.50.0	23656.11.1	23669.72.2	23683.33.3	23696.94.4	23710.55.5	23724.16.6	23737.77.7	4900
5000	24125.00.0	24138.88.8	24152.77.7	24166.66.6	24180.55.5	24194.44.4	24208.33.3	24222.22.2	5000
s. $ cts.	$ cts.	$ cts.	$ cts.	$ cts.	$ cts.	$ cts.	$ cts.	$ cts.	s.
1	.24.1	.24.1	.24.1	.24.2	.24.2	.24.2	.24.2	.24.2	1
2	.48.2	.48.3	.48.3	.48.3	.48.4	.48.4	.48.4	.48.4	2
3	.72.4	.72.4	.72.4	.72.5	.72.5	.72.6	.72.6	.72.7	3
4	.96.5	.96.5	.96.6	.96.7	.96·7	.96.8	.96.8	.96.9	4
5	1.20.6	1.20.7	1.20.8	1.20.8	1.20.9	1.21.0	1.21.0	1.21.1	5
6	1.44.7	1.44.8	1.44.9	1.45.0	1.45.1	1.45.2	1.45.2	1.45.3	6
7	1.68.9	1.69.0	1.69.1	1.69.2	1.69.3	1.69.4	1.69.4	1.69.5	7
8	1.93.0	1.93.1	1.93.2	1.93.3	1.93.4	1.93.5	1.93.7	1.93.8	8
9	2.17.1	2.17.2	2.17.4	2.17.5	2.17.6	2.17.7	2.17.9	2.18.0	9
10	2.41.2	2.41.4	2.41.5	2.41.7	2.41.8	2.41.9	2.42.1	2.42.2	10
11	2.65.4	2.65.5	2.65.7	2.65.8	2.66.0	2.66.1	2 66.3	2.66.4	11
12	2.89.5	2.89.7	2.89.8	2.90.0	2.90.2	2.90.3	2.90.5	2.90.7	12
13	3.13.6	3.13.8	3.14.0	3.14.2	3.14.3	3.14.5	3.14.7	3.14.9	13
14	3.37.7	3.37.9	3.38.1	3.38.3	3.38.5	3 38.7	3.38.9	3.39.1	14
15	3.61.9	3.62.1	3.62.3	3.62.5	3.62.7	3.62.9	3.63.1	3.63.3	15

STERLING EXCHANGE TABLES.

$8\frac{9}{16}$ TO 9 %

STERLING INTO DOLLARS AND CENTS.

Stg.	$8\frac{9}{16}$	$8\frac{5}{8}$	$8\frac{11}{16}$	$8\frac{3}{4}$	$8\frac{13}{16}$	$8\frac{7}{8}$	$8\frac{15}{16}$	9	Stg.
£	$ cts.	$ cts.	$ cts.	$ cts.	$ cts.	$ cts.	$ cts.	$ cts.	£
5100..	24607.50 0	24621.66.6	24635.83.3	24650.00.0	24664.16.6	21678.33 3	24692.50.0	24706.66.6	..5100
5200..	25090.00.0	25104.44.4	25118.88.8	25133.33.3	25147.77.7	25162.22.2	25176.66.6	25191.11.1	..5200
5300..	25572.50.0	25587.22.2	25601.94.4	25616.66.6	25631.38.8	25646.11.1	25660.83.3	25675.55.5	..5300
5400..	26055.00.0	26070.00.0	26085.00.0	26100.00.0	26115.0 1.0	26130.00.0	26145.00.0	26160.00.0	..5400
5500..	26537.50.0	26552.77.7	26568.05.5	26583.33.3	26598.61.1	26613.88.8	26629.16.6	26644.44.4	..5500
5600..	27020.00.0	27035.55.5	27051.11.1	27066.66.6	27082.22.2	27097.77.7	27113.33.3	27128.88.8	..5600
5700..	27502.50.0	27518.33.3	27534.16.6	27550.00.0	27565.83.3	27581.66.6	27597.50.0	27613.33.3	..5700
5800..	27985.00.0	28001.11.1	28017.22.2	28033.33.3	28049.44.4	28065.55.5	28081.66.6	28097.77.7	..5800
5900..	28467.50.0	28483.88.8	28500.27.7	28516.66.6	28533.05.5	28549.44.4	28565.83.3	28582.22.2	..5900
6000..	28950.00.0	28966.66.6	28983.33.3	29000.00.0	29016.66.6	29033.33.3	29050.00.0	29066.66.6	..6000
6100..	29432.50.0	29449.44.4	29466.38.8	29483.33.3	29500.27.7	29517.22.2	29534.16.6	29551.11.1	..6100
6200..	29915.00.0	29932.22.2	29949.44.4	29966.66.6	29983 88.8	30001.11.1	30018.33.3	30035.55.5	..6200
6300..	30397.50.0	30415.00.0	30432.50.0	30150.00 0	30467.50.0	30485.00.0	30502.50.0	30520.00.0	..6300
6400..	30880.00.0	30897.77.7	30915.55.5	30933.33 3	30951.11.1	30968.88.8	30986.66.6	31004.44.4	..6400
6500..	31362.50.0	31380.55.5	31398.61.1	31416.66.6	31134.72.2	31452.77.7	31470.83 3	31488.88 8	..6500
6600..	31845.00.0	31863.33 3	31881.66.6	31900.00 0	31918.33.3	31936.66.6	31955.00 0	31973.33.3	..6600
6700..	32327.50.0	32346.11.1	32364.72.2	32383.33.3	32401.91.4	32420.55.5	32439.16 6	32457.77.7	..6700
6800..	32810.00.0	32828.83.8	32847.77.7	32866.66.6	32885.55.5	32904.44.4	32923.33.3	32942.22.2	..6800
6900..	33292.50.0	33311.66.6	33330.83.3	33350.00.0	33369.16.6	33388.33 3	33407.50.0	33426.66.6	..6900
7000..	33775.00.0	33794.44.4	33813.88.8	33833.33.3	33852.77.7	33872.22.2	33891.66 6	33911.11.1	..7000
7100..	34257.50.0	34277.22.2	34296.94.4	31316.66.6	34336.38.8	34356.11.1	54375.83.3	34395.55.5	..7100
7200..	34740.00.0	34760.00.0	34780.00.0	34800.00.0	34820.00.0	34840.00.0	34860.00.0	34880.00.0	..7200
7300..	35222.50.0	35242.77.7	35263.05.5	35283.33.3	35303.61.1	35323.88 8	35314.16.6	35364 44.4	..7300
7400..	35705.00.0	35725.55.5	35746.11.1	35766.66.6	35787.22.2	35807.77.7	35828.33.3	35848 88.8	..7400
7500..	36187.50.0	36208.33.3	36229.16.6	36250.00.0	36270.83.3	36291.66.6	36312.50.0	36333 33.3	..7500
7600..	36670.00.0	36691.11.1	36712.22.2	36733.33.3	36754.44.4	36775.55.5	36796.66.6	36817.77.7	..7600
7700..	37152.50.0	37173.88.8	37195 27.7	37216.66 6	37238.05.5	37259.41.4	37280.83.3	37302.22.2	..7700
7800..	37635.00.0	37656.66.6	37678.33.3	37700.00.0	37721.66.6	37743.33.3	37765.00.0	37786 66.6	..7800
7900..	38117.50.0	38139.44.4	38161.38.8	38183.33.3	38205.27.7	38227.22.2	38249.16.6	38271.11.1	..7900
8000..	38600.00.0	38622.22.2	38644.44.4	38666 66.6	38688.88.8	38711.11.1	38733.33.3	38755.55.5	..8000
8100..	39082.50.0	39105.00.0	39127.50.0	39150.00.0	39172.50.0	39195.00.0	39217.50.0	39240.00.0	..8100
8200..	39565.00.0	39587.77.7	39610.55.5	39633.33.3	39656.11.1	39678.88.8	39701.66.6	39724.44.4	..8200
8300..	40047.50.0	40070.55.5	40093.61.1	40116.66.6	40139.72.2	40162.77.7	40185 83.3	40208.88.8	..8300
8400..	40530.00.0	40553.33.3	40576.66.6	40600.00.0	40623.33.3	40646.66.6	40670.00.0	40693.33 3	..8400
8500..	41012.50.0	41036 11.1	41059.72 2	41083.33.3	41106.94.4	41130.55.5	41154.16.6	41177.77.7	..8500
8600..	41495.00.0	41518.88.8	41542.77.7	41566.66.6	41590.55.5	41614.44.4	41638.33.3	41662.22.2	..8600
8700..	41977.50.0	42001.66.6	42025.83.3	42050.00.0	42074.16.6	42098.33.3	42122.50.0	42146.66.6	..8700
8800..	42460.00.0	42484.44.4	42508.88.8	42533.33.3	42557.77.7	42582.22.2	42606.66.6	42631.11.1	..8800
8900..	42942.50.0	42967.22.2	42991.94.4	43016.66.6	43041.38.8	43066.11.1	43090.83.3	43115.55.5	..8900
9000..	43425.00.0	43450.00.0	43475.00.0	43500.00 0	43525.00.0	43550.00.0	43575.00.0	43600.00.0	..9000
9100..	43907.50.0	43932.77.7	43958.05.5	43983.33.3	44008.61.1	44033 88.8	44059.16.6	44084.44 4	..9100
9200..	44390.00.0	44415.55.5	44441 11.1	44466.66.6	44492.22.2	44517.77.7	44543.33.3	44568.88.8	..9200
9300..	44872.50.0	44898.33.3	44924.16.6	44950.00.0	44975 83.3	45001.66.6	45027.50.0	45053.33.3	..9300
9400..	45355.00.0	45381.11.1	45407.22.2	45433.33.3	45459.44.4	45485.55.5	45511.66.6	45537.77.7	..9400
9500..	45837.50.0	45863.88.8	45890.27.7	45916.66.6	45943.05.5	45969.44.4	45995.83.3	46022.22.2	..9500
9600..	46320.00.0	46346.66.6	46373.33.3	46400.00.0	46426.66.6	46453.33.3	46480.00.0	46506.66.6	..9600
9700..	46802.50.0	46829.44.4	46856.38.8	46883.33.3	46910.27.7	46937.22.2	46964.16.6	46991.11.1	..9700
9800..	47285.00.0	47312.22.2	47339.44.4	47366.66.6	47393.88.8	47421.11.1	47448.33.3	47475.55.5	..9800
9900..	47767.50.0	47795.00.0	47822.50.0	47850.00.0	47877.50.0	47905.00.0	47932.50.0	47960.00.0	..9900
10000..	48250.00.0	48277.77.7	48305.55.5	48333.33.3	48361.11.1	48388.88.8	48416.66.6	48444.44.4	..10000

s.	$ cts.	$ cts.	$ cts.	$ cts.	$ cts.	$ cts.	$ cts.	$ cts.	s.
16..	3.86.0	3.86.2	3.86.4	3.86.7	3.86.9	3.87.1	3.87.3	3.87.5	.16
17..	4.10.1	4.10.4	4.10.6	4.10.8	4.11.1	4.11.3	4.11.5	4.11.8	.17
18..	4.34.2	4.34.5	4.34.7	4.36.0	4.35.2	4.35.5	4.35.7	4.36.0	.18
19..	4.58.4	4.58.6	4.58.9	4.59.2	4.59.4	4.59.7	4.59.9	4.60.2	.19
1d.	.02.0	.02.0	.02.0	.02.0	.02.0	.02 0	.02.0	.02.0	.. 1d.
2..	.04.0	.04.0	.04.0	.04.0	.04.0	.04.0	.04.0	.04.0	.. 2
3..	.06.0	.06.0	.06.0	.06.0	.06.0	.06.0	.06.0	.06.0	.. 3
4..	.08.0	.08.0	.08.0	.08.0	.08.0	.08.0	.08.0	.08.0	.. 4
5..	.10.0	.10.0	.10.0	.10.0	.10.0	.10.0	.10.0	.10.0	.. 5
6..	.12.0	.12.0	.12.0	.12.0	.12.0	.12.0	.12 1	.12.1	.. 6
7..	.14.0	.14.0	.14.0	.14.0	.14.1	.14.1	.14.1	.14.1	.. 7
8..	.16.0	.16.0	.16.1	.16.1	.16.1	.16.1	.16.1	.16.1	.. 8
9..	.18 0	.18.1	.18.1	.18.1	.18.1	.18.1	.18.1	.18.2	.. 9
10..	.20.1	.20.1	.20.1	.20.1	.20.1	.20.2	.20.2	.20.2	..10
11..	.22.1	.22.1	.22.1	.22.1	.22.2	.22.2	.22.2	.22.2	..11

STERLING EXCHANGE TABLES.

$9\frac{1}{16}$ TO $9\frac{1}{2}$%

STERLING INTO DOLLARS AND CENTS.

Stg.	$9\frac{1}{16}$	$9\frac{1}{8}$	$9\frac{3}{16}$	$9\frac{1}{4}$	$9\frac{5}{16}$	$9\frac{3}{8}$	$9\frac{7}{16}$	$9\frac{1}{2}$	Stg.
£	$ cts.	$ cts.	$ cts.	$ cts.	$ cts.	$ cts.	$ cts.	$ cts.	£
100..	484.72.2	485 00.0	485.27.7	485.55.5	485.83.3	486.11.1	486.38.8	486.66.6	..100
200..	969.44.4	970.00.0	970.55.5	971.11.1	971.66.6	972.22.2	972.77.7	973.33.3	..200
300..	1454.16.6	1455 00.0	1455.83.3	1456.66.6	1457·50.0	1458.33.3	1459.16.6	1460.00.0	..300
400..	1938.88.8	1940.00 0	1941.11.1	1942.22.2	1943.33.3	1944.44.4	1945.55.5	1946.66.6	..400
500..	2423.61.1	2425.00.0	2426.38.8	2427.77.7	2429.16.6	2430.55.5	2431.94.4	2433.33.3	..500
600..	2908.33.3	2910.00.0	2911.66.6	2913.33.3	2915.00.0	2916.66.6	2918.33.3	2920.00.0	..600
700..	3393 05.5	3395.00.0	3396.94.4	3398.88.8	3400.83.3	3402.77.7	3404.72.2	3406.66.6	..700
800..	3877.77.7	3880.00.0	3882.22.2	3884.44.4	3886.66.6	3888.88.8	3891.11.1	3893.33.3	..800
900..	4362.50.0	4365.00.0	4367.50.0	4370.00.0	4372.50.0	4375.00.0	4377.50.0	4380.00.0	..900
1000..	4847.22.2	4850.00.0	4852.77.7	4855.55.5	4858.33.3	4861.11.1	4863.88.8	4866.66.6	..1000
1100..	5331.94.4	5335.00.0	5338.05.5	5341.11.1	5344.16.6	5347.22.2	5350.27.7	5353.33.3	..1100
1200..	5816.66.6	5820.00.0	5823.33.3	5826.66.6	5830.00.0	5833.33.3	5836 66.6	5840.00.0	..1200
1300..	6301.38.8	6305.00.0	6308.61.1	6312.22.2	6315.83.3	6319.44.4	6323.05.5	6326.66.6	..1300
1400..	6786.11.1	6790.00.0	6793.88.8	6797.77.7	6801.66.6	6805.55.5	6809.44.4	6813.33.3	..1400
1500..	7270.83.3	7275.00.0	7279.16.6	7283.33.3	7287.50.0	7291.66.6	7295.83.3	7300.00.0	..1500
1600..	7755.55.5	7760.00.0	7764.44.4	7768.88.8	7773.33.3	7777.77.7	7782.22.2	7786.66.6	..1600
1700..	8240.27.7	8245.00.0	8249.72.2	8254.44.4	8259.16.6	8263.88.8	8268.61.1	8273.33.3	..1700
1800..	8725.00.0	8730.00.0	8735.00.0	8740.00.0	8745.00.0	8750.00.0	8755.00 0	8760.00.0	..1800
1900..	9209.72.2	9215.00.0	9220.27.7	9225.55.5	9230.83.3	9236.11.1	9241.38.8	9246.66.6	..1900
2000	9694.44.4	9700.00.0	9705.55.5	9711.11.1	9716.66.6	9722.22 2	9727.77.7	9733.33.3	..2000
2100..	10179.16.6	10185.00.0	10190.83.3	10196.66.6	10202.50.0	10208.33.3	10214.16.6	10220.00.0	..2100
2200..	10663.88.8	10670.00.0	10676.11.1	10682.22.2	10688.33.3	10694 44.4	10700.55.5	10706.66.6	..2200
2300..	11148.61.1	11155.00.0	11161.38 8	11167.77.7	11174.16.6	11180.55.5	11186.94.4	11193.33.3	..2300
2400..	11633.33.3	11640.00.0	11646.66.6	11653.33.3	11660.00.0	11666.66.6	11673.33.3	11680.00.0	..2400
2500..	12118.05.5	12125.00.0	12131.94.4	12138.88.8	12145.83.3	12152.77.7	12159.72.2	12166.66.6	..2500
2600..	12602.77.7	12610.00.0	12617.22.2	12624.44.4	12631.66.6	12638.88.8	12646.11.1	12653.33.3	..2600
2700..	13087.50.0	13095.00.0	13102.50.0	13110.00.0	13117.50.0	13125.00.0	13132.50.0	13140.00.0	..2700
2800..	13572.22.2	13580.00.0	13587.77.7	13595.55.5	13603.33.3	13611.11.1	13618.88.8	13626.66.6	..2800
2900..	14056.94.4	14065.00.0	14073.05.5	14081.11.1	14089.16.6	14097.22 2	14105.27.7	14113.33.3	..2900
3000..	14541.66.6	14550.00.0	14558.33.3	14566.66.6	14575.00.0	14583.33.3	14591.66.6	14600.00.0	..3000
3100.	15026.38.8	15035.00.0	15043.61.1	15052 22.2	15060.83.3	15069.44.4	15078.05.5	15086.66.6	..3100
3200..	15511.11.1	15520.00.0	15528.88.8	15537.77.7	15546.66.6	15555.55.5	15564.44.4	15573.33.3	..3200
3300..	15995.83.3	16005.00.0	16014.16.6	16023.33.3	16032.50.0	16041.66 6	16050.83.3	16060.00.0	..3300
3400..	16180.55.5	16490.00.0	16499.44.4	16508.88.8	16518.33.3	16527.77.7	16537.22.2	16546.66.6	..3400
3500..	16965.27.7	16975.00.0	16984.72.2	16994.44.4	17004.16.6	17013.88.8	17023.61.1	17033.33.3	..3500
3600..	17450.00.0	17460.00.0	17470.00.0	17480.00.0	17490.00.0	17500.00.0	17510.00.0	17520.00.0	..3600
3700..	17934.72.2	17945.00.0	17955.27.7	17965.55.5	17975.83.3	17986.11.1	17996.38.8	18006.66.6	..3700
3800..	18419.44.4	18430.00.0	18440.55.5	18451.11.1	18461.66.6	18472.22.2	18482.77.7	18493.33.3	..3800
3900..	18904.16.6	18915.00.0	18925.83.3	18936 66.6	18947.50.0	18958.33.3	18969.16.6	18980.00.0	..3900
4000..	19388.88.8	19400.00.0	19411.11.1	19422.22.2	19433.33.3	19444.44.4	19455.55.5	19466.66.6	..4000
4100..	19873.61.1	19885.00.0	19896.38.8	19907 77.7	19919.16.6	19930.55.5	19941.94.4	19953.33.3	..4100
4200..	20358.33.3	20370.00.0	20381.66.6	20393.33.3	20405.00.0	20416.66.6	20428.33.3	20440.00.0	..4200
4300..	20813.05.5	20855.00.0	20866.94.4	20878 88.8	20890.83.3	20902.77.7	20914.72.2	20926.66.6	..4300
4400..	21327.77.7	21340.00.0	21352.22.2	21364.44.4	21376.66.6	21388.88.8	21401.11.1	21413.33.3	..4400
4500..	21812.50.0	21825.00.0	21837.50.0	21850.00.0	21862.50.0	21875.00.0	21887.50.0	21900.00.0	..4500
4600..	22297.22.2	22310.00.0	22322.77.7	22335.55.5	22348.33.3	22361.11.1	22373.88.8	22386.66.6	..4600
4700..	22781.94.4	22795.00 0	22808.05.5	22821.11.1	22834.16.6	22847.22.2	22860.27.7	22873.33.3	..4700
4800..	23266.66.6	23280.00.0	23293.33.3	23306.66.6	23320.00.0	23333.33.3	23346.66.6	23360.00.0	..4800
4900..	23751.38.8	23765.00.0	23778.61.1	23792.22.2	23805.83.3	23819.44.4	23833.05.5	23846.66.6	..4900
5000..	24236.11.1	24250.00.0	24263.88.8	24277.77.7	24291.66.6	24305.55.5	24319.44.4	24333.33.3	..5000

s.	$ cts.	$ cts.	$ cts.	$ cts.	$ cts.	$ cts.	$ cts.	$ cts.	s.
1..	.24.2	.24.2	.24.3	.24.3	.24.3	.24.3	.24.3	.24.3	..1
2..	.48.5	.48.5	.48.5	.48.5	.48.6	.48.6	.48.6	.48.7	..2
3..	.72.7	.72.7	.72.8	.72.8	.72.9	.72.9	.72.9	.73.0	..3
4..	.96.9	.97.0	.97.0	.97.1	.97.2	.97.2	.97.3	.97.3	..4
5..	1.21.2	1.21.2	1.21.3	1.21.4	1.21.4	1.21.5	1.21.6	1.21.7	..5
6..	1.45.4	1.45.5	1.45.6	1.45.7	1.45.7	1.45.8	1.45.9	1.46.0	..6
7..	1.69.6	1.69.7	1.69.8	1.69.9	1.70.0	1.70.1	1.70.2	1.70.3	..7
8..	1.93 9	1.94.0	1.94.1	1.94.2	1.94.3	1.94.4	1.94.5	1.94.7	..8
9..	2.18.1	2.18.2	2.18.4	2.18.5	2.18.6	2.18.7	2.18.9	2.19.0	..9
10..	2.42 4	2.42.5	2.42.6	2.42.8	2.42.9	2.43.0	2.43.2	2.43.3	..10
11..	2.66.6	2.66.7	2.66.9	2.67.0	2.67.2	2.67.4	2 67.5	2.67.7	..11
12..	2.90.8	2.91.0	2.91.2	2.91.3	2.91.5	2.91.7	2.91.8	2.92.0	..12
13..	3.15.1	3.15.2	3.15.4	3.15.6	3.15.8	3.16.0	3.16.1	3.16.3	..13
14..	3.39.3	3.39.5	3.39.7	3.39.9	3.40.1	3 40.3	3.40.5	3.40.7	..14
15..	3.63.5	3.63.7	3.64.0	3.64.2	3.64.4	3.64.6	3.64.8	3.65.0	..15

Stg.	$9\frac{1}{16}$	$9\frac{1}{8}$	$9\frac{3}{16}$	$9\frac{1}{4}$	$9\frac{5}{16}$	$9\frac{3}{8}$	$9\frac{7}{16}$	$9\frac{1}{2}$	Stg.
£	$ cts.	$ cts.	$ cts.	$ cts.	$ cts.	$ cts.	$ cts.	$ cts.	£
5100	24720.83.3	24735.00.0	24749.16.6	24763.33.3	24777.50.0	24791.66.6	24805.83.3	24820.00.0	5100
5200	25205.55.5	25220.00.0	25234.44.4	25248.88.8	25263.33.3	25277.77.7	25292.22.2	25306.66.6	5200
5300	25690.27.7	25705.00.0	25719.73.2	25734.44.4	25749.16.6	25763.88.8	25778.61.1	25793.33.3	5300
5400	26175.00.0	26190.00.0	26205.00.0	26220.00.0	26235 00.0	26250.00.0	26265.00.0	26280.00.0	5400
5500	26659.72.2	26675.00.0	26690.27.7	26705.55.5	26720.83.3	26736.11.1	26751.38.8	26766.66.6	5500
5600	27144.44.4	27160.00.0	27175.55.5	27191.11.1	27206.66.6	27222.22.2	27237.77.7	27253.33.3	5600
5700	27629.16.6	27645.00.0	27660.83.3	27676.66.6	27692.50.0	27703.33.3	27724.16.6	27740.00.0	5700
5800	28113.88.8	28130.00.0	28146.11.1	28162.22.2	28178.33.3	28194.44.4	28210.55.5	28226.66.6	5800
5900	28598.61.1	28615.00.0	28631.38.8	28647.77.7	28664.16.6	28680.55.5	28696.94.4	28713.33.3	5900
6000	29083.33.3	29100.00.0	29116.66.6	29133.33.3	29150.00.0	29166.66.6	29183.33.3	29200.00.0	6000
6100	29568.05.5	29585.00.0	29601.94.4	29618.88.8	29635.83.3	29652.77.7	29669.72.2	29686.66.6	6100
6200	30052.77.7	30070.00.0	30087.22.2	30104.44.4	30121.66.6	30138.88.8	30156.11.1	30173.33.3	6200
6300	30537.50.0	30555.00.0	30572.50.0	30590.00.0	30607.50.0	30625.00.0	30642.50.0	30660.00.0	6300
6400	31022.22.2	31040.00.0	31057.77.7	31075.55.5	31093.33.3	31111.11.1	31128.88.8	31146.66.6	6400
6500	31506.94.4	31525.00.0	31543.05.5	31561.11.1	31579.16.6	31597.22.2	31615.27.7	31633.33.3	6500
6600	31991.66.6	32010.00.0	32028.33.3	32046.66.6	32065.00.0	32083.33.3	32101.66.6	32120.00.0	6600
6700	32476.38.8	32495.00.0	32513.61.1	32532.22.2	32550.83.3	32569.44.4	32588.05.5	32606.66.6	6700
6800	32961.11.1	32980.00.0	32998.88.8	33017.77.7	33036.66.6	33055.55.5	33074.44.4	33093.33.3	6800
6900	33445.83.3	33465.00.0	33484.16.6	33503.33.3	33522.50.0	33541.66.6	33560.83.3	33580.00.0	6900
7000	33930.55.5	33950.00.0	33969.44.4	33988.88.8	34008.33.3	34027.77.7	34047.22.2	34066.66.6	7000
7100	34415.27.7	34435.00.0	34454.72.2	34474.44.4	34494.16.6	34513.88.8	34533.61.1	34553.33.3	7100
7200	34900.00.0	34920.00.0	34940.00.0	34960.00.0	34980.00.0	35000.00.0	35020.00.0	35040.00.0	7200
7300	35384.72.2	35405.00.0	35425.27.7	35445.55.5	35465.83.3	35486.11.1	35506.38.8	35526.66.6	7300
7400	35869.44.4	35890.00.0	35910.55.5	35931.11.1	35951.66.6	35972.22.2	35992.77.7	36013.33.3	7400
7500	36354.16.6	36375.00.0	36395.83.3	36416.66.6	36437.50.0	36458.33.3	36479.16.6	36500.00.0	7500
7600	36838.88.8	36860.00.0	36881.11.1	36902.22.2	36923.33.3	36944.44.4	36965.55.5	36986.66.6	7600
7700	37323.61.1	37315.00.0	37366.38.8	37387.77.7	37409.16.6	37430.55.5	37451.94.4	37473.33.3	7700
7800	37803.33.3	37830.00.0	37851.66.6	37873.33.3	37895.00.0	37916.66.6	37938.33.3	37960.00.0	7800
7900	38293.05.5	38315.00.0	38336.94.4	38358.88.8	38380.83.3	38402.77.7	38424.72.2	38446.66.6	7900
8000	38777.77.7	38800.00.0	38822.22.2	38844.44.4	38866.66.6	38888.88.8	38911.11.1	38933.33.3	8000
8100	39262.50.0	39285.00.0	39307.50.0	39330.00.0	39352.50.0	39375.00.0	39397.50.0	39420.00.0	8100
8200	39747.22.2	39770.00.0	39792.77.7	39815.55.5	39838.33.3	39861.11.1	39883.88.8	39906.66.6	8200
8300	40231.94.4	40255.00.0	40278.05.5	40301.11.1	40324.16.6	40347.22.2	40370.27.7	40393.33.3	8300
8400	40716.66.6	40740.00.0	40763.33.3	40786.66.6	40810.00.0	40833.33.3	40856.66.6	40880.00.0	8400
8500	41201.38.8	41225.00.0	41248.61.1	41272.22.2	41295.83.3	41319.44.4	41343.05.5	41366.66.6	8500
8600	41686.11.1	41710.00.0	41733.88.8	41757.77.7	41781.66.6	41805.55.5	41829.44.4	41853.33.3	8600
8700	42170.83.3	42195.00.0	42219.16.6	42243.33.3	42267.50.0	42291.66.6	42315.83.3	42340.00.0	8700
8800	42655.55.5	42680.00.0	42704.44.4	42728.88.8	42753.33.3	42777.77.7	42802.22.2	42826.66.6	8800
8900	43140.27.7	43165.00.0	43189.72.2	43214.44.4	43239.16.6	43263.88.8	43288.61.1	43313.33.3	8900
9000	43625.00.0	43650.00.0	43675.00.0	43700.00.0	43725.00.0	43750.00.0	43775.00.0	43800.00.0	9000
9100	44109.72.2	44135.00.0	44160.27.7	44185.55.5	44210.83.3	44236.11.1	44261.38.8	44286.66.6	9100
9200	44594.44.4	44620.00.0	44645.55.5	44671.11.1	44696.66.6	44722.22.2	44747.77.7	44773.33.3	9200
9300	45079.16.6	45105.00.0	45130.83.3	45156.66.6	45182.50.0	45208.33.3	45234.16.6	45260.00.0	9300
9400	45563.88.8	45590.00.0	45616.11.1	45642.22.2	45668.33.3	45694.44.4	45720.55.5	45746.66.6	9400
9500	46048.61.1	46075.00.0	46101.38.8	46127.77.7	46154.16.6	46180.55.5	46206.94.4	46233.33.3	9500
9600	46533.33.3	46560.00.0	46586.66.6	46613.33.3	46640.00.0	46666.66.6	46693.33.3	46720.00.0	9600
9700	47018.05.5	47045.00.0	47071.94.4	47098.88.8	47125.83.3	47152.77.7	47179.72.2	47206.66.6	9700
9800	47502.77.7	47530.00.0	47557.22.2	47584.44.4	47611.66.6	47638.88.8	47666.11.1	47693.33.3	9800
9900	47987.50.0	48015.00.0	48042.50.0	48070.00.0	48097.50.0	48125.00.0	48152.50.0	48180.00.0	9900
10000	48472.22.2	48500.00.0	48527.77.7	48555.55.5	48583.33.3	48611.11.1	48638.88.8	48666.66.6	10000

s.	$ cts.	$ cts.	$ cts.	$ cts.	$ cts.	$ cts.	$ cts.	$ cts.	s.
16	3.87.8	3.88.0	3.88.3	3.88.5	3.88.7	3.88.9	3.89.1	3.89.3	.16
17	4.12.0	4.12.2	4.12.5	4.12.7	4.12.9	4.13.2	4.13.4	4.13.7	.17
18	4.36.2	4.36.5	4.36.7	4.37.0	4.37.2	4.37.5	4.37.7	4.38.0	.18
19	4.60.5	4.60.7	4.61.0	4.61.3	4.61.5	4.61.8	4.62.1	4.62.3	.19
1d	.02.0	.02.0	.02.0	.02.0	.02.0	.02.0	.02.0	.02.0	1d
2	.04.0	.04.0	.04.0	.04.0	.04.0	.04.0	.04.0	.04.0	2
3	.06.0	.06.1	.06.1	.06.1	.06.1	.06.1	.06.1	.06.1	3
4	.08.1	.08.1	.08.1	.08.1	.08.1	.08.1	.08.1	.08.1	4
5	.10.1	.10.1	.10.1	.10.1	.10.1	.10.1	.10.1	.10.1	5
6	.12.1	.12.1	.12.1	.12.1	.12.1	.12.1	.12.1	.12.2	6
7	.14.1	.14.1	.14.1	.14.2	.14.2	.14.2	.14.2	.14.2	7
8	.16.1	.16.2	.16.2	.16.2	.16.2	.16.2	.16.2	.16.2	8
9	.18.2	.18.2	.18.2	.18.2	.18.2	.18.2	.18.2	.18.2	9
10	.20.2	.20.2	.20.2	.20.2	.20.2	.20.2	.20.3	.20.3	.10
11	.22.2	.22.2	.22.2	.22.2	.22.3	.22.3	.22.3	.22.3	.11

STERLING EXCHANGE TABLES.

$9\frac{9}{16}$ TO 10 %

STERLING INTO DOLLARS AND CENTS.

Stg.	$9\frac{9}{16}$	$9\frac{5}{8}$	$9\frac{11}{16}$	$9\frac{3}{4}$	$9\frac{13}{16}$	$9\frac{7}{8}$	$9\frac{15}{16}$	10	Stg.
£	$ cts.	$ cts.	$ cts.	$ cts.	$ cts.	$ cts.	$ cts.	$ cts.	£
100..	486.94.4	487.22.2	487.50.0	487.77.7	488.05.5	488.33.3	488.61.1	488.88.8	100
200..	973.88.8	974.44.4	975.00.0	975.55.5	976.11.1	976.66.6	977.22.2	977.77.7	200
300..	1460.83.3	1461.66.6	1462.50.0	1463.33.3	1464.16.6	1465.00.0	1465.83.3	1466.66.6	300
400..	1947.77.7	1948.88.8	1950.00.0	1951.11.1	1952.22.2	1953.33.3	1954.44.4	1955.55.5	400
500..	2434.72.2	2436.11.1	2437.50.0	2438.88.8	2440.27.7	2441.66.6	2443.05.5	2444.44.4	500
600..	2921.66.6	2923.33.3	2925.00.0	2926.66.6	2928.33.3	2930.00.0	2931.66.6	2933.33.3	600
700..	3408.61.1	3410.55.5	3412.50.0	3414.44.4	3416.38.8	3418.33.3	3420.27.7	3422.22.2	700
800..	3895.55.5	3897.77.7	3900.00.0	3902.22.2	3904.44.4	3906.66.6	3908.88.8	3911.11.1	800
900..	4382.50.0	4385.00.0	4387.50.0	4390.00.0	4392.50.0	4395.00.0	4397.50.0	4400.00.0	900
1000..	4869.44.4	4872.22.2	4875.00.0	4877.77.7	4880.55.5	4883.33.3	4886.11.1	4888.88.8	1000
1100..	5356.38.8	5359.44.4	5362.50.0	5365.55.5	5368.61.1	5371.66.6	5374.72.2	5377.77.7	1100
1200..	5843.33.3	5846.66.6	5850.00.0	5853.33.3	5856.66.6	5860.00.0	5863.33.3	5866.66.6	1200
1300..	6330.27.7	6333.88.8	6337.50.0	6341.11.1	6344.72.2	6348.33.3	6351.94.4	6355.55.5	1300
1400..	6817.22.2	6821.11.1	6825.00.0	6828.88.8	6832.77.7	6836.66.6	6840.55.5	6844.44.4	1400
1500..	7304.16.6	7308.33.3	7312.50.0	7316.66.6	7320.83.3	7325.00.0	7329.16.6	7333.33.3	1500
1600..	7791.11.1	7795.55.5	7800.00.0	7804.44.4	7808.88.8	7813.33.3	7817.77.7	7822.22.2	1600
1700..	8278.05.5	8282.77.7	8287.50.0	8292.22.2	8296.94.4	8301.66.6	8306.38.8	8311.11.1	1700
1800..	8765.00.0	8770.00.0	8775.00.0	8780.00.0	8785.00.0	8790.00.0	8795.00.0	8800.00.0	1800
1900..	9251.94.4	9257.22.2	9262.50.0	9267.77.7	9273.05.5	9278.33.3	9283.61.1	9288.88.8	1900
2000..	9738.88.8	9744.44.4	9750.00.0	9755.55.5	9761.11.1	9766.66.6	9772.22.2	9777.77.7	2000
2100..	10225.83.3	10231.66.6	10237.50.0	10243.33.3	10249.16.6	10255.00.0	10260.83.3	10266.66.6	2100
2200..	10712.77.7	10718.88.8	10725.00.0	10731.11.1	10737.22.2	10743.33.3	10749.44.4	10755.55.5	2200
2300..	11199.72.2	11206.11.1	11212.50.0	11218.88.8	11225.27.7	11231.66.6	11238.05.5	11244.44.4	2300
2400..	11686.66.6	11693.33.3	11700.00.0	11706.66.6	11713.33.3	11720.00.0	11726.66.6	11733.33.3	2400
2500..	12173.61.1	12180.55.5	12187.50.0	12194.44.4	12201.38.8	12208.33.3	12215.27.7	12222.22.2	2500
2600..	12660.55.5	12667.77.7	12675.00.0	12682.22.2	12689.44.4	12696.66.6	12703.88.8	12711.11.1	2600
2700..	13147.50.0	13155.00.0	13162.50.0	13170.00.0	13177.50.0	13185.00.0	13192.50.0	13200.00.0	2700
2800..	13634.44.4	13642.22.2	13650.00.0	13657.77.7	13665.55.5	13673.33.3	13681.11.1	13688.88.8	2800
2900..	14121.38.8	14129.44.4	14137.50.0	14145.55.5	14153.61.1	14161.66.6	14169.72.2	14177.77.7	2900
3000..	14608.33.3	14616.66.6	14625.00.0	14633.33.3	14641.66.6	14650.00.0	14658.33.3	14666.66.6	3000
3100..	15095.27.7	15103.88.8	15112.50.0	15121.11.1	15129.72.2	15138.33.3	15146.94.4	15155.55.5	3100
3200..	15582.22.2	15591.11.1	15600.00.0	15608.88.8	15617.77.7	15626.66.6	15635.55.5	15644.44.4	3200
3300..	16069.16.6	16078.33.3	16087.50.0	16096.66.6	16105.83.3	16115.00.0	16124.16.6	16133.33.3	3300
3400..	16556.11.1	16565.55.5	16575.00.0	16584.44.4	16593.88.8	16603.33.3	16612.77.7	16622.22.2	3400
3500..	17043.05.5	17052.77.7	17062.50.0	17072.22.2	17081.94.4	17091.66.6	17101.38.8	17111.11.1	3500
3600..	17530.00.0	17540.00.0	17550.00.0	17560.00.0	17570.00.0	17580.00.0	17590.00.0	17600.00.0	3600
3700..	18016.94.4	18027.22.2	18037.50.0	18047.77.7	18058.05.5	18068.33.3	18078.61.1	18088.88.8	3700
3800..	18503.88.8	18514.44.4	18525.00.0	18535.55.5	18546.11.1	18556.66.6	18567.22.2	18577.77.7	3800
3900..	18990.83.3	19001.66.6	19012.50.0	19023.33.3	19034.16.6	19045.00.0	19055.83.3	19066.66.6	3900
4000..	19477.77.7	19488.88.8	19500.00.0	19511.11.1	19522.22.2	19533.33.3	19544.44.4	19555.55.5	4000
4100..	19964.72.2	19976.11.1	19987.50.0	19998.88.8	20010.27.7	20021.66.6	20033.05.5	20044.44.4	4100
4200..	20451.66.6	20463.33.3	20475.00.0	20486.66.6	20498.33.3	20510.00.0	20521.66.6	20533.33.3	4200
4300..	20938.61.1	20950.55.5	20962.50.0	20974.44.4	20986.38.8	20998.33.3	21010.27.7	21022.22.2	4300
4400..	21425.55.5	21437.77.7	21450.00.0	21462.22.2	21474.44.4	21486.66.6	21498.88.8	21511.11.1	4400
4500..	21912.50.0	21925.00.0	21937.50.0	21950.00.0	21962.50.0	21975.00.0	21987.50.0	22000.00.0	4500
4600..	22399.44.4	22412.22.2	22425.00.0	22437.77.7	22450.55.5	22463.33.3	22476.11.1	22488.88.8	4600
4700..	22886.38.8	22899.44.4	22912.50.0	22925.55.5	22938.61.1	22951.66.6	22964.72.2	22977.77.7	4700
4800..	23373.33.3	23386.66.6	23400.00.0	23413.33.3	23426.66.6	23440.00.0	23453.33.3	23466.66.6	4800
4900..	23860.27.7	23873.88.8	23887.50.0	23901.11.1	23914.72.2	23928.33.3	23941.94.4	23955.55.5	4900
5000..	24347.22.2	24361.11.1	24375.00.0	24388.88.8	24402.77.7	24416.66.6	24430.55.5	24444.44.4	5000
s.	$ cts.	$ cts.	$ cts.	$ cts.	$ cts.	$ cts.	$ cts.	$ cts.	s.
1..	.24.3	.24.4	.24.4	.24.4	.24.4	.24.4	.24.4	.24.4	1
2..	.48.7	.48.7	.48.7	.48.8	.48.8	.48.8	.48.9	.48.9	2
3..	.73.0	.73.1	.73.1	.73.1	.73.2	.73.2	.73.3	.73.3	3
4..	.97.4	.97.4	.97.5	.97.5	.97.6	.97.7	.97.7	.97.8	4
5..	1.21.7	1.21.8	1.21.9	1.21.9	1.22.0	1.22.1	1.22.1	1.22.2	5
6..	1.46.1	1.46.2	1.46.2	1.46.3	1.46.4	1.46.5	1.46.6	1.46.7	6
7..	1.70.4	1.70.5	1.70.6	1.70.7	1.70.8	1.70.9	1.71.0	1.71.1	7
8..	1.94.8	1.94.9	1.95.0	1.95.1	1.95.2	1.95.3	1.95.4	1.95.5	8
9..	2.19.1	2.19.2	2.19.4	2.19.5	2.19.6	2.19.7	2.19.9	2.20.0	9
10..	2.43.5	2.43.6	2.43.7	2.43.9	2.44.0	2.44.2	2.44.3	2.44.4	10
11..	2.67.8	2.68.0	2.68.1	2.68.3	2.68.4	2.68.6	2.68.7	2.68.9	11
12..	2.92.2	2.92.3	2.92.5	2.92.7	2.92.8	2.93.0	2.93.2	2.93.3	12
13..	3.16.5	3.16.7	3.16.9	3.17.0	3.17.2	3.17.4	3.17.6	3.17.8	13
14..	3.40.9	3.41.0	3.41.2	3.41.4	3.41.6	3.41.8	3.42.0	3.42.2	14
15..	3.65.2	3.65.4	3.65.6	3.65.8	3.66.0	3.66.2	3.66.4	3.66.7	15

STERLING EXCHANGE TABLES.

$9\frac{9}{16}$ TO 10 %

STERLING INTO DOLLARS AND CENTS.

Stg. £	$9\frac{9}{16}$ $ cts.	$9\frac{5}{8}$ $ cts.	$9\frac{11}{16}$ $ cts.	$9\frac{3}{4}$ $ cts.	$9\frac{13}{16}$ $ cts.	$9\frac{7}{8}$ $ cts.	$9\frac{15}{16}$ $ cts.	10 $ cts.	Stg. £
5100	24834.16,6	24848.33,3	24862.50,0	24876.66,6	24890.83,3	24905.00,0	24919.16,6	24933,33,3	5100
5200	25321.11,1	25335,55,5	25350.00,0	25364.44,4	25378,88,8	25393,33,3	25407.77,7	25422.22,2	5200
5300	25808.05,5	25822.77,7	25837.50,0	25852.22,2	25866.94,4	25881.66,6	25896.38,8	25911.11,1	5300
5400	26295.00,0	26310.00,0	26325.00,0	26340.00,0	26355,00,0	26370.00,0	26385.00,0	26400.00,0	5400
5500	26781.94,4	26797.22,2	26812.50 0	26827.77,7	26843.05,5	26858.33,3	26873.61,1	26888,88,8	5500
5600	27268.88,8	27284.44,4	27300.00,0	27315 55,5	27331.11,1	27346.66,6	27362.22,2	27377.77,7	5600
5700	27755.83,3	27771.66,6	27787.50,0	27803.33,3	27819.16,6	27835.00 0	27850.83,3	27866.66,6	5700
5800	28242.77,7	28258.88,8	28275.00,0	28291.11,1	28307.22 2	28323.33 3	28339.44,4	28355.55,5	5800
5900	28729.72,2	28746.11,1	28762.50,0	28778.88,8	28795,27 7	28811.66,6	28828.05,5	28844,44,4	5900
6000	29216.66,6	29233.33,3	29250.00,0	29266,66,6	29283,33,3	29300.00,0	29316.66,6	29333.33,3	6000
6100	29703.61,1	29730.55,5	29737.50,0	29754.44,4	29771.38 8	29788.33,3	29805.27.7	29832,22,2	6100
6200	30190.55,5	30207.77,7	30225.00,0	30242,22,2	30259,44,4	30276.66 6	30293.88,8	30311.11,1	6200
6300	30677.50 0	30695.00,0	30712.50,0	30730.00,0	30747.50,0	30765,00,0	30782.50,0	30800.00,0	6300
6400	31164.44,4	31182.22,2	31200.00,0	31217.77,7	31235,55,5	31253,33 3	31271.11,1	31288.88,8	6400
6500	31651.38,8	31669.44,4	31687.50,0	31705.55,5	31723.61 1	31741.66,6	31759.72,2	31777.77,7	6500
6600	32138.33,3	32156.66,6	32175.00,0	32193,33,3	32211.66,6	32230.00,0	32248.33,3	32266.66,6	6600
6700	32625.27,7	32643.88,8	32662,50,0	32681.11 1	32699.72,2	32718.33,3	32736.94,4	32755.55,5	6700
6800	33112.22,2	33131.11,1	33150.00,0	33168.88,8	33187.77,7	33206.66,6	33225.55,5	33244.44,4	6800
6900	33599.16,6	33618.33,3	33637.50,0	33656.66,6	33675.83,3	33695,00,0	33714.16 6	33733.33,3	6900
7000	34086.11,1	34105.55,5	34125.00,0	34144.44,4	34163.88,8	34183.33,3	34202.77,7	34222,22,2	7000
7100	34573.05,5	34592.77,7	34612.50,0	34632,22,2	34651.94,4	34671.66,6	34691.38 8	34711.11,1	7100
7200	35060 00,0	35080,00,0	35100.00,0	35120.00,0	35140.00,0	35160.00,0	35180.00 0	35200.00,0	7200
7300	35546.94,4	35567.22,2	35587.50,0	35607,77,7	35628.05,5	35648.33,3	35668 61,1	35688 88,8	7300
7400	36033.88,8	36054.44,4	36075,00,0	36095.55,5	36116.11,1	36136.66,6	36157.22,2	36177.77,7	7400
7500	36520.83,3	36541.66,6	36562.50,0	36583.33,3	36604.16,6	36625,00,0	36645.83,3	36666.66,6	7500
7600	37007.77,7	37028.88,8	37050,00,0	37071.11,1	37092.22,2	37113.33 3	37134,44,4	37155.55,5	7600
7700	37494.72,2	37516.11,1	37537.50,0	37558 88,8	37580.27,7	37601.66,6	37623.05,5	37644.44,4	7700
7800	37981.66,6	38003.33,3	38025.00,0	38046,66 6	38068.33,3	38090.00,0	38111.66,6	38133,33,3	7800
7900	38468.61,1	38490,55,5	38512 50,0	38534.44,4	38556,33,8	38578.33,3	38600,22 7	38622,22,2	7900
8000	38955.55,5	38977.77,7	39000.00,0	39022 23,2	39044.44,4	39066.66,6	39088.88,8	39111,11,1	8000
8100	39442.50,0	39465 00,0	39487.50,0	39510.00,0	39532.50,0	39555,00,0	39577.50,0	39600.00,0	8100
8200	39929.44,4	39952.22,2	39975,00,0	39997.77,7	40020.55,5	40043.33,3	40066,11,1	40088.88,8	8200
8300	40416.38,8	40439.44,4	40162.50,0	40485.55,5	40508.61,1	40531.66,6	40554.72,2	40577.77,7	8300
8400	40903.33,3	40926.66,6	40950 00 0	40973,33,3	40996.66,6	41020.00,0	41043.33,3	41066.66,6	8400
8500	41390 27.7	41413.88,8	41437.50,0	41461.11,1	41484.72,2	41508.33 3	41531.94,4	41555.55,5	8500
8600	41877.22,2	41901.11,1	41925.00,0	41948.88,8	41972.77,7	41996.66,6	42020.55,5	42044.44,4	8600
8700	42364.16,6	42388.33,3	42412.50,0	42436 66,6	42460.83 3	42485.00,0	42509.16,6	42533.33,3	8700
8800	42851.11,1	42875.55,5	42900.00,0	42924.44,4	42948.88,8	42973.33,3	42997.77,7	43022.22,2	8800
8900	43338 05,5	43362 77,7	43387.50,0	43412.22,2	43436.94,4	43461.66 6	43486.38,8	43511.11,1	8900
9000	43825.00,0	43850 00,0	43875.00 0	43900.00,0	43925.00,0	43950.00,0	43975.00,0	44000.00,0	9000
9100	44311.94,4	44337.22,2	44362.50,0	44387.77,7	44413.05,5	44438.33,3	44463.61,1	44488.88,8	9100
9200	44798.88,8	44824.44,4	44850.00,0	44875.55,5	44901.11,1	44926.66,6	44952,22,2	44977.77,7	9200
9300	45285.83,3	45311.66 6	45337.50 0	45363.33,3	45389.16,6	45415.00,0	45440.83,3	45466.66,6	9300
9400	45772.77,7	45798.88,8	45825.00,0	45851.11,1	45877.22 2	45903.33,3	45929.44,4	45955.55,5	9400
9500	46259.72,2	46286 11,1	46312.50 0	46338.88,8	46365.27,7	46391.66,6	46418.05,5	46444.44,4	9500
9600	46746.66,6	46773.33,3	46800.00 0	46826.66,6	46853,33,3	46880,00,0	46906.66,6	46933.33,3	9600
9700	47233.61,1	47260.55,5	47287.50,0	47314.44,4	47341.38,8	47368 33,3	47395.27,7	47422.22,2	9700
9800	47720.55,5	47747.77,7	47775.00,0	47802.22,2	47829.44 4	47856 66,6	47883.88,8	47911.11,1	9800
9900	48207.50,0	48235.00,0	48262.50 0	48290.00 0	48317.50 0	48345 00,0	48372.50,0	48400.00,0	9900
10000	48694.41,4	48722.22 2	48750.00 0	48777.77,7	48805.55,5	48833.33,3	48861.11,1	48888 88,8	10000

s.	$ cts.	$ cts.	$ cts.	$ cts.	$ cts.	$ cts.	$ cts.	$ cts.	s.
16	3.89,5	3.89,8	3.90,0	3.90,2	3.90,4	3.90,7	3.90,9	3.91,1	16
17	4.13,9	4.14,1	4.14,4	4.14,6	4.14,8	4.15,1	4.15,3	4.15,5	17
18	4.38,2	4.38,5	4.38,7	4.39,0	4.39,2	4.39,5	4.39,7	4.40,0	18
19	4.62,6	4.62,9	4.63,1	4.63,4	4.63,6	4.63,9	4.64,2	4.64,4	19
1d.	.02,0	.02,0	.02,0	.02,0	.02,0	.02 0	.02,0	.02,0	1d.
2	.04,0	.04,1	.04,1	.04,1	.04,1	.04,1	.04,1	.04,1	2
3	.06,1	.06,1	.06,1	.06,1	.06,1	.06,1	.06,1	.06,1	3
4	.08,1	.08,1	.08,1	.08,1	.08,1	.08,1	.08,1	.08,1	4
5	.10,1	.10,1	.10,1	.10,2	.10,2	.10,2	.10,2	.10,2	5
6	.12,2	.12,2	.12,2	.12,2	.12,2	.12 2	.12 2	.12,2	6
7	.14,2	.14 2	.14,2	.14,2	.14,2	.14,2	.14,2	.14 2	7
8	.16,2	.16,2	.16,2	.16,2	.16,2	.16,2	.16,2	.16,2	8
9	.18,3	.18,3	.18,3	.18,3	.18,3	.18,3	.18,3	.18,3	9
10	.20,3	.20,3	.20,3	.20,3	.20,3	.20,3	.20,3	.20,4	10
11	.22,3	.22,3	.22,3	.22,3	.22,4	.22,4	.22,4	.22,4	11

STERLING EXCHANGE TABLES.

$10\frac{1}{16}$ TO $10\frac{1}{2}$%

STERLING INTO DOLLARS AND CENTS.

Stg.	$10\frac{1}{16}$	$10\frac{1}{8}$	$10\frac{3}{16}$	$10\frac{1}{4}$	$10\frac{5}{16}$	$10\frac{3}{8}$	$10\frac{7}{16}$	$10\frac{1}{2}$	Stg.
£	$ cts.	$ cts.	$ cts.	$ cts.	$ cts.	$ cts.	$ cts.	$ cts.	£
100..	489.16.6	489.44.4	489.72.2	490.00.0	490.27.7	490.55.5	490.83.3	491.11.1	..100
200..	978.33.3	978.88.8	979.44.4	980.00.0	980.55.5	981.11.1	981.66.6	982.22.2	..200
300..	1467.50.0	1468.33.3	1469.16.6	1470.00.0	1470.83.3	1471.66.6	1472.50.0	1473.33.3	..300
400..	1956.66.6	1957.77.7	1958.88.8	1960.00.0	1961.11.1	1962.22.2	1963.33.3	1964.44.4	..400
500..	2445.83.3	2447.22.2	2448.61.1	2450.00.0	2451.38.8	2452.77.7	2454.16.6	2455.55.5	..500
600..	2935.00.0	2936.66.6	2938.33.3	2940.00.0	2941.66.6	2943.33.3	2945.00.0	2946.66.6	..600
700..	3424.16.6	3426.11.1	3428.05.5	3430.00.0	3431.94.4	3433.88.8	3435.83.3	3437.77.7	..700
800..	3913.33.3	3915.55.5	3917.77.7	3920.00.0	3922.22.2	3924.44.4	3926.66.6	3928.88.8	..800
900..	4402.50.0	4405.00.0	4407.50.0	4410.00.0	4412.50.0	4415.00.0	4417.50.0	4420.00.0	..900
1000..	4891.66.6	4894.44.4	4897.22.2	4900.00.0	4902.77.7	4905.55.5	4908.33.3	4911.11.1	..1000
1100..	5380.83.3	5383.88.8	5386.94.4	5390.00.0	5393.05.5	5396.11.1	5399.16.6	5402.22.2	..1100
1200..	5870.00.0	5873.33.3	5876.66.6	5880.00.0	5883.33.3	5886.66.6	5890.00.0	5893.33.3	..1200
1300..	6359.16.6	6362.77.7	6366.38.8	6370.00.0	6373.61.1	6377.22.2	6380.83.3	6384.44.4	..1300
1400..	6848.33.3	6852.22.2	6856.11.1	6860.00.0	6863.88.8	6867.77.7	6871.66.6	6875.55.5	..1400
1500..	7337.50.0	7341.66.6	7345.83.3	7350.00.0	7354.16.6	7358.33.3	7362.50.0	7366.66.6	..1500
1600..	7826.66.6	7831.11.1	7835.55.5	7840.00.0	7844.44.4	7848.88.8	7853.33.3	7857.77.7	..1600
1700..	8315.83.3	8320.55.5	8325.27.7	8330.00.0	8334.72.2	8339.44.4	8344.16.6	8348.88.8	..1700
1800..	8805.00.0	8810.00.0	8815.00.0	8820.00.0	8825.00.0	8830.00.0	8835.00.0	8840.00.0	..1800
1900..	9294.16.6	9299.44.4	9304.72.2	9310.00.0	9315.27.7	9320.55.5	9325.83.3	9331.11.1	..1900
2000..	9783.33.3	9788.88.8	9794.44.4	9800.00.0	9805.55.5	9811.11.1	9816.66.6	9822.22.2	..2000
2100..	10272.50.0	10278.33.3	10284.16.6	10290.00.0	10295.83.3	10301.66.6	10307.50.0	10313.33.3	..2100
2200..	10761.66.6	10767.77.7	10773.88.8	10780.00.0	10786.11.1	10792.22.2	10798.33.3	10804.44.4	..2200
2300..	11250.83.3	11257.22.2	11263.61.1	11270.00.0	11276.38.8	11282.77.7	11289.16.6	11295.55.5	..2300
2400..	11740.00.0	11746.66.6	11753.33.3	11760.00.0	11766.66.6	11773.33.3	11780.00.0	11786.66.6	..2400
2500..	12229.16.6	12236.11.1	12243.05.5	12250.00.0	12256.94.4	12263.88.8	12270.83.3	12277.77.7	..2500
2600..	12718.33.3	12725.55.5	12732.77.7	12740.00.0	12747.22.2	12754.44.4	12761.66.6	12768.88.8	..2600
2700..	13207.50.0	13215.00.0	13222.50.0	13230.00.0	13237.50.0	13245.00.0	13252.50.0	13260.00.0	..2700
2800..	13696.66.6	13704.44.4	13712.22.2	13720.00.0	13727.77.7	13735.55.5	13743.33.3	13751.11.1	..2800
2900..	14185.83.3	14193.88.8	14201.94.4	14210.00.0	14218.05.5	14226.11.1	14234.16.6	14242.22.2	..2900
3000..	14675.00.0	14683.33.3	14691.66.6	14700.00.0	14708.33.3	14716.66.6	14725.00.0	14733.33.3	..3000
3100..	15164.16.6	15172.77.7	15181.38.8	15190.00.0	15198.61.1	15207.22.2	15215.83.3	15224.44.4	..3100
3200..	15653.33.3	15662.22.2	15671.11.1	15680.00.0	15688.88.8	15697.77.7	15706.66.6	15715.55.5	..3200
3300..	16142.50.0	16151.66.6	16160.83.3	16170.00.0	16179.16.6	16188.33.3	16197.50.0	16206.66.6	..3300
3400..	16631.66.6	16641.11.1	16650.55.5	16660.00.0	16669.44.4	16678.88.8	16688.33.3	16697.77.7	..3400
3500..	17120.83.3	17130.55.5	17140.27.7	17150.00.0	17159.72.2	17169.44.4	17179.16.6	17188.88.8	..3500
3600..	17610.00.0	17620.00.0	17630.00.0	17640.00.0	17650.00.0	17660.00.0	17670.00.0	17680.00.0	..3600
3700..	18099.16.6	18109.44.4	18119.72.2	18130.00.0	18140.27.7	18150.55.5	18160.83.3	18171.11.1	..3700
3800..	18588.33.3	18598.88.8	18609.44.4	18620.00.0	18630.55.5	18641.11.1	18651.66.6	18662.22.2	..3800
3900..	19077.50.0	19088.33.3	19099.16.6	19110.00.0	19120.83.3	19131.66.6	19142.50.0	19153.33.3	..3900
4000..	19566.66.6	19577.77.7	19588.88.8	19600.00.0	19611.11.1	19622.22.2	19633.33.3	19644.44.4	..4000
4100..	20055.83.3	20067.22.2	20078.61.1	20090.00.0	20101.38.8	20112.77.7	20124.16.6	20135.55.5	..4100
4200..	20545.00.0	20556.66.6	20568.33.3	20580.00.0	20591.66.6	20603.33.3	20615.00.0	20626.66.6	..4200
4300..	21034.16.6	21046.11.1	21058.05.5	21070.00.0	21081.94.4	21093.88.8	21105.83.3	21117.77.7	..4300
4400..	21523.33.3	21535.55.5	21547.77.7	21560.00.0	21572.22.2	21584.44.4	21596.66.6	21608.88.8	..4400
4500..	22012.50.0	22025.00.0	22037.50.0	22050.00.0	22062.50.0	22075.00.0	22087.50.0	22100.00.0	..4500
4600..	22501.66.6	22514.44.4	22527.22.2	22540.00.0	22552.77.7	22565.55.5	22578.33.3	22591.11.1	..4600
4700..	22990.83.3	23003.88.8	23016.94.4	23030.00.0	23043.05.5	23056.11.1	23069.16.6	23082.22.2	..4700
4800..	23480.00.0	23493.33.3	23506.66.6	23520.00.0	23533.33.3	23546.66.6	23560.00.0	23573.33.3	..4800
4900..	23969.16.6	23982.77.7	23996.38.8	24010.00.0	24023.61.1	24037.22.2	24050.83.3	24064.44.4	..4900
5000..	24458.33.3	24472.22.2	24486.11.1	24500.00.0	24513.88.8	24527.77.7	24541.66.6	24555.55.5	..5000

s.	$ cts.	$ cts.	$ cts.	$ cts.	$ cts.	$ cts.	$ cts.	$ cts.	s.
1..	.24.4	.24.5	.24.5	.24.5	.24.5	.24.5	.24.5	.24.5	1
2..	.48.9	.48.9	.49.0	.49.0	.49.0	.49.0	.49.1	.49.1	2
3..	.73.4	.73.4	.73.4	.73.5	.73.5	.73.6	.73.6	.73.7	3
4..	.97.8	.97.9	.97.9	.98.0	.98.0	.98.1	.98.2	.98.2	4
5..	1.22.3	1.22.4	1.22.4	1.22.5	1.22.6	1.22.6	1.22.7	1.22.8	5
6..	1.46.7	1.46.8	1.46.9	1.47.0	1.47.1	1.47.2	1.47.2	1.47.3	6
7..	1.71.2	1.71.3	1.71.4	1.71.5	1.71.6	1.71.7	1.71.8	1.71.9	7
8..	1.95.7	1.95.8	1.95.9	1.96.0	1.96.1	1.96.2	1.96.3	1.96 4	8
9..	2.20.1	2.20.2	2.20.4	2.20.5	2.20.6	2.20.7	2.20.9	2.21.0	9
10..	2.44.6	2.44.7	2.44.9	2.45.0	2.45.1	2.45.3	2.45.4	2.45.5	10
11..	2.69.0	2.69.2	2.69.3	2.69.5	2.69.6	2.69.8	2 69.9	2.70.0	11
12..	2.93.5	2.93.7	2.93.8	2.94.0	2.94.2	2.94.3	2.94.5	2.94.7	12
13..	3.17.9	3.18.1	3.18.3	3.18.5	3.18.7	3.18.9	3.19.0	3.19.2	13
14..	3.42.4	3.42.6	3.42.8	3.43.0	3.43.2	3.43.4	3.43.6	3.43.8	14
15..	3.66.9	3.67.1	3.67.3	3.67.5	3.67.7	3.67.9	3.68.1	3.68.3	15

STERLING EXCHANGE TABLES.

$10\tfrac{1}{16}$ TO $10\tfrac{1}{2}\%$

STERLING INTO DOLLARS AND CENTS.

Stg. £	$10\tfrac{1}{16}$ $ cts.	$10\tfrac{1}{8}$ $ cts.	$10\tfrac{3}{16}$ $ cts.	$10\tfrac{1}{4}$ $ cts.	$10\tfrac{5}{16}$ $ cts.	$10\tfrac{3}{8}$ $ cts.	$10\tfrac{7}{16}$ $ cts.	$10\tfrac{1}{2}$ $ cts.	Stg. £
5100..	24947,50.0	24961.66,6	24975,83.3	24990,00.0	25004.16.6	25018.33.3	25032,50.0	25046,66.6	..5100
5200..	25436.66.6	25451.11.1	25465,55.5	25480.00.0	25494.44.4	25508.88.8	25523.33.3	25537.77.7	..5200
5300..	25925.83.3	25940.55.5	25955.27.7	25970,00.0	25984.72.2	25999.44.4	26014.16.6	26028.88.8	..5300
5400..	26415.00.0	26430,00.0	26445.00.0	26460.00.0	26475.00.0	26490.00.0	26505.00.0	26520.00.0	..5400
5500..	26904.16.6	26919.44 4	26934.72.2	26950,00.0	26965.27.7	26980.55.5	26995 83.3	27011.11.1	..5500
5600..	27393.33.3	27408.88.8	27424.44.4	27440.00.0	27455.55.5	27471.11.1	27486.66.6	27502.22.2	..5600
5700..	27882.50.0	27898.33.3	27914.16.6	27930.00.0	27945.83.3	27961.66.6	27977.50.0	27993.33.3	..5700
5800..	28371.66.6	28387.77.7	28403.88.8	28420.00.0	28436.11 1	28452.22.2	28468.33.3	28484.44 4	..5800
5900..	28860 83.3	28877.22.2	28893.61.1	28910.00.0	28926.38.8	28942.77.7	28959.16.6	28975.55.5	..5900
6000..	29350.00.0	29366,66.6	29383.33.3	29400.00.0	29416.66.6	29433.33.3	29450.00.0	29466.66.6	..6000
6100..	29839.16 6	29856.11.1	29873 05.5	29890.00.0	29906.94.4	29923.88.8	29940.83.3	29957.77.7	..6100
6200..	30328.33.3	30345 55.5	30362.77.7	30380.00.0	30397.22.2	30414.44 4	30431.66.6	30448.88.8	..6200
6300..	30817.50.0	30835.00.0	30852.50.0	30870.00 0	30887.50.0	30905.00.0	30922.50.0	30940.00.0	..6300
6400..	31306.66.6	31324.44.4	31342.22.2	31360.00.0	31377.77.7	31395.55.5	31413.33.3	31431.11.1	..6400
6500..	31795.83.3	31813.88.8	31831.94.4	31850.00.0	31868.05.5	31886.11.1	31904.16.6	31922.22.2	..6500
6600..	32285.00.0	32303.33.3	32321.66.6	32340.00.0	32358.33.3	32376.66.6	32395.00.0	32413.33.3	..6600
6700..	32774.16.6	32792.77.7	32811.38.8	32830.00.0	32848.61.1	32867.22.2	32885.83.3	32904.44.4	..6700
6800..	33263.33.3	33282.22.2	33301.11.1	33320.00.0	33338 88.8	33357.77.7	33376.66.6	33395.55.5	..6800
6900..	33752.50.0	33771.66.6	33790.83.3	33810.00.0	33829.16.6	33848 33.3	33867.50.0	33886.66.6	..6900
7000..	34241 66.6	34261.11.1	34280.55.5	34300.00.0	34319.44.4	34338.88.8	34358.33.3	34377.77.7	..7000
7100..	34730.83.3	34750.55.5	34770.27.7	34790.00.0	34809.72.2	34829.44 4	34849.16.6	34868.88.8	..7100
7200..	35220.00.0	35240.00.0	35260.00.0	35280.00.0	35300.00.0	35320.00.0	35340.00.0	35360.00.0	..7200
7300..	35709.16.6	35729.44.4	35749.72.2	35770.00.0	35790.27.7	35810.55.5	35830.83.3	35851.11.1	..7300
7400..	36198.33.3	36218.88.8	36239.44.4	36260.00.0	36280.55.5	36301.11.1	36321.66.6	36342.22.2	..7400
7500..	36687.50.0	36708.33.3	36729.16 6	36750.00.0	36770.83.3	36791.66.6	36812.50.0	36833.33.3	..7500
7600..	37176.66.6	37197.77.7	37218.88 8	37240.00.0	37261.11.1	37282 22.2	37303.33.3	37324.44.4	..7600
7700..	37665.83.3	37687.22.2	37708.61.1	37730.00.0	37751.38.8	37772.77.7	37794.16.6	37815.55.5	..7700
7800..	38155.00.0	38176.66.6	38198.33.3	38220.00.0	38241.66.6	38263.33.3	38285 00.0	38306.66.6	..7800
7900..	38644.16.6	38666.11.1	38688 05.5	38710.00.0	38731.94.4	38753.88.8	38775.83.3	38797.77.7	..7900
8000..	39133.33.3	39155.55.5	39177.77.7	39200.00.0	39222.22.2	39244.44 4	39266.66.6	39288.88.8	..8000
8100..	39622.50.0	39645.00.0	39667.50.0	39690.00.0	39712 50.0	39735.00.0	39757.50.0	39780.00.0	..8100
8200..	40111.66.6	40134.44.4	40157.22.2	40180.00.0	40202.77.7	40225.55.5	40248.33.3	40271.11.1	..8200
8300..	40600.83.3	40623.88.8	40646.94.4	40670.00.0	40693.05 5	40716.11.1	40739.16.6	40762.22.2	..8300
8400..	41090.00.0	41113.33.3	41136.66.6	41160 00.0	41183.33.3	41206.66.6	41230.00.0	41253.33.3	..8400
8500..	41579.16.6	41602.77.7	41626.38.8	41650.00.0	41673.61.1	41697.22.2	41720.83.3	41744.44.4	..8500
8600..	42068.33.3	42092.22.2	42116.11.1	42140 00.0	42163 88.8	42187.77.7	42211.66.6	42235.55.5	..8600
8700..	42557.50.0	42581.66.6	42605 83.3	42630.00.0	42654.16.6	42678.33.3	42702.50.0	42726.66.6	..8700
8800..	43046.66.6	43071.11.1	43095.55.5	43120.00.0	43144.44.4	43168.88 8	43193.33.3	43217.77.7	..8800
8900..	43535.83.3	43560.55.5	43585.27.7	43610.00.0	43634.72.2	43659.44 4	43684.16.6	43708.88.8	..8900
9000..	44025 00.0	44050 00.0	44075.00.0	44100.00.0	44125.00.0	44150.00.0	44175.00.0	44200.00.0	..9000
9100..	44514.16 6	44539.44.4	44564.72.2	44590.00 0	44615.27.7	44640.55.5	44665.83.3	44691.11.1	..9100
9200..	45003 33.3	45028.88.8	45054.44.4	45080.00.0	45105.55.5	45131.11.1	45156.66.6	45182.22.2	..9200
9300..	45492.50.0	45518.33.3	45544.16.6	45570.00 0	45595.83.3	45621.66.6	45647.50.0	45673.33.3	..9300
9400..	45981.66.6	46007.77.7	46033.88.8	46060.00.0	46086.11.1	46112.22.2	46138.33.3	46164.44.4	..9400
9500..	46470.83.3	46497.22.2	46523.61.1	46550.00.0	46576.38.8	46602.77.7	46629.16.6	46655.55.5	..9500
9600..	46960.00.0	46986.66.6	47013.33.3	47040.00.0	47066.66.6	47093.33.3	47120.00.0	47146.66.6	..9600
9700..	47449.16.6	47476.11.1	47503.05.5	47530.00.0	47556.94.4	47583.88 8	47610 83.3	47637.77.7	..9700
9800..	47938.33.3	47965.55.5	47992.77.7	48020.00.0	48047.22.2	48074.44.4	48101.66.6	48128.88.8	..9800
9900..	48427.50.0	48455.00.0	48482.50.0	48510 00.0	48537.50.0	48565.00.0	48592.50.0	48620 00.0	..9900
10000..	48916.66.6	48944.44.4	48972.22.2	49000.00.0	49027.77.7	49055.55.5	49083.33.3	49111.11.1	..10000

s.	$ cts.	$ cts.	$ cts.	$ cts.	$ cts.	$ cts.	$ cts.	$ cts.	s.
16..	3.91.3	3.91.6	3.91.8	3.92.0	3.92.2	3.92.4	3.92.7	3.92.9	..16
17..	4.15.8	4.16.0	4.16.3	4.16.5	4.16.7	4.17 0	4.17.2	4.17.4	..17
18..	4.40.2	4.40.5	4.40.7	4.41.0	4.41.2	4.41.5	4.41.7	4.42.0	..18
19..	4.64.7	4.65.0	4.65.2	4.65.5	4.65.8	4.66.0	4.66.3	4.66.5	..19
1d..	.02.0	.02.0	.02.0	.02.0	.02.0	.02.0	.02.0	.02.0	..1d.
2..	.04.1	.04.1	.04.1	.04.1	.04.1	.04.1	.04.1	.04.1	..2
3..	.06.1	.06.1	.06.1	.06.1	.06.1	.06.1	.06.1	.06.1	..3
4..	.08.1	.08.1	.08.2	.08.2	.08.2	.08 2	.08.2	.08.2	..4
5..	.10.2	.10.2	.10.2	.10.2	.10.2	.10.2	.10.2	.10.2	..5
6..	.12.2	.12.2	.12.2	.12.2	.12.2	.12.3	.12 3	.12.3	..6
7..	.14.3	.14 3	.14.3	.14.3	.14.3	.14.3	.14.3	.14.3	..7
8..	.16.3	.16.3	.16.3	.16.3	.16.3	.16.3	.16.3	.16.3	..8
9..	.18 3	.18.3	.18.4	.18.4	.18.4	.18.4	.18.4	.18.4	..9
10..	.20.4	.20.4	.20.4	.20.4	.20.4	.20.4	.20.4	.20.5	..10
11..	.22.4	.22.4	.22.4	.22.4	.22.5	.22.5	.22.5	.22.5	..11

STERLING EXCHANGE TABLES.

$10\frac{9}{16}$ TO 11 %

STERLING INTO DOLLARS AND CENTS.

Stg. £	$10\frac{9}{16}$ $ cts.	$10\frac{5}{8}$ $ cts.	$10\frac{11}{16}$ $ cts.	$10\frac{3}{4}$ $ cts.	$10\frac{13}{16}$ $ cts.	$10\frac{7}{8}$ $ cts.	$10\frac{15}{16}$ $ cts.	11 $ cts.	Stg. £
100..	491.38.8	491.66.6	491.94.4	492.22.2	492.50.0	492.77.7	493.05.5	493.33.3	..100
200..	982.77.7	983.33.3	983.88.8	984.44.4	985.00.0	985.55.5	986.11.1	986.66.6	..200
300..	1474.16.6	1475.00.0	1475.83.3	1476.66.6	1477.50.0	1478.33.3	1479.16.6	1480.00.0	..300
400..	1965.55.5	1966.66.6	1967.77.7	1968.88.8	1970.00.0	1971.11.1	1972.22.2	1973.33.3	..400
500..	2456.94.4	2458.33.3	2459.72.2	2461.11.1	2462.50.0	2463.88.8	2465.27.7	2466.66.6	..500
600..	2948.33.3	2950.00.0	2951.66.6	2953.33.3	2955.00.0	2956.66.6	2958.33.3	2960.00.0	..600
700..	3439.72.2	3441.66.6	3443.61.1	3445.55.5	3447.50.0	3449.44.4	3451.38.8	3453.33.3	..700
800..	3931.11.1	3933.33.3	3935.55.5	3937.77.7	3940.00.0	3942.22.2	3944.44.4	3946.66.6	..800
900..	4422.50.0	4425.00.0	4427.50.0	4430.00.0	4432.50.0	4435.00.0	4437.50.0	4440.00.0	..900
1000..	4913.88.8	4916.66.6	4919.44.4	4922.22.2	4925.00.0	4927.77.7	4930.55.5	4933.33.3	..1000
1100..	5405.27.7	5408.33.3	5411.38.8	5414.44.4	5417.50.0	5420.55.5	5423.61.1	5426.66.6	..1100
1200..	5896.66.6	5900.00.0	5903.33.3	5906.66.6	5910.00.0	5913.33.3	5916.66.6	5920.00.0	..1200
1300..	6388.05.5	6391.66.6	6395.27.7	6398.88.8	6402.50.0	6406.11.1	6409.72.2	6413.33.3	..1300
1400..	6879.44.4	6883.33.3	6887.22.2	6891.11.1	6895.00.0	6898.88.8	6902.77.7	6906.66.6	..1400
1500..	7370.83.3	7375.00.0	7379.16.6	7383.33.3	7387.50.0	7391.66.6	7395.83.3	7400.00.0	..1500
1600..	7862.22.2	7866.66.6	7871.11.1	7875.55.5	7880.00.0	7884.44.4	7888.88.8	7893.33.3	..1600
1700..	8353.61.1	8358.33.3	8363.05.5	8367.77.7	8372.50.0	8377.22.2	8381.94.4	8386.66.6	..1700
1800..	8845.00.0	8850.00.0	8855.00.0	8860.00.0	8865.00.0	8870.00.0	8875.00.0	8880.00.0	..1800
1900..	9336.38.8	9341.66.6	9346.94.4	9352.22.2	9357.50.0	9362.77.7	9368.05.5	9373.33.3	..1900
2000.	9827.77.7	9833.33.3	9838.88.8	9844.44.4	9850.00.0	9855.55.5	9861.11.1	9866.66.6	..2000
2100..	10319.16.6	10325.00.0	10330.83.3	10336.66.6	10342.50.0	10348.33.3	10354.16.6	10360.00.0	..2100
2200..	10810.55.5	10816.66.6	10822.77.7	10828.88.8	10835.00.0	10841.11.1	10847.22.2	10853.33.3	..2200
2300..	11301.94.4	11308.33.3	11314.72.2	11321.11.1	11327.50.0	11333.88.8	11340.27.7	11346.66.6	..2300
2400..	11793.33.3	11800.00.0	11806.66.6	11813.33.3	11820.00.0	11826.66.6	11833.33.3	11840.00.0	..2400
2500..	12284.72.2	12291.66.6	12298.61.1	12305.55.5	12312.50.0	12319.44.4	12326.38.8	12333.33.3	..2500
2600..	12776.11.1	12783.33.3	12790.55.5	12797.77.7	12805.00.0	12812.22.2	12819.44.4	12826.66.6	..2600
2700..	13267.50.0	13275.00.0	13282.50.0	13290.00.0	13297.50.0	13305.00.0	13312.50.0	13320.00.0	..2700
2800..	13758.88.8	13766.66.6	13774.44.4	13782.22.2	13790.00.0	13797.77.7	13805.55.5	13813.33.3	..2800
2900..	14250.27.7	14258.33.3	14266.38.8	14274.44.4	14282.50.0	14290.55.5	14298.61.1	14306.66.6	..2900
3000..	14741.66.6	14750.00.0	14758.33.3	14766.66.6	14775.00.0	14783.33.3	14791.66.6	14800.00.0	..3000
3100.	15233.05.5	15241.66.6	15250.27.7	15258.88.8	15267.50.0	15276.11.1	15284.72.2	15293.33.3	..3100
3200..	15724.44.4	15733.33.3	15742.22.2	15751.11.1	15760.00.0	15768.88.8	15777.77.7	15786.66.6	..3200
3300..	16215.83.3	16225.00.0	16234.16.6	16243.33.3	16252.50.0	16261.66.6	16270.83.3	16280.00.0	..3300
3400..	16707.22.2	16716.66.6	16726.11.1	16735.55.5	16745.00.0	16754.44.4	16763.88.8	16773.33.3	..3400
3500..	17198.61.1	17208.33.3	17218.05.5	17227.77.7	17237.50.0	17247.22.2	17256.94.4	17266.66.6	..3500
3600..	17690.00.0	17700.00.0	17710.00.0	17720.00.0	17730.00.0	17740.00.0	17750.00.0	17760.00.0	..3600
3700..	18181.38.8	18191.66.6	18201.94.4	18212.22.2	18222.50.0	18232.77.7	18243.05.5	18253.33.3	..3700
3800..	18672.77.7	18683.33.3	18693.88.8	18704.44.4	18715.00.0	18725.55.5	18736.11.1	18746.66.6	..3800
3900..	19164.16.6	19175.00.0	19185.83.3	19196.66.6	19207.50.0	19218.33.3	19229.16.6	19240.00.0	..3900
4000..	19655.55.5	19666.66.6	19677.77.7	19688.88.8	19700.00.0	19711.11.1	19722.22.2	19733.33.3	..4000
4100..	20146.94.4	20158.33.3	20169.72.2	20181.11.1	20192.50.0	20203.88.8	20215.27.7	20226.66.6	..4100
4200..	20638.33.3	20650.00.0	20661.66.6	20673.33.3	20685.00.0	20696.66.6	20708.33.3	20720.00.0	..4200
4300..	21129.72.2	21141.66.6	21153.61.1	21165.55.5	21177.50.0	21189.44.4	21201.38.8	21213.33.3	..4300
4400..	21621.11.1	21633.33.3	21645.55.5	21657.77.7	21670.00.0	21682.22.2	21694.44.4	21706.66.6	..4400
4500..	22112.50.0	22125.00.0	22137.50.0	22150.00.0	22162.50.0	22175.00.0	22187.50.0	22200.00.0	..4500
4600..	22603.88.8	22616.66.6	22629.44.4	22642.22.2	22655.00.0	22667.77.7	22680.55.5	22693.33.3	..4600
4700..	23095.27.7	23108.33.3	23121.38.8	23134.44.4	23147.50.0	23160.55.5	23173.61.1	23186.66.6	..4700
4800..	23586.66.6	23600.00.0	23613.33.3	23626.66.6	23640.00.0	23653.33.3	23666.66.6	23680.00.0	..4800
4900..	24078.05.5	24091.66.6	24105.27.7	24118.88.8	24132.50.0	24146.11.1	24159.72.2	24173.33.3	..4900
5000..	24569.44.4	24583.33.3	24597.22.2	24611.11.1	24625.00.0	24638.88.8	24652.77.7	24666.66.6	..5000

s.	$ cts.	$ cts.	$ cts.	$ cts.	$ cts.	$ cts.	$ cts.	$ cts.	s.
1..	.24.6	.24.6	.24.6	.24.6	.24.6	.24.6	.24.6	.24.7	..1
2..	.49.1	.49.2	.49.2	.49.2	.49.2	.49.3	.49.3	.49.3	..2
3..	.73.7	.73.7	.73.8	.73.8	.73.9	.73.9	.74.0	.74.0	..3
4..	.98.3	.98.3	.98.4	.98.4	.98.5	.98.5	.98.6	.98.7	..4
5..	1.22.8	1.22.9	1.23.0	1.23.0	1.23.1	1.23.2	1.23.3	1.23.3	..5
6..	1.47.4	1.47.5	1.47.6	1.47.7	1.47.7	1.47.8	1.47.9	1.48.0	..6
7..	1.72.0	1.72.1	1.72.2	1.72.3	1.72.4	1.72.5	1.72.6	1.72.7	..7
8..	1.96.5	1.96.7	1.96.8	1.96.9	1.97.0	1.97.1	1.97.2	1.97.3	..8
9..	2.21.1	2.21.2	2.21.4	2.21.5	2.21.6	2.21.7	2.21.9	2.22.0	..9
10..	2.45.7	2.45.8	2.46.0	2.46.1	2.46.2	2.46.4	2.46.5	2.46.7	..10
11..	2.70.3	2.70.4	2.70.6	2.70.7	2.70.9	2.71.0	2.71.2	2.71.3	..11
12..	2.94.8	2.95.0	2.95.2	2.95.3	2.95.5	2.95.7	2.95.8	2.96.0	..12
13..	3.19.4	3.19.6	3.19.8	3.19.9	3.20.1	3.20.3	3.20.5	3.20.7	..13
14..	3.44.0	3.44.2	3.44.4	3.44.5	3.44.7	3.44.9	3.45.1	3.45.3	..14
15..	3.68.5	3.68.7	3.68.9	3.69.2	3.69.4	3.69.6	3.69.8	3.70.0	..15

STERLING EXCHANGE TABLES.

10 9/16 TO 11 %

STERLING INTO DOLLARS AND CENTS.

Stg.	10⁹⁄₁₆	10⅝	10¹¹⁄₁₆	10¾	10¹³⁄₁₆	10⅞	10¹⁵⁄₁₆	11	Stg.
£	$ cts.	$ cts.	$ cts.	$ cts.	$ cts.	$ cts.	$ cts.	$ cts.	£
5100..	25060.83.3	25075.00.0	25089.16.6	25103.33.3	25117.50.0	25131.66.6	25145.83.3	25160.00.0	..5100
5200..	25552.22.2	25566.66.6	25581.11.1	25595.55.5	25610.00.0	25624.44.4	25638.88.8	25653.33.3	..5200
5300..	26043.61.1	26058.33.3	26073.05.5	26087.77.7	26102.50.0	26117.22.2	26131.94.4	26146.66.6	..5300
5400..	26535.00.0	26550.00.0	26565.00.0	26580.00.0	26595.00.0	26610.00.0	26625.00.0	26640.00.0	..5400
5500..	27026.38.8	27041.66.6	27056.94.4	27072.22.2	27087.50.0	27102.77.7	27118.05.5	27133.33.3	..5500
5600..	27517.77.7	27533.33.3	27548.88.8	27564.44.4	27580.00.0	27595.55.5	27611.11.1	27626.66.6	..5600
5700..	28009.16.6	28025.00.0	28040.83.3	28056.66.6	28072.50.0	28088.33.3	28104.16.6	28120.00.0	..5700
5800..	28500.55.5	28516.66.6	28532.77.7	28548.88.8	28565.00.0	28581.11.1	28597.22.2	28613.33.3	..5800
5900..	28991.94.4	29008.33.3	29024.72.2	29041.11.1	29057.50.0	29073.88.8	29090.27.7	29106.66.6	..5900
6000..	29483.33.3	29500.00.0	29516.66.6	29533.33.3	29550.00.0	29566.66.6	29583.33.3	29600.00.0	..6000
6100..	29974.72.2	29991.66.6	30008.61.1	30025.55.5	30042.50.0	30059.44.4	30076.38.8	30093.33.3	..6100
6200..	30466.11.1	30483.33.3	30500.55.5	30517.77.7	30535.00.0	30552.22.2	30569.44.4	30586.66.6	..6200
6300..	30957.50.0	30975.00.0	30992.50.0	31010.00.0	31027.50.0	31045.00.0	31062.50.0	31080.00.0	..6300
6400..	31448.88.8	31466.66.6	31484.44.4	31502.22.2	31520.00.0	31537.77.7	31555.55.5	31573.33.3	..6400
6500..	31940.27.7	31958.33.3	31976.38.8	31994.44.4	32012.50.0	32030.55.5	32048.61.1	32066.66.6	..6500
6600..	32431.66.6	32450.00.0	32468.33.3	32486.66.6	32505.00.0	32523.33.3	32541.66.6	32560.00.0	..6600
6700..	32923.05.5	32941.66.6	32960.27.7	32978.88.8	32997.50.0	33016.11.1	33034.72.2	33053.33.3	..6700
6800..	33414.44.4	33433.33.3	33452.22.2	33471.11.1	33490.00.0	33508.88.8	33527.77.7	33546.66.6	..6800
6900..	33905.83.3	33925.00.0	33944.16.6	33963.33.3	33982.50.0	34001.66.6	34020.83.3	34040.00.0	..6900
7000..	34397.22.2	34416.66.6	34436.11.1	34455.55.5	34475.00.0	34494.44.4	34513.88.8	34533.33.3	..7000
7100..	34888.61.1	34908.33.3	34928.05.5	34947.77.7	34967.50.0	34987.22.2	35006.94.4	35026.66.6	..7100
7200..	35380.00.0	35400.00.0	35420.00.0	35440.00.0	35460.00.0	35480.00.0	35500.00.0	35520.00.0	..7200
7300..	35871.38.8	35891.66.6	35911.94.4	35932.22.2	35952.50.0	35972.77.7	35993.05.5	36013.33.3	..7300
7400..	36362.77.7	36383.33.3	36403.88.8	36424.44.4	36445.00.0	36465.55.5	36486.11.1	36506.66.6	..7400
7500..	36854.16.6	36875.00.0	36895.83.3	36916.66.6	36937.50.0	36958.33.3	36979.16.6	37000.00.0	..7500
7600..	37345.55.5	37366.66.6	37387.77.7	37408.88.8	37430.00.0	37451.11.1	37472.22.2	37493.33.3	..7600
7700..	37836.94.4	37858.33.3	37879.72.2	37901.11.1	37922.50.0	37943.88.8	37965.27.7	37986.66.6	..7700
7800..	38328.33.3	38350.00.0	38371.66.6	38393.33.3	38415.00.0	38436.66.6	38458.33.3	38480.00.0	..7800
7900..	38819.72.2	38841.66.6	38863.61.1	38885.55.5	38907.50.0	38929.44.4	38951.38.8	38973.33.3	..7900
8000..	39311.11.1	39333.33.3	39355.55.5	39377.77.7	39400.00.0	39422.22.2	39444.44.4	39466.66.6	..8000
8100..	39802.50.0	39825.00.0	39847.50.0	39870.00.0	39892.50.0	39915.00.0	39937.50.0	39960.00.0	..8100
8200..	40293.88.8	40316.66.6	40339.44.4	40362.22.2	40385.00.0	40407.77.7	40430.55.5	40453.33.3	..8200
8300..	40785.27.7	40808.33.3	40831.36.8	40854.44.4	40877.50.0	40900.55.5	40923.61.1	40946.66.6	..8300
8400..	41276.66.6	41300.00.0	41323.33.3	41346.66.6	41370.00.0	41393.33.3	41416.66.6	41440.00.0	..8400
8500..	41768.05.5	41791.66.6	41815.27.7	41838.88.8	41862.50.0	41886.11.1	41909.72.2	41933.33.3	..8500
8600..	42259.44.4	42283.33.3	42307.22.2	42331.11.1	42355.00.0	42378.88.8	42402.77.7	42426.66.6	..8600
8700..	42750.83.3	42775.00.0	42799.16.6	42823.33.3	42847.50.0	42871.66.6	42895.83.3	42920.00.0	..8700
8800..	43242.22.2	43266.66.6	43291.11.1	43315.55.5	43340.00.0	43364.44.4	43388.88.8	43413.33.3	..8800
8900..	43733.61.1	43758.33.3	43783.05.5	43807.77.7	43832.50.0	43857.22.2	43881.94.4	43906.66.6	..8900
9000..	44225.00.0	44250.00.0	44275.00.0	44300.00.0	44325.00.0	44350.00.0	44375.00.0	44400.00.0	..9000
9100..	44716.38.8	44741.66.6	44766.94.4	44792.22.2	44817.50.0	44842.77.7	44868.05.5	44893.33.3	..9100
9200..	45207.77.7	45233.33.3	45258.88.8	45284.44.4	45310.00.0	45335.55.5	45361.11.1	45386.66.6	..9200
9300..	45699.16.6	45725.00.0	45750.83.3	45776.66.6	45802.50.0	45828.33.3	45854.16.6	45880.00.0	..9300
9400..	46190.55.5	46216.66.6	46242.77.7	46268.88.8	46295.00.0	46321.11.1	46347.22.2	46373.33.3	..9400
9500..	46681.94.4	46708.33.3	46734.72.2	46761.11.1	46787.50.0	46813.88.8	46840.27.7	46866.66.6	..9500
9600..	47173.33.3	47200.00.0	47226.66.6	47253.33.3	47280.00.0	47306.66.6	47333.33.3	47360.00.0	..9600
9700..	47664.72.2	47691.66.6	47718.61.1	47745.55.5	47772.50.0	47799.44.4	47826.38.8	47853.33.3	..9700
9800..	48156.11.1	48183.33.3	48210.55.5	48237.77.7	48265.00.0	48292.22.2	48319.44.4	48346.66.6	..9800
9900..	48647.50.0	48675.00.0	48702.50.0	48730.00.0	48757.50.0	48785.00.0	48812.50.0	48840.00.0	..9900
10000..	49138.88.8	49166.66.6	49194.44.4	49222.22.2	49250.00.0	49277.77.7	49305.55.5	49333.33.3	..10000

s.	$ cts.	$ cts.	$ cts.	$ cts.	$ cts.	$ cts.	$ cts.	$ cts.	s.
16..	3.93.1	3.93.3	3.93.5	3.93.8	3.94.0	3.94.3	3.94.4	3.94.7	..16
17..	4.17.7	4.17.9	4.18.1	4.18.4	4.18.6	4.18.9	4.19.1	4.19.3	..17
18..	4.42.2	4.42.5	4.42.7	4.43.0	4.43.2	4.43.5	4.43.7	4.44.0	..18
19..	4.66.8	4.67.1	4.67.3	4.67.6	4.67.9	4.68.1	4.68.4	4.68.7	..19
1d.	.02.0	.02.0	.02.0	.02.0	.02.0	.02.0	.02.0	.02.0	..1d.
2..	.04.1	.04.1	.04.1	.04.1	.04.1	.04.1	.04.1	.04.1	..2
3..	.06.1	.06.1	.06.1	.06.1	.06.1	.06.1	.06.2	.06.2	..3
4..	.08.2	.08.2	.08.2	.08.2	.08 2	.08.2	.08.2	.08.2	..4
5..	.10.2	.10.2	.10.2	.10.2	.10.3	.10.3	.10.3	.10.3	..5
6..	.12.3	.12.3	.12.3	.12.3	.12.3	.12.3	.12 3	.12.3	..6
7..	.14.3	.14.3	.14.3	.14.3	.14.4	.14.4	.14.4	.14.4	..7
8..	.16.4	.16.4	.16.4	.16.4	.16.4	.16.4	.16.4	.16.4	..8
9..	.18.4	.18.4	.18.4	.18.4	.18.5	.18.5	.18.5	.18.5	..9
10..	.20.5	.20.5	.20.5	.20.5	.20.5	.20.5	.20.5	.20.5	..10
11..	.22.5	.22.5	.22.5	.22.5	.22.6	.22.6	.22.6	.22.6	..11

STERLING EXCHANGE TABLES.

$11\frac{1}{16}$ TO $11\frac{1}{2}$%

STERLING INTO DOLLARS AND CENTS.

Stg. £	$11\frac{1}{16}$ $ cts.	$11\frac{1}{8}$ $ cts.	$11\frac{3}{16}$ $ cts.	$11\frac{1}{4}$ $ cts.	$11\frac{5}{16}$ $ cts.	$11\frac{3}{8}$ $ cts.	$11\frac{7}{16}$ $ cts.	$11\frac{1}{2}$ $ cts.	Stg. £
100..	493.61.1	493.88.8	494.16.6	494.44.4	494.72.2	495.00.0	495.27.7	495.55.5	..100
200..	987.22.2	987.77.7	988.33.3	988.88.8	989.44.4	990.00.0	990.55.5	991.11.1	..200
300..	1480.83.3	1481.66.6	1482.50 0	1483.33.3	1484.16.6	1485.00.0	1485.83.3	1486.66.6	..300
400..	1974.44.4	1975.55 5	1976.66.6	1977.77.7	1978.88.8	1980 00 0	1981.11.1	1982.22.2	..400
500..	2468.05.5	2469.44.4	2470.83.3	2472.22.2	2473.61.1	2475.00.0	2476.38.8	2477.77.7	..500
600..	2961.66.6	2963.33.3	2965.00.0	2966.66.6	2968.33.3	2970.00.0	2971.66.6	2973.33 3	..600
700..	3455.27.7	3457.22.2	3459.16.6	3461.11.1	3463.05.5	3465.00.0	3466.94.4	3468.88.8	..700
800..	3948.88.8	3951.11.1	3953.33.3	3955.55.5	3957.77.7	3960.00.0	3962.22.2	3964.44.4	..800
900..	4442.50.0	4445.00.0	4447.50.0	4450.00.0	4452.50.0	4455.00.0	4457.50.0	4460.00.0	..900
1000..	4936.11.1	4938.88.8	4941.66.6	4944.44.4	4947.22.2	4950.00.0	4952.77.7	4955.55.5	..1000
1100..	5429.72.2	5432.77.7	5435.83.3	5438.88.8	5441.94.4	5445.00.0	5448.05.5	5451.11.1	..1100
1200..	5923.33.3	5926.66.6	5930.00.0	5933.33.3	5936.66.6	5940.00.0	5943.33.3	5946.66.6	..1200
1300..	6416.94.4	6420.55.5	6424.16.6	6427.77.7	6431.38.8	6435.00.0	6438.61.1	6442.22.2	..1300
1400..	6910.55.5	6914.44.4	6918.33.3	6922.22.2	6926.11.1	6930.00.0	6933.88.8	6937.77.7	..1400
1500..	7404.16.6	7408.33.3	7412.50.0	7416.66.6	7420.83.3	7425.00 0	7429.16.6	7433.33.3	..1500
1600..	7897.77.7	7902.22.2	7906.66.6	7911.11.1	7915.55.5	7920.00.0	7924.44.4	7928.88.8	..1600
1700..	8391.38.8	8396.11.1	8400.83.3	8405.55.5	8410.27.7	8415.00.0	8419.72.2	8424.44.4	..1700
1800..	8885.00.0	8890.00.0	8895.00.0	8900.00.0	8905.00.0	8910.00.0	8915.00.0	8920.00.0	..1800
1900..	9378.61.1	9383.88.8	9389.16.6	9394.44.4	9399.72.2	9405.00.0	9410.27.7	9415.55.5	..1900
2000 .	9872.22.2	9877.77.7	9883.33.3	9888.88.8	9894.44.4	9900.00.0	9905.55.5	9911.11.1	..2000
2100..	10365.83.3	10371.66.6	10377.50.0	10383.33.3	10389.16.6	10395.00.0	10400.83.3	10406.66.6	..2100
2200..	10859.44.4	10865.55.5	10871.66.6	10877.77.7	10883.88.8	10890.00.0	10896.11.1	10902.22 2	..2200
2300..	11353.05.5	11359.44.4	11365.83 3	11372.22.2	11378.61.1	11385.00.0	11391.38.8	11397.77.7	..2300
2400..	11846.66.6	11853.33.3	11860.00 0	11866.66.6	11873.33.3	11880.00.0	11886.66.6	11893.33.3	..2400
2500..	12340.27.7	12347.22.2	12354.16.6	12361.11.1	12368.05.5	12375.00.0	12381.94.4	12388.88.8	..2500
2600..	12833.88.8	12841.11.1	12848.33 3	12855.55.5	12862 77.7	12870.00.0	12877.22.2	12884.44.4	..2600
2700..	13327.50.0	13335.00.0	13342.50 0	13350.00.0	13357.50.0	13365.00.0	13372.50.0	13380.00.0	..2700
2800..	13821.11.1	13828.88.8	13836.66.6	13844.44.4	13852.22.2	13860.00.0	13867.77.7	13875.55.5	..2800
2900..	14314.72.2	14322.77.7	14330.83.3	14338.88.8	14346.94.4	14355.00.0	14363.05.5	14371.11.1	..2900
3000..	14808.33.3	14816.66.6	14825.00.0	14833.33.3	14841.66.6	14850.00.0	14858.33.3	14866.66.6	3000
3100.	15301.94.4	15310.55.5	15319.16.6	15327.77.7	15336.38.8	15345.00.0	15353.61.1	15362.22.2	..3100
3200..	15795.55.5	15804.44.4	15813.33.3	15822.22.2	15831.11.1	15840.00.0	15848.88.8	15857.77.7	..3200
3300..	16289.16.6	16298.33.3	16307.50.0	16316.66.6	16325.83.3	16335.00.0	16344.16.6	16353.33.3	..3300
3400..	16782.77.7	16792.22.2	16801.66.6	16811.11.1	16820.55.5	16830.00.0	16839.44.4	16848.88.8	..3400
3500..	17276.38.8	17286.11.1	17295.83.3	17305.55.5	17315.27.7	17325.00.0	17334.72.2	17344.44.4	..3500
3600..	17770.00.0	17780.00.0	17790.00.0	17800.00.0	17810.00.0	17820.00.0	17830.00.0	17840.00.0	..3600
3700..	18263.61.1	18273.88.8	18284.16.6	18294.44.4	18304.72.2	18315.00.0	18325.27.7	18335.55.5	..3700
3800..	18757.22.2	18767.77.7	18778.33.3	18788.88.8	18799.44.4	18810.00.0	18820.55.5	18831.11.1	..3800
3900..	19250.83.3	19261.66.6	19272.50.0	19283.33.3	19294.16.6	19305.00.0	19315.83.3	19326.66 6	..3900
4000..	19744.44.4	19755.55.5	19766.66.6	19777.77.7	19788.88.8	19800.00.0	19811.11.1	19822.22.2	..4000
4100..	20238.05.5	20249.44.4	20260.83.3	20272.22.2	20283.61.1	20295.00.0	20306.38.8	20317.77.7	..4100
4200..	20731.66.6	20743.33.3	20755.00.0	20766.66.6	20778.33.3	20790.00.0	20801.66.6	20813.33.3	..4200
4300..	21225.27.7	21237.22.2	21249.16.6	21261.11.1	21273.05.5	21285.00.0	21296.94.4	21308.88.8	..4300
4400..	21718.88.8	21731.11.1	21743.33.3	21755 55.5	21767.77.7	21780.00.0	21792.22.2	21804.44.4	..4400
4500..	22212.50.0	22225.00.0	22237.50.0	22250.00.0	22262.50.0	22275.00.0	22287.50.0	22300.00.0	..4500
4600..	22706.11.1	22718.88.8	22731.66.6	22744.44.4	22757.22.2	22770.00.0	22782.77.7	22795.55.5	..4600
4700..	23199.72.2	23212.77.7	23225.83.3	23238.88.8	23251.94.4	23265.00.0	23278 05.5	23291 11.1	..4700
4800..	23693.33.3	23706.66.6	23720.00.0	23733.33.3	23746.66.6	23760.00.0	23773.33.3	23786.66.6	..4800
4900..	24186.94.4	24200.55.5	24214.16.6	24227.77.7	24241.38.8	24255.00.0	24268.61.1	24282.22.2	..4900
5000..	24680.55.5	24694.44.4	24708.33.3	24722.22.2	24736.11.1	24750.00.0	24763.88.8	24777.77.7	..5000

s.	$ cts.	$ cts.	$ cts.	$ cts.	$ cts.	$ cts.	$ cts.	$ cts.	s.
1..	.24.7	.24.7	.24.7	.24.7	.24.7	.24.7	.24.8	.24.8	..1
2..	.49 4	.49·4	.49.4	.49.4	.49.5	.49.5	.49.5	.49.5	..2
3..	.74.0	.74.1	.74.1	.74.2	.74.2	.74.2	.74.3	.74.3	..3
4..	.98.7	.98.8	.98.8	.98.9	.98.9	.99.0	.99.0	.99.1	..4
5..	1.23.4	1.23.5	1.23.6	1.23.6	1.23.7	1.23.7	1.23.8	1.23.9	..5
6..	1.48.1	1.48.2	1.48.2	1.48.3	1.48.4	1.48.5	1.48.6	1.48.7	..6
7..	1.72.8	1.72.9	1.72.9	1.73.0	1.73.1	1.73.2	1.73.3	1.73.4	..7
8..	1.97.4	1.97.5	1.97.7	1.97.8	1.97.9	1.98.0	1.98.1	1.98.2	..8
9..	2.22.1	2.22.2	2.22.4	2.22.5	2.22.6	2.22.7	2.22.9	2.23.0	..9
10..	2.46.8	2.46.9	2.47.1	2.47.2	2.47.4	2.47.5	2.47.6	2.47.8	..10
11..	2.71.5	2.71.6	2.71.8	2.71.9	2.72.1	2.72.2	2.72.4	2.72.5	..11
12..	2.96.2	2.96.3	2.96.5	2.96.7	2.96.8	2.97.0	2.97.2	2.97.3	..12
13..	3.20.8	3.21.0	3.21.2	3.21.4	3.21.6	3.21.7	3.21.9	3.22.1	..13
14..	3.45.5	3.45.7	3.45.9	3.46.1	3.46.3	3.46.5	3.46.7	3.46.9	..14
15..	3.70.2	3.70.4	3.70.6	3.70.8	3.71.0	3.71.2	3.71.4	3.71.7	..15

STERLING EXCHANGE TABLES.

$11\tfrac{1}{16}$ TO $11\tfrac{1}{2}$%

STERLING INTO DOLLARS AND CENTS.

Stg. £	$11\tfrac{1}{16}$ $ cts.	$11\tfrac{1}{8}$ $ cts.	$11\tfrac{3}{16}$ $ cts.	$11\tfrac{1}{4}$ $ cts.	$11\tfrac{5}{16}$ $ cts.	$11\tfrac{3}{8}$ $ cts.	$11\tfrac{7}{16}$ $ cts.	$11\tfrac{1}{2}$ $ cts.	Stg. £
5100	25174.16.6	25188.33.3	25202.50.0	25216.66.6	25230.83.3	25245.00.0	25259 16.6	25273.33.3	5100
5200	25667.77.7	25682 22.2	25696.66.6	25711.11.1	25725.55.5	25740.00.0	25754.44.4	25768.88.8	5200
5300	26161.38.8	26176.11.1	26190.83.3	26205.55.5	26220.27.7	26235.00.0	26249.72.2	26264.44.4	5300
5400	26655.00.0	26670.00.0	26685.00.0	26700.00.0	26715.00.0	26730.00.0	26745.00.0	26760.00.0	5400
5500	27148.61.1	27163.88.8	27179.16.6	27194.44.4	27209.72.2	27225.00.0	27240.27.7	27255.55.5	5500
5600	27642.22.2	27657.77.7	27673.33.3	27688.88.8	27704.44.4	27720.00.0	27735.55.5	27751.11.1	5600
5700	28135.83.3	28151.66.6	28167.50.0	28183.33.3	28199.16.6	28215.00.0	28230.83.3	28246.66.6	5700
5800	28629.44.4	28645.55.5	28661.66.6	28677.77.7	28693.88.8	28710.00.0	28726.11.1	28742.22.2	5800
5900	29123.05.5	29139.44.4	29155.83.3	29172.22.2	29188.61.1	29205.00.0	29221.38.8	29237.77.7	5900
6000	29616.66.6	29633.33.3	29650.00.0	29666.66.6	29683.33.3	29700.00.0	29716.66.6	29733.33.3	6000
6100	30110.27.7	30127.22.2	30144.16.6	30161.11.1	30178.05.5	30195.00.0	30211.94.4	30228.88.8	6100
6200	30603.88.8	30621.11.1	30638.33.3	30655.55.5	30672.77.7	30690.00.0	30707.22.2	30724.44.4	6200
6300	31097.50.0	31115.00.0	31132.50.0	31150.00.0	31167.50.0	31185.00.0	31202.50.0	31220.00.0	6300
6400	31591.11.1	31608.88.8	31626.66.6	31644.44.4	31662.22.2	31680.00.0	31697.77.7	31715.55.5	6400
6500	32084.72.2	32102.77.7	32120.83.3	32138.88.8	32156.94.4	32175.00.0	32193.05.5	32211.11.1	6500
6600	32578.33.3	32596.66.6	32615.00.0	32633.33.3	32651.66.6	32670.00.0	32688.33.3	32706.66.6	6600
6700	33071.94.4	33090.55.5	33109.16.6	33127.77.7	33146.38.8	33165.00.0	33183.61.1	33202.22.2	6700
6800	33565.55.5	33584.44.4	33603.33.3	33622.22.2	33641.11.1	33660.00.0	33678.88.8	33697.77.7	6800
6900	34059.16.6	34078.33.3	34097.50 0	34116.66.6	34135.83.3	34155.00.0	34174.16.6	34193.33.3	6900
7000	34552.77.7	34572.22.2	34591.66.6	34611.11.1	34630.55.5	34650.00.0	34669.44 4	34688.88.8	7000
7100	35046.38.8	35066.11.1	35085.83.3	35105.55.5	35125.27.7	35145.00.0	35164.72.2	35184.44.4	7100
7200	35540.00.0	35560.00.0	35580.00.0	35600.00.0	35620.00.0	35640.00.0	35660.00.0	35680.00.0	7200
7300	36033.61.1	36053.88.8	36074.16.6	36094.44.4	36114.72.2	36135.00.0	36155.27.7	36175.55.5	7300
7400	36527.22.2	36547.77.7	36568.33.3	36588.88.8	36609.44.4	36630.00.0	36650.55.5	36671.11.1	7400
7500	37020.83.3	37041.66.6	37062.50.0	37083.33.3	37104.16.6	37125.00.0	37145.83.3	37166.66.6	7500
7600	37514.44.4	37535.55.5	37556.66.6	37577.77.7	37598.88.8	37620.00.0	37641.11.1	37662.22.2	7600
7700	38008.05.5	38029.44.4	38050.83.3	38072.22.2	38093.61.1	38115.00.0	38136.38.8	38157.77.7	7700
7800	38501.66.6	38523.33.3	38545.00.0	38566.66.6	38588.33.3	38610.00.0	38631.66.6	38653.33.3	7800
7900	38995.27.7	39017.22.2	39039.16.6	39061.11.1	39083.05.5	39105.00.0	39126.94.4	39148.88.8	7900
8000	39488.88.8	39511.11.1	39533.33.3	39555.55.5	39577.77.7	39600.00.0	39622.22.2	39644.44.4	8000
8100	39982.50.0	40005.00.0	40027.50.0	40050.00.0	40072.50.0	40095.00.0	40117.50.0	40140.00.0	8100
8200	40476.11.1	40498.88.8	40521.66.6	40544.44.4	40567.22.2	40590.00.0	40612.77.7	40635.55.5	8200
8300	40969.72.2	40992.77.7	41015.83.3	41038.88.8	41061.94.4	41085.00.0	41108.05.5	41131.11.1	8300
8400	41463.33.3	41486.66.6	41510.00.0	41533.33.3	41556.66.6	41580.00.0	41603.33 3	41626.66.6	8400
8500	41956.94.4	41980.55.5	42004.16.6	42027.77.7	42051.38.8	42075.00.0	42098.61.1	42122.22.2	8500
8600	42450.55.5	42474.44.4	42498.33.3	42522.22.2	42546.11.1	42570.00.0	42593.88.8	42617.77.7	8600
8700	42944.16.6	42968.33.3	42992.50.0	43016.66.6	43040.83.3	43065.00.0	43089.16.6	43113.33.3	8700
8800	43437.77.7	43462.22.2	43486.66.6	43511.11.1	43535.55.5	43560.00.0	43584.44.4	43608.88 8	8800
8900	43931.38.8	43956.11.1	43980.83.3	44005.55.5	44030.27.7	44055.00.0	44079.72.2	44104.44.4	8900
9000	44425.00.0	44450.00.0	44475.00.0	44500.00.0	44525.00.0	44550.00.0	44575.00.0	44600.00.0	9000
9100	44918.61.1	44943.88.8	44969.16.6	44994.44.4	45019 72.2	45045.00.0	45070.27.7	45095.55.5	9100
9200	45412.22.2	45437.77.7	45463.33.3	45488.88 8	45514.44.4	45540.00.0	45565.55.5	45591.11.1	9200
9300	45905.83.3	45931.66.6	45957.50.0	45983.33.3	46009.16.6	46035.00.0	46060.83.3	46086.66.6	9300
9400	46399.44.4	46425.55.5	46451.66.6	46477.77.7	46503.88.8	46530.00.0	46556.11.1	46582.22.2	9400
9500	46893.05.5	46919.44.4	46945.83.3	46972.22.2	46998.61.1	47025.00.0	47051.38.8	47077.77.7	9500
9600	47386.66.6	47413.33.3	47440.00.0	47466.66.6	47493.33.3	47520.00.0	47546.66.6	47573.33.3	9600
9700	47880.27.7	47907.22.2	47934.16.6	47961.11.1	47988.05.5	48015.00.0	48041.94.4	48068.88.8	9700
9800	48373.88.8	48401.11.1	48428.33.3	48445.55.5	48482.77.7	48510.00.0	48537.22.2	48564.44.4	9800
9900	48867.50.0	48895.00.0	48922.50.0	48950.00.0	48977.50.0	49005.00.0	49032.50.0	49060.00.0	9900
10000	49361.11.1	49388.88.8	49416.66.6	49444.44.4	49472.22.2	49500.00.0	49527.77.7	49555.55.5	10000

s.	$ cts.	$ cts.	$ cts.	$ cts.	$ cts.	$ cts.	$ cts.	$ cts.	s.
10..	3.94.9	3.95.1	3.95.3	3.95.5	3.95.8	3.96.0	3.96.2	3.96.4	.16
17..	4.19.6	4.19.8	4.20.0	4.20.3	4.20.5	4.20.7	4.21.0	4.21.2	.17
18..	4.44.2	4.44.5	4.44.7	4.45.0	4.45.2	4.45.5	4.45.7	4.46.0	.18
19..	4.08.9	4.69.2	4.69.4	4.69.7	4.70.0	4.70.2	4.70.5	4.70.8	.19
1d.	.02.0	.02.0	.02.0	.02.1	.02.1	.02.1	.02.1	.02.1	1d.
2..	.04.1	.04.1	.04.1	.04.1	.04.1	.04.1	.04.1	.04.1	2
3..	.06.2	.06.2	.06.2	.06.2	.06.2	.06.2	.06.2	.06.2	3
4..	.08.2	.08.2	.08.2	.08.2	.08 2	.08.2	.08.2	.08.2	4
5..	.10.3	.10.3	.10.3	.10.3	.10.3	.10.3	.10.3	.10.3	5
6..	.12.3	.12.3	.12.3	.12.4	.12.4	.12.4	.12.4	.12.4	6
7..	.14.4	.14.4	.14.4	.14.4	.14.4	.14.4	.14.4	.14.4	7
8..	.16.4	.16.5	.16.5	.16.5	.16.5	.16.5	.16.5	.16.5	8
9..	.18.5	.18.5	.18.5	.18.5	.18.5	.18.6	.18.6	.18.6	9
10..	.20.6	20.6	.20.6	.20.6	.20.6	.20.6	.20.6	.20.6	.10
11..	.22.6	.22.6	.22.6	.22.7	.22.7	.22.7	.22.7	.22.7	.11

STERLING EXCHANGE TABLES.

11 9/16 TO 12 %

STERLING INTO DOLLARS AND CENTS.

Stg.	11 9/16	11 5/8	11 11/16	11 3/4	11 13/16	11 7/8	11 15/16	12	Stg.
£	$ cts.	$ cts.	$ cts.	$ cts.	$ cts.	$ cts.	$ cts.	$ cts.	£
100..	495.83.3	496.11.1	496.38.8	496.66.6	496.94.4	497.22.2	497.50.0	497.77.7	..100
200..	991.66.6	992.22.2	992.77.7	993.33.3	993.88.8	994.44.4	995.00.0	995.55.5	..200
300..	1487.50.0	1488.33.3	1489.16.6	1490.00.0	1490.83.3	1491.66.6	1492.50.0	1493.33.3	..300
400..	1983.33.3	1984.44.4	1985.55.5	1986.66.6	1987.77.7	1988.88.8	1990.00.0	1991.11.1	..400
500..	2479.16.6	2480.55.5	2481.94.4	2483.33.3	2484.72.2	2486.11.1	2487.50.0	2488.88.8	..500
600..	2975.00.0	2976.66.6	2978.33.3	2980.00.0	2981.66.6	2983.33.3	2985.00.0	2986.66.6	..600
700..	3470.83.3	3472.77.7	3474.72.2	3476.66.6	3478.61.1	3480.55.5	3482.50.0	3484.44.4	..700
800..	3966.66.6	3968.88.8	3971.11.1	3973.33.3	3975.55.5	3977.77.7	3980.00.0	3982.22.2	..800
900..	4462.50.0	4465.00.0	4467.50.0	4470.00.0	4472.50.0	4475.00.0	4477.50.0	4480.00.0	..900
1000..	4958.33.3	4961.11.1	4963.88.8	4966.66.6	4969.44.4	4972.22.2	4975.00.0	4977.77.7	..1000
1100..	5454.16.6	5457.22.2	5460.27.7	5463.33.3	5466.38.8	5469.44.4	5472.50.0	5475.55.5	..1100
1200..	5950.00.0	5953.33.3	5956.66.6	5960.00.0	5963.33.3	5966.66.6	5970.00.0	5973.33.3	..1200
1300..	6445.83.3	6449.44.4	6453.05.5	6456.66.6	6460.27.7	6463.88.8	6467.50.0	6471.11.1	..1300
1400..	6941.66.6	6945.55.5	6949.44.4	6953.33.3	6957.22.2	6961.11.1	6965.00.0	6968.88.8	..1400
1500..	7437.50.0	7441.66.6	7445.83.3	7450.00.0	7454.16.6	7458.33.3	7462.50.0	7466.66.6	..1500
1600..	7933.33.3	7937.77.7	7942.22.2	7946.66.6	7951.11.1	7955.55.5	7960.00.0	7964.44.4	..1600
1700..	8429.16.6	8433.88.8	8438.61.1	8443.33.3	8448.05.5	8452.77.7	8457.50.0	8462.22.2	..1700
1800..	8925.00.0	8930.00.0	8935.00.0	8940.00.0	8945.00.0	8950.00.0	8955.00.0	8960.00.0	..1800
1900..	9420.83.3	9426.11.1	9431.38.8	9436.66.6	9441.94.4	9447.22.2	9452.50.0	9457.77.7	..1900
2000..	9916.66.6	9922.22.2	9927.77.7	9933.33.3	9938.88.8	9944.44.4	9950.00.0	9955.55.5	..2000
2100..	10412.50.0	10418.33.3	10424.16.6	10430.00.0	10435.83.3	10441.66.6	10447.50.0	10453.33.3	..2100
2200..	10908.33.3	10914.44.4	10920.55.5	10926.66.6	10932.77.7	10938.88.8	10945.00.0	10951.11.1	..2200
2300..	11404.16.6	11410.55.5	11416.94.4	11423.33.3	11429.72.2	11436.11.1	11442.50.0	11448.88.8	..2300
2400..	11900.00.0	11906.66.6	11913.33.3	11920.00.0	11926.66.6	11933.33.3	11940.00.0	11946.66.6	..2400
2500..	12395.83.3	12402.77.7	12409.72.2	12416.66.6	12423.61.1	12430.55.5	12437.50.0	12444.44.4	..2500
2600..	12891.66.6	12898.88.8	12906.11.1	12913.33.3	12920.55.5	12927.77.7	12935.00.0	12942.22.2	..2600
2700..	13387.50.0	13395.00.0	13402.50.0	13410.00.0	13417.50.0	13425.00.0	13432.50.0	13440.00.0	..2700
2800..	13883.33.3	13891.11.1	13898.88.8	13906.66.6	13914.44.4	13922.22.2	13930.00.0	13937.77.7	..2800
2900..	14379.16.6	14387.22.2	14395.27.7	14403.33.3	14411.38.8	14419.44.4	14427.50.0	14435.55.5	..2900
3000..	14875.00.0	14883.33.3	14891.66.6	14900.00.0	14908.33.3	14916.66.6	14925.00.0	14933.33.3	..3000
3100..	15370.83.3	15379.44.4	15388.05.5	15396.66.6	15405.27.7	15413.88.8	15422.50.0	15431.11.1	..3100
3200..	15866.66.6	15875.55.5	15884.44.4	15893.33.3	15902.22.2	15911.11.1	15920.00.0	15928.88.8	..3200
3300..	16362.50.0	16371.66.6	16380.83.3	16390.00.0	16399.16.6	16408.33.3	16417.50.0	16426.66.6	..3300
3400..	16858.33.3	16867.77.7	16877.22.2	16886.66.6	16896.11.1	16905.55.5	16915.00.0	16924.44.4	..3400
3500..	17354.16.6	17363.88.8	17373.61.1	17383.33.3	17393.05.5	17402.77.7	17412.50.0	17422.22.2	..3500
3600..	17850.00.0	17860.00.0	17870.00.0	17880.00.0	17890.00.0	17900.00.0	17910.00.0	17920.00.0	..3600
3700..	18345.83.3	18356.11.1	18366.38.8	18376.66.6	18386.94.4	18397.22.2	18407.50.0	18417.77.7	..3700
3800..	18841.66.6	18852.22.2	18862.77.7	18873.33.3	18883.88.8	18894.44.4	18905.00.0	18915.55.5	..3800
3900..	19337.50.0	19348.33.3	19359.16.6	19370.00.0	19380.83.3	19391.66.6	19402.50.0	19413.33.3	..3900
4000..	19833.33.3	19844.44.4	19855.55.5	19866.66.6	19877.77.7	19888.88.8	19900.00.0	19911.11.1	..4000
4100..	20329.16.6	20340.55.5	20351.94.4	20363.33.3	20374.72.2	20386.11.1	20397.50.0	20408.88.8	..4100
4200..	20825.00.0	20836.66.6	20848.33.3	20860.00.0	20871.66.6	20883.33.3	20895.00.0	20906.66.6	..4200
4300..	21320.83.3	21332.77.7	21344.72.2	21356.66.6	21368.61.1	21380.55.5	21392.50.0	21404.44.4	..4300
4400..	21816.66.6	21828.88.8	21841.11.1	21853.33.3	21865.55.5	21877.77.7	21890.00.0	21902.22.2	..4400
4500..	22312.50.0	22325.00.0	22337.50.0	22350.00.0	22362.50.0	22375.00.0	22387.50.0	22400.00.0	..4500
4600..	22808.33.3	22821.11.1	22833.88.8	22846.66.6	22859.44.4	22872.22.2	22885.00.0	22897.77.7	..4600
4700..	23304.16.6	23317.22.2	23330.27.7	23343.33.3	23356.38.8	23369.44.4	23382.50.0	23395.55.5	..4700
4800..	23800.00.0	23813.33.3	23826.66.6	23840.00.0	23853.33.3	23866.66.6	23880.00.0	23893.33.3	..4800
4900..	24295.83.3	24309.44.4	24323.05.5	24336.66.6	24350.27.7	24363.88.8	24377.50.0	24391.11.1	..4900
5000..	24791.66.6	24805.55.5	24819.44.4	24833.33.3	24847.22.2	24861.11.1	24875.00.0	24888.88.8	..5000

s.	$ cts.	$ cts.	$ cts.	$ cts.	$ cts.	$ cts.	$ cts.	$ cts.	s.
1..	.24.8	.24.8	.24.8	.24.8	.24.8	.24.9	.24.9	.24.9	..1
2..	.49.6	.49.6	.49.6	.49.7	.49.7	.49.7	.49.7	.49.8	..2
3..	.74.4	.74.4	.74.4	.74.5	.74.5	.74.6	.74.6	.74.7	..3
4..	.99.2	.99.2	.99.3	.99.3	.99.4	.99.4	.99.5	.99.5	..4
5..	1.23.9	1.24.0	1.24.1	1.24.2	1.24.2	1.24.3	1.24.4	1.24.4	..5
6..	1.48.7	1.48.8	1.48.9	1.49.0	1.49.1	1.49.2	1.49.2	1.49.3	..6
7..	1.73.5	1.73.6	1.73.7	1.73.8	1.73.9	1.74.0	1.74.1	1.74.2	..7
8..	1.98.3	1.98.4	1.98.5	1.98.7	1.98.8	1.98.9	1.99.0	1.99.1	..8
9..	2.23.1	2.23.2	2.23.4	2.23.5	2.23.6	2.23.7	2.23.9	2.24.0	..9
10..	2.47.9	2.48.0	2.48.2	2.48.3	2.48.5	2.48.6	2.48.7	2.48.9	..10
11..	2.72.7	2.72.9	2.73.0	2.73.2	2.73.3	2.73.5	2.73.6	2.73.8	..11
12..	2.97.5	2.97.7	2.97.8	2.98.0	2.98.2	2.98.3	2.98.5	2.98.7	..12
13..	3.22.3	3.22.5	3.22.6	3.22.8	3.23.0	3.23.2	3.23.4	3.23.5	..13
14..	3.47.1	3.47.3	3.47.5	3.47.7	3.47.9	3.48.0	3.48.2	3.48.4	..14
15..	3.71.9	3.72.1	3.72.3	3.72.5	3.72.7	3.72.9	3.73.1	3.73.3	..15

STERLING EXCHANGE TABLES.

11 9/16 TO 12 %

STERLING INTO DOLLARS AND CENTS.

Stg.	11 9/16	11 5/8	11 11/16	11 3/4	11 13/16	11 7/8	11 15/16	12	Stg.
£	$ cts.	$ cts.	$ cts.	$ cts.	$ cts.	$ cts.	$ cts.	$ cts.	£
5100..	25287.50.0	25301.66.6	25315.83.3	25330.00.0	25344.16.6	25358.33.3	25372.50.0	25336.66.6	..5100
5200..	25783.33.3	25797.77.7	25812.22.2	25826.66.6	25841.11.1	25855.55.5	25870.00.0	25884.44.4	..5200
5300..	26279.16.6	26293.88.8	26308.61.1	26323.33.3	26338.05.5	26352.77.7	26367.50.0	26382.22.2	..5300
5400..	26775.00.0	26790.00.0	26805.00.0	26820.00.0	26835.00.0	26850.00.0	26865.00.0	26880.00.0	..5400
5500..	27270.83.3	27286.11.1	27301.38.8	27316.66.6	27331.94.4	27347.22.2	27362.50.0	27377.77.7	..5500
5600..	27766.66.6	27782.22.2	27797.77.7	27813.33.3	27828.88.8	27844.44.4	27860.00.0	27875.55.5	..5600
5700..	28262.50.0	28278.33.3	28294.16.6	28310.00.0	28325.83.3	28341.66.6	28357.50.0	28373.33.3	..5700
5800..	28758.33.3	28774.44.4	28790.55.5	28806.66.6	28822.77.7	28838.88.8	28855.00.0	28871.11.1	..5800
5900..	29254.16.6	29270.55.5	29286.94.4	29303.33.3	29319.72.2	29336.11.1	29352.50.0	29368.88.8	..5900
6000..	29750.00.0	29766.66.6	29783.33.3	29800.00.0	29816.66.6	29833.33.3	29850.00.0	29866.66.6	..6000
6100..	30245.83.3	30262.77.7	30279.72.2	30296.66.6	30313.61.1	30330.55.5	30347.50.0	30364.44.4	..6100
6200..	30741.66.6	30758.88.8	30776.11.1	30793.33.3	30810.55.5	30827.77.7	30845.00.0	30862.22.2	..6200
6300..	31237.50.0	31255.00.0	31272.50.0	31290.00.0	31307.50.0	31325.00.0	31342.50.0	31360.00.0	..6300
6400..	31733.33.3	31751.11.1	31768.88.8	31786.66.6	31804.44.4	31822.22.2	31840.00.0	31857.77.7	..6400
6500..	32229.16.6	32247.22.2	32265.27.7	32283.33.3	32301.38.8	32319.44.4	32337.50.0	32355.55.5	..6500
6600..	32725.00.0	32743.33.3	32761.66.6	32780.00.0	32798.33.3	32816.66.6	32835.00.0	32853.33.3	..6600
6700..	33220.83.3	33239.44.4	33258.05.5	33276.66.6	33295.27.7	33313.88.8	33332.50.0	33351.11.1	..6700
6800..	33716.66.6	33735.55.5	33754.44.4	33773.33.3	33792.22.2	33811.11.1	33830.00.0	33848.88.8	..6800
6900..	34212.50.0	34231.66.6	34250.83.3	34270.00.0	34289.16.6	34308.33.3	34327.50.0	34346.66.6	..6900
7000..	34708.33.3	34727.77.7	34747.22.2	34766.66.6	34786.11.1	34805.55.5	34825.00.0	34844.44.4	..7000
7100..	35204.16.6	35223.88.8	35243.61.1	35263.33.3	35283.05.5	35302.77.7	35322.50.0	35342.22.2	..7100
7200..	35700.00.0	35720.00.0	35740.00.0	35760.00.0	35780.00.0	35800.00.0	35820.00.0	35840.00.0	..7200
7300..	36195.83.3	36216.11.1	36236.38.8	36256.66.6	36276.94.4	36297.22.2	36317.50.0	36337.77.7	..7300
7400..	36691.66.6	36712.22.2	36732.77.7	36753.33.3	36773.88.8	36794.44.4	36815.00.0	36835.55.5	..7400
7500..	37187.50.0	37208.33.3	37229.16.6	37250.00.0	37270.83.3	37291.66.6	37312.50.0	37333.33.3	..7500
7600..	37683.33.3	37704.44.4	37725.55.5	37746.66.6	37767.77.7	37788.88.8	37810.00.0	37831.11.1	..7600
7700..	38179.16.6	38200.55.5	38221.94.4	38243.33.3	38264.72.2	38286.11.1	38307.50.0	38328.88.8	..7700
7800..	38675.00.0	38696.66.6	38718.33.3	38740.00.0	38761.66.6	38783.33.3	38805.00.0	38826.66.6	..7800
7900..	39170.83.3	39192.77.7	39214.72.2	39236.66.6	39258.61.1	39280.55.5	39302.50.0	39324.44.4	..7900
8000..	39666.66.6	39688.88.8	39711.11.1	39733.33.3	39755.55.5	39777.77.7	39800.00.0	39822.22.2	..8000
8100..	40162.50.0	40185.00.0	40207.50.0	40230.00.0	40252.50.0	40275.00.0	40297.50.0	40320.00.0	..8100
8200..	40658.33.3	40681.11.1	40703.88.8	40726.66.6	40749.44.4	40772.22.2	40795.00.0	40817.77.7	..8200
8300..	41154.16.6	41177.22.2	41200.27.7	41223.33.3	41246.38.8	41269.44.4	41292.50.0	41315.55.5	..8300
8400..	41650.00.0	41673.33.3	41696.66.6	41720.00.0	41743.33.3	41766.66.6	41790.00.0	41813.33.3	..8400
8500..	42145.83.3	42169.44.4	42193.05.5	42216.66.6	42240.27.7	42263.88.8	42287.50.0	42311.11.1	..8500
8600..	42641.66.6	42665.55.5	42689.44.4	42713.33.3	42737.22.2	42761.11.1	42785.00.0	42808.88.8	..8600
8700..	43137.50.0	43161.66.6	43185.83.3	43210.00.0	43234.16.6	43258.33.3	43282.50.0	43306.66.6	..8700
8800..	43633.33.3	43657.77.7	43682.22.2	43706.66.6	43731.11.1	43755.55.5	43780.00.0	43804.44.4	..8800
8900..	44129.16.6	44153.88.8	44178.61.1	44203.33.3	44228.05.5	44252.77.7	44277.50.0	44302.22.2	..8900
9000..	44625.00.0	44650.00.0	44675.00.0	44700.00.0	44725.00.0	44750.00.0	44775.00.0	44800.00.0	..9000
9100..	45120.83.3	45146.11.1	45171.38.8	45196.66.6	45221.94.4	45247.22.2	45272.50.0	45297.77.7	..9100
9200..	45616.66.6	45642.22.2	45667.77.7	45693.33.3	45718.88.8	45744.44.4	45770.00.0	45795.55.5	..9200
9300..	46112.50.0	46138.33.3	46164.16.6	46190.00.0	46215.83.3	46241.66.6	46267.50.0	46293.33.3	..9300
9400..	46608.33.3	46634.44.4	46660.55.5	46686.66.6	46712.77.7	46738.88.8	46765.00.0	46791.11.1	..9400
9500..	47104.16.6	47130.55.5	47156.94.4	47183.33.3	47209.72.2	47236.11.1	47262.50.0	47288.88.8	..9500
9600..	47600.00.0	47626.66.6	47653.33.3	47680.00.0	47706.66.6	47733.33.3	47760.00.0	47786.66.6	..9600
9700..	48095.83.3	48122.77.7	48149.72.2	48176.66.6	48203.61.1	48230.55.5	48257.50.0	48284.44.4	..9700
9800..	48591.66.6	48618.88.8	48646.11.1	48673.33.3	48700.55.5	48727.77.7	48755.00.0	48782.22.2	..9800
9900..	49087.50.0	49115.00.0	49142.50.0	49170.00.0	49197.50.0	49225.00.0	49252.50.0	49280.00.0	..9900
10000..	49583.33.3	49611.11.1	49638.88.8	49666.66.6	49694.44.4	49722.22.2	49750.00.0	49777.77.7	..10000

s.	$ cts.	$ cts.	$ cts.	$ cts.	$ cts.	$ cts.	$ cts.	$ cts.	s.
16..	3.96.7	3.96.9	3.97.1	3.97.3	3.97.5	3.97.8	3.98.0	3.98.2	..16
17..	4.21.4	4.21.7	4.21.9	4.22.2	4.22.4	4.22.6	4.22.9	4.23.1	..17
18..	4.46.2	4.46.5	4.46.7	4.47.0	4.47.2	4.47.5	4.47.7	4.48.0	..18
19..	4.71.0	4.71.3	4.71.6	4.71.8	4.72.1	4.72.4	4.72.6	4.72.9	..19
1d..	.02.1	.02.1	.02.1	.02.1	.02.1	.02.1	.02.1	.02.1	..1d.
2..	.04.1	.04.1	.04.1	.04.1	.04.1	.04.1	.04.1	.04.1	..2
3..	.06.2	.06.2	.06.2	.06.2	.06.2	.06.2	.06.2	.06.2	..3
4..	.08.3	.08.3	.08.3	.08.3	.08.3	.08.3	.04.3	.08.3	..4
5..	.10.3	.10.3	.10.3	.10.3	.10.3	.10.3	.10.4	.10.4	..5
6..	.12.4	.12.4	.12.4	.12.4	.12.4	.12.4	.12.4	.12.4	..6
7..	.14.5	.14.5	.14.5	.14.5	.14.5	.14.5	.14.5	.14.5	..7
8..	.16.5	.16.5	.16.5	.16.5	.16.6	.16.6	.16.6	.16.6	..8
9..	.18.6	.18.6	.18.6	.18.6	.18.6	.18.6	.18.6	.18.7	..9
10..	.20.6	.20.7	.20.7	.20.7	.20.7	.20.7	.20.7	.20.7	..10
11..	.22.7	.22.7	.22.7	.22.8	.22.8	.22.8	.22.8	.22.8	..11

STERLING EXCHANGE TABLES.

$12\frac{1}{16}$ TO $12\frac{1}{2}$%

STERLING INTO DOLLARS AND CENTS.

Stg.	$12\frac{1}{16}$	$12\frac{1}{8}$	$12\frac{3}{16}$	$12\frac{1}{4}$	$12\frac{5}{16}$	$12\frac{3}{8}$	$12\frac{7}{16}$	$12\frac{1}{2}$	Stg.
£	$ cts.	$ cts.	$ cts.	$ cts.	$ cts.	$ cts.	$ cts.	$ cts.	£
100..	498.05.5	498.33.3	498.61.1	498.88.8	499.16.6	499.44.4	499.72.2	500.00.0	..100
200..	996.11.1	996.66.6	997.22.2	997.77.7	998.33.3	998.88.8	999.44.4	1000.00.0	..200
300..	1494.16.6	1495.00.0	1495.83.3	1496.66.6	1497.50.0	1498.33.3	1499.16.6	1500.00.0	..300
400..	1992.22.2	1993.33.3	1994.44.4	1995.55.5	1996.66.6	1997.77.7	1998.88.8	2000.00.0	..400
500..	2490.27.7	2491.66.6	2493.05.5	2494.44.4	2495.83.3	2497.22.2	2498.61.1	2500.00.0	..500
600..	2988.33.3	2990.00.0	2991.66.6	2993.33.3	2995.00.0	2996.66.6	2998.33.3	3000.00.0	..600
700..	3486.38.8	3488.33.3	3490.27.7	3492.22.2	3494.16.6	3496.11.1	3498.05.5	3500.00.0	..700
800..	3984.44.4	3986.66.6	3988.88.8	3991.11.1	3993.33.3	3995.55.5	3997.77.7	4000.00.0	..800
900..	4482.50.0	4485.00.0	4487.50.0	4490.00.0	4492.50.0	4495.00.0	4497.50.0	4500.00.0	..900
1000..	4980.55.5	4983.33.3	4986.11.1	4988.88.8	4991.66.6	4994.44.4	4997.22.2	5000.00.0	..1000
1100..	5478.61.1	5481.66.6	5484.72.2	5487.77.7	5490.83.3	5493.88.8	5496.94.4	5500.00.0	..1100
1200..	5976.66.6	5980.00.0	5983.33.3	5986.66.6	5990.00.0	5993.33.3	5996.66.6	6000.00.0	..1200
1300..	6474.72.2	6478.33.3	6481.94.4	6485.55.5	6489.16.6	6492.77.7	6496.38.8	6500.00.0	..1300
1400..	6972.77.7	6976.66.6	6980.55.5	6984.44.4	6988.33.3	6992.22.2	6996.11.1	7000.00.0	..1400
1500..	7470.83.3	7475.00.0	7479.16.6	7483.33.3	7487.50.0	7491.66.6	7495.83.3	7500.00.0	..1500
1600..	7968.88.8	7973.33.3	7977.77.7	7982.22.2	7986.66.6	7991.11.1	7995.55.5	8000.00.0	..1600
1700..	8466.94.4	8471.66.6	8476.38.8	8481.11.1	8485.83.3	8490.55.5	8495.27.7	8500.00.0	..1700
1800..	8965.00.0	8970.00.0	8975.00.0	8980.00.0	8985.00.0	8990.00.0	8995.00.0	9000.00.0	..1800
1900..	9463.05.5	9468.33.3	9473.61.1	9478.88.8	9484.16.6	9489.44.4	9494.72.2	9500.00.0	..1900
2000..	9961.11.1	9966.66.6	9972.22.2	9977.77.7	9983.33.3	9988.88.8	9994.44.4	10000.00.0	..2000
2100..	10459.16.6	10465.00.0	10470.83.3	10476.66.6	10482.50.0	10488.33.3	10494.16.6	10500.00.0	..2100
2200..	10957.22.2	10963.33.3	10969.44.4	10975.55.5	10981.66.6	10987.77.7	10993.88.8	11000.00.0	..2200
2300..	11455.27.7	11461.66.6	11468.05.5	11474.44.4	11480.83.3	11487.22.2	11493.61.1	11500.00.0	..2300
2400..	11953.33.3	11960.00.0	11966.66.6	11973.33.3	11980.00.0	11986.66.6	11993.33.3	12000.00.0	..2400
2500..	12451.38.8	12458.33.3	12465.27.7	12472.22.2	12479.16.6	12486.11.1	12493.05.5	12500.00.0	..2500
2600..	12949.44.4	12956.66.6	12963.88.8	12971.11.1	12978.33.3	12985.55.5	12992.77.7	13000.00.0	..2600
2700..	13447.50.0	13455.00.0	13462.50.0	13470.00.0	13477.50.0	13485.00.0	13492.50.0	13500.00.0	..2700
2800..	13945.55.5	13953.33.3	13961.11.1	13968.88.8	13976.66.6	13984.44.4	13992.22.2	14000.00.0	..2800
2900..	14443.61.1	14451.66.6	14459.72.2	14467.77.7	14475.83.3	14483.88.8	14491.94.4	14500.00.0	..2900
3000..	14941.66.6	14950.00.0	14958.33.3	14966.66.6	14975.00.0	14983.33.3	14991.66.6	15000.00.0	..3000
3100.	15439.72.2	15448.33.3	15456.94.4	15465.55.5	15474.16.6	15482.77.7	15491.38.8	15500.00.0	..3100
3200..	15937.77.7	15946.66.6	15955.55.5	15964.44.4	15973.33.3	15982.22.2	15991.11.1	16000.00.0	..3200
3300..	16435.83.3	16445.00.0	16454.16.6	16463.33.3	16472.50.0	16481.66.6	16490.83.3	16500.00.0	..3300
3400..	16933.88.8	16943.33.3	16952.77.7	16962.22.2	16971.66.6	16981.11.1	16990.55.5	17000.00.0	..3400
3500..	17431.94.4	17441.66.6	17451.38.8	17461.11.1	17470.83.3	17480.55.5	17490.27.7	17500.00.0	..3500
3600..	17930.00.0	17940.00.0	17950.00.0	17960.00.0	17970.00.0	17980.00.0	17990.00.0	18000.00.0	..3600
3700..	18428.05.5	18438.33.3	18448.61.1	18458.88.8	18469.16.6	18479.44.4	18489.72.2	18500.00.0	..3700
3800..	18926.11.1	18936.66.6	18947.22.2	18957.77.7	18968.33.3	18978.88.8	18989.44.4	19000.00.0	..3800
3900..	19424.16.6	19435.00.0	19445.83.3	19456.66.6	19467.50.0	19478.33.3	19489.16.6	19500.00.0	..3900
4000..	19922.22.2	19933.33.3	19944.44.4	19955.55.5	19966.66.6	19977.77.7	19988.88.8	20000.00.0	..4000
4100..	20420.27.7	20431.66.6	20443.05.5	20454.44.4	20465.83.3	20477.22.2	20488.61.1	20500.00.0	..4100
4200..	20918.33.3	20930.00.0	20941.66.6	20953.33.3	20965.00.0	20976.66.6	20988.33.3	21000.00.0	..4200
4300..	21416.38.8	21428.33.3	21440.27.7	21452.22.2	21464.16.6	21476.11.1	21488.05.5	21500.00.0	..4300
4400..	21914.44.4	21926.66.6	21938.88.8	21951.11.1	21963.33.3	21975.55.5	21987.77.7	22000.00.0	..4400
4500..	22412.50.0	22425.00.0	22437.50.0	22450.00.0	22462.50.0	22475.00.0	22487.50.0	22500.00.0	..4500
4600..	22910.55.5	22923.33.3	22936.11.1	22948.88.8	22961.66.6	22974.44.4	22987.22.2	23000.00.0	..4600
4700..	23408.61.1	23421.66.6	23434.72.2	23447.77.7	23460.83.3	23473.88.8	23486.94.4	23500.00.0	..4700
4800..	23906.66.6	23920.00.0	23933.33.3	23946.66.6	23960.00.0	23973.33.3	23986.66.6	24000.00 0	..4800
4900..	24404.72.2	24418.33.3	24431.94.4	24445.55.5	24459.16.6	24472.77.7	24486.38.8	24500.00.0	..4900
5000..	24902.77.7	24916.66.6	24930.55.5	24944.44.4	24958.33.3	24972.22.2	24986.11.1	25000.00.0	..5000

s.	$ cts.	$ cts.	$ cts.	$ cts.	$ cts.	$ cts.	$ cts.	$ cts.	s.
1..	.24.9	.24.9	.24.9	.24.9	.24.9	.25.0	.25.0	.25.0	..1
2..	.49.8	.49.8	.49.9	.49.9	.49.9	.49.9	.50.0	.50.0	..2
3..	.74.7	.74.7	.74.8	.74.8	.74.9	.74.9	.74.9	.75.0	..3
4..	.99.6	.99.7	.90.7	.99.8	.99.8	.99.9	.99.9	1.00.0	..4
5..	1.24.5	1.24.6	1.24.6	1.24.7	1.24.8	1.24.9	1.24.9	1.25.0	..5
6..	1.49.4	1.49.5	1.49.6	1.49.7	1.49.7	1.49.8	1.49.9	1.50.0	..6
7..	1.74.3	1.74.4	1.74.5	1.74.6	1.74.7	1.74.8	1.74.9	1.75.0	..7
8..	1.99.2	1.99.3	1.99.4	1.99.5	1.99.7	1.99.8	1.99.9	2.00.0	..8
9..	2.24.1	2.24.2	2.24.4	2.24.5	2.24.6	2.24.7	2.24.9	2.25.0	..9
10..	2.49.0	2.49.2	2.49.3	2.49.4	2.49.6	2.49.7	2.49.9	2.50.0	..10
11..	2.73.9	2.74.1	2.74.2	2.74.4	2.74.5	2.74.7	2.74.8	2.75.0	..11
12..	2.98.9	2.99.0	2.99.2	2.99.3	2.99.5	2.99.7	2.99.8	3.00.0	..12
13..	3.23.7	3.23.9	3.24.1	3.24.3	3.24.4	3.24.6	3.24.8	3.25.0	..13
14..	3.48.6	3.48.8	3.49.0	3.49.2	3.49.4	3.49.6	3.49.8	3.50.0	..14
15..	3.73.5	3.73.7	3.73.9	3.74.2	3.74.4	3.74.6	3.74.8	3.75.0	..15

STERLING EXCHANGE TABLES.

$12\frac{1}{16}$ TO $12\frac{1}{2}$%

STERLING INTO DOLLARS AND CENTS.

Stg.	$12\frac{1}{16}$	$12\frac{1}{8}$	$12\frac{3}{16}$	$12\frac{1}{4}$	$12\frac{5}{16}$	$12\frac{3}{8}$	$12\frac{7}{16}$	$12\frac{1}{2}$	Stg.
£	$ cts.	$ cts.	$ cts.	$ cts.	$ cts.	$ cts.	$ cts.	$ cts.	£
5100..	25400.83.3	25415.00.0	25429.16.6	25443.33.3	25457.50.0	25471.66.6	25485.83.3	25500.00.0	..5100
5200..	25898.88.8	25913.33.3	25927.77.7	25942.22.2	25956.66.6	25971.11.1	25985.55.5	26000.00.0	..5200
5300..	26396.94.4	26411.66.6	26426.38.8	26441.11.1	26455.83.3	26470.55.5	26485.27.7	26500.00.0	..5300
5400..	26895.00.0	26910.00.0	26925.00.0	26940.00.0	26955.00.0	26970.00.0	26985.00.0	27000.00.0	..5400
5500..	27393.05.5	27408.33.3	27423.61.1	27438.88.8	27454.16.6	27469.44.4	27484.72.2	27500.00.0	..5500
5600..	27891.11.1	27906.66.6	27922.22.2	27937.77.7	27953.33.3	27968.88.8	27984.44.4	28000.00.0	..5600
5700..	28389.16.6	28405.00.0	28420.83.3	28436.66.6	28452.50.0	28468.33.3	28484.16.6	28500.00.0	..5700
5800..	28887.22.2	28903.33.3	28919.44.4	28935.55.5	28951.66.6	28967.77.7	28983.88.8	29000.00.0	..5800
5900..	29385.27.7	29401.66.6	29418.05.5	29434.44.4	29450.83.3	29467.22.2	29483.61.1	29500.00.0	..5900
6000..	29883.33.3	29900.00.0	29916.66.6	29933.33.3	29950.00.0	29966.66.6	29983.33.3	30000.00.0	..6000
6100..	30381.38.8	30398.33.3	30415.27.7	30432.22.2	30449.16.6	30466.11.1	30483.05.5	30500.00.0	..6100
6200..	30879.44.4	30896.66.6	30913.88.8	30931.11.1	30948.33.3	30965.55.5	30982.77.7	31000.00.0	..6200
6300..	31377.50.0	31395.00.0	31412.50.0	31430.00.0	31447.50.0	31465.00.0	31482.50.0	31500.00.0	..6300
6400..	31875.55.5	31893.33.3	31911.11.1	31928.88.8	31946.66.6	31964.44.4	31982.22.2	32000.00.0	..6400
6500..	32373.61.1	32391.66.6	32409.72.2	32427.77.7	32445.83.3	32463.88.8	32481.94.4	32500.00.0	..6500
6600..	32871.66.6	32890.00.0	32908.33.3	32926.66.6	32945.00.0	32963.33.3	32981.66.6	33000.00.0	..6600
6700..	33369.72.2	33388.33.3	33406.94.4	33425.55.5	33444.16.6	33462.77.7	33481.38.8	33500.00.0	..6700
6800..	33867.77.7	33886.66.6	33905.55.5	33924.44.4	33943.33.3	33962.22.2	33981.11.1	34000.00.0	..6800
6900..	34365.83.3	34385.00.0	34404.16.6	34423.33.3	34442.50.0	34461.66.6	34480.83.3	34500.00.0	..6900
7000..	34863.88.8	34883.33.3	34902.77.7	34922.22.2	34941.66.6	34961.11.1	34980.55.5	35000.00.0	..7000
7100..	35361.94.4	35381.66.6	35401.38.8	35421.11.1	35440.83.3	35460.55.5	35480.27.7	35500.00.0	..7100
7200..	35860.00.0	35880.00.0	35900.00.0	35920.00.0	35940.00.0	35960.00.0	35980.00.0	36000.00.0	..7200
7300..	36358.05.5	36378.33.3	36398.61.1	36418.88.8	36439.16.6	36459.44.4	36479.72.2	36500.00.0	..7300
7400..	36856.11.1	36876.66.6	36897.22.2	36917.77.7	36938.33.3	36958.88.8	36979.44.4	37000.00.0	..7400
7500..	37354.16.6	37375.00.0	37395.83.3	37416.66.6	37437.50.0	37458.33.3	37479.16.6	37500.00.0	..7500
7600..	37852.22.2	37873.33.3	37894.44.4	37915.55.5	37936.66.6	37957.77.7	37978.88.8	38000.00.0	..7600
7700..	38350.27.7	38371.66.6	38393.05.5	38414.44.4	38435.83.3	38457.22.2	38478.61.1	38500.00.0	..7700
7800..	38848.33.3	38870.00.0	38891.66.6	38913.33.3	38935.00.0	38956.66.6	38978.33.3	39000.00.0	..7800
7900..	39346.38.8	39368.33.3	39390.27.7	39412.22.2	39434.16.6	39456.11.1	39478.05.5	39500.00.0	..7900
8000..	39844.44.4	39866.66.6	39888.88.8	39911.11.1	39933.33.3	39955.55.5	39977.77.7	40000.00.0	..8000
8100..	40342.50.0	40365.00.0	40387.50.0	40410.00.0	40432.50.0	40455.00.0	40477.50.0	40500.00.0	..8100
8200..	40840.55.5	40863.33.3	40886.11.1	40908.88.8	40931.66.6	40954.44.4	40977.22.2	41000.00.0	..8200
8300..	41338.61.1	41361.66.6	41384.72.2	41407.77.7	41430.83.3	41453.88.8	41476.94.4	41500.00.0	..8300
8400..	41836.66.6	41860.00.0	41883.33.3	41906.66.6	41930.00.0	41953.33.3	41976.66.6	42000.00.0	..8400
8500..	42334.72.2	42358.33.3	42381.94.4	42405.55.5	42429.16.6	42452.77.7	42476.38.8	42500.00.0	..8500
8600..	42832.77.7	42856.66.6	42880.55.5	42904.44.4	42928.33.3	42952.22.2	42976.11.1	43000.00.0	..8600
8700..	43330.83.3	43355.00.0	43379.16.6	43403.33.3	43427.50.0	43451.66.6	43475.83.3	43500.00.0	..8700
8800..	43828.88.8	43853.33.3	43877.77.7	43902.22.2	43926.66.6	43951.11.1	43975.55.5	44000.00.0	..8800
8900..	44326.94.4	44351.66.6	44376.38.8	44401.11.1	44425.83.3	44450.55.5	44475.27.7	44500.00.0	..8900
9000..	44825.00.0	44850.00.0	44875.00.0	44900.00.0	44925.00.0	44950.00.0	44975.00.0	45000.00.0	..9000
9100..	45323.05.5	45348.33.3	45373.61.1	45398.88.8	45424.16.6	45449.44.4	45474.72.2	45500.00.0	..9100
9200..	45821.11.1	45846.66.6	45872.22.2	45897.77.7	45923.33.3	45948.88.8	45974.44.4	46000.00.0	..9200
9300..	46319.16.6	46345.00.0	46370.83.3	46396.66.6	46422.50.0	46448.33.3	46474.16.6	46500.00.0	..9300
9400..	46817.22.2	46843.33.3	46869.44.4	46895.55.5	46921.66.6	46947.77.7	46973.88.8	47000.00.0	..9400
9500..	47315.27.7	47341.66.6	47368.05.5	47394.44.4	47420.83.3	47447.22.2	47473.61.1	47500.00.0	..9500
9600..	47813.33.3	47840.00.0	47866.66.6	47893.33.3	47920.00.0	47946.66.6	47973.33.3	48000.00.0	..9600
9700..	48311.38.8	48338.33.3	48365.27.7	48392.22.2	48419.16.6	48446.11.1	48473.05.5	48500.00.0	..9700
9800..	48809.44.4	48836.66.6	48863.88.8	48891.11.1	48918.33.3	48945.55.5	48972.77.7	49000.00.0	..9800
9900..	49307.50.0	49335.00.0	49362.50.0	49390.00.0	49417.50.0	49445.00.0	49472.50.0	49500.00.0	..9900
10000..	49805.55.5	49833.33.3	49861.11.1	49888.88.8	49916.66.6	49944.44.4	49972.22.2	50000.00.0	..10000

s.	$ cts.	$ cts.	$ cts.	$ cts.	$ cts.	$ cts.	$ cts.	$ cts.	s.
16..	3.98.4	3.98.7	3.98.9	3.99.1	3.99.3	3.99.5	3.99.8	4.00.0	..16
17..	4.23.3	4.23.6	4.23.8	4.24.0	4.24.3	4.24.5	4.24.8	4.25.0	..17
18..	4.48.2	4.48.5	4.48.7	4.49.0	4.49.2	4.49.5	4.49.7	4.50.0	..18
19..	4.73.1	4.73.4	4.73.7	4.73.9	4.74.2	4.74.5	4.74.7	4.75.0	..19
1d..	.02.1	.02.1	.02.1	.02.1	.02.1	.02.1	.02.1	.02.1	..1d.
2..	.04.1	.04.1	.04.1	.04.1	.04.1	.04.2	.04.2	.04.2	..2
3..	.06.2	.06.2	.06.2	.06.2	.06.2	.06.2	.06.2	.06.2	..3
4..	.08.3	.08.3	.08.3	.08.3	.08.3	.08.3	.08.3	.08.3	..4
5..	.10.4	.10.4	.10.4	.10.4	.10.4	.10.4	.10.4	.10.4	..5
6..	.12.4	.12.5	.12.5	.12.5	.12.5	.12.5	.12 5	.12.6	..6
7..	.14.5	.14.5	.14.5	.14.5	.14.5	.14.6	.14.6	.14.6	..7
8..	.16.6	.16.6	.16.6	.16.6	.16.6	.16.6	.16.6	.16.7	..8
9..	.18.7	.18.7	.18.7	.18.7	.18.7	.18.7	.18.7	.18.7	..9
10..	.20.7	.20.8	.20.8	.20.8	.20.8	.20.8	.20.8	.20.8	..10
11..	.22.8	.22.8	.22.8	.22.9	.22.9	.22.9	.22.9	.22.9	..11

STERLING EXCHANGE TABLES.

8 9/16 TO 9%

DOLLARS AND CENTS INTO STERLING.

Doll'rs	8 9/16			8⅝			8 11/16			8¾			8 13/16			8⅞			8 15/16			9			Doll'rs
	£	s.	d.	£	s.	d.	£	s.	d.	£	s.	d.	£	s.	d.	£	s.	d.	£	s.	d.	£	s.	d.	
1..		4	1¾		4	1¾		4	1¾		4	1¾		4	1¾		4	1½		4	1½		4	1½	.. 1
2..		8	3½		8	3½		8	3¼		8	3¼		8	3¼		8	3¼		8	3¼		8	3	.. 2
3..		12	5¼		12	5¼		12	5		12	5		12	5		12	4¾		12	4¾		12	4½	.. 3
4.		16	7		16	6¾		16	6¾		16	6½		16	6½		16	6½		16	6¼		16	6¼	.. 4
5..	1	0	8¾	1	0	8½	1	0	8½	1	0	8¼	1	0	8¼	1	0	8	1	0	7¾	1	0	7¾	.. 5
6..	1	4	10½	1	4	10¼	1	4	10	1	4	10	1	4	9¾	1	4	9½	1	4	9½	1	4	9¼	.. 6
7..	1	9	0¼	1	9	0	1	8	11¾	1	8	11½	1	8	11½	1	8	11¼	1	8	11	1	8	10¾	.. 7
8..	1	13	2	1	13	1¾	1	13	1½	1	13	1¼	1	13	1	1	13	0¾	1	13	0½	1	13	0¼	.. 8
9..	1	17	3¾	1	17	3½	1	17	3¼	1	17	3	1	17	2¾	1	17	2½	1	17	2	1	17	1¾	.. 9
10..	2	1	6½	2	1	5	2	1	4¾	2	1	4½	2	1	4¼	2	1	4	2	1	3¾	2	1	3½	.. 10
20..	4	2	10¾	4	2	10¼	4	2	9¾	4	2	9	4	2	8½	4	2	8	4	2	7½	4	2	6¾	.. 20
30..	6	4	4¼	6	4	3¼	6	4	2½	6	4	1¾	6	4	0¾	6	4	0	6	3	11	6	3	10¼	.. 30
40..	8	5	9¾	8	5	8½	8	5	7¼	8	5	6¼	8	5	5	8	5	4	8	5	2¾	8	5	1¾	.. 40
50..	10	7	3	10	7	1½	10	7	0¼	10	6	10¾	10	6	9¼	10	6	8	10	6	6½	10	6	5	.. 50
60..	12	8	8½	12	8	6¾	12	8	5	12	8	3¼	12	8	1½	12	8	0	12	7	10¼	12	7	8½	.. 60
70..	14	10	1¾	14	9	11¾	14	9	9¾	14	9	7¾	14	9	5¼	14	9	3¾	14	9	2	14	9	0	.. 70
80..	16	11	7¼	16	11	5	16	11	2¾	16	11	0½	16	10	10¼	16	10	7¾	16	10	5½	16	10	3¼	.. 80
90..	18	13	0¾	18	12	10	18	12	7½	18	12	5	18	12	2½	18	11	11¾	18	11	9¼	18	11	6¾	.. 90
100..	20	14	6	20	14	3¼	20	14	0¼	20	13	9½	20	13	6¾	20	13	3¾	20	13	1	20	12	10	..100
110..	22	15	11½	22	15	8¼	22	15	5¼	22	15	2	22	14	11	22	14	7¾	22	14	4¾	22	14	1½	..110
120..	24	17	5	24	17	1½	24	16	10	24	16	6½	24	16	3¼	24	15	11¾	24	15	8¼	24	15	5	..120
130..	26	18	10½	26	18	6½	26	18	3	26	17	11¼	26	17	7½	26	17	3¾	26	17	0	26	16	8¼	..130
140..	29	0	3¾	28	19	11¾	28	19	7¾	28	19	3¾	28	18	11¾	28	18	7¾	28	18	3¾	28	17	11¾	..140
150..	31	1	9¼	31	1	4¾	31	1	0½	31	0	8¼	31	0	4	30	19	11¾	30	19	7½	30	19	3¼	. 150
160..	33	3	2½	33	2	10	33	2	5½	33	2	0¾	33	1	8¼	33	1	3¾	33	0	11¼	33	0	6½	..160
170..	35	4	8	35	4	3	35	3	10¼	35	3	5¼	35	3	0½	35	2	7¾	35	2	2¾	35	1	10	..170
180..	37	6	1¼	37	5	8¼	37	5	3	37	4	10	37	4	4¾	37	3	11¾	37	3	6½	37	3	1½	..180
190..	39	7	6¾	39	7	1¼	39	6	8	39	6	2½	39	5	9	39	5	3¾	39	4	10¼	39	4	4¾	..190
200..	41	9	0¼	41	8	6½	41	8	0¾	41	7	7	41	7	1¼	41	6	7½	41	6	2	41	5	8¼	..200
210..	43	10	5½	43	9	11½	43	9	5½	43	8	11½	43	8	5½	43	7	11½	43	7	5¾	43	6	11¾	..210
220..	45	11	11	45	11	4¾	45	10	10½	45	10	4¼	45	9	9¾	45	9	3½	45	8	9¼	45	8	3	..220
230..	47	13	4½	47	12	9¾	47	12	3¼	47	11	8¾	47	11	2¼	47	10	7½	47	10	1	47	9	6½	..230
240..	49	14	9¾	49	14	3	49	13	8	49	13	1¼	49	12	6½	49	11	11½	49	11	4¾	49	10	10	..240
250..	51	16	3¼	51	15	8	51	15	1	51	14	5¾	51	13	10¾	51	13	3½	51	12	8½	51	12	1¼	..250
260..	53	17	8¾	53	17	1¼	53	16	5¾	53	15	10¼	53	15	3	53	14	7½	53	14	0	53	13	4¾	..260
270..	55	19	2	55	18	6¼	55	17	10½	55	17	3	55	16	7¼	55	15	11½	55	15	3¾	55	14	8¼	..270
280..	58	0	7½	57	19	11½	57	19	3½	57	18	7½	57	17	11½	57	17	3½	57	16	7½	57	15	11½	..280
290..	60	2	1	60	1	4½	60	0	8¼	60	0	0	59	19	3¾	59	18	7½	59	17	11¼	59	17	3	..290
300..	62	3	6¼	62	2	9¾	62	2	1	62	1	4½	62	0	8	61	19	11½	61	19	3	61	18	6¼	..300
310..	64	4	11¾	64	4	2¾	64	3	6	64	2	9	64	2	0¼	64	1	3½	64	0	6½	63	19	9¾	..310
320..	66	6	5	66	5	8	66	4	10¾	66	4	1¾	66	3	4½	66	2	7½	66	1	10¼	66	1	1¼	..320
330..	68	7	10½	68	7	1	68	6	3¾	68	5	6¼	68	4	8¾	68	3	11½	68	3	2	68	2	4½	..330
340..	70	9	4	70	8	6½	70	7	8½	70	6	10¾	70	6	1	70	5	3½	70	4	5¾	70	3	8	..340
350..	72	10	9¼	72	9	11¼	72	9	1¼	72	8	3¼	72	7	5¼	72	6	7¼	72	5	9½	72	4	11½	..350
360..	74	12	2¾	74	11	4½	74	10	6¼	74	9	7¾	74	8	9½	74	7	11¼	74	7	1	74	6	2¾	..360
370..	76	13	8¼	76	12	9½	76	11	11	76	11	0½	76	10	1¾	76	9	3¼	76	8	4¾	76	7	6¼	..370
380..	78	15	1½	78	14	2¾	78	13	3¾	78	12	5	78	11	6¼	78	10	7½	78	9	8½	78	8	9¾	..380
390..	80	16	7	80	15	7¾	80	14	8¾	80	13	9½	80	12	10½	80	11	11¼	80	11	0¼	80	10	1	..390
400..	82	18	0½	82	17	1	82	16	1½	82	15	2	82	14	2¾	82	13	3¼	82	12	4	82	11	4½	..400
410..	84	19	5¾	84	18	6	84	17	6¼	84	16	6½	84	15	7	84	14	7¼	84	13	7½	84	12	8	..410
420..	87	0	11¼	86	19	11½	86	18	11¼	86	17	11½	86	16	11¼	86	15	11¼	86	14	11¼	86	13	11¼	..420
430..	89	2	4½	89	1	4¼	89	0	4	88	19	3¾	88	18	3½	88	17	3¼	88	16	3	88	15	2¾	..430
440..	91	3	10	91	2	9½	91	1	8¾	91	0	8¼	90	19	7¾	90	18	7¼	90	17	6¾	90	16	6¼	..440
450..	93	5	3½	93	4	2½	93	3	1¾	93	2	0¾	93	1	0	92	19	11¼	92	18	10¼	92	17	9½	. 450
460..	95	6	8¾	95	5	7¾	95	4	6½	95	3	5½	95	2	4¼	95	1	3¼	95	0	2	94	19	1	..460
470..	97	8	2¼	97	7	0¾	97	5	11¼	97	4	10	97	3	8½	97	2	7¼	97	1	5¾	97	0	4½	..470
480..	99	9	7¾	99	8	6	99	7	4¼	99	6	2½	99	5	0¾	99	3	11¼	99	2	9½	99	1	7¾	..480
490..	101	11	1	101	9	11	101	8	9	101	7	7	101	6	5	101	5	3	101	4	1¼	101	2	11¼	..490
500..	103	12	6½	103	11	4¼	103	10	1¾	103	8	11½	103	7	9¼	103	6	7	103	5	4¾	103	4	2¾	..500

Cents.	0		1		2		3		4		5		6		7		8		9		Cents.
	s.	d.	s.	d.	s.	d.	s.	d.	s.	d.	s.	d.	s.	d.	s.	d.	s.	d.	s.	d.	
0 ...		0		0½		1		1½		2		2½		3		3½		4		4½	... 0
10 ...		5		5½		6		6½		7		7½		8		8½		9		9½	... 10
20 ...		10		10½		11		11½	1	0	1	0½	1	1	1	1½	1	2	1	2½	... 20
30 ...	1	3	1	3½	1	4	1	4½	1	5	1	5½	1	6	1	6½	1	6¾	1	7¼	... 30
40 ...	1	7¾	1	8¼	1	8¾	1	9¼	1	9¾	1	10¼	1	10¾	1	11¼	1	11¾	2	0¼	... 40

STERLING EXCHANGE TABLES.

8 9/16 TO 9%

DOLLARS AND CENTS INTO STERLING.

Doll's	8 9/16 £ s. d.	8 5/8 £ s. d.	8 11/16 £ s. d.	8 3/4 £ s. d.	8 13/16 £ s. d.	8 7/8 £ s. d.	8 15/16 £ s. d.	9 £ s. d.	Doll's
510	105 13 11¾	105 12 9¼	105 11 6¾	105 10 4¼	105 9 1½	105 7 11	105 6 8½	105 5 6	510
520	107 15 5½	107 14 2½	107 12 11½	107 11 8¾	107 10 5¾	107 9 3	107 8 0¼	107 6 9½	520
530	109 16 10¾	109 15 7½	109 14 4½	109 13 1¼	109 11 10	109 10 7	109 9 4	109 8 1	530
540	111 18 4	111 17 0¾	111 15 9¼	111 14 5¾	111 13 2½	111 11 11	111 10 7¾	111 9 4½	540
550	113 19 9½	113 18 5¾	113 17 2	113 15 10¼	113 14 6¾	113 13 3	113 11 11¼	113 10 7¾	550
560	116 1 3	115 19 11	115 18 7	115 17 3	115 15 11	115 14 7	115 13 3	115 11 11	560
570	118 2 8½	118 1 4	117 19 11¾	117 18 7½	117 17 3¼	117 15 11	117 14 6¾	117 13 2½	570
580	120 4 1¾	120 2 9¼	120 1 4½	120 0 0	119 18 7½	119 17 3	119 15 10½	119 14 6	580
590	122 5 7¼	122 4 2¼	122 2 9½	122 1 4½	121 19 11¾	121 18 7	121 17 2¼	121 15 9¼	590
600	124 7 0½	124 5 7½	124 4 2¾	124 2 9	124 1 4	123 19 11	123 18 5¾	123 17 0¾	600
610	126 8 6	126 7 0½	126 5 7	126 4 1¾	126 2 8¼	126 1 3	125 19 9½	125 18 4½	610
620	128 9 11½	128 8 5¾	128 7 0	128 5 6½	128 4 0½	128 2 6¾	128 1 1¼	127 19 7½	620
630	130 11 4¾	130 9 10¾	130 8 4¾	130 6 10¾	130 5 4¾	130 3 10¾	130 2 5	130 0 11	630
640	132 12 10¼	132 11 4	132 9 9½	132 8 3¼	132 6 9	132 5 2¾	132 3 8½	132 2 2½	640
650	134 14 3½	134 12 9	134 11 2½	134 9 7¾	134 8 1¼	134 6 6¾	134 5 0¼	134 3 5¾	650
660	136 15 9	136 14 2¼	136 12 7½	136 11 0½	136 9 5½	136 7 10¾	136 6 4	136 4 9¼	660
670	138 17 2½	138 15 7¼	138 14 0	138 12 5	138 10 9¾	138 9 2¾	138 7 7¾	138 6 0¾	670
680	140 18 7¾	140 17 0½	140 15 5	140 13 9½	140 12 2	140 10 6¾	140 8 11½	140 7 4	680
690	143 0 1¼	142 18 6½	142 16 9¾	142 15 2	142 13 6½	142 11 10¾	142 10 3	142 8 7½	690
700	145 1 6¾	144 19 10½	144 18 2½	144 16 6½	144 14 10¾	144 13 2¾	144 11 6¾	144 9 11	700
710	147 3 0	147 1 3¾	146 19 7½	146 17 11¼	146 16 3	146 14 6¾	146 12 10½	146 11 2¼	710
720	149 4 5½	149 2 8¾	149 1 0¼	148 19 3¾	148 17 7¼	148 15 10¾	148 14 2¼	148 12 5¾	720
730	151 5 11	151 4 2	151 2 5¼	151 0 8½	150 18 11½	150 17 2¾	150 15 6	150 13 9¼	730
740	153 7 4¼	153 5 7	153 3 10	153 2 0¾	153 0 3¾	152 18 6¾	152 16 9½	152 15 0½	740
750	155 8 9¾	155 7 0¼	155 5 2¾	155 3 5½	155 1 8	154 19 10¾	154 18 1¼	154 16 4	750
760	157 10 3	157 8 5¼	157 6 7¾	157 4 10	157 3 0¼	157 1 2½	156 19 5	156 17 7½	760
770	159 11 8½	159 9 10½	159 8 0½	159 6 2½	159 4 4½	159 2 6½	159 0 8¾	158 18 10¾	770
780	161 13 2	161 11 3½	161 9 5¼	161 7 7	161 5 8¾	161 3 10½	161 2 0	161 0 2¼	780
790	163 14 7¼	163 12 8¾	163 10 10¼	163 8 11½	163 7 1	163 5 2½	163 3 4	163 1 5½	790
800	165 16 0¾	165 14 1¾	165 12 3	165 10 4¼	165 8 5¼	165 6 6½	165 4 7¾	165 2 9	800
810	167 17 6¼	167 15 7	167 13 7¾	167 11 8¾	167 9 9½	167 7 10½	167 5 11½	167 4 0½	810
820	169 18 11½	169 17 0	169 15 0¾	169 13 1¼	169 11 1¾	169 9 2½	169 7 3¼	169 5 3¾	820
830	172 0 5	171 18 5¼	171 16 5½	171 14 5¾	171 12 6	171 10 6½	171 8 6¾	171 6 7¼	830
840	174 1 10½	173 19 10¼	173 17 10¼	173 15 10¼	173 13 10½	173 11 10½	173 9 10½	173 7 10¾	840
850	176 3 3¾	176 1 3½	175 19 3¼	175 17 3	175 15 2¾	175 13 2½	175 11 2¼	175 9 2	850
860	178 4 9¼	178 2 8½	178 0 8	177 18 7½	177 16 7	177 14 6½	177 12 6	177 10 5½	860
870	180 6 2½	180 4 1¾	180 2 0¾	180 0 0	179 17 11¼	179 15 10½	179 13 9¾	179 11 9	870
880	182 7 8	182 5 6¾	182 3 5¾	182 1 4½	181 19 3½	181 17 2½	181 15 1¼	181 13 0¼	880
890	184 9 1½	184 7 0	184 4 10½	184 2 9	184 0 7¾	183 18 6½	183 16 5	183 14 3¾	890
900	186 10 6¾	186 8 5	186 6 3¼	186 4 1¾	186 2 0	185 19 10¼	185 17 8¾	185 15 7¼	900
910	188 12 0¼	188 9 10¼	188 7 8¼	188 5 6¼	188 3 4¼	188 1 2¼	187 19 0½	187 16 10½	910
920	190 13 5¾	190 11 3¼	190 9 1	190 6 10¾	190 4 8¼	190 2 6¼	190 0 4¼	189 18 2	920
930	192 14 11	192 12 8½	192 10 5¾	192 8 3¼	192 6 0¾	192 3 10¼	192 1 7¾	191 19 5½	930
940	194 16 4½	194 14 1½	194 11 10¾	194 9 7¾	194 7 5	194 5 2¼	194 2 11½	194 0 8¾	940
950	196 17 10	196 15 6¾	196 13 3½	196 11 0½	196 8 9¼	196 6 6¼	196 4 3¼	196 2 0¼	950
960	198 19 3¼	198 16 11¾	198 14 8½	198 12 5	198 10 1½	198 7 10¼	198 5 7	198 3 3¾	960
970	201 0 8¾	200 18 5	200 16 1¼	200 13 9½	200 11 5¾	200 9 2¼	200 6 10½	200 4 7	970
980	203 2 2	202 19 10	202 17 6	202 15 2	202 12 10	202 10 6½	202 8 2¼	202 5 10½	980
990	205 3 7½	205 1 3¼	204 18 11	204 16 6¾	204 14 2½	204 11 10¼	204 9 6	204 7 1¾	990
1000	207 5 1	207 2 8¼	207 0 3¾	206 17 11¼	206 15 6¾	206 13 2¼	206 10 9¾	206 8 5¼	1000
2000	414 10 1¾	414 5 4¾	414 0 7½	413 15 10¼	413 11 1¼	413 6 4¼	413 1 7½	412 16 10½	2000
3000	621 15 2¾	621 8 1	621 0 11¼	620 13 9½	620 6 8	619 19 6½	619 12 5¼	619 5 3¾	3000
4000	829 0 3¾	828 10 9¼	828 1 3	827 11 8¾	827 2 2½	826 12 8¾	826 3 2¾	825 13 9¼	4000
5000	1036 5 4¾	1035 13 5½	1035 1 6¾	1034 9 7¾	1033 17 9¼	1033 5 10¾	1032 14 0½	1032 2 2½	5000
6000	1243 10 5½	1242 16 2	1242 1 10½	1241 7 7	1240 13 4	1239 19 1	1239 4 10¼	1238 10 7¾	6000
7000	1450 15 6½	1449 18 10½	1449 2 2	1448 5 6¼	1447 8 10½	1446 12 3¼	1445 15 8	1444 19 1	7000
8000	1658 0 7½	1657 1 6½	1656 2 5¾	1655 3 5½	1654 4 5½	1653 5 5½	1652 6 5¾	1651 7 6¼	8000
9000	1865 5 8¼	1864 4 3	1863 2 9½	1862 1 4½	1860 19 11¾	1859 18 7½	1858 17 3½	1857 15 11½	9000
10000	2072 10 9¼	2071 6 11¼	2070 3 1¼	2068 19 3¾	2067 15 6½	2066 11 9¾	2065 8 1¼	2064 4 4¾	10000

Cents.	0 s. d.	1 s. d.	2 s. d.	3 s. d.	4 s. d.	5 s. d.	6 s. d.	7 s. d.	8 s. d.	9 s. d.	Cents.
50 ...	2 0¾	2 1¼	2 1¾	2 2¼	2 2¾	2 3¼	2 3¾	2 4¼	2 4¾	2 5¼	... 50
60 ...	2 5¾	2 6¼	2 6¾	2 7¼	2 7¾	2 8¼	2 8¾	2 9¼	2 9¾	2 10¼	... 60
70 ...	2 10¾	2 11¼	2 11¾	3 0¼	3 0¾	3 1¼	3 1¾	3 2¼	3 2¾	3 3¼	... 70
80 ...	3 3¾	3 4¼	3 4¾	3 5¼	3 5¾	3 6¼	3 6¾	3 7¼	3 7¾	3 8¼	... 80
90 ...	3 8¾	3 9¼	3 9¾	3 10¼	3 10¾	3 11¼	3 11¾	4 0¼	4 0¾	4 1¼	... 90

STERLING EXCHANGE TABLES.

9 1/16 TO 9 1/2 %

DOLLARS AND CENTS INTO STERLING.

Doll'rs	9 1/16			9 1/8			9 3/16			9 1/4		
	£	s.	d.	£	s.	d.	£	s.	d.	£	s.	d.
1..		4	1½		4	1½		4	1½		4	1½
2..		8	3		8	3		8	3		8	2¾
3..		12	4½		12	4½		12	4¼		12	4¼
4..		16	6		16	6		16	5¾		16	5¾
5..	1	0	7½	1	0	7½	1	0	7¼	1	0	7¼
6..	1	4	9	1	4	9	1	4	8¾	1	4	8½
7..	1	8	10½	1	8	10½	1	8	10¼	1	8	10
8..	1	13	0	1	13	0	1	12	11¾	1	12	11½
9..	1	17	1½	1	17	1½	1	17	1	1	17	0¾
10..	2	1	3	2	1	2¾	2	1	2½	2	1	2¼
20..	4	2	6¼	4	2	5¾	4	2	5¼	4	2	4½
30..	6	3	9½	6	3	8½	6	3	7¾	6	3	6¾
40..	8	5	0½	8	4	11½	8	4	10¼	8	4	9
50..	10	6	3¾	10	6	2¼	10	6	0¾	10	5	11½
60..	12	7	6¾	12	7	5	12	7	3¼	12	7	1¾
70..	14	8	10	14	8	8	14	8	6	14	8	4
80..	16	10	1	16	9	10¾	16	9	8½	16	9	6¼
90..	18	11	4¼	18	11	1½	18	10	11	18	10	8½
100..	20	12	7¼	20	12	4½	20	12	1½	20	11	10¾
110..	22	13	10½	22	13	7¼	22	13	4¼	22	13	1
120..	24	15	1½	24	14	10¼	24	14	6¾	24	14	3½
130..	26	16	4¾	26	16	1	26	15	9¼	26	15	5¾
140..	28	17	7¾	28	17	3¾	28	16	11¾	28	16	8
150..	30	18	11	30	18	6¾	30	18	2½	30	17	10¼
160..	33	0	2	32	19	9½	32	19	5	32	19	0½
170..	35	1	5¼	35	1	0¼	35	0	7½	35	0	2¾
180..	37	2	8¼	37	2	3¼	37	1	10	37	1	5
190..	39	3	11½	39	3	6	39	3	0¾	39	2	7¼
200..	41	5	2½	41	4	9	41	4	3¼	41	3	9½
210..	43	6	5¾	43	5	11¾	43	5	5¾	43	4	11¾
220..	45	7	9	45	7	2½	45	6	8¼	45	6	2¼
230..	47	9	0	47	8	5½	47	7	11	47	7	4½
240..	49	10	3	49	9	8¼	49	9	1½	49	8	6¾
250..	51	11	6¼	51	10	11¼	51	10	4	51	9	9
260..	53	12	9¼	53	12	2	53	11	6½	53	10	11¼
270..	55	14	0½	55	13	4¾	55	12	9¼	55	12	1½
280..	57	15	3½	57	14	7¾	57	13	11¾	57	13	3¾
290..	59	16	6¾	59	15	10½	59	15	2¼	59	14	6
300..	61	17	10	61	17	1¼	61	16	4¾	61	15	8½
310..	63	19	1	63	18	4¼	63	17	7½	63	16	10¾
320..	66	0	4	65	19	7	65	18	10	65	18	1
330..	68	1	7¼	68	0	10	68	0	0½	67	19	3¼
340..	70	2	10¼	70	2	0¾	70	1	3	70	0	5½
350..	72	4	1½	72	3	3½	72	2	5½	72	1	7¾
360..	74	5	4¾	74	4	6½	74	3	8¼	74	2	10
370..	76	6	7¾	76	5	9¼	76	4	10¾	76	4	0¼
380..	78	7	11	78	7	0	78	6	1¼	78	5	2½
390..	80	9	2	80	8	3	80	7	4	80	6	5
400..	82	10	5¼	82	9	5¾	82	8	6½	82	7	7¼
410..	84	11	8¼	84	10	8¾	84	9	9	84	8	9½
420..	86	12	11½	86	11	11½	86	10	11½	86	9	11¾
430..	88	14	2½	88	13	2¼	88	12	2¼	88	11	2
440..	90	15	5¾	90	14	5¼	90	13	4¾	90	12	4¼
450..	92	16	8¾	92	15	8	92	14	7¼	92	13	6½
460..	94	18	0	94	16	11	94	15	9¾	94	14	8¾
470..	96	19	3	96	18	1¾	96	17	0½	96	15	11¼
480..	99	0	6¼	98	19	4½	98	18	3	98	17	1½
490..	101	1	9¼	101	0	7½	100	19	5½	100	18	3¾
500..	103	3	0½	103	1	10¼	103	0	8	102	19	6

9 5/16			9 3/8			9 7/16			9 1/2			Doll'rs
£	s.	d.	£	s.	d.	£	s.	d.	£	s.	d.	
	4	1½		4	1¼		4	1¼		4	1¼	.. 1
	8	2¾		8	2¾		8	2¾		8	2¾	.. 2
	12	4¼		12	4		12	4		12	4	.. 3
	16	5½		16	5½		16	5¼		16	5¼	.. 4
1	0	7	1	0	6¾	1	0	6¾	1	0	6½	.. 5
1	4	8½	1	4	8¼	1	4	8	1	4	8	.. 6
1	8	9¾	1	8	9½	1	8	9½	1	8	9¼	.. 7
1	12	11¼	1	12	11	1	12	10¾	1	12	10½	.. 8
1	17	0½	1	17	0¼	1	17	0	1	16	11¾	.. 9
2	1	2	2	1	1¾	2	1	1½	2	1	1¼	.. 10
4	2	4	4	2	3½	4	2	2¾	4	2	2¼	.. 20
6	3	6	6	3	5¼	6	3	4¼	6	3	3½	.. 30
8	4	8	8	4	6¾	8	4	5¾	8	4	4½	.. 40
10	5	10	10	5	8½	10	5	7¼	10	5	5¾	.. 50
12	7	0	12	6	10¼	12	6	8½	12	6	7	.. 60
14	8	2	14	8	0	14	7	10	14	7	8	.. 70
16	9	4	16	9	1¾	16	8	11½	16	8	9¼	.. 80
18	10	6	18	10	3½	18	10	1	18	9	10½	.. 90
20	11	8	20	11	5¼	20	11	2¼	20	10	11½	..100
22	12	10	22	12	6¾	22	12	3¾	22	12	0¾	..110
24	14	0	24	13	8½	24	13	5¼	24	13	1¾	..120
26	15	2	26	14	10¼	26	14	6½	26	14	3	..130
28	16	4	28	16	0	28	15	8	28	15	4	..140
30	17	6	30	17	1¾	30	16	9½	30	16	5¼	..150
32	18	8	32	18	3½	32	17	11	32	17	6½	..160
34	19	10	34	19	5¼	34	19	0¼	34	18	7½	..170
37	1	0	37	0	6¾	37	0	1¾	36	19	8¾	..180
39	2	2	39	1	8½	39	1	3¼	39	0	9¾	..190
41	3	4	41	2	10¼	41	2	4¾	41	1	11	..200
43	4	6	43	4	0	43	3	6	43	3	0¼	..210
45	5	8	45	5	1¾	45	4	7½	45	4	1¼	..220
47	6	10	47	6	3½	47	5	9	47	5	2½	..230
49	8	0	49	7	5¼	49	6	10¼	49	6	3½	..240
51	9	2	51	8	6¾	51	7	11¾	51	7	4¾	..250
53	10	4	53	9	8½	53	9	1¼	53	8	6	..260
55	11	6	55	10	10¼	55	10	2¾	55	9	7	..270
57	12	8	57	12	0	57	11	4	57	10	8¼	..280
59	13	10	59	13	1¾	59	12	5½	59	11	9¼	..290
61	15	0	61	14	3½	61	13	7	61	12	10½	..300
63	16	2	63	15	5¼	63	14	8½	63	13	11¾	..310
65	17	4	65	16	6¾	65	15	9¾	65	15	0¾	..320
67	18	6	67	17	8½	67	16	11¼	67	16	2	..330
69	19	8	69	18	10¼	69	18	0¾	69	17	3	..340
72	0	9¾	72	0	0	71	19	2¼	71	18	4½	..350
74	1	11¾	74	1	1¾	74	0	3½	73	19	5½	..360
76	3	1¾	76	2	3½	76	1	5	76	0	6½	..370
78	4	3¾	78	3	5¼	78	2	6½	78	1	7¾	..380
80	5	5¾	80	4	6¾	80	3	7¾	80	2	8¾	..390
82	6	7¾	82	5	8½	82	4	9¼	82	3	10	..400
84	7	9¾	84	6	10¼	84	5	10¾	84	4	11¼	..410
86	8	11¾	86	8	0	86	7	0¼	86	6	0¼	..420
88	10	1¾	88	9	1¾	88	8	1½	88	7	1½	..430
90	11	3¾	90	10	3½	90	9	3	90	8	2½	..440
92	12	5¾	92	11	5¼	92	10	4½	92	9	3¾	..450
94	13	7¾	94	12	6¾	94	11	5¾	94	10	5	..460
96	14	9¾	96	13	8½	96	12	7¼	96	11	6	..470
98	15	11¾	98	14	10¼	98	13	8¾	98	12	7¼	..480
100	17	1¾	100	16	0	100	14	10¼	100	13	8¼	..490
102	18	3¾	102	17	1¾	102	15	11½	102	14	9½	..500

Cents.	0		1		2		3		4		5		6		7		8		9		Cents.
	s.	d.	s.	d.	s.	d.	s.	d.	s.	d.	s.	d.	s.	d.	s.	d.	s.	d.	s.	d.	
0 ...		0		0½		1		1½		2		2½		3		3½		4		4½	... 0
10 ...		5		5½		6		6½		7		7½		8		8½		9		9½	... 10
20 ...		10		10½		10¾		11¼		11¾	1	0¼	1	0¾	1	1¼	1	1¾	1	2¼	... 20
30 ...	1	2¾	1	3¼	1	3¾	1	4¼	1	4¾	1	5¼	1	5¾	1	6¼	1	6¾	1	7¼	... 30
40 ...	1	7¾	1	8¼	1	8¾	1	9¼	1	9¾	1	10¼	1	10¾	1	11¼	1	11¾	2	0¼	... 40

STERLING EXCHANGE TABLES.

$9\frac{1}{16}$ TO $9\frac{1}{2}$%

DOLLARS AND CENTS INTO STERLING.

Doll's	$9\frac{1}{16}$	$9\frac{1}{8}$	$9\frac{3}{16}$	$9\frac{1}{4}$	$9\frac{5}{16}$	$9\frac{3}{8}$	$9\frac{7}{16}$	$9\frac{1}{2}$	Doll's
	£ s. d.	£ s. d.	£ s. d.	£ s. d.	£ s. d.	£ s. d.	£ s. d.	£ s. d.	
510	105 4 3¼	105 3 1	105 1 10¾	105 0 8¼	104 19 5¾	104 18 3½	104 17 1	104 15 10¾	510
520	107 5 6¾	107 4 4	107 3 1¼	107 1 10½	107 0 7¾	106 19 5¼	106 18 2½	106 16 11¾	520
530	109 6 9¾	109 5 6¾	109 4 3¾	109 3 0¾	109 1 9¾	109 0 6¾	108 19 4	108 18 1	530
540	111 8 1	111 6 9¾	111 5 6¼	111 4 3	111 2 11¾	111 1 8½	111 0 5¼	110 19 2¼	540
550	113 9 4	113 8 0½	113 6 9	113 5 5¼	113 4 1¾	113 2 10¼	113 1 6¾	113 0 3¼	550
560	115 10 7¼	115 9 3¼	115 7 11½	115 6 7¾	115 5 3¾	115 4 0	115 2 8¼	115 1 4½	560
570	117 11 10¼	117 10 6¼	117 9 2	117 7 10	117 6 5¾	117 5 1¾	117 3 9¾	117 2 5½	570
580	119 13 1½	119 11 9	119 10 4½	119 9 0¼	119 7 7¾	119 6 3½	119 4 11	119 3 6¾	580
590	121 14 4½	121 12 11¾	121 11 7¼	121 10 2½	121 8 9¾	121 7 5¼	121 6 0½	121 4 8	590
600	123 15 7¾	123 14 2¾	123 12 9¾	123 11 4¾	123 9 11¾	123 8 6¾	123 7 2	123 5 9	600
610	125 16 10¾	125 15 5½	125 14 0¼	125 12 7	125 11 1¾	125 9 8½	125 8 3¼	125 6 10¼	610
620	127 18 2	127 16 8½	127 15 2¾	127 13 9¼	127 12 3¾	127 10 10¼	127 9 4¾	127 7 11¼	620
630	129 19 5	129 17 11¼	129 16 5½	129 14 11½	129 13 5¾	129 12 0	129 10 6¼	129 9 0½	630
640	132 0 8¼	131 19 2	131 17 8	131 16 1¾	131 14 7¾	131 13 1¾	131 11 7¾	131 10 1¾	640
650	134 1 11¼	134 0 5	133 18 10½	133 17 4¼	133 15 9¾	133 14 3½	133 12 9	133 11 2¾	650
660	136 3 2½	136 1 7¾	136 0 1	135 18 6½	135 16 11¾	135 15 5¼	135 13 10½	135 12 4	660
670	138 4 5¾	138 2 10¾	138 1 3¾	137 19 8¾	137 18 1¾	137 16 6¾	137 15 0	137 13 5	670
680	140 5 8¾	140 4 2½	140 2 6¼	140 0 11	139 19 3¾	139 17 8½	139 16 1½	139 14 6¼	680
690	142 7 0	142 5 5¼	142 3 8¾	142 2 1¼	142 0 5¾	141 18 10¼	141 17 2¾	141 15 7½	690
700	144 8 3	144 6 7¼	144 4 11¼	144 3 3½	144 1 7⅝	144 0 0	143 18 4¼	143 16 8½	700
710	146 9 6½	146 7 10	146 6 2	146 4 5¾	146 2 9¾	146 1 1¾	145 19 6¼	145 17 9¾	710
720	148 10 9¼	148 9 0¾	148 7 4½	148 5 8	148 3 11¾	148 2 3½	148 0 7	147 18 10¾	720
730	150 12 0½	150 10 3¾	150 8 7	150 6 10½	150 5 1¾	150 3 5¼	150 1 8½	150 0 0	730
740	152 13 3¾	152 11 6½	152 9 9½	152 8 0¾	152 6 3¾	152 4 6¾	152 2 10	152 1 1¼	740
750	154 14 6¾	154 12 9½	154 11 0¼	154 9 3	154 7 5¾	154 5 8½	154 3 11½	154 2 2¼	750
760	156 15 9¾	156 14 0¼	156 12 2¾	156 10 5¼	156 8 7¾	156 6 10¼	156 5 0¾	156 3 3½	760
770	158 17 1	158 15 3	158 13 5¼	158 11 7½	158 9 9¾	158 8 0	158 6 2¼	158 4 4½	770
780	160 18 4	160 16 6	160 14 7¾	160 12 9¾	160 10 11¾	160 9 1¾	160 7 3¾	160 5 5¾	780
790	162 19 7¼	162 17 8¾	162 15 10½	162 14 0	162 12 1¾	162 10 3½	162 8 5¼	162 6 7	790
800	165 0 10¼	164 18 11¾	164 17 1	164 15 2¼	164 13 3¾	164 11 5¼	164 9 6½	164 7 8	800
810	167 2 1½	167 0 2½	166 18 3½	166 16 4¾	166 14 5¾	166 12 6¾	166 10 8	166 8 9¼	810
820	169 3 4½	169 1 5¼	168 19 6	168 17 7	168 15 7¾	168 13 8½	168 11 9½	168 9 10¼	820
830	171 4 7¾	171 2 8¼	171 0 8¾	170 18 9¼	170 16 9¾	170 14 10¼	170 12 10¾	170 10 11¼	830
840	173 5 10¾	173 3 11	173 1 11¼	172 19 11½	172 17 11¾	172 16 0	172 14 0¼	172 12 0¾	840
850	175 7 2	175 5 1¾	175 3 1¾	175 1 1¾	174 19 1¾	174 17 1¾	174 15 1¾	174 13 1¾	850
860	177 8 5	177 6 4¾	177 4 4¼	177 2 4	177 0 3¾	176 18 3½	176 16 3¼	176 14 3	860
870	179 9 8¼	179 7 7½	179 5 7	179 3 6¼	179 1 5¾	178 19 5¼	178 17 4½	178 15 4	870
880	181 10 11¼	181 8 10½	181 6 9½	181 4 8½	181 2 7¾	181 0 6¾	180 18 6	180 16 5¼	880
890	183 12 2½	183 10 1¼	183 8 0	183 5 10¾	183 3 9¾	183 1 8½	182 19 7½	182 17 6½	890
900	185 13 5½	185 11 4	185 9 2½	185 7 1¼	185 4 11⅝	185 2 10½	185 0 9	184 18 7½	900
910	187 14 8¾	187 12 7	187 10 5¼	187 8 3½	187 6 1¾	187 4 0	187 1 10¼	186 19 8¾	910
920	189 15 11¾	189 13 9¾	189 11 7¾	189 9 5¾	189 7 3¾	189 5 1¾	189 2 11¾	189 0 9¾	920
930	191 17 3	191 15 0¾	191 12 10¼	191 10 8	191 8 5¾	191 6 3½	191 4 1¼	191 1 11	930
940	193 18 6	193 16 3½	193 14 0¾	193 11 10¼	193 9 7¾	193 7 5¼	193 5 2½	193 3 0¼	940
950	195 19 9¼	195 17 6¼	195 15 3½	195 13 0½	195 10 9¾	195 8 6¾	195 6 4	195 4 1¼	950
960	198 1 0½	197 18 9¼	197 16 6	197 14 2¾	197 11 11¾	197 9 8½	197 7 5½	197 5 2½	960
970	200 2 3½	200 0 0	199 17 8½	199 15 5	199 13 1¾	199 10 10¼	199 8 7	199 6 3½	970
980	202 3 6½	202 1 3	201 18 11	201 16 7½	201 14 3¾	201 12 0	201 9 8¼	201 7 4¾	980
990	204 4 9¾	204 2 5¾	204 0 1¾	203 17 9¾	203 15 5¾	203 13 1¾	203 10 9¾	203 8 6	990
1000	206 6 0¾	206 3 8½	206 1 4¼	205 19 0	205 16 7⅝	205 14 3½	205 11 11¼	205 9 7	1000
2000	412 12 1¾	412 7 5	412 2 8½	411 17 11¾	411 13 3¼	411 8 6¾	411 3 10½	410 19 2¼	2000
3000	618 18 2¾	618 11 1½	618 4 0¾	617 16 11¾	617 9 11	617 2 10¼	616 15 9¾	616 8 9¼	3000
4000	825 4 3½	824 14 10¼	824 5 4¾	823 15 11¾	823 6 6⅝	822 17 1¾	822 7 9	821 18 4¼	4000
5000	1031 10 4½	1030 18 6¾	1030 6 9	1029 14 11½	1029 3 2¼	1028 11 5¼	1027 19 8¼	1027 7 11½	5000
6000	1237 16 5¼	1237 3 3¼	1236 8 1¼	1235 13 11½	1234 19 10	1234 5 8½	1233 11 7½	1232 17 6½	6000
7000	1444 2 6¼	1443 5 11¾	1442 9 5½	1441 12 11½	1440 16 5½	1440 0 0	1439 3 6¾	1438 7 1½	7000
8000	1650 8 7¼	1649 9 8¼	1648 10 9¾	1647 11 11½	1646 13 1¼	1645 14 3½	1644 15 6	1643 16 8½	8000
9000	1856 14 8	1855 13 4¾	1854 12 2	1853 10 11¼	1852 9 9	1851 8 6¾	1850 7 5¼	1849 6 3¾	9000
10000	2063 0 9	2061 17 1½	2060 13 6¼	2059 9 11¼	2058 6 4½	2057 2 10¼	2055 19 4½	2054 15 10¾	10000

Cents.	0	1	2	3	4	5	6	7	8	9	Cents.
	s. d.	s. d.	s. d.	s. d.	s. d.	s. d.	s. d.	s. d.	s. d.	s. d.	
50 ...	2 0¾	2 1¼	2 1¾	2 2¼	2 2¾	2 3¼	2 3¾	2 4¼	2 4¾	2 5¼	... 50
60 ...	2 5¾	2 6¼	2 6¾	2 7¼	2 7¾	2 8¼	2 8½	2 9	2 9½	2 10	... 60
70 ...	2 10½	2 11	2 11½	3 0	3 0½	3 1	3 1½	3 2	3 2½	3 3	... 70
80 ...	3 3½	3 4	3 4½	3 5	3 5½	3 6	3 6½	3 7	3 7½	3 8	... 80
90 ...	3 8½	3 9	3 9½	3 10	3 10½	3 11	3 11½	4 0	4 0½	4 1	... 90

STERLING EXCHANGE TABLES.

9 9/16 TO 10%

DOLLARS AND CENTS INTO STERLING.

Doll'rs	9 9/16 £ s. d.	9⅝ £ s. d.	9 11/16 £ s. d.	9¾ £ s. d.	9 13/16 £ s. d.	9⅞ £ s. d.	9 15/16 £ s. d.	10 £ s. d.	Doll'rs
1..	4 1¼	4 1½	4 1½	4 1¼	4 1¼	4 1¼	4 1	4 1	.. 1
2..	8 2½	8 2½	8 2½	8 2½	8 2¼	8 2¼	8 2¼	8 2¼	.. 2
3..	12 3¾	12 3¾	12 3¾	12 3½	12 3½	12 3½	12 3¼	12 3¼	.. 3
4..	16 5¼	16 5	16 5	16 4¾	16 4¾	16 4½	16 4½	16 4¼	.. 4
5..	1 0 6½	1 0 6¼	1 0 6¼	1 0 6	1 0 5¾	1 0 5¾	1 0 5½	1 0 5½	.. 5
6..	1 4 7¾	1 4 7½	1 4 7½	1 4 7¼	1 4 7	1 4 7	1 4 6¾	1 4 6½	.. 6
7..	1 8 9	1 8 8¾	1 8 8½	1 8 8½	1 8 8¼	1 8 8	1 8 7¾	1 8 7¾	.. 7
8..	1 12 10¼	1 12 10	1 12 9¾	1 12 9½	1 12 9½	1 12 9¼	1 12 9	1 12 8¾	.. 8
9..	1 16 11½	1 16 11¼	1 16 11	1 16 10¾	1 16 10½	1 16 10¼	1 16 10	1 16 9¾	.. 9
10..	2 1 0¾	2 1 0½	2 1 0¼	2 1 0	2 0 11¾	2 0 11½	2 0 11¼	2 0 11	.. 10
20..	4 2 1¾	4 2 1¼	4 2 0½	4 2 0	4 1 11½	4 1 11	4 1 10½	4 1 9¾	.. 20
30..	6 3 2½	6 3 1¾	6 3 1	6 3 0	6 2 11¼	6 2 10½	6 2 9½	6 2 8¾	.. 30
40..	8 4 3½	8 4 2¼	8 4 1¼	8 4 0	8 3 11	8 3 9¾	8 3 8¾	8 3 7¾	.. 40
50..	10 5 4¼	10 5 3	10 5 1½	10 5 0¼	10 4 10¾	10 4 9½	10 4 8	10 4 6½	.. 50
60..	12 6 5¼	12 6 3½	12 6 1¾	12 6 0¼	12 5 10½	12 5 8¾	12 5 7¼	12 5 5½	.. 60
70..	14 7 6	14 7 4	14 7 2¼	14 7 0¼	14 6 10¼	14 6 8¼	14 6 6¼	14 6 4¼	.. 70
80..	16 8 7	16 8 4¾	16 8 2½	16 8 0¼	16 7 10	16 7 7¾	16 7 5½	16 7 3¼	.. 80
90..	18 9 7¾	18 9 5¼	18 9 2¾	18 9 0¼	18 8 9¾	18 8 7¼	18 8 4¾	18 8 2¼	.. 90
100..	20 10 8¾	20 10 6	20 10 3	20 10 0¼	20 9 9½	20 9 6¾	20 9 4	20 9 1	..100
110..	22 11 9½	22 11 6½	22 11 3½	22 11 0¼	22 10 9¼	22 10 6¼	22 10 3	22 10 0	..110
120..	24 12 10½	24 12 7	24 12 3¾	24 12 0¼	24 11 9	24 11 5½	24 11 2¼	24 10 10	..120
130..	26 13 11¼	26 13 7¾	26 13 4	26 13 0¼	26 12 8¾	26 12 5	26 12 1½	26 11 9¾	..130
140..	28 15 0¼	28 14 8¼	28 14 4¼	28 14 0½	28 13 8½	28 13 4½	28 13 0¾	28 12 8¾	..140
150..	30 16 1	30 15 8¾	30 15 4½	30 15 0½	30 14 8¼	30 14 4	30 13 11¾	30 13 7¾	..150
160..	32 17 2	32 16 9½	32 16 5	32 16 0½	32 15 8	32 15 3½	32 14 11	32 14 6½	..160
170..	34 18 2¾	34 17 10	34 17 5¼	34 17 0½	34 16 7¾	34 16 3	34 15 10¼	34 15 5½	..170
180..	36 19 3¾	36 18 10½	36 18 5½	36 18 0½	36 17 7½	36 17 2½	36 16 9½	36 16 4¼	..180
190..	39 0 4½	38 19 11¼	38 19 5¾	38 19 0½	38 18 7¼	38 18 2	38 17 8½	38 17 3¼	..190
200..	41 1 5½	41 0 11¾	41 0 6¼	41 0 0½	40 19 7	40 19 1¼	40 18 7¾	40 18 2¼	..200
210..	43 2 6¼	43 2 0¼	43 1 6½	43 1 0½	43 0 6¾	43 0 0¾	42 19 7	42 19 1	..210
220..	45 3 7¼	45 3 1	45 2 6¾	45 2 0½	45 1 6½	45 1 0¼	45 0 6¼	45 0 0	..220
230..	47 4 8	47 4 1½	47 3 7	47 3 0¾	47 2 6¼	47 1 11¾	47 1 5¼	47 0 11	..230
240..	49 5 8¾	49 5 2	49 4 7½	49 4 0¾	49 3 6	49 2 11¼	49 2 4½	49 1 9¾	..240
250..	51 6 9¾	51 6 2¾	51 5 7¾	51 5 0¾	51 4 5¾	51 3 10¾	51 3 3¾	51 2 8¾	..250
260..	53 7 10½	53 7 3¼	53 6 8	53 6 0¾	53 5 5½	53 4 10¼	53 4 3	53 3 7¾	..260
270..	55 8 11½	55 8 4	55 7 8¼	55 7 0¾	55 6 5¼	55 5 9½	55 5 2	55 4 6½	..270
280..	57 10 0¼	57 9 4½	57 8 8½	57 8 0¾	57 7 5	57 6 9	57 6 1¼	57 5 5½	..280
290	59 11 1¼	59 10 5	59 9 9	59 9 0¾	59 8 4¾	59 7 8½	59 7 0½	59 6 4¼	..290
300..	61 12 2	61 11 5¾	61 10 9¼	61 10 0¾	61 9 4½	61 8 8	61 7 11¾	61 7 3¼	..300
310..	63 13 3	63 12 6½	63 11 9½	63 11 0¾	63 10 4¼	63 9 7½	63 8 10¾	63 8 2¼	..310
320..	65 14 3¾	65 13 6¾	65 12 9¾	65 12 1	65 11 4	65 10 7	65 9 10	65 9 1	..320
330..	67 15 4¾	67 14 7½	67 13 10¼	67 13 1	67 12 3¾	67 11 6½	67 10 9¼	67 10 0	. 330
340..	69 16 5½	69 15 8	69 14 10½	69 14 1	69 13 3½	69 12 6	69 11 8½	69 10 11	..340
350..	71 17 6½	71 16 8½	71 15 10¾	71 15 1	71 14 3¼	71 13 5¼	71 12 7½	71 11 9¾	..350
360..	73 18 7¼	73 17 9¼	73 16 11	73 16 1	73 15 3	73 14 4¾	73 13 6¾	73 12 8¾	..360
370..	75 19 8¼	75 18 9¾	75 17 11½	75 17 1	75 16 2¾	75 15 4¼	75 14 6	75 13 7¾	..370
380..	78 0 9	77 19 10¼	77 18 11¾	77 18 1	77 17 2½	77 16 3¾	77 15 5¼	77 14 6½	..380
390..	80 1 10	80 0 11	80 0 0	79 19 1	79 18 2¼	79 17 3½	79 16 4¾	79 15 5½	..390
400..	82 2 10¾	82 1 11½	82 1 0¼	82 0 1	81 19 2	81 18 2¾	81 17 3½	81 16 4¼	..400
410..	84 3 11¾	84 3 0¼	84 2 0½	84 1 1¼	84 0 1¾	83 19 2¼	83 18 2¾	83 17 3¼	..410
420..	86 5 0½	86 4 0¾	86 3 1	86 2 1¼	86 1 1½	86 0 1¾	85 19 2	85 18 2¼	..420
430..	88 6 1½	88 5 1¼	88 4 1¼	88 3 1¼	88 2 1¼	88 1 1	88 0 1	87 19 1	..430
440..	90 7 2¼	90 6 2	90 5 1½	90 4 1¼	90 3 1	90 2 0½	90 1 0¼	90 0 0	..440
450..	92 8 3¼	92 7 2½	92 6 1¾	92 5 1¼	92 4 0¾	92 3 0	92 1 11½	92 0 11	..450
460..	94 9 4	94 8 3	94 7 2¼	94 6 1¼	94 5 0½	94 3 11½	94 2 10¾	94 1 9¾	..460
470..	96 10 4¾	96 9 3¾	96 8 2½	96 7 1¼	96 6 0¼	96 4 11	96 3 9¾	96 2 8¾	..470
480..	98 11 5¾	98 10 4¼	98 9 2¾	98 8 1¼	98 7 0	98 5 10½	98 4 9	98 3 7¾	..480
490..	100 12 6½	100 11 4¾	100 10 3	100 9 1¼	100 7 11¾	100 6 10	100 5 8¼	100 4 6½	..490
500..	102 13 7½	102 12 5½	102 11 3¼	102 10 1¼	102 8 11¼	102 7 9¼	102 6 7½	102 5 5½	..500

Cents.	0 s. d.	1 s. d.	2 s. d.	3 s. d.	4 s. d.	5 s. d.	6 s. d.	7 s. d.	8 s. d.	9 s. d.	Cents.
0 ...	0 0	0 0½	0 1	0 1½	0 2	0 2½	0 3	0 3½	0 4	0 4½	... 0
10 ...	0 5	0 5½	0 6	0 6½	0 7	0 7½	0 7¾	0 8¼	0 8¾	0 9¼	... 10
20 ...	0 9¾	0 10¼	0 10¾	0 11¼	0 11¾	1 0¼	1 0¾	1 1¼	1 1¾	1 2¼	... 20
30 ...	1 2¾	1 3¼	1 3¾	1 4¼	1 4¾	1 5¼	1 5¾	1 6¼	1 6¾	1 7¼	... 30
40 ...	1 7¾	1 8¼	1 8¾	1 9¼	1 9¾	1 10¼	1 10¾	1 11¼	1 11½	2 0	... 40

STERLING EXCHANGE TABLES.

9 9/16 TO 10%

DOLLARS AND CENTS INTO STERLING.

Dol'rs	9 9/16			9⅝			9 11/16			9¾			9 13/16			9⅞			9 15/16			10			Dol'rs
	£	s.	d.	£	s.	d.	£	s.	d.	£	s.	d.	£	s.	d.	£	s.	d.	£	s.	d.	£	s.	d.	
510	104	14	8½	104	13	6	104	12	3¾	104	11	1½	104	9	11	104	8	8¾	104	7	6½	104	6	4¼	510
520	106	15	9¼	106	14	6½	106	13	4	106	12	1½	106	10	10¾	106	9	8¼	106	8	5¾	106	7	3¼	520
530	108	16	10	108	15	7¼	108	14	4¼	108	13	1½	108	11	10½	108	10	7¾	108	9	5	108	8	2¼	530
540	110	17	11	110	16	7¾	110	15	4½	110	14	1½	110	12	10¼	110	11	7¼	110	10	4¼	110	9	1	540
550	112	18	11¾	112	17	8¼	112	16	5	112	15	1½	112	13	10	112	12	6¾	112	11	3½	112	10	0	550
560	115	0	0¾	114	18	9	114	17	5¼	114	16	1½	114	14	9¾	114	13	6½	114	12	2½	114	10	11	560
570	117	1	1½	116	19	9½	116	18	5½	116	17	1½	116	15	9½	116	14	5¾	116	13	1¾	116	11	9¾	570
580	119	2	2½	119	0	10	118	19	5¾	118	18	1½	118	16	9¼	118	15	5	118	14	1	118	12	8¾	580
590	121	3	3¼	121	1	10¾	121	0	6¼	120	19	1½	120	17	9	120	16	4½	120	15	0	120	13	7¾	590
600	123	4	4¼	123	2	11¼	123	1	6½	123	0	1¾	122	18	8¾	122	17	4	122	15	11¼	122	14	6½	600
610	125	5	5	125	4	0	125	2	6¾	125	1	1¾	124	19	8½	124	18	3½	124	16	10½	124	15	5½	610
620	127	6	6	127	5	0½	127	3	7	127	2	1¾	127	0	8¼	126	19	3	126	17	9¾	126	16	4¼	620
630	129	7	6¾	129	6	1	129	4	7½	129	3	1¾	129	1	8	129	0	2½	128	18	8¾	128	17	3¼	630
640	131	8	7¾	131	7	1¾	131	5	7¾	131	4	1¾	131	2	7¾	131	1	2	130	19	8	130	18	2¼	640
650	133	9	8½	133	8	2¼	133	6	8	133	5	1¾	133	3	7½	133	2	1½	133	0	7¼	132	19	1	650
660	135	10	9½	135	9	2¾	135	7	8¼	135	6	1¾	135	4	7¼	135	3	0¾	135	1	6½	135	0	0	660
670	137	11	10¼	137	10	3½	137	8	8½	137	7	1¾	137	5	7	137	4	0¼	137	2	6½	137	0	11	670
680	139	12	11	139	11	4	139	9	9	139	8	1¾	139	6	6¾	139	4	11¾	139	3	4¾	139	1	9¾	680
690	141	14	0	141	12	4½	141	10	9¼	141	9	2	141	7	6½	141	5	11¼	141	4	4	141	2	8¾	690
700	143	15	0¾	143	13	5¼	143	11	9½	143	10	2	143	8	6¼	143	6	10¾	143	5	3¼	143	3	7¾	700
710	145	16	1¾	145	14	5¾	145	12	9¾	145	11	2	145	9	6	145	7	10¼	145	6	2½	145	4	6½	710
720	147	17	2½	147	15	6¼	147	13	10¼	147	12	2	147	10	5¾	147	8	9¾	147	7	1½	147	5	5½	720
730	149	18	3½	149	16	7	149	14	10½	149	13	2	149	11	5½	149	9	9¼	149	8	0¾	149	6	4½	730
740	151	19	4¼	151	17	7½	151	15	10¾	151	14	2	151	12	5¼	151	10	8½	151	9	0	151	7	3½	740
750	154	0	5¼	153	18	8¼	153	16	11	153	15	2	153	13	5	153	11	8	153	9	11	153	8	2¼	750
760	156	1	6	155	19	8¾	155	17	11½	155	16	2	155	14	4¾	155	12	7½	155	10	10¼	155	9	1	760
770	158	2	7	158	0	9¼	157	18	11¾	157	17	2	157	15	4½	157	13	7	157	11	9½	157	10	0	770
780	160	3	7¾	160	1	10	160	0	0	159	18	2¼	159	16	4¼	159	14	6½	159	12	8¾	159	10	11	780
790	162	4	8¾	162	2	10½	162	1	0¼	161	19	2¼	161	17	4	161	16	6	161	13	8	161	11	9¾	790
800	164	5	9½	164	3	11	164	2	0½	164	0	2¼	163	18	3¾	163	16	5½	163	14	7	163	12	8¾	800
810	166	6	10½	166	4	11¾	166	3	1	166	1	2¼	165	19	3½	165	17	4¾	165	15	6¼	165	13	7¾	810
820	168	7	11¼	168	6	0¼	168	4	1½	168	2	2¼	168	0	3¼	167	18	4¼	167	16	5½	167	14	6½	820
830	170	9	0¼	170	7	0¾	170	5	1½	170	3	2¼	170	1	3	169	19	3¾	169	17	4¾	169	15	5½	830
840	172	10	1	172	8	1½	172	6	1¾	172	4	2¼	172	2	2¾	172	0	3¼	171	18	3¾	171	16	4½	840
850	174	11	2	174	9	2	174	7	2¼	174	5	2¼	174	3	2½	174	1	2¾	173	19	3	173	17	3¼	850
860	176	12	2¾	176	10	2½	176	8	2½	176	6	2¼	176	4	2¼	176	2	2¼	176	0	2¼	175	8	2¼	860
870	178	13	3¾	178	11	3¼	178	9	2¾	178	7	2½	178	5	2	178	3	1¾	178	1	1½	177	19	1	870
880	180	14	4½	180	12	3¾	180	10	3	180	8	2½	180	6	1¾	180	4	1¼	180	2	0½	180	0	0	880
890	182	15	5½	182	13	4¼	182	11	3½	182	9	2½	182	7	1½	182	5	0½	182	2	11¾	182	0	11	890
900	184	16	6¼	184	14	5	184	12	3¾	184	10	2½	184	8	1¼	184	6	0	184	3	11	184	1	9¾	900
910	186	17	7	186	15	5½	186	13	4	186	11	2½	186	9	1	186	6	11½	186	4	10¼	186	2	8¾	910
920	188	18	8	188	16	6¼	188	14	4¼	188	12	2½	188	10	0¾	188	7	11	188	5	9¼	188	3	7¾	920
930	190	19	8¾	190	17	6¾	190	15	4½	190	13	2½	190	11	0½	190	8	10½	190	6	8½	190	4	6½	930
940	193	0	9¾	192	18	7¼	192	16	5	192	14	2½	192	12	0¼	192	9	10	192	7	7¾	192	5	5½	940
950	195	1	10½	194	19	8	194	17	5¼	194	16	2½	194	13	0	194	10	9½	194	8	7	194	6	4½	950
960	197	2	11¼	197	0	8½	196	18	5½	196	16	2¾	196	13	11¾	196	11	9	196	9	6	196	7	3¼	960
970	199	4	0¼	199	1	9	198	19	5¾	198	17	2¾	198	14	11½	198	12	8¼	198	10	5¼	198	8	2¼	970
980	201	5	1¼	201	2	9¾	201	0	6¼	200	18	2¾	200	15	11¼	200	13	7¾	200	11	4¼	200	9	1	980
990	203	6	2	203	3	10¼	203	1	6½	202	19	2¾	202	16	11	202	14	7¼	202	12	3¾	202	10	0	990
1000	205	7	3	205	4	10¾	205	2	6¾	205	0	2¾	204	17	10¾	204	15	6¾	204	13	2¾	204	10	11	1000
2000	410	14	6	410	9	9¾	410	5	1½	410	0	5½	409	15	9½	409	11	1½	409	6	5¼	409	1	9¾	2000
3000	616	1	8¾	615	14	8½	615	7	8¼	615	0	8¼	614	13	8¼	614	6	8¼	613	19	8½	613	12	8¾	3000
4000	821	8	11¾	820	19	7¼	820	10	3	820	0	11	819	11	7	819	2	3	818	12	11¼	818	3	7¾	4000
5000	1026	16	2¾	1026	4	6¼	1025	12	9¾	1025	1	1¾	1024	9	5½	1023	17	9¾	1023	6	2	1022	14	6½	5000
6000	1232	3	5¾	1231	9	5	1230	15	4½	1230	1	4½	1229	7	4½	1228	13	4½	1227	19	5	1227	5	5½	6000
7000	1437	10	8½	1436	14	3¾	1435	17	11½	1435	1	7¼	1434	5	3¼	1433	8	11¼	1432	12	7¾	1431	16	4½	7000
8000	1642	17	11½	1641	19	2¾	1641	0	6¼	1640	1	10	1639	3	2	1638	4	6	1637	5	10½	1636	7	3¼	8000
9000	1848	5	2½	1847	4	1½	1846	3	1	1845	2	0¾	1844	1	0½	1843	0	0¾	1841	19	1¼	1840	18	2¼	9000
10000	2053	12	5½	2052	9	0½	2051	5	7¾	2050	2	3½	2048	18	11¼	2047	15	7½	2046	12	4¼	2045	9	1¼	10000

Cents.	0		1		2		3		4		5		6		7		8		9		Cents.
	s.	d.	s.	d.	s.	d.	s.	d.	s.	d.	s.	d.	s.	d.	s.	d.	s.	d.	s.	d.	
50 ...	2	0½	2	1	2	1½	2	2	2	2½	2	3	2	3½	2	4	2	4½	2	5	... 50
60 ...	2	5½	2	6	2	6½	2	7	2	7½	2	8	2	8½	2	9	2	9½	2	10	... 60
70 ...	2	10½	2	11	2	11½	3	0	3	0½	3	1	3	1½	3	2	3	2½	3	2¾	... 70
80 ...	3	3¼	3	3¾	3	4¼	3	4¾	3	5¼	3	5¾	3	6¼	3	6¾	3	7¼	3	7¾	... 80
90 ...	3	8¼	3	8¾	3	9¼	3	9¾	3	10¼	3	10¾	3	11¼	3	11¾	4	0¼	4	0¾	... 90

STERLING EXCHANGE TABLES.

10 1/16 to 10 1/2 %

DOLLARS AND CENTS INTO STERLING.

Doll'rs	10 1/16			10⅛			10 3/16			10¼			10 5/16			10⅜			10 7/16			10½			Doll'rs
	£	s.	d.	£	s.	d.	£	s.	d.	£	s.	d.	£	s.	d.	£	s.	d.	£	s.	d.	£	s.	d.	
1..		4	1		4	1		4	1		4	1		4	1		4	1		4	1		4	0¾	.. 1
2..		8	2¼		8	2		8	2		8	2		8	2		8	1¾		8	1¾		8	1¾	.. 2
3..		12	3¼		12	3		12	3		12	3		12	2¾		12	2¾		12	2¾		12	2½	.. 3
4..		16	4¼		16	4¼		16	4		16	4		16	3¾		16	3¾		16	3½		16	3½	.. 4
5..	1	0	5¼	1	0	5¼	1	0	5	1	0	5	1	0	4¾	1	0	4½	1	0	4½	1	0	4½	.. 5
6..	1	4	6½	1	4	6¼	1	4	6	1	4	5¾	1	4	5¾	1	4	5½	1	4	5½	1	4	5¼	.. 6
7..	1	8	7½	1	8	7¼	1	8	7	1	8	6¾	1	8	6¾	1	8	6½	1	8	6¼	1	8	6	.. 7
8..	1	12	8½	1	12	8¼	1	12	8	1	12	7¾	1	12	7½	1	12	7½	1	12	7¼	1	12	7	.. 8
9..	1	16	9½	1	16	9¼	1	16	9	1	16	8¾	1	16	8½	1	16	8¼	1	16	8	1	16	7¾	.. 9
10..	2	0	10¾	2	0	10¼	2	0	10	2	0	9¾	2	0	9½	2	0	9¼	2	0	9	2	0	8¾	.. 10
20..	4	1	9¼	4	1	8¾	4	1	8¼	4	1	7½	4	1	7	4	1	6½	4	1	6	4	1	5½	.. 20
30..	6	2	8	6	2	7	6	2	6¼	6	2	5½	6	2	4½	6	2	3¾	6	2	3	6	2	2	.. 30
40..	8	3	6½	8	3	5½	8	3	4¼	8	3	3¼	8	3	2	8	3	1	8	2	11¾	8	2	10¾	.. 40
50..	10	4	5¼	10	4	3¾	10	4	2¼	10	4	1	10	3	11½	10	3	10¼	10	3	8¾	10	3	7½	.. 50
60..	12	5	3¾	12	5	2	12	5	0½	12	4	10¾	12	4	9	12	4	7½	12	4	5¾	12	4	4¼	.. 60
70..	14	6	2½	14	6	0½	14	5	10½	14	5	8½	14	5	6¾	14	5	4¾	14	5	2¾	14	5	0¾	.. 70
80..	16	7	1	16	6	10¾	16	6	8½	16	6	6¼	16	6	4¼	16	6	2	16	5	11¾	16	5	9½	.. 80
90..	18	7	11¾	18	7	9¼	18	7	6¾	18	7	4¼	18	7	1¾	18	6	11¼	18	6	8¾	18	6	6¼	.. 90
100..	20	8	10¼	20	8	7½	20	8	4¾	20	8	2	20	7	11¼	20	7	8½	20	7	5¾	20	7	3	..100
110..	22	9	9	22	9	5¾	22	9	2¾	22	8	11¾	22	8	8¾	22	8	5¾	22	8	2½	22	7	11½	..110
120..	24	10	7½	24	10	4¼	24	10	0¾	24	9	9½	24	9	6¼	24	9	3	24	8	11½	24	8	8¼	..120
130..	26	11	6¼	26	11	2½	26	10	11	26	10	7¼	26	10	3¾	26	10	0¼	26	9	8½	26	9	5	..130
140..	28	12	4¾	28	12	1	28	11	9	28	11	5¼	28	11	1¼	28	10	9¼	28	10	5½	28	10	1½	..140
150..	30	13	3½	30	12	11¼	30	12	7	30	12	3	30	11	10¾	30	11	6½	30	11	2¼	30	10	10¼	. 150
160..	32	14	2	32	13	9¾	32	13	5¼	32	13	0¾	32	12	8¼	32	12	3¾	32	11	11½	32	11	7	..160
170..	34	15	0¾	34	14	8	34	14	3½	34	13	10½	34	13	5¾	34	13	1	34	12	8½	34	12	3¾	..170
180..	36	15	11¼	36	15	6½	36	15	1½	36	14	8½	36	14	3½	36	13	10¼	36	13	5¼	36	13	0½	..180
190..	38	16	10	38	16	4¾	38	15	11½	38	15	6	38	15	0¾	38	14	7½	38	14	2¼	38	13	9	..190
200..	40	17	8½	40	17	3	40	16	9½	40	16	4	40	15	10¼	40	15	4¾	40	14	11¼	40	14	5¾	..200
210..	42	18	7¼	42	18	1½	42	17	7½	42	17	1¾	42	16	8	42	16	2	42	15	8¼	42	15	2½	..210
220..	44	19	5¾	44	18	11¾	44	18	5½	44	17	11½	44	17	5½	44	16	11¼	44	16	5¼	44	15	11¼	..220
230..	47	0	4½	46	19	10	46	19	3¾	46	18	9¼	46	18	3	46	17	8½	46	17	2¼	46	16	7¾	..230
240..	49	1	3	49	0	8½	49	0	1¾	48	19	7	48	19	0½	48	18	5¾	48	17	11¼	48	17	4½	..240
250..	51	2	1¾	51	1	6¾	51	0	11¾	51	0	5	50	19	10	50	19	3	50	18	8	50	18	1¼	..250
260..	53	3	0½	53	2	5¼	53	1	10	53	1	2¾	53	0	7½	53	0	0¼	52	19	5	52	18	10	..260
270..	55	3	11	55	3	3½	55	2	8	55	2	0½	55	1	5	55	0	9½	55	0	2	54	19	6½	..270
280..	57	4	9¾	57	4	1¾	57	3	6	57	2	10¼	57	2	2½	57	1	6¾	57	0	11	57	0	3¼	..280
290..	59	5	8¼	59	5	0¼	59	4	4½	59	3	8	59	3	0	59	2	4	59	1	8	59	1	0	..290
300..	61	6	7	61	5	10½	61	5	2¼	61	4	6	61	3	9½	61	3	1¼	61	2	5	61	1	8¾	..300
310..	63	7	5½	63	6	9	63	6	0¼	63	5	3¾	63	4	7	63	3	10½	63	3	2	63	2	5¼	..310
320..	65	8	4¼	65	7	7½	65	6	10¾	65	6	1½	65	5	4½	65	4	7¾	65	3	10¾	65	3	2	..320
330..	67	9	2¾	67	8	5½	67	7	8½	67	6	11¼	67	6	2	67	5	5	67	4	7¾	67	3	10¾	..330
340..	69	10	1½	69	9	4	69	8	6½	69	7	9	69	6	11¾	69	6	2¼	69	5	4¾	69	4	7½	..340
350..	71	11	0	71	10	2¼	71	9	4½	71	8	6¾	71	7	9¼	71	6	11½	71	6	1¾	71	5	4	..350
360..	73	11	10¾	73	11	0¾	73	10	2¾	73	9	4¾	73	8	6¾	73	7	8¾	73	6	10¾	73	6	0¾	..360
370..	75	12	9¼	75	11	11	75	11	0¾	75	10	2½	75	9	4¼	75	8	6	75	7	7¾	75	6	9½	..370
380..	77	13	8	77	12	9¼	77	11	10¾	77	11	0¼	77	10	1¾	77	9	3¼	77	8	4¾	77	7	6¼	..380
390..	79	14	6½	79	13	7¾	79	12	8¾	79	11	10	79	10	11¼	79	10	0½	79	9	1½	79	8	2¾	..390
400..	81	15	5¼	81	14	6	81	13	7	81	12	7¾	81	11	8¾	81	10	9¾	81	9	10½	81	8	11½	..400
410..	83	16	3¾	83	15	4½	83	14	5	83	13	5¾	83	12	6¼	83	11	7	83	10	7½	83	9	8¼	..410
420..	85	17	2½	85	16	2¾	85	15	3	85	14	3¼	85	13	3¾	85	12	4¼	85	11	4½	85	10	5	..420
430..	87	18	1	87	17	1½	87	16	1¼	87	15	1¼	87	14	1¼	87	13	1¼	87	12	1½	87	11	1½	..430
440..	89	18	11¾	89	17	11½	89	16	11¼	89	15	11	89	14	10¾	89	13	10½	89	12	10½	89	11	10¼	..440
450..	91	19	10¼	91	18	9¾	91	17	9¼	91	16	8¾	91	15	8¼	91	14	7¾	91	13	7½	91	12	7	. 450
460..	94	0	9	93	19	8¼	93	18	7½	93	17	6½	93	16	5¾	93	15	5	93	14	4¼	93	13	3¾	..460
470..	96	1	7½	96	0	6½	95	19	5½	95	18	4½	95	17	3¼	95	16	2¼	95	15	1¼	95	14	0¼	..470
480..	98	2	6¼	98	1	5	98	0	3½	97	19	2¼	97	18	1	97	16	11½	97	15	10¼	97	14	9	..480
490..	100	3	5	100	2	3½	100	1	1½	100	0	0	99	18	10½	99	17	8¾	99	16	7¼	99	15	5¾	..490
500..	102	4	3½	102	3	1½	102	1	11¾	102	0	9¾	101	19	8	101	18	6	101	17	4¼	101	16	2½	..500

Cents.	0		1		2		3		4		5		6		7		8		9		Cents.
	s.	d.	s.	d.	s.	d.	s.	d.	s.	d.	s.	d.	s.	d.	s.	d.	s.	d.	s.	d.	
0 ...		0		0½		1		1½		2		2½		3		3½		4		4½	... 0
10 ...		5		5½		6		6½		6¾		7¼		7¾		8¼		8¾		9¼	... 10
20 ...		9¾		10¼		10¾		11¼		11¾	1	0¼	1	0¾	1	1¼	1	1¾	1	2¼	... 20
30 ...	1	2¾	1	3¼	1	3¾	1	4¼	1	4¾	1	5¼	1	5¾	1	6	1	6½	1	7	... 30
40 ...	1	7½	1	8	1	8½	1	9	1	9½	1	10	1	10½	1	11	1	11½	2	0	... 40

STERLING EXCHANGE TABLES.

$10\frac{1}{16}$ TO $10\frac{1}{2}$%

DOLLARS AND CENTS INTO STERLING.

Dol'rs	$10\frac{1}{16}$	$10\frac{1}{8}$	$10\frac{3}{16}$	$10\frac{1}{4}$	$10\frac{5}{16}$	$10\frac{3}{8}$	$10\frac{7}{16}$	$10\frac{1}{2}$	Dol'rs
	£ s. d.	£ s. d.	£ s. d.	£ s. d.	£ s. d.	£ s. d.	£ s. d.	£ s. d.	
510	104 5 2¼	104 4 0	104 2 9¾	104 1 7½	104 0 5½	103 19 3¼	103 18 1¼	103 16 11	510
520	106 6 0¾	106 4 10½	106 3 7¾	106 2 5½	106 1 3	106 0 0¼	105 18 10¼	105 17 7¾	520
530	108 6 11½	108 5 8¾	108 4 6	108 3 3¼	108 2 0½	108 0 9¾	107 19 7	107 18 4½	530
540	110 7 10	110 6 7	110 5 4	110 4 1	110 2 10	110 1 7	110 0 4	109 19 1¼	540
550	112 8 8¾	112 7 5½	112 6 2	112 4 10¾	112 3 7½	112 2 4¼	112 1 1	111 19 9¾	550
560	114 9 7¼	114 8 3¾	114 7 0½	114 5 8¾	114 4 5	114 3 1½	114 1 10	114 0 6½	560
570	116 10 6	116 9 2	116 7 10½	116 6 6¼	116 5 2½	116 3 10¾	116 2 7	116 1 3¼	570
580	118 11 4½	118 10 0½	118 8 8¼	118 7 4¼	118 6 0	118 4 8	118 3 4	118 2 0	580
590	120 12 3¼	120 10 10¾	120 9 6¼	120 8 2	120 6 9½	120 5 5¼	120 4 1	120 2 8½	590
600	122 13 1¾	122 11 9	122 10 4½	122 8 11¾	122 7 7	122 6 2½	122 4 9¾	122 3 5¼	600
610	124 14 0½	124 12 7½	124 11 2½	124 9 9½	124 8 4¾	124 6 11¾	124 5 6¾	124 4 2	610
620	126 14 11	126 13 5¾	126 12 0½	126 10 7¼	126 9 2¼	126 7 9	126 6 3¾	126 4 10¾	620
630	128 15 9¾	128 14 4¼	128 12 10¾	128 11 5¼	128 9 11¾	128 8 6¼	128 7 0¾	128 5 7¼	630
640	130 16 8¼	130 15 2½	130 13 8¾	130 12 3	130 10 9¼	130 9 3½	130 7 9¾	130 6 4	640
650	132 17 7	132 16 1	132 14 6¾	132 13 0¾	132 11 6¾	132 10 0¾	132 8 6¾	132 7 0¾	650
660	134 18 5½	134 16 11¼	134 15 4¾	134 13 10½	134 12 4¼	134 10 10	134 9 3¾	134 7 9½	660
670	136 19 4¼	136 17 9½	136 16 3	136 14 8¼	136 13 1¾	136 11 7¼	136 10 0½	136 8 6	670
680	139 0 2¾	138 18 8	138 17 1	138 15 6	138 13 11¼	138 12 4¼	138 10 9¼	138 9 2¾	680
690	141 1 1½	140 19 6¼	140 17 11	140 16 4	140 14 8¾	140 13 1¾	140 11 6½	140 9 11½	690
700	143 2 0	143 0 4¾	142 18 9¼	142 17 1¾	142 15 6¼	142 13 11	142 12 3½	142 10 8¼	700
710	145 2 10¾	145 1 3	144 19 7¼	144 17 11½	144 16 3¾	144 14 8¼	144 13 0½	144 11 4¾	710
720	147 3 9½	147 2 1¼	147 0 5¼	146 18 9¼	146 17 1¼	146 15 5¼	146 13 9½	146 12 1½	720
730	149 4 8	149 2 11¾	149 1 3½	148 19 7	148 17 10¾	148 16 2¼	148 14 6½	148 12 10¼	730
740	151 5 6¾	151 3 10	151 2 1½	151 0 5	150 18 8¼	150 16 11¾	150 15 3¼	150 13 7	740
750	153 6 5¼	153 4 8½	153 2 11½	153 1 2¾	152 19 6	152 17 9	152 16 0¼	152 14 3½	750
760	155 7 4	155 5 6¾	155 3 9½	155 2 0½	155 0 3½	154 18 6¼	154 16 9¼	154 15 0¼	760
770	157 8 2½	157 6 5	157 4 7¾	157 2 10¼	157 1 1	156 19 3½	156 17 6¼	156 15 9	770
780	159 9 1¼	159 7 3½	159 5 5¾	159 3 8	159 1 10½	159 0 0¾	158 18 3¼	158 16 5¾	780
790	161 9 11¾	161 8 1¾	161 6 3¾	161 4 6	161 2 8	161 0 10	160 19 0¼	160 17 2¼	790
800	163 10 10½	163 9 0¼	163 7 2	163 5 3¾	163 3 5½	163 1 7¼	162 19 9¼	162 17 11	800
810	165 11 9	165 9 10½	165 8 0	165 6 1½	165 4 3	165 2 4½	165 0 6	164 18 7¾	810
820	167 12 7¾	167 10 8¾	167 8 10	167 6 11¼	167 5 0½	167 3 1¾	167 1 3	166 19 4½	820
830	169 13 6¼	169 11 7¼	169 9 8	169 7 9	169 5 10	169 3 11	169 2 0	169 0 1	830
840	171 14 5	171 12 5½	171 10 6¼	171 8 6¾	171 6 7½	171 4 8¼	171 2 9	171 0 9¾	840
850	173 15 3½	173 13 4	173 11 4½	173 9 4¾	173 7 5	173 5 5½	173 3 6	173 1 6½	850
860	175 16 2¼	175 14 2¼	175 12 2¼	175 10 2½	175 8 2½	175 6 2¾	175 4 3	175 2 3¼	860
870	177 17 0¾	177 15 0½	177 13 0½	177 11 0¼	177 9 0	177 7 0	177 5 0	177 2 11¾	870
880	179 17 11½	179 15 11	179 13 10½	179 11 10	179 9 9¾	179 7 9¼	179 5 8¾	179 3 8½	880
890	181 18 10	181 16 9¼	181 14 8½	181 12 7¾	181 10 7¼	181 8 6½	181 6 5¾	181 4 5¼	890
900	183 19 8¾	183 17 7¾	183 15 6¾	183 13 5¾	183 11 4¾	183 9 3¾	183 7 2¾	183 5 2	900
910	186 0 7¼	185 18 6	185 16 4¾	185 14 3½	185 12 2¼	185 10 1	185 7 11¾	185 5 10½	910
920	188 1 6	187 19 4½	187 17 2¾	187 15 1¼	187 12 11¾	187 10 10¼	187 8 8¾	187 6 7¼	920
930	190 2 4½	190 0 2¾	189 18 0¾	189 15 11	189 13 9¼	189 11 7½	189 9 5¾	189 7 4	930
940	192 3 3¼	192 1 1	191 18 11	191 16 8¾	191 14 6½	191 12 4¾	191 10 2¾	191 8 0¾	940
950	194 4 1¾	194 1 11½	193 19 9	193 17 6½	193 15 4¼	193 13 2	193 10 11½	193 8 9¼	950
960	196 5 0½	196 2 9¾	196 0 7	195 18 4½	195 16 1¾	195 13 11¼	195 11 8½	195 9 6	960
970	198 5 11¼	198 3 8¼	198 1 5¼	197 19 2¼	197 16 11¼	197 14 8½	197 12 5½	197 10 2¾	970
980	200 6 9¾	200 4 6½	200 2 3¼	200 0 0	199 17 8¾	199 15 5¾	199 13 2½	199 10 11½	980
990	202 7 8½	202 5 4¾	202 3 1¼	202 0 9¾	201 18 6¼	201 16 3	201 13 11½	201 11 8	990
1000	204 8 7	204 6 3½	204 3 11¼	204 1 7½	203 19 3¾	203 17 0	203 14 8½	203 12 4¾	1000
2000	408 17 2	408 12 6¼	408 7 10¾	408 3 3¼	407 18 7¾	407 14 0¼	407 9 4¾	407 4 9½	2000
3000	613 5 9	612 18 9½	612 11 10	612 4 10¾	611 17 11½	611 11 0¼	611 4 1¼	610 17 2¼	3000
4000	817 14 4	817 5 0¾	816 15 9½	816 6 6¼	815 17 3½	815 8 0½	814 18 9¾	814 9 7	4000
5000	1022 2 11¼	1021 11 4	1020 19 8¾	1020 8 2	1019 16 7¼	1019 5 0¾	1018 13 6¼	1018 2 0	5000
6000	1226 11 6¼	1225 17 7¼	1225 3 8¼	1224 9 9½	1223 15 11	1223 2 0¾	1222 8 2½	1221 14 4¾	6000
7000	1431 0 1¼	1430 3 10¼	1429 7 7¾	1428 11 5¼	1427 15 3	1426 19 1	1426 2 11	1425 6 9½	7000
8000	1635 8 8¼	1634 10 1½	1633 11 7	1632 13 0¾	1631 14 6¾	1630 16 1	1629 17 7½	1628 19 2¼	8000
9000	1839 17 3¼	1838 16 4¾	1837 15 6½	1836 14 8¼	1835 13 10½	1834 13 1¼	1833 12 4	1832 11 7	9000
10000	2044 5 10¼	2043 2 8	2041 19 5¾	2040 16 4	2039 13 2½	2038 10 1¼	2037 7 0¼	2036 3 11¾	10000

Cents.	0	1	2	3	4	5	6	7	8	9	Cents.
	s. d.	s. d.	s. d.	s. d.	s. d.	s. d.	s. d.	s. d.	s. d.	s. d.	
50 ...	2 0½	2 1	2 1½	2 2	2 2½	2 3	2 3½	2 4	2 4½	2 5	... 50
60 ...	2 5½	2 6	2 6½	2 6¾	2 7¼	2 7¾	2 8¼	2 8¾	2 9¼	2 9¾	... 60
70 ...	2 10¼	2 10¾	2 11¼	2 11¾	3 0¼	3 0¾	3 1¼	3 1¾	3 2¼	3 2¾	... 70
80 ...	3 3¼	3 3¾	3 4¼	3 4¾	3 5¼	3 5¾	3 6	3 6½	3 7	3 7½	... 80
90 ...	3 8	3 8½	3 9	3 9½	3 10	3 10½	3 11	3 11½	4 0	4 0½	... 90

EXPLANATIONS TO PART III.

AMERICAN OR DOMESTIC EXCHANGE TABLES,

AND

BROKERAGES OF $\frac{1}{32}$, $\frac{1}{16}$, AND $\frac{1}{8}$ OF 1% FOR STERLING AND AMERICAN EXCHANGE.

These tables present an entirely new feature, and are nothing more nor less than a set of calculations for transactions in Exchange on New York or other important American cities.

The arrangement of the form is by thousands, from $1000 to $100000 ($1000 to $50000 on left hand and $51000 to $100000 on right-hand side of each sheet); but by a removal of the decimal point the amounts can be adjusted to units, tens, or hundreds; thus, $45000 at $\frac{3}{16}$ premium = $45084 37, $4500 = $4508.43, $450 = $450.84, and $15 = $45.08.

Example : To find value of $13000 New York draft at $\frac{3}{16}$ discount (see pages 120 and 121) :—

$$\$13000 \; \tfrac{3}{16} = \$12975.62$$
$$\$1300 \; \tfrac{3}{16} = \$1297.56$$

To find the value of $41000 New York draft at $\frac{3}{16}$ premium (see pages 126 and 127) :—

$$\$41000 \; \tfrac{3}{16} = \$41230.62$$
$$\$4100 \; \tfrac{3}{16} = \$4123 06$$

The Brokerages will be found of great use in the handling of large amounts where the assistance of a broker is required. The same system as in the preceding tables is carried out, Sterling Exchange in amounts from £100 to £10000 on left-hand and American Exchange in amounts from $1000 to $100000 on right-hand sheet.

Example : To find Brokerage of $\frac{1}{32}$ on £5000 (see page 128) = $6.94.

To find Brokerage of $\frac{1}{32}$ on $100000 (see page 129) = $31.25.

AMERICAN EXCHANGE TABLES,

CALCULATED AT FRACTIONS OF

$\frac{1}{16}$ TO $\frac{1}{2}$ % DISCOUNT

TO FACILITATE TRANSACTIONS IN

AMERICAN OR DOMESTIC EXCHANGE,

Par.	$\frac{1}{16}$	$\frac{1}{8}$	$\frac{3}{16}$	$\frac{1}{4}$	$\frac{5}{16}$	$\frac{3}{8}$	$\frac{7}{16}$	$\frac{1}{2}$	Par.
$	$ cts.	$ cts	$ cts.	$ cts.	$ cts.	$ cts.	$ cts.	$ cts.	$
1000	999.37	998.75	998.12	997.50	996.87	996.25	995.62	995.00	.. 1000
2000	1998.75	1997.50	1996.25	1995.00	1993.75	1992.50	1991.25	1990.00	.. 2000
3000	2998.12	2996.25	2994.37	2992.50	2990.62	2988.75	2986.87	2985.00	.. 3000
4000	3997.50	3995.00	3992.50	3990.00	3987.50	3985.00	3982.50	3980.00	.. 4000
5000	4996.87	4993.75	4990.62	4987.50	4984.37	4981.25	4978.12	4975.00	.. 5000
6000	5996.25	5992.50	5988.75	5985.00	5981.25	5977.50	5973.75	5970.00	.. 6000
7000	6995.62	6991.25	6986.87	6982.50	6978.12	6973.75	6969.37	6965.00	.. 7000
8000	7995.00	7990.00	7985.00	7980.00	7975.00	7970.00	7965.00	7960.00	.. 8000
9000	8994.37	8988.75	8983.12	8977.50	8971.87	8966.25	8960.62	8955.00	.. 9000
10000	9993.75	9987.50	9981.25	9975.00	9968.75	9962.50	9956.25	9950.00	.. 10000
11000	10993.12	10986.25	10979.37	10972.50	10965.62	10958.75	10951.87	10945.00	.. 11000
12000	11992.50	11985.00	11977.50	11970.00	11962.50	11955.00	11947.50	11940.00	.. 12000
13000	12991.87	12983.75	12975.62	12967.50	12959.37	12951.25	12943.12	12935.00	.. 13000
14000	13991.25	13982.50	13973.75	13965.00	13956.25	13947.50	13938.75	13930.00	.. 14000
15000	14990.62	14981.25	14971.87	14962.50	14953.12	14943.75	14934.37	14925.00	.. 15000
16000	15990.00	15980.00	15970.00	15960.00	15950.00	15940.00	15930.00	15920.00	.. 16000
17000	16989.37	16978.75	16968.12	16957.50	16946.87	16936.25	16925.62	16915.00	.. 17000
18000	17988.75	17977.50	17966.25	17955.00	17913.75	17932.50	17921.25	17910.00	.. 18000
19000	18988.12	18976.25	18964.37	18952.50	18940.62	18928.75	18916.87	18905.00	.. 19000
20000	19987.50	19975.00	19962.50	19950.00	19937.50	19925.00	19912.50	19900.00	.. 20000
21000	20986.87	20973.75	20960.62	20947.50	20934.37	20921.25	20908.12	20895.00	.. 21000
22000	21986.25	21972.50	21958.75	21945.00	21931.25	21917.50	21903.75	21890.00	.. 22000
23000	22985.62	22971.25	22956.87	22942.50	22928.12	22913.75	22899.37	22885.00	.. 23000
24000	23985.00	23970.00	23955.00	23940.00	23925.00	23910.00	23895.00	23880.00	.. 24000
25000	24984.37	24968.75	24953.12	24937.50	24921.87	24906.25	24890.62	24875.00	.. 25000
26000	25983.75	25967.50	25951.25	25935.00	25918.75	25902.50	25886.25	25870.00	.. 26000
27000	26983.12	26966.25	26949.37	26932.50	26915.62	26898.75	26881.87	26865.00	.. 27000
28000	27982.50	27965.00	27947.50	27930.00	27912.50	27895.00	27877.50	27860.00	.. 28000
29000	28981.87	28963.75	28945.62	28927.50	28909.37	28891.25	28873.12	28855.00	.. 29000
30000	29981.25	29962.50	29943.75	29925.00	29906.25	29887.50	29868.75	29850.00	.. 30000
31000	30980.62	30961.25	30941.87	30922.50	30903.12	30883.75	30864.37	30845.00	.. 31000
32000	31980.00	31960.00	31940.00	31920.00	31900.00	31880.00	31860.00	31840.00	.. 32000
33000	32979.37	32958.75	32938.12	32917.50	32896.87	32876.25	32855.62	32835.00	.. 33000
34000	33978.75	33957.50	33936.25	33915.00	33893.75	33872.50	33851.25	33830.00	.. 34000
35000	34978.12	34956.25	34934.37	34912.50	34890.62	34868.75	34846.87	34825.00	.. 35000
36000	35977.50	35955.00	35932.50	35910.00	35887.50	35865.00	35842.50	35820.00	. 36000
37000	36976.87	36953.75	36930.62	36907.50	36884.37	36861.25	36838.12	36815.00	.. 37000
38000	37976.25	37952.50	37928.75	37905.00	37881.25	37857.50	37833.75	37810.00	.. 38000
39000	38975.62	38951.25	38926.87	38902.50	38878.12	38853.75	38829.37	38805.00	.. 39000
40000	39975.00	39950.00	39925.00	39900.00	39875.00	39850.00	39825.00	39800.00	.. 40000
41000	40974.37	40948.75	40923.12	40897.50	40871.87	40846.25	40820.62	40795.00	.. 41000
42000	41973.75	41947.50	41921.25	41895.00	41868.75	41842.50	41816.25	41790.00	.. 42000
43000	42973.12	42946.25	42919.37	42892.50	42865.62	42838.75	42811.87	42785.00	.. 43000
44000	43972.50	43945.00	43917.50	43890.00	43862.50	43835.00	43807.50	43780.00	.. 44000
45000	44971.87	44943.75	44915.62	44887.50	44859.37	44831.25	44803.12	44775.00	.. 45000
46000	45971.25	45942.50	45913.75	45885.00	45856.25	45827.50	45798.75	45770.00	.. 46000
47000	46970.62	46941.25	46911.87	46882.50	46853.12	46823.75	46794.37	46765.00	.. 47000
48000	47970.00	47940.00	47910.00	47880.00	47850.00	47820.00	47790.00	47760.00	.. 48000
49000	48969.37	48938.75	48908.12	48877.50	48846.87	48816.25	48785.62	48755.00	.. 49000
50000	49968.75	49937.50	49906.25	49875.00	49843.75	49812.50	49781.25	49750.00	.. 50000

AMERICAN EXCHANGE TABLES.

CALCULATED AT FRACTIONS OF

$\frac{1}{16}$ TO $\frac{1}{2}$% DISCOUNT

TO FACILITATE TRANSACTIONS IN

AMERICAN OR DOMESTIC EXCHANGE.

Par.	$\frac{1}{16}$	$\frac{1}{8}$	$\frac{3}{16}$	$\frac{1}{4}$	$\frac{5}{16}$	$\frac{3}{8}$	$\frac{7}{16}$	$\frac{1}{2}$	Par.
$	$ cts.	$ cts.	$ cts.	$ cts.	$ cts.	$ cts.	$ cts.	$ cts.	$
51000	50968.12	50936.25	50904.37	50872.50	50840.62	50808.75	50776.87	50745.00	51000
52000	51967.50	51935.00	51902.50	51870.00	51837.50	51805.00	51772.50	51740.00	52000
53000	52966.87	52933.75	52900.62	52867.50	52834.37	52801.25	52768.12	52735.00	53000
54000	53966.25	53932.50	53898.75	53865.00	53831.25	53797.50	53763.75	53730.00	54000
55000	54965.62	54931.25	54896.87	54862.50	54828.12	54793.75	54759.37	54725.00	55000
56000	55965.00	55930.00	55895.00	55860.00	55825.00	55790.00	55755.00	55720.00	56000
57000	56964.37	56928.75	56893.12	56857.50	56821.87	56786.25	56750.62	56715.00	57000
58000	57963.75	57927.50	57891.25	57855.00	57818.75	57782.50	57746.25	57710.00	58000
59000	58963.12	58926.25	58889.37	58852.50	58815.62	58778.75	58741.87	58705.00	59000
60000	59962.50	59925.00	59887.50	59850.00	59812.50	59775.00	59737.50	59700.00	60000
61000	60961.87	60923.75	60885.62	60847.50	60809.37	60771.25	60733.12	60695.00	61000
62000	61961.25	61922.50	61883.75	61845.00	61806.25	61767.50	61728.75	61690.00	62000
63000	62960.62	62921.25	62881.87	62842.50	62803.12	62763.75	62724.37	62685.00	63000
64000	63960.00	63920.00	63880.00	63840.00	63800.00	63760.00	63720.00	63680.00	64000
65000	64959.37	64918.75	64878.12	64837.50	64796.87	64756.25	64715.62	64675.00	65000
66000	65958.75	65917.50	65876.25	65835.00	65793.75	65752.50	65711.25	65670.00	66000
67000	66958.12	66916.25	66874.37	66832.50	66790.62	66748.75	66706.87	66665.00	67000
68000	67957.50	67915.00	67872.50	67830.00	67787.50	67745.00	67702.50	67660.00	68000
69000	68956.87	68913.75	68870.62	68827.50	68784.37	68741.25	68698.12	68655.00	69000
70000	69956.25	69912.50	69868.75	69825.00	69781.25	69737.50	69693.75	69650.00	70000
71000	70955.62	70911.25	70866.87	70822.50	70778.12	70733.75	70689.37	70645.00	71000
72000	71955.00	71910.00	71865.00	71820.00	71775.00	71730.00	71685.00	71640.00	72000
73000	72954.37	72908.75	72863.12	72817.50	72771.87	72726.25	72680.62	72635.00	73000
74000	73953.75	73907.50	73861.25	73815.00	73768.75	73722.50	73676.25	73630.00	74000
75000	74953.12	74906.25	74859.37	74812.50	74765.62	74718.75	74671.87	74625.00	75000
76000	75952.50	75905.00	75857.50	75810.00	75762.50	75715.00	75667.50	75620.00	76000
77000	76951.87	76903.75	76855.62	76807.50	76759.37	76711.25	76663.12	76615.00	77000
78000	77951.25	77902.50	77853.75	77805.00	77756.25	77707.50	77658.75	77610.00	78000
79000	78950.62	78901.25	78851.87	78802.50	78753.12	78703.75	78654.37	78605.00	79000
80000	79950.00	79900.00	79850.00	79800.00	79750.00	79700.00	79650.00	79600.00	80000
81000	80949.37	80898.75	80848.12	80797.50	80746.87	80696.25	80645.62	80595.00	81000
82000	81948.75	81897.50	81846.25	81795.00	81743.75	81692.50	81641.25	81590.00	82000
83000	82948.12	82896.25	82844.37	82792.50	82740.62	82688.75	82636.87	82585.00	83000
84000	83947.50	83895.00	83842.50	83790.00	83737.50	83685.00	83632.50	83580.00	84000
85000	84946.87	84893.75	84840.62	84787.50	84734.37	84681.25	84628.12	84575.00	85000
86000	85946.25	85892.50	85838.75	85785.00	85731.25	85677.50	85623.75	85570.00	86000
87000	86945.62	86891.25	86836.87	86782.50	86728.12	86673.75	86619.37	86565.00	87000
88000	87945.00	87890.00	87835.00	87780.00	87725.00	87670.00	87615.00	87560.00	88000
89000	88944.37	88888.75	88833.12	88777.50	88721.87	88666.25	88610.62	88555.00	89000
90000	89943.75	89887.50	89831.25	89775.00	89718.75	89662.50	89606.25	89550.00	90000
91000	90943.12	90886.25	90829.37	90772.50	90715.62	90658.75	90601.87	90545.00	91000
92000	91942.50	91885.00	91827.50	91770.00	91712.50	91655.00	91597.50	91540.00	92000
93000	92941.87	92883.75	92825.62	92767.50	92709.37	92651.25	92593.12	92535.00	93000
94000	93941.25	93882.50	93823.75	93765.00	93706.25	93647.50	93588.75	93530.00	94000
95000	94940.62	94881.25	94821.87	94762.50	94703.12	94643.75	94584.37	94525.00	95000
96000	95940.00	95880.00	95820.00	95760.00	95700.00	95640.00	95580.00	95520.00	96000
97000	96939.37	96878.75	96818.12	96757.50	96696.87	96636.25	96575.62	96515.00	97000
98000	97938.75	97877.50	97816.25	97755.00	97693.75	97632.50	97571.25	97510.00	98000
99000	98938.12	98876.25	98814.37	98752.50	98690.62	98628.75	98566.87	98505.00	99000
100000	99937.50	99875.00	99812.50	99750.00	99687.50	99625.00	99562.50	99500.00	100000

AMERICAN EXCHANGE TABLES,

CALCULATED AT FRACTIONS OF

$\frac{9}{16}$ TO 1% DISCOUNT

TO FACILITATE TRANSACTIONS IN

AMERICAN OR DOMESTIC EXCHANGE.

Par.	$\frac{9}{16}$	$\frac{5}{8}$	$\frac{11}{16}$	$\frac{3}{4}$	$\frac{13}{16}$	$\frac{7}{8}$	$\frac{15}{16}$	1	Par.
$	$ cts.	$ cts.	$ cts.	$ cts.	$ cts.	$ cts.	$ cts.	$ cts.	$
1000	994.37	993.75	993.12	992.50	991 87	991.25	990.62	990 00	1000
2000	1988.75	1987.50	1986.25	1985.00	1983.75	1982 50	1981.25	1980.00	2000
3000	2983.12	2981.25	2979.37	2977.50	2975.62	2973.75	2971.87	2970.00	3000
4000	3977.50	3975.00	3972.50	3970.00	3967.50	3965.00	3962.50	3960.00	4000
5000	4971.87	4968.75	4965.62	4962.50	4959.37	4956.25	4953.12	4950.00	5000
6000	5966.25	5962.50	5958.75	5955 00	5951.25	5947.50	5943.75	5940 00	6000
7000	6960.62	6956.25	6951.87	6947.50	6943.12	6938.75	6934.37	6930.00	7000
8000	7955.00	7950.00	7945.00	7940.00	7935.00	7930.00	7925.00	7920.00	8000
9000	8949.37	8943.75	8938.12	8932.50	8926.87	8921.25	8915.62	8910.00	9000
10000	9943.75	9937.50	9931.25	9925.00	9918.75	9912.50	9906.25	9900.00	10000
11000	10938.12	10931.25	10924.37	10917.50	10910.62	10903.75	10896.87	10890.00	11000
12000	11932.50	11925.00	11917.50	11910.00	11902.50	11895.00	11887.50	11880.00	12000
13000	12926.87	12918.75	12910.62	12902.50	12894.37	12886.25	12878.12	12870.00	13000
14000	13921.25	13912.50	13903.75	13895.00	13886.25	13877.50	13868.75	13860.00	14000
15000	14915.62	14906.25	14896.87	14887.50	14878.12	14868.75	14859.37	14850.00	15000
16000	15910.00	15900.00	15890.00	15880.00	15870.00	15860.00	15850.00	15840.00	16000
17000	16904.37	16893.75	16883.12	16872.50	16861.87	16851.25	16840.62	16830.00	17000
18000	17898.75	17887.50	17876.25	17865.00	17853.75	17842.50	17831.25	17820.00	18000
19000	18893.12	18881.25	18869.37	18857.50	18845.62	18833.75	18821.87	18810.00	19000
20000	19887.50	19875.00	19862.50	19850.00	19837.50	19825.00	19812.50	19800.00	20000
21000	20881.87	20868.75	20855.62	20842.50	20829.37	20816.25	20803.12	20790.00	21000
22000	21876.25	21862.50	21848.75	21835.00	21821.25	21807.50	21793.75	21780.00	22000
23000	22870.62	22856.25	22841 87	22827.50	22813.12	22798.75	22784.37	22770.00	23000
24000	23865.00	23850.00	23835.00	23820.00	23805.00	23790.00	23775.00	23760.00	24000
25000	24859.37	24843.75	24828.12	24812.50	24796.87	24781.25	24765.62	24750.00	25000
26000	25853.75	25837.50	25821.25	25805.00	25788.75	25772 50	25756.25	25740.00	26000
27000	26848.12	26831.25	26814.37	26797.50	26780.62	26763.75	26746.87	26730.00	27000
28000	27842.50	27825.00	27807.50	27790.00	27772.50	27755.00	27737.50	27720.00	28000
29000	28836.87	28818.75	28800.62	28782.50	28764.37	28746.25	28728.12	28710.00	29000
30000	29831.25	29812.50	29793.75	29775.00	29756.25	29737.50	29718.75	29700.00	30000
31000	30825.62	30806.25	30786.87	30767.50	30748.12	30728.75	30709.37	30690.00	31000
32000	31820.00	31800.00	31780.00	31760.00	31740.00	31720.00	31700.00	31680.00	32000
33000	32814.37	32793.75	32773.12	32752.50	32731.87	32711.25	32690.62	32670.00	33000
34000	33808.75	33787.50	33766.25	33745.00	33723.75	33702.50	33681.25	33660.00	34000
35000	34803.12	34781.25	34759.37	34737.50	34715.62	34693.75	34671.87	34650.00	35000
36000	35797.50	35775.00	35752.50	35730 00	35707.50	35685.00	35662.50	35640.00	36000
37000	36791.87	36768.75	36745.62	36722.50	36699.37	36676.25	36653.12	36630.00	37000
38000	37786.25	37762.50	37738.75	37715.00	37691.25	37667.50	37643.75	37620.00	38000
39000	38780.62	38756.25	38731.87	38707.50	38683.12	38658.75	38634.37	38610.00	39000
40000	39775.00	39750.00	39725.00	39700.00	39675.00	39650.00	39625.00	39600.00	40000
41000	40769.37	40743.75	40718.12	40692.50	40666.87	40641.25	40615.62	40590.00	41000
42000	41763.75	41737.50	41711.25	41685.00	41658.75	41632.50	41606.25	41580.00	42000
43000	42758.12	42731.25	42704.37	42677.50	42650.62	42623.75	42596.87	42570.00	43000
44000	43752.50	43725.00	43697.50	43670.00	43642.50	43615.00	43587.50	43560.00	44000
45000	44746.87	44718.75	44690.62	44662.50	44634.37	44606.25	44578.12	44550.00	45000
46000	45741.25	45712.50	45683.75	45655.00	45626.25	45597.50	45568.75	45540.00	46000
47000	46735.62	46706.25	46676.87	46647.50	46618.12	46588.75	46559.37	46530.00	47000
48000	47730.00	47700.00	47670.00	47640.00	47610.00	47580.00	47550.00	47520.00	48000
49000	48724.37	48693.75	48663.12	48632.50	48601.87	48571.25	48540.62	48510.00	49000
50000	49718.75	49687.50	49656.25	49625.00	49593.75	49562.50	49531.25	49500.00	50000

AMERICAN EXCHANGE TABLES.

CALCULATED AT FRACTIONS OF

⁹⁄₁₆ ᵀᴼ 1% DISCOUNT

TO FACILITATE TRANSACTIONS IN

AMERICAN OR DOMESTIC EXCHANGE.

Par.	⁹⁄₁₆	⅝	¹¹⁄₁₆	¾	¹³⁄₁₆	⅞	¹⁵⁄₁₆	1	Par.
$	$ cts.	$ cts.	$ cts.	$ cts.	$ cts.	$ cts.	$ cts.	$ cts.	$
51000 ..	50713.12	50681.25	50649.37	50617.50	50585.62	50553.75	50521.87	50490.00	.. 51000
52000 ..	51707.50	51675.00	51642.50	51610.00	51577.50	51545.00	51512.50	51480.00	.. 52000
53000 ..	52701.87	52668.75	52635.62	52602.50	52569.37	52536.25	52503.12	52470.00	.. 53000
54000 ..	53696.25	53662.50	53628.75	53595.00	53561.25	53527.50	53493.75	53460.00	.. 54000
55000 ..	54690.62	54656.25	54621.87	54587.50	54553.12	54518.75	54484.37	54450.00	.. 55000
56000 ..	55685.00	55650.00	55615.00	55580.00	55545.00	55510.00	55475.00	55440.00	.. 56000
57000 ..	56679.37	56643.75	56608.12	56572.50	56536.87	56501.25	56465.62	56430.00	.. 57000
58000 ..	57673.75	57637.50	57601.25	57565.00	57528.75	57492.50	57456.25	57420.00	.. 58000
59000 ..	58668.12	58631.25	58594.37	58557.50	58520.62	58483.75	58446.87	58410.00	.. 59000
60000 ..	59662.50	59625.00	59587.50	59550.00	59512.50	59475.00	59437.50	59400.00	.. 60000
61000 ..	60656.87	60618.75	60580.62	60542.50	60504.37	60466.25	60428.12	60390.00	.. 61000
62000 ..	61651.25	61612.50	61573.75	61535.00	61496.25	61457.50	61418.75	61380.00	.. 62000
63000 ..	62645.62	62606.25	62566.87	62527.50	62488.12	62448.75	62409.37	62370.00	.. 63000
64000 ..	63640.00	63600.00	63560.00	63520.00	63480.00	63440.00	63400.00	63360.00	.. 64000
65000 ..	64634.37	64593.75	64553.12	64512.50	64471.87	64431.25	64390.62	64350.00	.. 65000
66000	65628.75	65587.50	65546.25	65505.00	65463.75	65422.50	65381.25	65340.00	.. 66000
67000	66623.12	66581.25	66539.37	66497.50	66455.62	66413.75	66371.87	66330.00	.. 67000
68000	67617.50	67575.00	67532.50	67490.00	67447.50	67405.00	67362.50	67320.00	.. 68000
69000	68611.87	68568.75	68525.62	68482.50	68439.37	68396.25	68353.12	68310.00	.. 69000
70000	69606.25	69562.50	69518.75	69475.00	69431.25	69387.50	69343.75	69300.00	.. 70000
71000 ..	70600.62	70556.25	70511.87	70467.50	70423.12	70378.75	70334.37	70290.00	.. 71000
72000 ..	71595.00	71550.00	71505.00	71460.00	71415.00	71370.00	71325.00	71280.00	.. 72000
73000 ..	72589.37	72543.75	72498.12	72452.50	72406.87	72361.25	72315.62	72270.00	.. 73000
74000 ..	73583.75	73537.50	73491.25	73445.00	73398.75	73352.50	73306.25	73260.00	.. 74000
75000 ..	74578.12	74531.25	74484.37	74437.50	74390.62	74343.75	74296.87	74250.00	.. 75000
76000 ..	75572.50	75525.00	75477.50	75430.00	75382.50	75335.00	75287.50	75240.00	.. 76000
77000 ..	76566.87	76518.75	76470.62	76422.50	76374.37	76326.25	76278.12	76230.00	.. 77000
78000 ..	77561.25	77512.50	77463.75	77415.00	77366.25	77317.50	77268.75	77220.00	.. 78000
79000 ..	78555.62	78506.25	78456.87	78407.50	78358.12	78308.75	78259.37	78210.00	.. 79000
80000 ..	79550 00	79500.00	79450.00	79400.00	79350.00	79300.00	79250.00	79200.00	.. 80000
81000 ..	80544.37	80493.75	80443.12	80392.50	80341.87	80291.25	80240.62	80190.00	.. 81000
82000 ..	81538.75	81487.50	81436.25	81385.00	81333.75	81282.50	81231.25	81180.00	.. 82000
83000 ..	82533.12	82481.25	82429.37	82377.50	82325.62	82273.75	82221.87	82170.00	.. 83000
84000 ..	83527.50	83475.00	83422.50	83370.00	83317.50	83265.00	83212.50	83160.00	.. 84000
85000 ..	84521.87	84468.75	84415.62	84362.50	84309.37	84256.25	84203.12	84150.00	.. 85000
86000 ..	85516.25	85462.50	85408.75	85355.00	85301.25	85247.50	85193.75	85140.00	.. 86000
87000 ..	86510.62	86456.25	86401.87	86347.50	86293.12	86238.75	86184.37	86130.00	.. 87000
88000 .	87505.00	87450.00	87395.00	87340.00	87285.00	87230.00	87175.00	87120.00	.. 88000
89000 ..	88499.37	88443.75	88388.12	88332.50	88276.87	88221.25	88165.62	88110.00	.. 89000
90000 ..	89493.75	89437.50	89381.25	89325.00	89268.75	89212.50	89156.25	89100.00	.. 90000
91000 ..	90488.12	90431.25	90374.37	90317.50	90260.62	90203.75	90146.87	90090.00	.. 91000
92000 ..	91482.50	91425.00	91367.50	91310.00	91252.50	91195.00	91137.50	91080.00	.. 92000
93000 ..	92476.87	92418.75	92360.62	92302.50	92244.37	92186.25	92128.12	92070.00	.. 93000
94000 ..	93471.25	93412.50	93353.75	93295.00	93236.25	93177.50	93118.75	93060.00	.. 94000
95000 ..	94465.62	94406.25	94346.87	94287.50	94228.12	94168.75	94109.37	94050.00	.. 95000
96000 ..	95460.00	95400.00	95340.00	95280.00	95220.00	95160.00	95100.00	95040.00	.. 96000
97000 ..	96454.37	96393.75	96333.12	96272.50	96211.87	96151.25	96090.62	96030.00	.. 97000
98000 ..	97448.75	97387.50	97326.25	97265.00	97203.75	97142.50	97081.25	97020.00	.. 98000
99000 ..	98443.12	98381.25	98319.37	98257.50	98195.62	98133.75	98071.87	98010.00	.. 99000
100000 ..	99437.50	99375.00	99312.50	99250.00	99187.50	99125.00	99062.50	99000.00	..100000

AMERICAN EXCHANGE TABLES,

CALCULATED AT FRACTIONS OF

$\frac{1}{16}$ TO $\frac{1}{2}$ % PREMIUM

TO FACILITATE TRANSACTIONS IN

AMERICAN OR DOMESTIC EXCHANGE,

Par.	$\frac{1}{16}$	$\frac{1}{8}$	$\frac{3}{16}$	$\frac{1}{4}$	$\frac{5}{16}$	$\frac{3}{8}$	$\frac{7}{16}$	$\frac{1}{2}$	Par.
$	$ cts.	$ cts.	$ cts.	$ cts.	$ cts.	$ cts.	$ cts.	$ cts.	$
1000 .	1000.62	1001.25	1001.87	1002.50	1003.12	1003.75	1004.37	1005.00	.. 1000
2000 ..	2001.25	2002.50	2003.75	2005.00	2006.25	2007.50	2008.75	2010.00	.. 2000
3000 ..	3001.87	3003.75	3005.62	3007.50	3009.37	3011.25	3013.12	3015.00	.. 3000
4000 ..	4002.50	4005.00	4007.50	4010.00	4012.50	4015.00	4017.50	4020.00	.. 4000
5000 ..	5003.12	5006.25	5009.37	5012.50	5015.62	5018.75	5021.87	5025.00	.. 5000
6000 ..	6003.75	6007.50	6011.25	6015.00	6018.75	6022.50	6026.25	6030.00	.. 6000
7000 ..	7004.37	7008.75	7013.12	7017.50	7021.87	7026.25	7030.62	7035.00	.. 7000
8000 ..	8005.00	8010.00	8015.00	8020.00	8025.00	8030.00	8035.00	8040.00	.. 8000
9000 ..	9005.62	9011.25	9016.87	9022.50	9028.12	9033.75	9039.37	9045.00	.. 9000
10000 ..	10006.25	10012.50	10018.75	10025.00	10031.25	10037.50	10043.75	10050.00	.. 10000
11000 ..	11006.87	11013.75	11020.62	11027.50	11034.37	11041.25	11048 12	11055.00	.. 11000
12000 ..	12007.50	12015.00	12022.50	12030.00	12037.50	12045.00	12052.50	12060.00	.. 12000
13000 ..	13008.12	13016.25	13024.37	13032.50	13040.62	13048 75	13056.87	13065.00	.. 13000
14000 ..	14008.75	14017.50	14026.25	14035.00	14043.75	14052.50	14061.25	14070.00	.. 14000
15000 ..	15009.37	15018.75	15028.12	15037.50	15046.87	15056.25	15065.62	15075 00	.. 15000
16000 ..	16010.00	16020.00	16030.00	16040.00	16050.00	16060.00	16070.00	16080.00	.. 16000
17000 ..	17010.62	17021.25	17031.87	17042.50	17053.12	17063.75	17074.37	17085.00	.. 17000
18000 ..	18011.25	18022.50	18033.75	18045.00	18056.25	18067.50	18078.75	18090.00	.. 18000
19000 ..	19011.87	19023.75	19035.62	19047.50	19059.37	19071.25	19083.12	19095.00	.. 19000
20000 ..	20012.50	20025.00	20037.50	20050.00	20062.50	20075.00	20087.50	20100.00	.. 20000
21000 ..	21013.12	21026.25	21039.37	21052.50	21065.62	21078.75	21091.87	21105.00	.. 21000
22000 ..	22013.75	22027.50	22041.25	22055.00	22068.75	22082.50	22096.25	22110.00	.. 22000
23000 ..	23014.37	23028.75	23043.12	23057.50	23071 87	23086.25	23100.62	23115.00	.. 23000
24000 ..	24015.00	24030.00	24045.00	24060.00	24075.00	24090.00	24105.00	24120.00	.. 24000
25000 ..	25015 62	25031.25	25046.87	25062.50	25078 12	25093.75	25109.37	25125.00	.. 25000
26000 ..	26016.25	26032.50	26048.75	26065.00	26081.25	26097.50	26113.75	26130.00	.. 26000
27000 ..	27016.87	27033.75	27050.62	27067.50	27084.37	27101.25	27118.12	27135.00	.. 27000
28000 ..	28017.50	28035 00	28052.50	28070.00	28087.50	28105.00	28122.50	28140.00	.. 28000
29000 ..	29018.12	29036.25	29054.37	29072.50	29090.62	29108.75	29126.87	29145.00	.. 29000
30000 ..	30018.75	30037.50	30056.25	30075.00	30093.75	30112.50	30131.25	30150.00	.. 30000
31000 ..	31019.37	31038.75	31058.12	31077 50	31096 87	31116.25	31135.62	31155.00	.. 31000
32000 ..	32020.00	32040.00	32060.00	32080.00	32100.00	32120.00	32140.00	32160.00	.. 32000
33000 ..	33020.62	33041.25	33061.87	33082.50	33103 12	33123.75	33144.37	33165.00	.. 33000
34000 ..	34021.25	34042.50	34063.75	34085.00	34106.25	34127.50	34148.75	34170.00	.. 34000
35000 ..	35021.87	35043.75	35065.62	35087.50	35109.37	35131.25	35153.12	35175.00	.. 35000
36000 ..	36022.50	36045.00	36067.50	36090.00	36112.50	36135.00	36157.50	36180.00	. 36000
37000 ..	37023.12	37046.25	37069.37	37092.50	37115.62	37138.75	37161.87	37185.00	. 37000
38000 ..	38023.75	38047.50	38071.25	38095.00	38118.75	38142.50	38166.25	38190.00	.. 38000
39000 ..	39024.37	39048.75	39073.12	39097.50	39121 87	39146.25	39170.62	39195.00	.. 39000
40000 ..	40025.00	40050.00	40075.00	40100.00	40125.00	40150.00	40175.00	40200.00	.. 40000
41000 ..	41025.62	41051.25	41076.87	41102.50	41128.12	41153.75	41179.37	41205.00	.. 41000
42000 ..	42026.25	42052.50	42078.75	42105.00	42131.25	42157.50	42183.75	42210.00	.. 42000
43000 ..	43026.87	43053.75	43080.62	43107.50	43134.37	43161.25	43188.12	43215.00	.. 43000
44000 ..	44027.50	44055.00	44082.50	44110.00	44137.50	44165.00	44192.50	44220.00	.. 44000
45000 ..	45028.12	45056.25	45084.37	45112.50	45140.62	45168.75	45196.87	45225.00	.. 45000
46000 ..	46028.75	46057.50	46086.25	46115.00	46143.75	46172.50	46201.25	46230.00	.. 46000
47000 ..	47029.37	47058.75	47088.12	47117.50	47146.87	47176.25	47205.62	47235.00	.. 47000
48000 ..	48030.00	48060.00	48090.00	48120 00	48150.00	48180.00	48210.00	48240.00	.. 48000
49000 ..	49030.62	49061.25	49091.87	49122.50	49153.12	49183.75	49214 37	49245.00	.. 49000
50000 ..	50031.25	50062.50	50093.75	50125.00	50156.25	50187.50	50218.75	50250.00	.. 50000

AMERICAN EXCHANGE TABLES.

CALCULATED AT FRACTIONS OF

$\frac{1}{16}$ TO $\frac{1}{2}$ % PREMIUM

TO FACILITATE TRANSACTIONS IN

AMERICAN OR DOMESTIC EXCHANGE.

Par.	$\frac{1}{16}$	$\frac{1}{8}$	$\frac{3}{16}$	$\frac{1}{4}$	$\frac{5}{16}$	$\frac{3}{8}$	$\frac{7}{16}$	$\frac{1}{2}$	Par.
$	$ cts.	$ cts.	$ cts.	$ cts.	$ cts.	$ cts.	$ cts.	$ cts.	$
51000 ..	51031.87	51063.75	51095.62	51127.50	51159.37	51191.25	51223.12	51255.00	.. 51000
52000 ..	52032.50	52065.00	52097.50	52130.00	52162.50	52195.00	52227.50	52260.00	.. 52000
53000 ..	53033.12	53066.25	53099.37	53132.50	53165.62	53198.75	53231.87	53265.00	.. 53000
54000 ..	54033.75	54067.50	54101.25	54135.00	54168.75	54202.50	54236.25	54270.00	.. 54000
55000 ...	55034.37	55068.75	55103.12	55137.50	55171.87	55206.25	55240.62	55275.00	.. 55000
56000 ..	56035.00	56070.00	56105.00	56140.00	56175.00	56210.00	56245.00	56280.00	.. 56000
57000 ..	57035.62	57071.25	57106.87	57142.50	57178.12	57213.75	57249.37	57285.00	.. 57000
58000 ..	58036.25	58072.50	58108.75	58145.00	58181.25	58217.50	58253.75	58290.00	.. 58000
59000 ..	59036.87	59073.75	59110.62	59147.50	59184.37	59221.25	59258.12	59295.00	.. 59000
60000 ..	60037.50	60075.00	60112.50	60150.00	60187.50	60225.00	60262.50	60300.00	.. 60000
61000 ..	61038.12	61076.25	61114.37	61152.50	61190.62	61228.75	61266.87	61305.00	.. 61000
62000 ..	62038.75	62077.50	62116.25	62155.00	62193.75	62232.50	62271.25	62310.00	.. 62000
63000 ..	63039.37	63078.75	63118.12	63157.50	63196.87	63236.25	63275.62	63315.00	.. 63000
64000 ..	64040.00	64080.00	64120.00	64160.00	64200.00	64240.00	64280.00	64320.00	.. 64000
65000 ..	65040.62	65081.25	65121.87	65162.50	65203.12	65243.75	65284.37	65325.00	.. 65000
66000 ..	66041.25	66082.50	66123.75	66165.00	66206.25	66247.50	66288.75	66330.00	.. 66000
67000 ..	67041.87	67083.75	67125.62	67167.50	67209.37	67251.25	67293.12	67335.00	.. 67000
68000 ..	68042.50	68085.00	68127.50	68170.00	68212.50	68255.00	68297.50	68340.00	.. 68000
69000 ..	69043.12	69086.25	69129.37	69172.50	69215.62	69258.75	69301.87	69345.00	.. 69000
70000 ..	70043.75	70087.50	70131.25	70175.00	70218.75	70262.50	70306.25	70350.00	.. 70000
71000 ..	71044.37	71088.75	71133.12	71177.50	71221.87	71266.25	71310.62	71355.00	.. 71000
72000 ..	72045.00	72090.00	72135.00	72180.00	72225.00	72270.00	72315.00	72360.00	.. 72000
73000 ..	73045.62	73091.25	73136.87	73182.50	73228.12	73273.75	73319.37	73365.00	.. 73000
74000 ..	74046.25	74092.50	74138.75	74185.00	74231.25	74277.50	74323.75	74370.00	.. 74000
75000 ..	75046.87	75093.75	75140.62	75187.50	75234.37	75281.25	75328.12	75375.00	.. 75000
76000 ..	76047.50	76095.00	76142.50	76190.00	76237.50	76285.00	76332.50	76380.00	.. 76000
77000 ..	77048.12	77096.25	77144.37	77192.50	77240.62	77288.75	77336.87	77385.00	.. 77000
78000 ..	78048.75	78097.50	78146.25	78195.00	78243.75	78292.50	78341.25	78390.00	.. 78000
79000 ..	79049.37	79098.75	79148.12	79197.50	79246.87	79296.25	79345.62	79395.00	.. 79000
80000 ..	80050.00	80100.00	80150.00	80200.00	80250.00	80300.00	80350.00	80400.00	.. 80000
81000 ..	81050.62	81101.25	81151.87	81202.50	81253.12	81303.75	81354.37	81405.00	.. 81000
82000 ..	82051.25	82102.50	82153.75	82205.00	82256.25	82307.50	82358.75	82410.00	.. 82000
83000 ..	83051.87	83103.75	83155.62	83207.50	83259.37	83311.25	83363.12	83415.00	.. 83000
84000 ..	84052.50	84105.00	84157.50	84210.00	84262.50	84315.00	84367.50	84420.00	.. 84000
85000 ..	85053.12	85106.25	85159.37	85212.50	85265.62	85318.75	85371.87	85425.00	.. 85000
86000 ..	86053.75	86107.50	86161.25	86215.00	86268.75	86322.50	86376.25	86430.00	.. 86000
87000 ..	87054.37	87108.75	87163.12	87217.50	87271.87	87326.25	87380.62	87435.00	.. 87000
88000 ..	88055.00	88110.00	88165.00	88220.00	88275.00	88330.00	88385.00	88440.00	.. 88000
89000 ..	89055.62	89111.25	89166.87	89222.50	89278.12	89333.75	89389.37	89445.00	.. 89000
90000 ..	90056.25	90112.50	90168.75	90225.00	90281.25	90337.50	90393.75	90450.00	.. 90000
91000 ..	91056.87	91113.75	91170.62	91227.50	91284.37	91341.25	91398.12	91455.00	.. 91000
92000 ..	92057.50	92115.00	92172.50	92230.00	92287.50	92345.00	92402.50	92460.00	.. 92000
93000 ..	93058.12	93116.25	93174.37	93232.50	93290.62	93348.75	93406.87	93465.00	.. 93000
94000 ..	94058.75	94117.50	94176.25	94235.00	94293.75	94352.50	94411.25	94470.00	.. 94000
95000 ..	95059.37	95118.75	95178.12	95237.50	95296.87	95356.25	95415.63	95475.00	.. 95000
96000 ..	96060.00	96120.00	96180.00	96240.00	96300.00	96360.00	96420.00	96480.00	.. 96000
97000 ..	97060.62	97121.25	97181.87	97242.50	97303.12	97363.75	97424.37	97485.00	.. 97000
98000 ..	98061.25	98122.50	98183.75	98245.00	98306.25	98367.50	98428.75	98490.00	.. 98000
99000 ..	99061.87	99123.75	99185.62	99247.50	99309.37	99371.25	99433.12	99495.00	.. 99000
100000 ..	100062.50	100125.00	100187.50	100250.00	100312.50	100375.00	100437.50	100500.00	.. 100000

AMERICAN EXCHANGE TABLES,

CALCULATED AT FRACTIONS OF

$\frac{9}{16}$ TO 1% PREMIUM

TO FACILITATE TRANSACTIONS IN

AMERICAN OR DOMESTIC EXCHANGE,

Par.	9/16	5/8	11/16	3/4	13/16	7/8	15/16	1	Par.
$	$ cts.	$ cts.	$ cts.	$ cts.	$ cts.	$ cts.	$ cts.	$ cts.	$
1000 .	1005.62	1006.25	1006.87	1007.50	1008.12	1008.75	1009.37	1010.00	.. 1000
2000 ..	2011.25	2012.50	2013.75	2015.00	2016.25	2017.50	2018.75	2020.00	.. 2000
3000 ..	3016.87	3018.75	3020.62	3022.50	3024.37	3026.25	3028.12	3030.00	.. 3000
4000 ..	4022.50	4025.00	4027.50	4030.00	4032.50	4035.00	4037.50	4040.00	.. 4000
5000 ..	5028.12	5031.25	5034.37	5037.50	5040.62	5043.75	5046.87	5050.00	.. 5000
6000 ..	6033.75	6037.50	6041.25	6045.00	6048.75	6052.50	6056.25	6060.00	.. 6000
7000 ..	7039.37	7043.75	7048.12	7052.50	7056.87	7061.25	7065.62	7070.00	.. 7000
8000 ..	8045.00	8050.00	8055.00	8060.00	8065.00	8070.00	8075.00	8080.00	.. 8000
9000 ..	9050.62	9056.25	9061.87	9067.50	9073.12	9078.75	9084.37	9090.00	.. 9000
10000 ..	10056.25	10062.50	10068.75	10075.00	10081.25	10087.50	10093.75	10100.00	.. 10000
11000 ..	11061.87	11068.75	11075.62	11082.50	11089.37	11096.25	11103.12	11110.00	.. 11000
12000 ..	12067.50	12075.00	12082.50	12090.00	12097.50	12105.00	12112.50	12120.00	.. 12000
13000 ..	13073.12	13081.25	13089.37	13097.50	13105.62	13113.75	13121.87	13130.00	.. 13000
14000 ..	14078.75	14087.50	14096.25	14105.00	14113.75	14122.50	14131.25	14140.00	.. 14000
15000 ..	15084.37	15093.75	15103.12	15112.50	15121.87	15131.25	15140.62	15150.00	.. 15000
16000 ..	16090.00	16100.00	16110.00	16120.00	16130.00	16140.00	16150.00	16160.00	.. 16000
17000 ..	17095.62	17106.25	17116.87	17127.50	17138.12	17148.75	17159.37	17170.00	.. 17000
18000 ..	18101.25	18112.50	18123.75	18135.00	18146.25	18157.50	18168.75	18180.00	.. 18000
19000 ..	19106.87	19118.75	19130.62	19142.50	19154.37	19166.25	19178.12	19190.00	.. 19000
20000 ..	20112.50	20125.00	20137.50	20150.00	20162.50	20175.00	20187.50	20200.00	.. 20000
21000 ..	21118.12	21131.25	21144.37	21157.50	21170.62	21183.75	21196.87	21210.00	.. 21000
22000 ..	22123.75	22137.50	22151.25	22165.00	22178.75	22192.50	22206.25	22220.00	.. 22000
23000 ..	23129.37	23143.75	23158.12	23172.50	23186.87	23201.25	23215.62	23230.00	.. 23000
24000 ..	24135.00	24150.00	24165.00	24180.00	24195.00	24210.00	24225.00	24240.00	.. 24000
25000 ..	25140.62	25156.25	25171.87	25187.50	25203.12	25218.75	25234.37	25250.00	.. 25000
26000 ..	26146.25	26162.50	26178.75	26195.00	26211.25	26227.50	26243.75	26260.00	.. 26000
27000 ..	27151.87	27168.75	27185.62	27202.50	27219.37	27236.25	27253.12	27370.00	.. 27000
28000 ..	28157.50	28175.00	28192.50	28210.00	28227.50	28245.00	28262.50	28280.00	.. 28000
29000 ..	29163.12	29181.25	29199.37	29217.50	29235.62	29253.75	29271.87	29290.00	.. 29000
30000 ..	30168.75	30187.50	30206.25	30225.00	30243.75	30262.50	30281.25	30300.00	.. 30000
31000 ..	31174.37	31193.75	31213.12	31232.50	31251.87	31271.25	31290.62	31310.00	.. 31000
32000 ..	32180.00	32200.00	32220.00	32240.00	32260.00	32280.00	32300.00	32320.00	.. 32000
33000 ..	33185.62	33206.25	33226.87	33247.50	33268.12	33288.75	33309.37	33330.00	.. 33000
34000 ..	34191.25	34212.50	34233.75	34255.00	34276.25	34297.50	34318.75	34340.00	.. 34000
35000 ..	35196.87	35218.75	35240.62	35262.50	35284.37	35306.25	35328.12	35350.00	.. 35000
36000 ..	36202.50	36225.00	36247.50	36270.00	36292.50	36315.00	36337.50	36360.00	. 36000
37000 ..	37208.12	37231.25	37254.37	37277.50	37300.62	37323.75	37346.87	37370.00	.. 37000
38000 ..	38213.75	38237.50	38261.25	38285.00	38308.75	38332.50	38356.25	38380.00	.. 38000
39000 ..	39219.37	39243.75	39268.12	39292.50	39316.87	39341.25	39365.62	39390.00	.. 39000
40000 ..	40225.00	40250.00	40275.00	40300.00	40325.00	40350.00	40375.00	40400.00	.. 40000
41000 ..	41230.62	41256.25	41281.87	41307.50	41333.12	41358.75	41384.37	41410.00	.. 41000
42000 ..	42236.25	42262.50	42288.75	42315.00	42341.25	42367.50	42393.75	42420.00	.. 42000
43000 ..	43241.87	43268.75	43295.62	43322.50	43349.37	43376.25	43403.12	43430.00	.. 43000
44000 ..	44247.50	44275.00	44302.50	44330.00	44357.50	44385.00	44412.50	44440.00	.. 44000
45000 ..	45253.12	45281.25	45309.37	45337.50	45365.62	45393.75	45421.87	45450.00	.. 45000
46000 ..	46258.75	46287.50	46316.25	46345.00	46373.75	46402.50	46431.25	46460.00	.. 46000
47000 ..	47264.37	47293.75	47323.12	47352.50	47381.87	47411.25	47440.62	47470.00	.. 47000
48000 ..	48270.00	48300.00	48330.00	48360.00	48390.00	48420.00	48450.00	48480.00	.. 48000
49000 ..	49275.62	49306.25	49336.87	49367.50	49398.12	49428.75	49459.37	49490.00	.. 49000
50000 ..	50281.25	50312.50	50343.75	50375.00	50406.25	50437.50	50468.75	50500.00	.. 50000

AMERICAN EXCHANGE TABLES.

CALCULATED AT FRACTIONS OF

$\frac{9}{16}$ TO 1% PREMIUM

TO FACILITATE TRANSACTIONS IN

AMERICAN OR DOMESTIC EXCHANGE.

Par.	$\frac{9}{16}$	$\frac{5}{8}$	$\frac{11}{16}$	$\frac{3}{4}$	$\frac{13}{16}$	$\frac{7}{8}$	$\frac{15}{16}$	1	Par.
$	$ cts.	$ cts.	$ cts.	$ cts.	$ cts.	$ cts.	$ cts.	$ cts.	$
51000 ..	51286.87	51318.75	51350.62	51382.50	51414.37	51446.25	51478.12	51510.00	.. 51000
52000 ..	52292.50	52325.00	52357.50	52390.00	52422.50	52455.00	52487.50	52520.00	.. 52000
53000 ..	53298.12	53331.25	53364.37	53397.50	53430.62	53463.75	53496.87	53530.00	.. 53000
54000 ..	54303.75	54337.50	54371.25	54405.00	54438.75	54472.50	54506.25	54540.00	.. 54000
55000 ..	55309.37	55343.75	55378.12	55412.50	55446.87	55481.25	55515.62	55550.00	.. 55000
56000 ..	56315.00	56350.00	56385.00	56420.00	56455.00	56490.00	56525.00	56560.00	.. 56000
57000 ..	57320.62	57356.25	57391.87	57427.50	57463.12	57498.75	57534.37	57570.00	.. 57000
58000 ..	58326.25	58362.50	58398.75	58435.00	58471.25	58507.50	58543.75	58580.00	.. 58000
59000 ..	59331.87	59368.75	59405.62	59442.50	59479.37	59516.25	59553.12	59590.00	.. 59000
60000 ..	60337.50	60375.00	60412.50	60450.00	60487.50	60525.00	60562.50	60600.00	.. 60000
61000 ..	61343.12	61381.25	61419.37	61457.50	61495.62	61533.75	61571.87	61610.00	.. 61000
62000 ..	62348.75	62387.50	62426.25	62465.00	62503.75	62542.50	62581.25	62620.00	.. 62000
63000 ..	63354.37	63393.75	63433.12	63472.50	63511.87	63551.25	63590.62	63630.00	.. 63000
64000 ..	64360.00	64400.00	64440.00	64480.00	64520.00	64560.00	64600.00	64640.00	.. 64000
65000 ..	65365.62	65406.25	65446.87	65487.50	65528.12	65568.75	65609.37	65650.00	.. 65000
66000 ..	66371.25	66412.50	66453.75	66495.00	66536.25	66577.50	66618.75	66660.00	.. 66000
67000 ..	67376.87	67418.75	67460.62	67502.50	67544.37	67586.25	67628.12	67670.00	.. 67000
68000 ..	68382.50	68425.00	68467.50	68510.00	68552.50	68595.00	68637.50	68680.00	.. 68000
69000 ..	69388.12	69431.25	69474.37	69517.50	69560.62	69603.75	69646.87	69690.00	.. 69000
70000 ..	70393.75	70437.50	70481.25	70525.00	70568.75	70612.50	70656.25	70700.00	.. 70000
71000 ..	71399.37	71443.75	71488.12	71532.50	71576.87	71621.25	71665.62	71710.00	.. 71000
72000 ..	72405.00	72450.00	72495.00	72540.00	72585.00	72630.00	72675.00	72720.00	.. 72000
73000 ..	73410.62	73456.25	73501.87	73547.50	73593.12	73638.75	73684.37	73730.00	.. 73000
74000 ..	74416.25	74462.50	74508.75	74555.00	74601.25	74647.50	74693.75	74740.00	.. 74000
75000 ..	75421.87	75468.75	75515.62	75562.50	75609.37	75656.25	75703.12	75750.00	.. 75000
76000 ..	76427.50	76475.00	76522.50	76570.00	76617.50	76665.00	76712.50	76760.00	.. 76000
77000 ..	77433.12	77481.25	77529.37	77577.50	77625.62	77673.75	77721.87	77770.00	.. 77000
78000 ..	78438.75	78487.50	78536.25	78585.00	78633.75	78682.50	78731.25	78780.00	.. 78000
79000 ..	79444.37	79493.75	79543.12	79592.50	79641.87	79691.25	79740.62	79790.00	.. 79000
80000 ..	80450.00	80500.00	80550.00	80600.00	80650.00	80700.00	80750.00	80800.00	.. 80000
81000 ..	81455.62	81506.25	81556.87	81607.50	81658.12	81708.75	81759.37	81810.00	.. 81000
82000 ..	82461.25	82512.50	82563.75	82615.00	82666.25	82717.50	82768.75	82820.00	.. 82000
83000 ..	83466.87	83518.75	83570.62	83622.50	83674.37	83726.25	83778.12	83830.00	.. 83000
84000 ..	84472.50	84525.00	84577.50	84630.00	84682.50	84735.00	84787.50	84840.00	.. 84000
85000 ..	85478.12	85531.25	85584.37	85637.50	85690.62	85743.75	85796.87	85850.00	.. 85000
86000 ..	86483.75	86537.50	86591.25	86645.00	86698.75	86752.50	86806.25	86860.00	.. 86000
87000 ..	87489.37	87543.75	87598.12	87652.50	87706.87	87761.25	87815.62	87870.00	.. 87000
88000 ..	88495.00	88550.00	88605.00	88660.00	88715.00	88770.00	88825.00	88880.00	.. 88000
89000 ..	89500.62	89556.25	89611.87	89667.50	89723.12	89778.75	89834.37	89890.00	.. 89000
90000 ..	90506.25	90562.50	90618.75	90675.00	90731.25	90787.50	90843.75	90900.00	.. 90000
91000 ..	91511.87	91568.75	91625.62	91682.50	91739.37	91796.25	91853.12	91910.00	.. 91000
92000 ..	92517.50	92575.00	92632.50	92690.00	92747.50	92805.00	92862.50	92920.00	.. 92000
93000 ..	93523.12	93581.25	93639.37	93697.50	93755.62	93813.75	93871.87	93930.00	.. 93000
94000 ..	94528.75	94587.50	94646.25	94705.00	94763.75	94822.50	94881.25	94940.00	.. 94000
95000 ..	95534.37	95593.75	95653.12	95712.50	95771.87	95831.25	95890.62	95950.00	.. 95000
96000 ..	96540.00	96600.00	96660.00	96720.00	96780.00	96840.00	96900.00	96960.00	.. 96000
97000 ..	97545.62	97606.25	97666.87	97727.50	97788.12	97848.75	97909.37	97970.00	.. 97000
98000 ..	98551.25	98612.50	98673.75	98735.00	98796.25	98857.50	98918.75	98980.00	.. 98000
99000 ..	99556.87	99618.75	99680.62	99742.50	99804.37	99866.25	99928.12	99990.00	.. 99000
100000 ..	100562.50	100625.00	100687.50	100750.00	100812.50	100875.00	100937.50	101000.00	.. 100000

BROKERAGES.

$$\tfrac{1}{32}, \ \tfrac{1}{16} \ \text{AND} \ \tfrac{1}{8} \ \text{OF} \ 1\%$$

STERLING INTO DOLLARS AND CENTS.

Stg. £	$\tfrac{1}{32}$ $ cts.	$\tfrac{1}{16}$ $ cts.	$\tfrac{1}{8}$ $ cts.	Stg. £	$\tfrac{1}{32}$ $ cts.	$\tfrac{1}{16}$ $ cts.	$\tfrac{1}{8}$ $ cts.
100 ..	.13.8	.27.7	.55.5	5100 ..	7.08.3	14.16.6	28.33.3
200 ..	.27.7	.55.5	1.11.1	5200 ..	7.22.2	14.44.4	28.88.8
300 ..	.41 6	.83.3	1.66.6	5300 ..	7.36.1	14.72.2	29.44.4
400 ..	.55.5	1.11.1	2.22.2	5400 ..	7.50.0	15.00.0	30.00.0
500 ..	.69.4	1.38.8	2.77.7	5500 ..	7.63.8	15.27.7	30.55.5
600 ..	.83.3	1.66.6	3.33.3	5600 ..	7.77.7	15.55.5	31.11.1
700 ..	.97.2	1.94.4	3.88.8	5700 ..	7.91.6	15.83.3	31.66.6
800 ..	1.11.1	2.22.2	4.44.4	5800 ..	8 05.5	16.11.1	32.22.2
900 ..	1.25.0	2.50.0	5.00.0	5900 ..	8.19.4	16.38.8	32.77.7
1000 ..	1.38.8	2.77.7	5.55.5	6000 ..	8.33.3	16.66.6	33.33.3
1100 ..	1.52.7	3.05.5	6.11.1	6100 ..	8.47.2	16.94.4	33.88.8
1200 ..	1.66.6	3.33.3	6.66.6	6200 ..	8.61.1	17.22.2	34.44.4
1300 ..	1.80.5	3.61.1	7.22.2	6300 ..	8.75.0	17.50.0	35.00.0
1400 ..	1.94.4	3.88.8	7.77.7	6400 ..	8.88.8	17.77.7	35.55.5
1500 ..	2.08.3	4.16.6	8.33.3	6500 ..	9.02.7	18.05.5	36.11.1
1600 ..	2.22.2	4.44.4	8.88.8	6600 ..	9.16.6	18.33.3	36.66.6
1700 ..	2.36.1	4.72.2	9.44.4	6700 ..	9.30.5	18.61.1	37 22.2
1800 ..	2.50.0	5.00.0	10.00.0	6800 ..	9.44.4	18.88.8	37.77.7
1900 ..	2.63.8	5.27.7	10.55.5	6900 ..	9.58.3	19.16.6	38.33.3
2000 ..	2.77.7	5.55.5	11.11.1	7000 ..	9.72.2	19.44.4	38.88.8
2100 ..	2 91.6	5.83.3	11.66.6	7100 ..	9.86.1	19.72.2	39.44.4
2200 ..	3.05.5	6.11.1	12.22.2	7200 ..	10.00.0	20.00.0	40.00.0
2300 ..	3.19.4	6.38.8	12.77.7	7300 ..	10.13.8	20.27.7	40.55.5
2400 ..	3.33.3	6.66.6	13.33.3	7400 ..	10.27.7	20.55.5	41.11.1
2500 ..	3.47.2	6.94.4	13.88.8	7500 ..	10.41.6	20.83.3	41.66.6
2600 ..	3.61.1	7.22.2	14.44.4	7600 ..	10.55.5	21.11.1	42.22.2
2700 ..	3.75.0	7.50.0	15.00.0	7700 ..	10.69.4	21.38.8	42.77.7
2800 ..	3.88.8	7.77.7	15.55.5	7800 ..	10.83.3	21.66.6	43.33.3
2900 ..	4.02.7	8.05.5	16.11.1	7900 ..	10.97.2	21.94.4	43.88.8
3000 ..	4.16.6	8.33.3	16.66.6	8000 ..	11.11.1	22.22.2	44.44.4
3100 ..	4.30.5	8.61.1	17.22.2	8100 ..	11.25.0	22.50.0	45.00.0
3200 ..	4.44.4	8.88.8	17.77.7	8200 ..	11.38.8	22.77.7	45.55.5
3300 ..	4.58.3	9.16.6	18.33.3	8300 ..	11.52.7	23.05.5	46.11.1
3400 ..	4.72.2	9.44.4	18.88.8	8400 ..	11.66.6	23.33.3	46.66.6
3500 ..	4.86.1	9.72.2	19.44.4	8500 .	11.80.5	23.61.1	47.22.2
3600 ..	5.00.0	10.00.0	20.00.0	8600 ..	11.94.4	23.88.8	47.77.7
3700 ..	5.13.8	10.27.7	20.55.5	8700 ..	12.08.3	24.16 6	48.33.3
3800 ..	5.27.7	10.55.5	21.11.1	8800 ..	12.22.2	24.44.4	48.88.8
3900 ..	5.41.6	10.83.3	21.66.6	8900 ..	12.36.1	24.72.2	49.44.4
4000 ..	5.55.5	11.11.1	22.22.2	9000 ..	12.50.0	25.00.0	50.00.0
4100 ..	5.69.4	11.38.8	22.77.7	9100 ..	12.63.8	25.27.7	50.55.5
4200 ..	5.83.3	11.66.6	23.33.3	9200 ..	12.77.7	25.55.5	51.11.1
4300 ..	5.97.2	11.94.4	23.88.8	9300 ..	12.91.6	25.83 3	51.66.6
4400 ..	6.11.1	12.22.2	24.44.4	9400 ..	13.05.5	26.11.1	52.22.2
4500 ..	6.25.0	12.50.0	25.00.0	9500 ..	13.19.4	26.38.8	52.77.7
4600 ..	6.38.8	12.77.7	25 55.5	9600 ..	13 33.3	26.66.6	53.33.3
4700 ..	6.52.7	13.05.5	26 11.1	9700 ..	13 47.2	26.94.4	53.88.8
4800 ..	6.66.6	13.33.3	26.66.6	9800 ..	13.61.1	27.22.2	54.44.4
4900 ..	6.80.5	13 61.1	27.22.2	9900 ..	13.75.0	27.50.0	55.00.0
5000 ..	6.94.4	13.88.8	27.77.7	10000 ..	13.88.8	27.77.7	55.55.5

BROKERAGES.

$\frac{1}{32}, \frac{1}{16},$ AND $\frac{1}{8}$ OF 1%

AMERICAN OR DOMESTIC EXCHANGE.

Par.	$\frac{1}{32}$	$\frac{1}{16}$	$\frac{1}{8}$	Par.	$\frac{1}{32}$	$\frac{1}{16}$	$\frac{1}{8}$
$	$ cts.	$ cts.	$ cts.	$	$ cts.	$ cts.	$ cts.
1000 ..	.31.2	.62.5	1.25.0	51000 ..	15.93.7	31.87.5	63.75.0
2000 ..	.62.5	1.25.0	2.50.0	52000 ..	16.25.0	32.50.0	65.00.0
3000 ..	.93.7	1.87.5	3.75.0	53000 ..	16.56.2	33.12.5	66.25.0
4000 ..	1.25.0	2.50.0	5.00.0	54000 ..	16.87.5	33.75.0	67.50.0
5000 ..	1.56.2	3.12.5	6.25.0	55000 ..	17.18.7	34.37.5	68.75.0
6000 ..	1.87.5	3.75.0	7.50.0	56000 ..	17.50.0	35.00.0	70.00.0
7000 ..	2.18.7	4.37.5	8.75.0	57000 ..	17.81.2	35.62.5	71.25.0
8000 ..	2.50.0	5.00.0	10.00.0	58000 ..	18.12.5	36.25.0	72.50.0
9000 ..	2.81.2	5.62.5	11.25.0	59000 ..	18.43.7	36.87.5	73.75.0
10000 ..	3.12.5	6.25.0	12.50.0	60000 ..	18.75.0	37.50.0	75.00.0
11000 ..	3.43.7	6.87.5	13.75.0	61000 ..	19.06.2	38.12.5	76.25.0
12000 ..	3.75.0	7.50.0	15.00.0	62000 ..	19.37.5	38.75.0	77.50.0
13000 ..	4.06.2	8.12.5	16.25.0	63000 ..	19.68.7	39.37.5	78.75.0
14000 ..	4.37.5	8.75.0	17.50.0	64000 ..	20.00.0	40.00.0	80.00.0
15000 ..	4.68.7	9.37.5	18.75.0	65000	20.31.2	40.62.5	81.25.0
16000 ..	5.00.0	10.00.0	20.00.0	66000 ..	20.62.5	41.25.0	82.50.0
17000 ..	5.31.2	10.62.5	21.25.0	67000 ..	20.93.7	41.87.5	83.75.0
18000 ..	5.62.5	11.25.0	22.50.0	68000 ..	21.25.0	42.50.0	85.00.0
19000 ..	5.93.7	11.87.5	23.75.0	69000 ..	21.56.2	43.12.5	86.25.0
20000 ..	6.25.0	12.50.0	25.00.0	70000 ..	21.87.5	43.75.0	87.50.0
21000 ..	6.56.2	13.12.5	26.25.0	71000 ..	22.18.7	44.37.5	88.75.0
22000 ..	6.87.5	13.75.0	27.50.0	72000 ..	22.50.0	45.00.0	90.00.0
23000 ..	7.18.7	14.37.5	28.75.0	73000 ..	22.81.2	45.62.5	91.25.0
24000 ..	7.50.0	15.00.0	30.00.0	74000 ..	23.12.5	46.25.0	92.50.0
25000 ..	7.81.2	15.62.5	31.25.0	75000 ..	23.43.7	46.87.5	93.75.0
26000 ..	8.12.5	16.25.0	32.50.0	76000 ..	23.75.0	47.50.0	95.00.0
27000 ..	8.43.7	16.87.5	33.75.0	77000 ..	24.06.2	48.12.5	96.25.0
28000 ..	8.75.0	17.50.0	35.00.0	78000 ..	24.37.5	48.75.0	97.50.0
29000 ..	9.06.2	18.12.5	36.25.0	79000 ..	24.68.7	49.37.5	98.75.0
30000 .	9.37.5	18.75.0	37.50.0	80000 ..	25.00.0	50.00.0	100.00.0
31000 ..	9.68.7	19.37.5	38.75.0	81000 ..	25.31.2	50.62.5	101.25.0
32000 ..	10.00.0	20.00.0	40.00.0	82000 ..	25.62.5	51.25.0	102.50.0
33000 ..	10.31.2	20.62.5	41.25 0	83000 ..	25.93.7	51.87.5	103.75.0
34000 ..	10.62.5	21.25.0	42.50.0	84000 ..	26.25.0	52.50.0	105.00.0
35000 ..	10.93.7	21.87.5	43.75.0	85000 ..	26.56.2	53.12.5	106.25.0
36000 ..	11.25.0	22.50.0	45.00.0	86000 ..	26.87.5	53.75.0	107.50.0
37000 ..	11.56.2	23.12.5	46.25.0	87000 ..	27.18.7	54.37.5	108.75.0
38000 ..	11.87.5	23.75.0	47.50.0	88000 ..	27.50.0	55.00.0	110.00 0
39000 ..	12.18.7	24.37.5	48.75.0	89000 ..	27.81.2	55.62.5	111.25.0
40000 ..	12.50.0	25.00.0	50.00.0	90000 ..	28.12.5	56.25.0	112.50.0
41000 .	12.81.2	25.62.5	51.25.0	91000 ..	28.43.7	56.87.5	113.75.0
42000 ..	13.12.5	26.25.0	52.50.0	92000 ..	28.75.0	57.50.0	115.00 0
43000 ..	13.43.7	26.87.5	53.75.0	93000 ..	29.06.2	58.12.5	116.25.0
44000 ..	13.75.0	27.50.0	55.00.0	94000 ..	29.37.5	58.75.0	117.50.0
45000 ..	14.06.2	28.12.5	56.25.0	95000 ..	29.68.7	59.37.5	118.75.0
46000 ..	14.37.5	28.75.0	57.50.0	96000 ..	30.00.0	60.00.0	120.00.0
47000 ..	14.68.7	29.37.5	58.75.0	97000 ..	30.31.2	60.62.5	121.25.0
48000 ..	15.00.0	30.00.0	60.00.0	98000 ..	30.62.5	61.25.0	122.50.0
49000 ..	15.31.2	30.62.5	61.25.0	99000 ..	30.93.7	61.87.5	123.75 0
50000 ..	15.62.5	31.25.0	62.50.0	100000 ..	31.25.0	62.50.0	125.00.0

www.ingramcontent.com/pod-product-compliance
Lightning Source LLC
Chambersburg PA
CBHW030617270326
41927CB00007B/1210